Progress in Commutative Algebra 1

Progress in Commutative Algebra 1

Combinatorics and Homology

edited by
Christopher Francisco
Lee Klingler
Sean Sather-Wagstaff
Janet C. Vassilev

De Gruyter

Mathematics Subject Classification 2010
13D02, 13D40, 05E40, 13D45, 13D22, 13H10, 13A35, 13A15, 13A05, 13B22, 13F15

ISBN 978-3-11-025034-3
e-ISBN 978-3-11-025040-4

Library of Congress Cataloging-in-Publication Data
A CIP catalog record for this book has been applied for at the Library of Congress.

Bibliographic information published by the Deutsche Nationalbibliothek
The Deutsche Nationalbibliothek lists this publication in the Deutsche Nationalbibliografie; detailed bibliographic data are available in the Internet at http://dnb.dnb.de.

Typesetting: Da-TeX Gerd Blumenstein, Leipzig, www.da-tex.de
Printing: Hubert & Co. GmbH & Co. KG, Göttingen
∞ Printed on acid-free paper

Printed in Germany

www.degruyter.com

Preface

This collection of papers in commutative algebra stemmed out of the 2009 Fall Southeastern American Mathematical Society Meeting which contained three special sessions in the field:

- Special Session on Commutative Ring Theory, a Tribute to the Memory of James Brewer, organized by Alan Loper and Lee Klingler;
- Special Session on Homological Aspects of Module Theory, organized by Andy Kustin, Sean Sather-Wagstaff, and Janet Vassilev; and
- Special Session on Graded Resolutions, organized by Chris Francisco and Irena Peeva.

Much of the commutative algebra community has split into two camps, for lack of a better word: the Noetherian camp and the non-Noetherian camp. Most researchers in commutative algebra identify with one camp or the other, though there are some notable exceptions to this. We had originally intended this to be a Proceedings Volume for the conference as the sessions had a nice combination of both Noetherian and non-Noetherian talks. However, the project grew into two Volumes with invited papers that are blends of survey material and new research. We hope that members from the two camps will read each others' papers and that this will lead to increased mathematical interaction between the camps.

As the title suggests, this volume, Progress in Commutative Algebra I, contains combinatorial and homological surveys. Contributions to this volume are written by speakers in the second and third sessions. To make the volume more complete, we have complemented these papers by articles on three topics which should be of broad interest: Boij–Söderburg theory, the current status of the homological conjectures and crepant resolutions. The collection represents the current trends in two of the most active areas of commutative algebra. Of course, the divisions we have outlined here are slightly artificial, given the interdependencies between these areas. For instance, much of combinatorial commutative algebra focuses on the topic of resolutions, a homological topic. So, we have not officially divided the volume into two parts.

The combinatorial papers document some of the increasing focus in commutative algebra recently on the interaction between algebra and combinatorics. Specifically, one can use combinatorial techniques to investigate resolutions and other algebraic structures as with the papers of Fløystad on Boij–Söderburg theory, of Geramita, Harbourne and Migliore and of Cooper on Hilbert Functions, of Clark on Minimal Poset Resolutions and of Mermin on Simplicial Resolutions. One can also utilize algebraic invariants to understand combinatorial structures like graphs, hypergraphs, and simplicial complexes such as in the paper of Morey and Villarreal on Edge Ideals.

Homological techniques have become indispensable tools for the study of Noetherian rings. These ideas have yielded amazing levels of interaction with other fields like algebraic topology (via differential graded techniques as well as the foundations of homological algebra), analysis (via the study of D-modules), and combinatorics (as described in the previous paragraph). The homological articles we have included in this volume relate mostly to how homological techniques help us better understand rings and singularities both Noetherian and non-Noetherian such as in the papers by Roberts, Yao, Hummel and Leuschke.

Enjoy!

March 2012

Sean Sather-Wagstaff
Chris Francisco
Lee Klingler
Janet C. Vassilev

Table of Contents

Progress in Commutative Algebra 1, 1–54

Boij–Söderberg Theory: Introduction and Survey

Gunnar Fløystad

Abstract. Boij–Söderberg theory describes the Betti diagrams of graded modules over the polynomial ring, up to multiplication by a rational number. Analog Eisenbud–Schreyer theory describes the cohomology tables of vector bundles on projective spaces up to rational multiple. We give an introduction and survey of these newly developed areas.

Keywords. Betti Diagrams, Cohomology of Vector Bundles, Cohen–Macaulay Modules, Pure Resolutions, Supernatural Bundles.

2010 Mathematics Subject Classification. Primary: 13D02, 14F05; Secondary: 13C14, 14N99.

Introduction

In November 2006 M. Boij and J. Söderberg put out on the arXiv a preprint "Graded Betti numbers of Cohen–Macaulay modules and the multiplicity conjecture". The paper concerned resolutions of graded modules over the polynomial ring $S = \Bbbk[x_1, \ldots, x_n]$ over a field $\Bbbk$. It put forth two striking conjectures on the form of their resolutions. These conjectures and their subsequent proofs have put the greatest floodlight on our understanding of resolutions over polynomial rings since the inception of the field in 1890. In this year David Hilbert published his syzygy theorem stating that a graded ideal over the polynomial ring in n variables has a resolution of length less than or equal to n. Resolutions of modules both over the polynomial ring and other rings have since then been one of the pivotal topics of algebraic geometry and commutative algebra, and more generally in the field of associative algebras.

For the next half a year after Boij and Söderberg put out their conjectures, they were incubating in the mathematical community, and probably not so much exposed to attacks. The turning point was the conference at MSRI, Berkeley in April 2007 in honor of David Eisenbud 60th birthday, where the conjectures became a topic of discussion.

For those familiar with resolutions of graded modules over the polynomial ring, a complete classification of their numerical invariants, the graded Betti numbers (β_{ij}), seemed a momentous task, completely out of reach (and still does). Perhaps the central idea of Boij and Söderberg is this: We do not try to determine if (β_{ij}) are the graded Betti numbers of a module, but let us see if we can determine if $m \cdot (\beta_{ij})$ are the graded Betti numbers of a module if m is some big integer.

This is the idea of *stability* which has been so successful in stable homotopy theory in algebraic topology and rational divisor theory in algebraic geometry.

Another way to phrase the idea of Boij and Söderberg is that we do not determine the graded Betti numbers (β_{ij}) but rather the positive rays $t \cdot (\beta_{ij})$ where t is a positive rational number. It is easy to see that these rays form a cone in a suitable vector space over the rational numbers.

The conjectures of Boij and Söderberg considered the cone B of such diagrams coming from modules of codimension c with the shortest possible length of resolution, c. This is the class of Cohen–Macaulay modules. The first conjecture states precisely what the extremal rays of the cone B are. The diagrams on these rays are called *pure diagrams*. They are the possible Betti diagrams of *pure resolutions*,

$$S(-d_0)^{\beta_{0,d_0}} \leftarrow S(-d_1)^{\beta_{1,d_1}} \leftarrow \cdots \leftarrow S(-d_c)^{\beta_{c,d_c}} \tag{0.1}$$

of graded modules, where the length c is equal to the codimension of the module. To prove this conjecture involved two tasks. The first is to show that there are vectors on these rays which actually are Betti diagrams of modules. The second is to show that these rays account for *all* the extremal rays in the cone B, in the sense that any Betti diagram is a positive rational combination of vectors on these rays. This last part was perhaps what people found most suspect. Eisenbud has said that his immediate reaction was that this could not be true.

Boij and Söderberg made a second conjecture giving a refined description of the cone B. There is a partial order on the pure diagrams, and in any chain in this partial order the pure diagrams are linearly independent. Pure diagrams in a chain therefore generate a simplicial cone. Varying over the different chains we then get a simplicial fan of Betti diagrams. The refinement of the conjectures states that the realization of this simplicial fan is the positive cone B. In this way each Betti diagram lies on a unique minimal face of the simplicial fan, and so we get a strong uniqueness statement on how to write the Betti diagram of a module.

After the MSRI conference in April 2007, Eisenbud and the author independently started to look into the existence question, to construct *pure resolutions* whose Betti diagram is a pure diagram. They came up with the construction of the $\mathrm{GL}(n)$-equivariant resolution described in Subsection 3.1. Jerzy Weyman was instrumental in proving the exactness of this resolution and the construction was published in a joint paper in September 2007 on the arXiv, [11]. In the same paper also appeared another construction of pure resolutions described in Subsection 3.5.

After this success D. Eisenbud and F.-O. Schreyer went on to work on the other part of the conjectures. And in December 2007 they published on the arXiv the proof of the second part of the conjectures of Boij and Söderberg, [12]. But at least two more interesting things appeared in this paper. They gave a construction of pure resolutions that worked in all characteristics. The constructions above [11], work only in characteristic zero. But most startling, they discovered a surprising duality with cohomology tables of algebraic vector bundles on projective spaces. And fairly parallel to the proof of the second Boij–Söderberg conjecture they were able to give a complete description of all cohomology tables of algebraic vector bundles on projective spaces, up to multiplication by a positive rational number.

In the wake of this a range of papers have followed, most of which are discussed in this survey. But one thing still needs to be addressed. What enticed Boij and Söderberg to come up with their conjectures? The origin here lies in an observation by Huneke and Miller from 1985, that if a Cohen–Macaulay quotient ring of $A = S/I$ has pure resolution (0.1) (so $d_0 = 0$ and $\beta_{0,d_0} = 1$), then its multiplicity $e(A)$ is equal to the surprisingly simple expression

$$\frac{1}{c!} \cdot \prod_{i=1}^{c} d_i.$$

This led naturally to consider resolutions $F_\bullet$ of Cohen–Macaulay quotient rings $A = S/I$ in general. In this case one has in each homological term F_i in the resolution a maximal twist $S(-a_i)$ (so a_i is minimal) and a minimal twist $S(-b_i)$ occurring. The *multiplicity conjecture* of Herzog, Huneke and Srinivasan, see [25] and [26], stated that the multiplicity of the quotient ring A is in the following range

$$\frac{1}{c!}\prod_{i=1}^{c} a_i \leq e(A) \leq \frac{1}{c!}\prod_{i=1}^{c} b_i.$$

Over the next two decades a substantial number of papers were published on this treating various classes of rings, and also various generalizations of this conjecture. But efforts in general did not succeed because of the lack of a strong enough understanding of the (numerical) structure of resolutions. Boij and Söderberg's central idea is to see the above inequalities as a projection of convexity properties of the Betti diagrams of graded Cohen–Macaulay modules: The pure diagrams generate the extremal rays in the cone of Betti diagrams.

Notation. The graded Betti numbers $\beta_{ij}(M)$ of a finitely generated module M are indexed by $i = 0, \dots, n$ and $j \in \mathbb{Z}$. Only a finite number of these are nonzero. By a *diagram* we shall mean a collection of rational numbers (β_{ij}), indexed as above, with only a finite number of them being nonzero.

The organization of this paper is as follows. In Section 1 we give the important notions, like the graded Betti numbers of a module, pure resolutions and Cohen–Macaulay modules. Such modules have certain linear constraints on their graded Betti numbers, the Herzog–Kühl equations, giving a subspace L^{HK} of the space of diagrams. We define the positive cone B in L^{HK} of Betti diagrams of Cohen–Macaulay modules. An important technical convenience is that we fix a "window" on the diagrams, considering Betti diagrams where the β_{ij} are nonzero only in a finite range of indices (i, j). This makes the Betti diagrams live in a finite dimensional vector space. Then we present the Boij–Söderberg conjectures. We give both the algorithmic version, concerning the decomposition of Betti diagrams, and the geometric version in terms of fans.

In Section 2 we define the simplicial fan Σ of diagrams. The goal is to show that its realization is the positive cone B, and to do this we study the exterior facets of Σ. The main work of this section is to find the equations of these facets. They are the key to the duality with algebraic vector bundles, and the form of their equations are derived from suitable pairings between Betti diagrams and cohomology tables of vector bundles. The positivity of the pairings proves that the cone B is contained in the realization of Σ, which is one part of the conjectures.

The other part, that Σ is contained in B, is shown in Section 3 by providing the existence of pure resolutions. We give in Subsection 3.1 the construction of the equivariant pure resolution of [11], in Subsection 3.4 the characteristic free resolution of [12], and in Subsection 3.5 the second construction of [11]. For cohomology of vector bundles, the bundles with supernatural cohomology play the analog role of pure resolutions. In Subsection 3.2 we give the equivariant construction of supernatural bundles, and in Subsection 3.3 the characteristic free construction of [12].

In Section 4 we first consider the cohomology of vector bundles on projective spaces, and give the complete classification of such tables up to multiplication by a positive rational number. The argument runs analogous to what we do for Betti diagrams. We define the positive cone of cohomology tables C, and the simplicial fan of tables Γ. We compute the equations of the exterior facets of Γ which again are derived from the pairings between Betti diagrams and cohomology tables. The positivity of these pairings show that $C \subseteq \Gamma$, and the existence of supernatural bundles that $\Gamma \subseteq C$, showing the desired equality $C = \Gamma$.

Section 5 considers extensions of the previous results. First in Subsection 5.1 we get the classification of graded Betti numbers of *all* modules up to positive rational multiples. For cohomology of coherent sheaves there is not yet a classification, but in Subsection 5.2 the procedure to decompose cohomology tables of vector bundles is extended to cohomology tables of coherent sheaves. However this procedure involves an infinite number of steps, so this decomposition involves an infinite sum.

Section 6 gives more results that have followed in the wake of the conjectures and their proofs. The ultimate goal, to classify Betti diagrams of modules (not just up to rational multiple) is considered in Subsection 6.1, and consists mainly of examples of diagrams which are or are not the Betti diagrams of modules. So far we have considered $\Bbbk[x_1, \ldots, x_n]$ to be standard graded, i.e. each $\deg x_i = 1$. In Subsection 6.2 we consider other gradings and multigradings on the x_i. Subsection 6.3 considers the partial order on pure diagrams, so essential in defining the simplicial fan Σ. In Subsection 6.4 we inform on computer packages related to Boij–Söderberg theory, and in Subsection 6.5 we give some important open problems.

1 The Boij–Söderberg Conjectures

We work over the standard graded polynomial ring $S = \Bbbk[x_1, \dots, x_n]$. For a graded module M over S, we denote by M_d its graded piece of degree d, and by $M(-r)$ the module where degrees are shifted so that $M(-r)_d = M_{d-r}$.

Note. We shall always assume our modules to be finitely generated and graded.

1.1 Resolutions and Betti Diagrams

A natural approach to understand such modules is to understand their numerical invariants. The most immediate of these is of course the Hilbert function:

$$h_M(d) = \dim_{\Bbbk} M_d.$$

Another set of invariants is obtained by considering its minimal graded free resolution:

$$F_0 \leftarrow F_1 \leftarrow F_2 \leftarrow \cdots \leftarrow F_l. \tag{1.1}$$

Here each F_i is a graded free S-module $\bigoplus_{j \in \mathbb{Z}} S(-j)^{\beta_{ij}}$.

Example 1.1. Let $S = \Bbbk[x, y]$ and M be the quotient ring $S/(x^2, xy, y^3)$. Then its minimal resolution is

$$S \xleftarrow{[x^2\ xy\ y^3]} S(-2)^2 \oplus S(-3) \xleftarrow{\begin{bmatrix} y & 0 \\ -x & y^2 \\ 0 & -x \end{bmatrix}} S(-3) \oplus S(-4).$$

The multiple $\beta_{i,j}$ of the term $S(-j)$ in the ith homological part F_i of the resolution, is called the ith graded Betti number of degree j. These Betti numbers constitute another natural set of numerical invariants, and the ones that are the topic of the present notes. By the resolution (1.1) we see that the graded Betti number determine the Hilbert function of M. In fact the dimension $\dim_{\Bbbk} M_d$ is the alternating sum $\sum (-1)^i \dim_{\Bbbk}(F_i)_d$. The Betti numbers are however more refined numerical invariants of graded modules than the Hilbert function.

Example 1.2. Let M' be the quotient ring $S/(x^2, y^2)$. Its minimal free resolution is

$$S \leftarrow S(-2)^2 \leftarrow S(-4).$$

Then M of Example 1.1 and M' have the same Hilbert functions, but their graded Betti numbers are different.

The Betti numbers are usually displayed in an array. The immediate natural choice is to put $\beta_{i,j}$ in the ith column and jth row, so the diagram of Example 1.1 would be:

$$\begin{array}{c} \\ 0 \\ 1 \\ 2 \\ 3 \\ 4 \end{array} \begin{array}{c} \begin{array}{ccc} 0 & 1 & 2 \end{array} \\ \begin{bmatrix} 1 & 0 & 0 \\ 0 & 0 & 0 \\ 0 & 2 & 0 \\ 0 & 1 & 1 \\ 0 & 0 & 1 \end{bmatrix} \end{array}$$

However, to reduce the number of rows, one uses the convention that the ith column is shifted i steps up. Thus $\beta_{i,j}$ is put in the ith column and the $j-i$th row. Alternatively, $\beta_{i,i+j}$ is put in the ith column and jth row. So the diagram above is displayed as:

$$\begin{array}{c} \\ 0 \\ 1 \\ 2 \end{array} \begin{array}{c} \begin{array}{ccc} 0 & 1 & 2 \end{array} \\ \begin{bmatrix} 1 & 0 & 0 \\ 0 & 2 & 1 \\ 0 & 1 & 1 \end{bmatrix} \end{array} \tag{1.2}$$

A Betti diagram has columns indexed by $0, \dots, n$ and rows indexed by elements of $\mathbb{Z}$, but any Betti diagram (of a finitely generated graded module) is nonzero only in a finite number of rows. Our goal is to understand the possible Betti diagrams that can occur for Cohen–Macaulay modules. This objective seems however as of yet out of reach. The central idea of Boij–Söderberg theory is rather to describe Betti diagrams up to a multiple by a rational number. I.e. we do not determine if a diagram β is a Betti diagram of a module, but we will determine if $q\beta$ is a Betti diagram for some positive rational number q. By Hilbert's syzygy theorem we know that the length l of the resolution (1.1) is $\leq n$. Thus we consider Betti diagrams to live in the $\mathbb{Q}$-vector space $\mathbb{D} = \bigoplus_{j \in \mathbb{Z}} \mathbb{Q}^{n+1}$, with the β_{ij} as coordinate functions. An element in this vector space, a collection of rational numbers $(\beta_{ij})_{i=0,\dots,n, j \in \mathbb{Z}}$ where only a finite number is nonzero, is called a *diagram*.

1.2 The Positive Cone of Betti Diagrams

We want to make our Betti diagram live in a finite dimensional vector space, so we fix a "window" in $\mathbb{D}$ as follows. Let $c \leq n$ and $\mathbb{Z}^{c+1}_{\deg}$ be the set of strictly increasing integer sequences $(a_0, \dots, a_c)$ in $\mathbb{Z}^{c+1}$. Such an element is called a *degree sequence.* Then $\mathbb{Z}^{c+1}_{\deg}$ is a partially ordered set with $\mathbf{a} \leq \mathbf{b}$ if $a_i \leq b_i$ for all $i = 0, \dots, c$.

Definition 1.3. For $\mathbf{a}, \mathbf{b}$ in $\mathbb{Z}^{c+1}_{\deg}$ let $\mathbb{D}(\mathbf{a}, \mathbf{b})$ be the set of diagrams $(\beta_{ij})_{i=0,\dots,n, j \in \mathbb{Z}}$ such that β_{ij} may be nonzero only in the range $0 \leq i \leq c$ and $a_i \leq j \leq b_i$.

We see that $\mathbb{D}(\mathbf{a}, \mathbf{b})$ is simply the $\mathbb{Q}$-vector space with basis elements indexed by the pairs (i, j) in the range above determined by $\mathbf{a}$ and $\mathbf{b}$. The diagram of Example 1.1, displayed above in (1.2), lives in the window $\mathbb{D}(\mathbf{a}, \mathbf{b})$ with $\mathbf{a} = (0, 1, 2)$ and $\mathbf{b} = (0, 3, 4)$ (or $\mathbf{a}$ any triple $\leq (0, 1, 2)$ and $\mathbf{b}$ any triple $\geq (0, 3, 4)$).

If the module M has codimension c, equivalently its Krull dimension is $n - c$, the depth of M is $\leq n - c$. By the Auslander–Buchsbaum theorem, [8], the length of the resolution is $l \geq c$. To make things simple we assume that l has its smallest possible value $l = c$ or equivalently that M has depth equal to the dimension $n - c$. This gives the class of Cohen–Macaulay (CM) modules.

Definition 1.4. Let $\mathbf{a}$ and $\mathbf{b}$ be in $\mathbb{Z}_{\deg}^{c+1}$.

- $L(\mathbf{a}, \mathbf{b})$ is the $\mathbb{Q}$-vector subspace of the window $\mathbb{D}(\mathbf{a}, \mathbf{b})$ spanned by the Betti diagrams of CM-modules of codimension c, whose Betti diagrams are in this window.
- $B(\mathbf{a}, \mathbf{b})$ is the set of nonnegative rays spanned by such Betti diagrams.

Lemma 1.5. *$B(\mathbf{a}, \mathbf{b})$ is a cone.*

Proof. We must show that if β_1 and β_2 are in $B(\mathbf{a}, \mathbf{b})$ then $q_1\beta_1 + q_2\beta_2$ is in $B(\mathbf{a}, \mathbf{b})$ for all positive rational numbers q_1 and q_2.

This is easily seen to be equivalent to the following: Let M_1 and M_2 be CM-modules of codimension c with Betti diagrams in $\mathbb{D}(\mathbf{a}, \mathbf{b})$. Show that $c_1\beta(M_1) + c_2\beta(M_2)$ is in $B(\mathbf{a}, \mathbf{b})$ for all natural numbers c_1 and c_2. But this linear combination is clearly the Betti diagram of the CM-module $M_1^{c_1} \oplus M_2^{c_2}$ of codimension c. And clearly this linear combination is still in the window $\mathbb{D}(\mathbf{a}, \mathbf{b})$. □

Our main objective is to describe this cone.

1.3 Herzog–Kühl Equations

Now given a resolution (1.1) of a module M, there are natural relations its Betti numbers β_{ij} must fulfil. First of all if the codimension $c \geq 1$, then clearly the alternating sum of the ranks of the F_i must be zero. I.e.

$$\sum_{i,j} (-1)^i \beta_{ij} = 0.$$

When the codimension $c \geq 2$ we get more numerical restrictions. Since M has dimension $n - c$, its Hilbert series is of the form $h_M(t) = \frac{p(t)}{(1-t)^{n-c}}$, where $p(t)$ is some polynomial. This may be computed as the alternating sum of the Hilbert series of each of the terms in the resolution (1.1):

$$h_M(t) = \frac{\sum_j \beta_{0j} t^j}{(1-t)^n} - \frac{\sum_j \beta_{1j} t^j}{(1-t)^n} + \cdots + (-1)^l \frac{\sum_j \beta_{lj} t^j}{(1-t)^n}.$$

Multiplying with $(1-t)^n$ we get

$$(1-t)^c p(t) = \sum_{i,j} (-1)^i \beta_{ij} t^j.$$

Differentiating this successively and setting $t = 1$, gives the equations

$$\sum_{i,j} (-1)^i j^p \beta_{ij} = 0, \quad p = 0, \ldots, c-1. \tag{1.3}$$

These equations are the *Herzog–Kühl equations* for the Betti diagram $\{\beta_{ij}\}$ of a module of codimension c.

We denote by $L^{\mathrm{HK}}(\mathbf{a}, \mathbf{b})$ the $\mathbb{Q}$-linear subspace of diagrams in $\mathbb{D}(\mathbf{a}, \mathbf{b})$ fulfilling the Herzog–Kühl equations (1.3). Note that $L(\mathbf{a}, \mathbf{b})$ is a subspace of $L^{\mathrm{HK}}(\mathbf{a}, \mathbf{b})$. We shall show that these spaces are equal.

1.4 Pure resolutions

Now we shall consider a particular case of the resolution (1.1). Let $\mathbf{d} = (d_0, \ldots, d_l)$ be a strictly increasing sequence of integers, a degree sequence. The resolution (1.1) is *pure* if it has the form

$$S(-d_0)^{\beta_{0,d_0}} \leftarrow S(-d_1)^{\beta_{1,d_1}} \leftarrow \cdots \leftarrow S(-d_l)^{\beta_{l,d_l}}.$$

By a *pure diagram* (of type $\mathbf{d}$) we shall mean a diagram such that for each column i there is only one nonzero entry β_{i,d_i}, and the d_i form an increasing sequence. We see that a pure resolution gives a pure Betti diagram.

When M is CM of codimension c, the Herzog–Kühl equations give the following set of equations

$$\begin{bmatrix} 1 & -1 & \cdots & (-1)^c \\ d_0 & -d_1 & \cdots & (-1)^c d_c \\ \vdots & & & \vdots \\ d_0^{c-1} & -d_1^{c-1} & \cdots & (-1)^c d_c^{c-1} \end{bmatrix} \begin{bmatrix} \beta_{0,d_0} \\ \beta_{1,d_1} \\ \vdots \\ \beta_{c,d_c} \end{bmatrix}.$$

This is a $c \times (c+1)$ matrix of maximal rank. Hence there is only a one-dimensional $\mathbb{Q}$-vector space of solutions. The solutions may be found by computing the maximal minors which are Vandermonde determinants and we find

$$\beta_{i,d_i} = (-1)^i \cdot t \cdot \prod_{k \neq i} \frac{1}{(d_k - d_i)}$$

where $t \in \mathbb{Q}$. When $t > 0$ all these are positive. Let $\pi(\mathbf{d})$ be the diagram which is the smallest integer solution to the equations above. As we shall see pure resolutions and pure diagrams play a central role in the description of Betti diagrams up to rational multiple.

1.5 Linear Combinations of Pure Diagrams

The rays generated by the $\pi(\mathbf{d})$ turn out to be exactly the extremal rays in the cone $B(\mathbf{a}, \mathbf{b})$. Thus any Betti diagram is a positive linear combination of pure diagrams. Let us see how this works in an example.

Example 1.6. If the diagram of Example 1.1

$$\beta = \begin{matrix} 0 \\ 1 \\ 2 \end{matrix} \begin{bmatrix} 1 & 0 & 0 \\ 0 & 2 & 1 \\ 0 & 1 & 1 \end{bmatrix}$$

is a positive linear combination of pure diagrams $\pi(\mathbf{d})$, the only possibilities for these diagrams are

$$\pi(0,2,3) = \begin{bmatrix} 1 & 0 & 0 \\ 0 & 3 & 2 \\ 0 & 0 & 0 \end{bmatrix}, \quad \pi(0,2,4) = \begin{bmatrix} 1 & 0 & 0 \\ 0 & 2 & 0 \\ 0 & 0 & 1 \end{bmatrix}, \quad \pi(0,3,4) = \begin{bmatrix} 1 & 0 & 0 \\ 0 & 0 & 0 \\ 0 & 4 & 3 \end{bmatrix}.$$

Note that by the natural partial order on degree sequences we have

$$(0,2,3) < (0,2,4) < (0,3,4).$$

To find this linear combination we proceed as follows. Take the largest positive multiple c_1 of $\pi(0,2,3)$ such that $\beta - c_1\pi(0,2,3)$ is still nonnegative. We see that $c_1 = 1/2$ and get

$$\beta_1 = \beta - \frac{1}{2}\pi(0,2,3) = \begin{bmatrix} 1/2 & 0 & 0 \\ 0 & 1/2 & 0 \\ 0 & 1 & 1 \end{bmatrix}.$$

Then take the largest possible multiple c_2 of $\pi(0,2,4)$ such that $\beta_1 - c_2\pi(0,2,4)$ is nonnegative. We see that $c_2 = 1/4$ and get

$$\beta_2 = \beta - \frac{1}{2}\pi(0,2,3) - \frac{1}{4}\pi(0,2,4) = \begin{bmatrix} 1/4 & 0 & 0 \\ 0 & 0 & 0 \\ 0 & 1 & 3/4 \end{bmatrix}.$$

Taking the largest multiple c_3 of $\pi(0,3,4)$ such that $\beta_2 - c_3\pi(0,3,4)$ is nonnegative, we see that $c_3 = 1/4$ and the last expression becomes the zero diagram. Thus we get β as a positive rational combination of pure diagrams

$$\beta = \frac{1}{2}\pi(0,2,3) + \frac{1}{4}\pi(0,2,4) + \frac{1}{4}\pi(0,3,4).$$

The basic part of Boij–Söderberg theory says that this procedure will always work: It gives a *nonnegative* linear combination of pure diagrams. We proceed to develop this in more detail. With $\mathbb{Z}^{c+1}_{\text{deg}}$ equipped with the natural partial order, we get for $\mathbf{a}, \mathbf{b} \in \mathbb{Z}^{c+1}_{\text{deg}}$ the interval $[\mathbf{a}, \mathbf{b}]_{\text{deg}}$ consisting of all degree sequences $\mathbf{d}$ with $\mathbf{a} \le \mathbf{d} \le \mathbf{b}$. The diagrams $\pi(\mathbf{d})$ where $\mathbf{d} \in [\mathbf{a}, \mathbf{b}]_{\text{deg}}$ are the pure diagrams in the window determined by $\mathbf{a}$ and $\mathbf{b}$.

Example 1.7. If $\mathbf{a} = (0, 2, 3)$ and $\mathbf{b} = (0, 3, 4)$, the vector space $\mathbb{D}(\mathbf{a}, \mathbf{b})$ consists of the diagrams which may be nonzero in the positions marked by $*$ below.

$$\begin{bmatrix} * & 0 & 0 \\ 0 & * & * \\ 0 & * & * \end{bmatrix},$$

and so is five-dimensional. The Herzog–Kühl equations for the diagrams ($c = 2$) are the following two equations

$$\beta_{0,0} - (\beta_{1,2} + \beta_{1,3}) + (\beta_{2,3} + \beta_{2,4}) = 0,$$
$$0 \cdot \beta_{0,0} - (2\beta_{1,2} + 3\beta_{1,3}) + (3\beta_{2,3} + 4\beta_{2,4}) = 0.$$

These are linearly independent and so $L^{\text{HK}}(\mathbf{a}, \mathbf{b})$ will be three-dimensional. On the other hand the diagrams $\pi(0, 2, 3), \pi(0, 2, 4)$ and $\pi(0, 3, 4)$ are clearly linearly independent in this vector space and so they form a basis for it. This is a general phenomenon.

The linear space $L^{\text{HK}}(\mathbf{a}, \mathbf{b})$ (and as will turn out $L(\mathbf{a}, \mathbf{b})$) may be described as follows.

Proposition 1.8. *Given any maximal chain*

$$\mathbf{a} = \mathbf{d}^1 < \mathbf{d}^2 < \cdots < \mathbf{d}^r = \mathbf{b}$$

in $[\mathbf{a}, \mathbf{b}]_{\text{deg}}$. *The associated pure diagrams*

$$\pi(\mathbf{d}^1), \pi(\mathbf{d}^2), \ldots, \pi(\mathbf{d}^r)$$

form a basis for $L^{\text{HK}}(\mathbf{a}, \mathbf{b})$. *The length of such a chain, and hence the dimension of the latter vector space is* $r = 1 + \sum(b_i - a_i)$.

Proof. Let β be a solution of the HK-equations contained in the window $\mathbb{D}(\mathbf{a}, \mathbf{b})$. The vectors $\mathbf{d}^1$ and $\mathbf{d}^2$ differ in one coordinate, suppose it is the ith coordinate, so $\mathbf{d}^1 = (\ldots, d_i^1, \ldots)$ and $\mathbf{d}^2 = (\ldots, d_i^1 + 1, \ldots)$. Let c_1 be such that $\beta_1 = \beta - c_1\pi(\mathbf{d}^1)$ is zero in position (i, d_i^1). Then β_1 is contained in the window $\mathbb{D}(\mathbf{d}^2, \mathbf{b})$ and $\mathbf{d}^2, \ldots, \mathbf{d}^r$

is a maximal chain in $[\mathbf{d}^2, \mathbf{b}]_{\text{deg}}$. We may proceed by induction and in the end get β_{r-1} contained in $[\mathbf{b}, \mathbf{b}]_{\text{deg}}$. Then β_{r-1} is pure and so is a multiple of $\pi(\mathbf{d}^r)$. In conclusion

$$\beta = \sum_{i=1}^{r} c_i \pi(\mathbf{d}^i).$$

To see that the $\pi(\mathbf{d}^i)$ are linearly independent, note that $\pi(\mathbf{d}^1)$ is not a linear combination of the $\pi(\mathbf{d}^i)$ for $i \geq 2$ since $\pi(\mathbf{d}^1)$ is nonzero in position (i, d_i^1) while the $\pi(\mathbf{d}^i)$ for $i \geq 2$ are zero in this position. Hence a dependency must involve only $\pi(\mathbf{d}^i)$ for $i \geq 2$. But then we may proceed by induction. □

1.6 The Boij–Söderberg Conjectures

The first part of the original Boij–Söderberg conjectures states the following.

Theorem 1.9. *For every degree sequence* $\mathbf{d}$, *a strictly increasing sequence of integers* $(d_0, \ldots, d_c)$, *there exists a Cohen–Macaulay module* M *of codimension* c *with pure resolution of type* $\mathbf{d}$.

We shall in Section 3 give an overview of the constructions of such resolutions, making the conjecture a theorem.

Corollary 1.10. *The linear space* $L(\mathbf{a}, \mathbf{b})$ *is equal to* $L^{\text{HK}}(\mathbf{a}, \mathbf{b})$.

Proof. The diagram $\pi(\mathbf{d})$ may now be realized, up to multiplication by a scalar, as the Betti diagram of a Cohen–Macaulay module. □

The second part of the Boij–Söderberg conjectures says the following.

Theorem 1.11. *Let* M *be a Cohen–Macaulay module of codimension* c *with Betti diagram* $\beta(M)$ *in* $\mathbb{D}(\mathbf{a}, \mathbf{b})$. *There is a unique chain*

$$\mathbf{d}^1 < \mathbf{d}^2 < \cdots < \mathbf{d}^r$$

in $[\mathbf{a}, \mathbf{b}]_{\text{deg}}$ *such that* $\beta(M)$ *is uniquely a linear combination*

$$c_1\pi(\mathbf{d}^1) + c_2\pi(\mathbf{d}^2) + \cdots + c_r\pi(\mathbf{d}^r),$$

where the c_i *are positive rational numbers.*

Remark 1.12. When M is any graded module of codimension $\geq c$, the same essentially holds true, but one must allow degree sequences $\mathbf{d}^i$ in $\mathbb{Z}^p_{\text{deg}}$ where p ranges over $c+1, \ldots, n+1$. See Subsection 5.1.

Remark 1.13. Combining this with Theorem 1.9 we see that there are modules M_i with pure resolution of type $\mathbf{d}^i$ such that for suitable multiples p and p_i then M^p and $\bigoplus_i M_i^{p_i}$ have the same Betti diagram.

1.7 Algorithmic Interpretation

As a consequence of Theorem 1.11 we get a simple algorithm to find this unique decomposition, which is the way we did it in Example 1.6. This algorithm, with interesting consequences, is presented in [12, Section 1]. For a diagram β, for each i let d_i be the minimal j such that β_{ij} is nonzero. This gives a sequence $\underline{\mathbf{d}}(\beta) = (d_0, d_1, \ldots, d_c)$, the *lower bound* of β.

Example 1.14. Below the nonzero positions of β is indicated by $*$'s.

$$\begin{array}{c|ccccc} 0 & * & * & 0 & 0 & 0 \\ 1 & * & * & * & 0 & 0 \\ 2 & 0 & * & * & * & * \\ 3 & 0 & 0 & 0 & * & * \end{array}$$

Then $\underline{\mathbf{d}}(\beta) = (0, 1, 3, 5, 6)$.

There is a pure Betti diagram $\pi(\underline{\mathbf{d}}(\beta))$ and let $c(\beta) > 0$ be the maximal number such that $\beta' = \beta - c(\beta)\pi(\underline{\mathbf{d}}(\beta))$ is nonnegative.

Let M be a Cohen–Macaulay module. The algorithm is now as follows.

1. Let $\beta = \beta(M)$ and $i = 1$.
2. Compute $\underline{\mathbf{d}}^i := \underline{\mathbf{d}}(\beta)$ and $c^i := c(\beta)$. Then $\underline{\mathbf{d}}^i$ will be a strictly increasing sequence. Let $\beta := \beta - c^i\pi(\underline{\mathbf{d}}^i)$.
3. If β is nonzero let $i := i + 1$ and continue with Step 2. Otherwise stop.

The output will then be the unique positive linear combination

$$\beta(M) = c^1\pi(\underline{\mathbf{d}}^1) + c^2\pi(\underline{\mathbf{d}}^2) + \cdots + c^r\pi(\underline{\mathbf{d}}^r).$$

1.8 Geometric Interpretation

Since for any chain $D : \mathbf{d}^1 < \mathbf{d}^2 < \cdots < \mathbf{d}^r$ in $[\mathbf{a}, \mathbf{b}]_{\deg}$ the Betti diagrams $\pi(\mathbf{d}^1), \ldots, \pi(\mathbf{d}^r)$ are linearly independent diagrams in $\mathbb{D}(\mathbf{a}, \mathbf{b})$, their positive rational linear combinations give a simplicial cone $\sigma(D)$ in $\mathbb{D}(\mathbf{a}, \mathbf{b})$, which actually is in the subspace $L^{\mathrm{HK}}(\mathbf{a}, \mathbf{b})$. Two such cones will intersect along another such cone, which is the content of the following.

Proposition 1.15. *The set of simplicial cones $\sigma(D)$ where D ranges over all chains $\mathbf{d}^1 < \cdots < \mathbf{d}^r$ in $[\mathbf{a}, \mathbf{b}]_{\deg}$ form a simplicial fan, which we denote as $\Sigma(\mathbf{a}, \mathbf{b})$.*

Proof. Let D be a chain like above and E another chain $\mathbf{e}^1 < \cdots < \mathbf{e}^s$ in $[\mathbf{a}, \mathbf{b}]_{\deg}$. We shall show that $\sigma(D)$ and $\sigma(E)$ intersect in $\sigma(D \cap E)$. So consider

$$\beta = \sum c_i \pi(\mathbf{d}^i) = \sum c_i' \pi(\mathbf{e}^i)$$

in the intersection. By omitting elements in the chain we may assume all c_i and c_i' positive. Then the lower bound of β which we denoted $\underline{\mathbf{d}}(\beta)$, will be $\mathbf{d}^1$. But it will also be $\mathbf{e}^1$, and so $\mathbf{e}^1 = \mathbf{d}^1$. Assume say that $c_1 \le c_1'$. Let $\beta' = \beta - c_1\pi(\mathbf{d}^1)$. Then β' is in $\sigma(D \backslash \{\mathbf{d}^1\})$ and in $\sigma(E)$. By induction on the sum of the cardinalities of D and E, we get that β' is in $\sigma(D \cap E \backslash \{\mathbf{d}^1\})$ and so β is in $\sigma(D \cap E)$. □

We now get the following description of the positive cone $B(\mathbf{a}, \mathbf{b})$.

Theorem 1.16. (i) *The realization of the fan* $\Sigma(\mathbf{a}, \mathbf{b})$ *is contained in the positive cone* $B(\mathbf{a}, \mathbf{b})$.

(ii) *The positive cone* $B(\mathbf{a}, \mathbf{b})$ *is contained in the realization of the fan* $\Sigma(\mathbf{a}, \mathbf{b})$.

In conclusion the realization of the fan $\Sigma(\mathbf{a}, \mathbf{b})$ *is equal to the positive cone* $B(\mathbf{a}, \mathbf{b})$.

It may seem overly pedantic to express it in this way but the reason should be clear from the proof.

Proof. Part (i) is equivalent to the first part of the Boij–Söderberg conjectures, Theorem 1.9. Part (ii) is equivalent to the second part of the Boij–Söderberg conjectures, Theorem 1.11. □

2 The Exterior Facets of the Boij–Söderberg Fan and Their Supporting Hyperplanes

In order to prove Theorem 1.11, which is equivalent to part b. of Theorem 1.16, we must describe the exterior facets of the Boij–Söderberg fan $\Sigma(\mathbf{a}, \mathbf{b})$ and their supporting hyperplanes.

2.1 The Exterior Facets

Let $D : \mathbf{d}^1 < \cdots < \mathbf{d}^r$ be a maximal chain in $[\mathbf{a}, \mathbf{b}]_{\deg}$. The positive rational linear combinations of the pure diagrams $\pi(\mathbf{d}^1), \ldots, \pi(\mathbf{d}^r)$ is a maximal simplicial cone $\sigma(D)$ in the Boij–Söderberg fan $\Sigma(\mathbf{a}, \mathbf{b})$. The facets of the cone $\sigma(D)$ are the cones $\sigma(D \backslash \{\mathbf{d}^i\})$ for $i = 1, \ldots, r$. We call such a facet *exterior* if it is on only one simplicial cone in the fan $\Sigma(\mathbf{a}, \mathbf{b})$.

Example 2.1. Let $\mathbf{a} = (0, 1, 3)$ and $\mathbf{b} = (0, 3, 4)$. The Hasse diagram of the poset $[\mathbf{a}, \mathbf{b}]_{\deg}$ is the diagram.

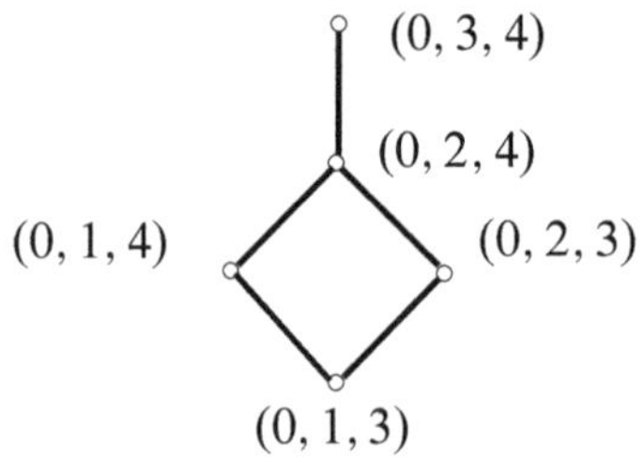

There are two maximal chains in this diagram

$$D: \quad (0,1,3) < (0,2,3) < (0,2,4) < (0,3,4)$$
$$E: \quad (0,1,3) < (0,1,4) < (0,2,4) < (0,3,4)$$

so the realisation of the Boij–Söderberg fan consists of the union of two simplicial cones of dimension four. We intersect this transversally with a hyperplane to get a three-dimension picture of this as the union of two tetrahedra. (The vertices are labelled by the pure diagrams on their rays.)

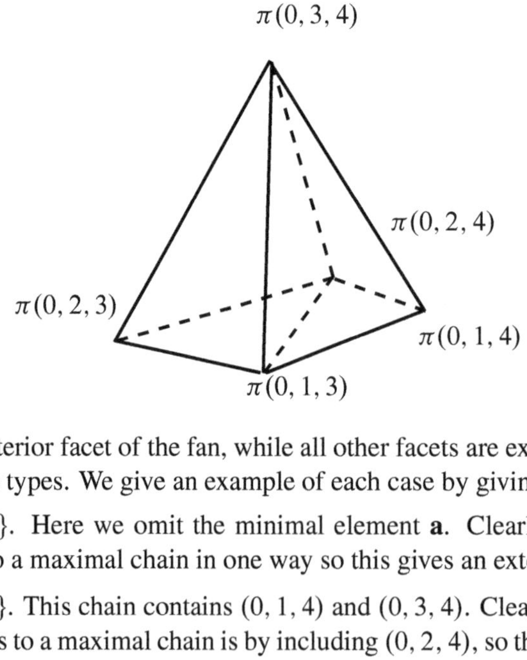

There is one interior facet of the fan, while all other facets are exterior. The exterior facets are of three types. We give an example of each case by giving the chain.

(i) $D \backslash \{(0,1,3)\}$. Here we omit the minimal element **a**. Clearly this can only be completed to a maximal chain in one way so this gives an exterior facet.

(ii) $E \backslash \{(0,2,4)\}$. This chain contains (0, 1, 4) and (0, 3, 4). Clearly the only way to complete this to a maximal chain is by including $(0,2,4)$, so this gives an exterior facet.

(iii) $D \backslash \{(0,2,4)\}$. This contains $(0,2,3)$ and $(0,3,4)$. When completing this to a maximal chain clearly one must first increase the last 3 in $(0,2,3)$ to 4, giving $(0,2,4)$. So D is the only maximal chain containing this.

The following tells that these three types are the only ways of getting exterior facets.

Proposition 2.2. *Let D be a maximal chain in $[\mathbf{a}, \mathbf{b}]_{\deg}$ and $f \in D$. Then $\sigma(D \backslash \{f\})$ is an exterior facet iff one of the following holds:*

(i) *f is either* **a** *or* **b**.

(ii) *The degree sequences of f^- and f^+ immediately before and after f in D differ in exactly one position. So for some r we have*

$$f^- = (\ldots, r-1, \ldots), \quad f = (\ldots, r, \ldots), \quad f^+ = (\ldots, r+1, \ldots).$$

(iii) *The degree sequences of f^- and f^+ immediately before and after f in D differ in exactly two adjacent positions such that in these two positions there is an integer r such that*

$$f^- = (\ldots, r-1, r, \ldots), \quad f = (\ldots, r-1, r+1, \ldots),$$
$$f^+ = (\ldots, r, r+1, \ldots).$$

In Case (iii) we denote the exterior facet by $\mathbf{facet}(f, \tau)$ where τ is the position of the number $r-1$ in f.

Proof. That these cases give exterior facets is immediate as in the discussion of the example above. That this is the only way to achieve exterior facets is also easy to verify. □

2.2 The Supporting Hyperplanes

If σ is full dimensional simplicial cone in a vector space L, each facet of σ is contained in a unique hyperplane, which is the kernel of a nonzero linear functional $h : L \to k$.

We shall apply this to the cones $\sigma(D)$ in $L^{\mathrm{HK}}(\mathbf{a}, \mathbf{b})$, and find the equations of the hyperplanes H defining the exterior facets of $\sigma(D)$. Actually we consider the inclusion $\sigma(D) \subseteq L^{\mathrm{HK}}(\mathbf{a}, \mathbf{b}) \subseteq \mathbb{D}(\mathbf{a}, \mathbf{b})$ and rather find a hyperplane H' in $\mathbb{D}(\mathbf{a}, \mathbf{b})$ with $H = H' \cap L^{\mathrm{HK}}(\mathbf{a}, \mathbf{b})$. The equation of such a hyperplane is not unique up to constant however. Since $L^{\mathrm{HK}}(\mathbf{a}, \mathbf{b})$ is cut out by the Herzog–Kühl equations, we may add any linear combinations of these equations, say ℓ, and get a new equation $h'' = h' + \ell$ defining another hyperplane $H'' \subseteq \mathbb{D}(\mathbf{a}, \mathbf{b})$ which still intersects $L^{\mathrm{HK}}(\mathbf{a}, \mathbf{b})$ in H. In Cases (i) and (ii) of Proposition 2.2 there turns out to be a unique natural choice for the hyperplane, while in Case (iii) there are two distinguished hyperplanes.

Example 2.3. We continue Example 2.1 and look at the various types of exterior facets of Proposition 2.2.

(i) In the chain

$$D: \quad (0,1,3) < (0,2,3) < (0,2,4) < (0,3,4)$$

if we look at the facet of $\sigma(D)$ we get by removing $(0,1,3)$, the natural equation for a hyperplane in $\mathbb{D}(\mathbf{a}, \mathbf{b})$ is $\beta_{1,1} = 0$. This hyperplane contains $\pi(0,2,3)$, $\pi(0,2,4)$, and $\pi(0,3,4)$, but it does not contain $\pi(0,1,3)$. We may get other equations by adding linear combinations of the Herzog–Kühl equations but this equation is undoubtedly the simplest one.

(ii) In the chain

$$E: \ (0,1,3) < (0,1,4) < (0,2,4) < (0,3,4)$$

if we consider the facet of $\sigma(D)$ we get by removing $(0,2,4)$, the natural equation for a supporting hyperplane is $\beta_{1,2} = 0$.

(iii) In the case that we remove $f = (0,2,4)$ from D things are more refined. There turns out to be two linear functionals on $\mathbb{D}(\mathbf{a},\mathbf{b})$ which define two distinguished hyperplanes, called respectively the upper and lower hyperplanes. We will represent the equation of a hyperplane in $\mathbb{D}(\mathbf{a},\mathbf{b})$ by giving the coefficients of the β_{ij}. To describe the upper hyperplane note that the Betti diagram of the sequence $f^+ = (0,3,4)$ immediately after f is

$$\begin{array}{r} -1 \\ 0 \\ 1 \\ 2 \end{array}\begin{bmatrix} 0 & 0 & 0 \\ 1^* & 0 & 0 \\ 0 & 0^- & 0^- \\ 0 & 4^+ & 3^+ \end{bmatrix}$$

The nonzero entries of the diagram have been additionally labelled with $*$, $+$ and $+$. Similarly the nonzero positions of $\pi(f^-)$ will be labelled by $*, -$ and $-$. Thinking of the Betti diagram as stretching infinitely upwards and downwards, the zeros in the diagram for f^+ are divided into an upper and lower part. The equation of the upper hyperplane, the *upper equation* will have possible nonzero values only in the upper part of f^+ (marked with normalsized $*$'s):

$$\begin{array}{r|lll} -2 & * & * & * \\ -1 & * & * & * \\ 0 & 0^* & * & * \\ 1 & 0 & *^- & *^- \\ 2 & 0 & 0^+ & 0^+ \end{array}\ . \tag{2.1}$$

Remark 2.4. The choice of facet equation $\beta_{ij} = 0$ for exterior facets of type (i) and (ii) and of the upper equation for exterior facets of type (iii) is further justified in the last paragraph of Subsection 5.1. The Betti diagrams of *all* graded modules whose Betti diagram is in the window $\mathbb{D}(\mathbf{a},\mathbf{b})$, generate a full-dimensional cone in this window. The exterior facet types above have corresponding larger facets in this cone, and the equations above give the unique (up to scalar) equations of these larger facets.

Before proceeding to find the upper hyperplane equation, we note the following which says that the choice of window bounds $\mathbf{a}$ and $\mathbf{b}$ does not have any essential effect on the exterior facets, and that the exterior facets of type (iii) essentially only depend on the f omitted and not on the chain.

Lemma 2.5. *Consider facets of type* (iii) *in Proposition* 2.2.

(i) *If D and E are two maximal chains in $[\mathbf{a},\mathbf{b}]_{\deg}$ which both contain the subsequence $f^- < f < f^+$, the exterior facets $\sigma(D\backslash\{f\})$ and $\sigma(E\backslash\{f\})$ define the same hyperplane in $L^{\mathrm{HK}}(\mathbf{a},\mathbf{b})$.*

(ii) *Let $\mathbf{a}' \leq \mathbf{a} \leq \mathbf{b} \leq \mathbf{b}'$ and suppose D' is a maximal chain in $[\mathbf{a}',\mathbf{b}']_{\deg}$ restricting to D in $[\mathbf{a},\mathbf{b}]_{\deg}$. If H' in $\mathbb{D}(\mathbf{a}',\mathbf{b}')$ is a hyperplane defining $\sigma(D'\backslash\{f\})$, then $H' \cap \mathbb{D}(\mathbf{a},\mathbf{b})$ is a hyperplane defining $\sigma(D\backslash\{f\})$.*

Proof. As for part (ii) $\sigma(D\backslash\{f\})$ is a subset of $\sigma(D'\backslash\{f\})$ and so is contained in H'. But H' does not contain $L^{\mathrm{HK}}(\mathbf{a},\mathbf{b})$ since H' does not contain $\pi(f)$ which is contained in this linear space, and so $H' \cap L^{\mathrm{HK}}(\mathbf{a},\mathbf{b})$ is a hyperplane in $L^{\mathrm{HK}}(\mathbf{a},\mathbf{b})$. Thus $H' \cap \mathbb{D}(\mathbf{a},\mathbf{b})$ is a hyperplane in $\mathbb{D}(\mathbf{a},\mathbf{b})$ defining $\sigma(D\backslash\{f\})$.

For part (i) note that if D is the chain

$$\mathbf{a} = \mathbf{d}^1 < \cdots < \mathbf{d}^{p-1} = f^- < f < f^+ = \mathbf{d}^{p+1} < \cdots < \mathbf{d}^r = \mathbf{b}$$

the space L^- spanned by $\pi(\mathbf{d}^1),\ldots,\pi(\mathbf{d}^{p-1})$ is by Proposition 1.8 equal to $L^{\mathrm{HK}}(\mathbf{a}, f^-)$, and so depends only on $\mathbf{a}$ and f^-. Similarly L^+ spanned by $\pi(\mathbf{d}^{p+1})$, $\ldots,\pi(\mathbf{d}^r)$ is equal to $L^{\mathrm{HK}}(f^+,\mathbf{b})$. The hyperplane of $\sigma(D\backslash\{f\})$ is then spanned by L^- and L^+, which depends only on f, f^+, f^-, $\mathbf{a}$ and $\mathbf{b}$. □

Example 2.6. Let us return to Example 2.3 to find the hyperplane equations when we remove $f = (0,2,4)$ from D. By the previous proposition we may as well assume that $\mathbf{a}$ is some tuple with small coordinates and $\mathbf{b}$ is a tuple with large coordinates.

The upper hyperplane equation h_{up}, which has the form given in (2.1), does not vanish on $\pi(f)$ but will, by Lemma 2.5 vanish on $\pi(g)$ when $g < f$. In particular it vanishes on

$$\pi(f^-) = \pi(0,2,3) = \begin{matrix} -1 \\ 0 \\ 1 \end{matrix}\begin{bmatrix} 0 & 0 & 0 \\ 1^* & 0 & 0 \\ 0 & 3^- & 2^- \end{bmatrix},$$

and so the coefficients of h_{up} must have the form

$$\begin{array}{r|ccc} -2 & * & * & * \\ -1 & * & * & * \\ 0 & 0 & * & * \\ 1 & 0 & 2\alpha & -3\alpha \end{array}$$

where α is some nonzero constant, which we may as well take to be $\alpha = 1$. Also h_{up} must vanish on

$$\pi(0,1,3) = \begin{matrix} 0 \\ 1 \end{matrix}\begin{bmatrix} 2 & 3 & 0 \\ 0 & 0 & 1 \end{bmatrix}.$$

This shows that the coefficients of h_{up} must be

$$\begin{array}{r|ccc} -2 & * & * & * \\ -1 & * & * & * \\ 0 & 0 & 1 & * \\ 1 & 0 & 2 & -3 \end{array}.$$

We may continue with an element just before $(0,1,3)$ in a maximal chain, say $(0,1,2)$. Since

$$\pi(0,1,2) = 0\begin{bmatrix} 1 & 2 & 1 \end{bmatrix}.$$

we get that the coefficients of h_{up} are

$$\begin{array}{r|ccc} -2 & * & * & * \\ -1 & * & * & * \\ 0 & 0 & 1 & -2 \\ 1 & 0 & 2 & -3 \end{array}.$$

In this way we may continue and h_{up} will be uniquely determined in all positions in the window determined by **a** and **b**. We find that the coefficients of h_{up} are given by the diagram:

$$\begin{array}{r|ccc} -3 & 3 & -2 & 1 \\ -2 & 2 & -1 & 0 \\ -1 & 1 & 0 & -1 \\ 0 & 0^* & 1 & -2 \\ 1 & 0 & 2^- & -3^- \\ 2 & 0 & 0^+ & 0^+ \end{array}.$$

In order to find the lower equation, we may in a similar way consider the diagram of $\pi(f^-)$

$$\begin{array}{r} -1 \\ 0 \\ 1 \\ 2 \end{array}\begin{bmatrix} 0 & 0 & 0 \\ 1^* & 0 & 0 \\ 0 & 3^- & 2^- \\ 0 & 0^+ & 0^+ \end{bmatrix}.$$

Again thinking of the Betti diagram as stretching infinitely upwards and downwards, the positions with zero are divided into an upper and a lower part. There is a unique hyperplane defined by a linear form h_{low} which may have nonzero entries only in the lower part of the diagram of $\pi(f^-)$. We find that the coefficients of h_{low} are given by the following:

$$\begin{array}{c|ccc} 0 & 0^* & 0 & 0 \\ 1 & -1 & 0^- & 0^- \\ 2 & -2 & 3^+ & -4^+ \\ 3 & -3 & 4 & -5 \\ 4 & -4 & 5 & -6 \\ 5 & -5 & 6 & -7 \end{array}.$$

Proposition 2.7. *Let $f^- < f < f^+$ be the degree sequence as in part 3 of Proposition 2.2. There is a unique hyperplane in $\mathbb{D}(\mathbf{a}, \mathbf{b})$, the upper hyperplane, that contains* $\mathbf{facet}(f, \tau)$ *and whose equation has coefficient zero of $\beta_{i,j}$ for all $j \geq f_i^+$.*

Proof. This is done as in the example by choosing any chain $f^- = \mathbf{d}^{p-1} > \mathbf{d}^{p-2} > \cdots > \mathbf{d}^1 = \mathbf{a}$ and making the equation of the hyperplane vanish on the elements of this chain. Lemma 2.5 shows that we get the same hyperplane equation independent of the choice of chain. □

A regular feature of the equations is that the diagonals from lower left to upper right have the same absolute values but alternating signs in the range where they are nonzero.

Lemma 2.8. *Let b_{ij} be the coefficient of β_{ij} in the upper equation h_{up}. If $j < f_i^+$ then $b_{i+1,j} = -b_{i,j}$.*

Proof. Both h_{up} and h_{low} are equations of the same hyperplane in the subspace $L^{\text{HK}}(\mathbf{a}, \mathbf{b})$. A linear combination of them, in our examples $h_{\text{up}} + h_{\text{low}}$, then vanishes on this space and so must be a linear combination of the Herzog–Kühl equations (1.3). But looking at these equations we see that the coefficient of $\beta_{i+1,j}$ and $\beta_{i,j}$ always have the same absolute value but different signs. □

What are the explicit forms of the facet equations, i.e. what determines the numbers occurring in these equations? We are interested in this because each supporting hyperplane H defines a halfspace H^+ and the intersection of all these halfspaces is a positive cone contained in the Boij–Söderberg fan $\Sigma(\mathbf{a}, \mathbf{b})$. We will be able to show that each Betti diagram of a module is in all the positive halfspaces. This shows that the positive cone $B(\mathbf{a}, \mathbf{b})$ is contained in the realization of $\Sigma(\mathbf{a}, \mathbf{b})$, so we obtain part b. of Theorem 1.11.

The numbers in the example above are too simple to make any deductions as to what governs them in general. A more sophisticated example is the following.

Example 2.9. The upper equation of **facet**$((-1,0,2,3),1)$ has coefficients:

$$U: \begin{array}{r|rrrr} -4 & 21 & -12 & 5 & 0 \\ -3 & 12 & -5 & 0 & 3 \\ -2 & 5 & 0 & -3 & 4 \\ -1 & 0^* & 3^- & -4^- & 3 \\ 0 & 0 & 0^+ & 0^+ & 0^* \end{array}$$

where in U the superscripts $*$ and $+$ indicate the nonzero parts of $\pi(f^+)$, while the $*$ and $-$ indicate the nonzero parts of $\pi(f^-)$. The polynomial ring in this case is $S = \Bbbk[x_1, x_2, x_3]$. Eisenbud and Schreyer, [12], recognised the numbers in this diagram as the Hilbert functions of the homology modules of the complex

$$E : E^0 = S(1)^5 \xrightarrow{d} S(2)^3 = E^1 \tag{2.2}$$

for a general map d.

The homology table of this complex is:

d	-3	-2	-1	0	1	2	3	4	5	6
$\dim_{\Bbbk}(H^1E)_d$	0	3	4	3	0	0	0	0	0	0
$\dim_{\Bbbk}(H^0E)_d$	0	0	0	0	0	5	12	21	32	45.

Two features of this cohomology table that we note are the following. The dimensions of $(H^0E)_d$ are the values of the Hilbert polynomial of H^0E for $d \geq 1$. This Hilbert polynomial is

$$5\binom{d+3}{2} - 3\binom{d+4}{2} = (d-1)(d+3).$$

This polynomial also gives the dimensions of H^1E in the degrees $d = -2, -1, 0$ but with opposite sign. Note also that the roots of this polynomial are 1 and -3 which are the negatives of the first and last entry in the degree sequence $(-1, 0, 2, 3)$ that we consider. In fact the lower and upper facet equations are now fairly simple to describe.

Given a sequence $\mathbf{z} : z_1 > z_2 > \cdots > z_{c-1}$ of integers. It gives a polynomial

$$p(d) = \prod_{i=1}^{c-1}(d - z_i).$$

Let $H(\mathbf{z})$ be the diagram in $\mathbb{D}$ such that:

- The value in position $(0, d)$ is $p(-d)$.
- The entries in positions $(i+1, d)$ and (i, d) for $i = 0, \dots, c-1$ have the same absolute values but opposite signs.

Example 2.10. When $c = 3$ and $p(d) = (d-1)(d+3)$ the diagram $H(\mathbf{z})$ is the following rotated 90° counterclockwise

	6	5	4	3	2	1	0	−1	−2	−3	
...	5	0	−3	−4	−3	0	5	12	21	32	...
...	−12	−5	0	3	4	3	0	−5	−12	−21	...
...	21	12	5	0	−3	−4	−3	0	5	12	...
...	−32	−21	−12	−5	0	3	4	3	0	−5	...

Now given a **facet**(f,τ) of the Boij–Söderberg fan. Associated to the sequence of integers

$$\hat{f} : -f_0 > -f_1 > \cdots > -f_{\tau-1} > -f_{\tau+2} > \cdots > -f_c$$

we get a $H(\hat{f})$. Let $U(f,\tau)$ be the diagram we get by making all entries of $H(\hat{f})$ on and below the positions occupied by $\pi(f^+)$ equal to zero. Explicitly $U(f,\tau)_{ij} = H(\hat{f})_{ij}$ for $j < f_i^+$ and $U(f,\tau)_{ij} = 0$ otherwise. The associated linear form is then:

$$\sum_{\substack{i<\tau \\ d<f_i}} (-1)^i \beta_{i,d}\, p(-d) + \sum_{\substack{i=\tau \\ d\leq f_\tau}} (-1)^i \beta_{i,d}\, p(-d) + \sum_{\substack{i>\tau \\ d<f_i}} (-1)^i \beta_{i,d}\, p(-d). \quad (2.3)$$

Proposition 2.11. *The upper equation h_{up} of* **facet**(f,τ) *has coefficients given by the diagram $U(f,\tau)$. The coefficients of the lower facet equation is $H(\hat{f}) - U(f,\tau)$.*

Proof. First note that $h_{\mathrm{up}}(\pi(f))$ is nonzero. If the degree sequence $f' \geq f^+$ then clearly $h_{\mathrm{up}}(\pi(f')) = 0$. When $f' \leq f^-$ it is shown in [12, Theorem 7.1] that $h_{\mathrm{up}}(\pi(f')) = 0$. □

2.3 Pairings of Vector Bundles and Resolutions

In order to prove Proposition 2.11 we had to show that the hyperplane equation h_{up} given by $U(f,\tau)$ is positive on $\pi(f)$ and vanishes on the other $\pi(f')$. With the explicit forms we have for all these expressions this could be done with numerical calculations. However to prove Theorem 1.11 we need to show that the form given by h_{up} is nonnegative on all Betti diagrams of Cohen–Macaulay modules.

In order to prove this positivity we must go beyond the numerics. It then appears that if β is a Betti diagram, the linear functional determined by $U(f,\tau)$ evaluated on β arises from a pairing between a Betti diagram and the cohomology table of a vector bundle. This is the fruitful viewpoint which enables us to show the desired positivity.

Example 2.12. Going back to the complex (2.2), if we sheafify this complex to get a complex of direct sums of line bundles on the projective plane $\mathbb{P}^2$

$$\tilde{E} : \mathcal{O}_{\mathbb{P}^2}(1)^5 \xrightarrow{\tilde{d}} \mathcal{O}_{\mathbb{P}^2}(2)^3,$$

the map $\tilde{d}$ is surjective and so the only nonvanishing homology is $\mathcal{E} = H^0(\tilde{E})$. The table below is the cohomology table of the vector bundle $\mathcal{E}$.

d	$\cdots$	-6	-5	-4	-3	-2	-1	0	1	2	3	4	5	6	$\cdots$
$\dim_{\Bbbk} H^2\mathcal{E}(d)$	$\cdots$	21	12	5	0	0	0	0	0	0	0	0	0	0	$\cdots$
$\dim_{\Bbbk} H^1\mathcal{E}(d)$	$\cdots$	0	0	0	0	3	4	3	0	0	0	0	0	0	$\cdots$
$\dim_{\Bbbk} H^0\mathcal{E}(d)$	$\cdots$	0	0	0	0	0	0	0	0	5	12	21	32	45	$\cdots$

and the values are the absolute values of $(d-1)(d+3)$. This is an example of a bundle with supernatural cohomology as we now define.

Definition 2.13. Let

$$z_1 > z_2 > \cdots > z_m$$

be a sequence of integers. A vector bundle $\mathcal{E}$ on the projective space $\mathbb{P}^m$ has *supernatural cohomology* if:

(i) The Hilbert polynomial is $\chi\mathcal{E}(d) = \frac{r^0}{m!} \cdot \prod_{i=1}^m (d - z_i)$ for a constant r^0 (which must be the rank of $\mathcal{E}$).

(ii) For each d let i be such that $z_i > d > z_{i+1}$. Then

$$H^i\mathcal{E}(d) = \begin{cases} \frac{r^0}{m!} \cdot \prod_{i=1}^m |d - z_i|, & z_i > d > z_{i+1}, \\ 0, & \text{otherwise.} \end{cases}$$

The sequence $z_1 > z_2 > \cdots > z_m$ is called the *root sequence* of the bundle $\mathcal{E}$.

In particular we see that for each d there is at most one nonvanishing cohomology group. We show in Section 3 that for any sequence $\mathbf{z}$ of strictly decreasing integers such a vector bundle exists.

Remark 2.14. The naturality of the notion of supernatural cohomology for a vector bundle, may be seen from the fact that it is equivalent to its Tate resolution, see Subsection 6.5, being pure, i.e. each cohomological term in the Tate resolution, a free module over the exterior algebra, being generated in a single degree.

Proposition 2.11 and the explicit form (2.3) just before it, may now be translated to the following.

Proposition 2.15. *For a* **facet**$(f, \tau,)$ *let* $\mathcal{E}$ *be a vector bundle on* $\mathbb{P}^{c-1}$ *with supernatural cohomology and root sequence*

$$-f_0 > -f_1 > \cdots > -f_{\tau-1} > -f_{\tau+2} > \cdots > -f_c.$$

Let $\gamma_{i,d} = H^i \mathcal{E}(d)$ and $\gamma_{\le i,d}$ the alternating sum $\gamma_{0,d} - \gamma_{1,d} + \cdots + (-1)^i \gamma_{i,d}$. The upper facet equation $h_{\text{up}}(\beta)$ is defined by the linear form

$$\sum_{\substack{i<\tau \\ d<f_i}} (-1)^i \beta_{i,d}\gamma_{\le i,-d} + \sum_{d\le f_\tau} (-1)^\tau \beta_{\tau,d}\gamma_{\le\tau,-d} + \sum_{d<f_{\tau+1}} (-1)^{\tau+1}\beta_{\tau+1,d}\gamma_{\le\tau,-d}$$
$$+ \sum_{\substack{i>\tau+1 \\ d<f_i}} (-1)^i \beta_{i,d}\gamma_{\le i-2,-d}.$$

To understand this as a special case of the upcoming (2.4), we may note the following.

- $\gamma_{\le i,-d} = 0$ when $i < \tau$ and $d \ge f_i$.
- $\gamma_{\le \tau-1,-d} = 0$ when $d > f_\tau$.
- $\gamma_{\le i-2,-d} = 0$ when $i > \tau + 1$ and $d \ge f_i$.

By studying many examples and a leap of insight, Eisenbud and Schreyer defined for any integer e and $0 \le \tau \le n-1$ a pairing $\langle \beta, \gamma\rangle_{e,\tau}$ between diagrams and cohomology tables as the expression

$$\sum_{\substack{i<\tau, \\ d\in\mathbb{Z}}} (-1)^i \beta_{i,d}\gamma_{\le i,-d} + \sum_{d\le e} (-1)^\tau \beta_{\tau,d}\gamma_{\le\tau,-d} + \sum_{d>e} (-1)^\tau \beta_{\tau,d}\gamma_{\le\tau-1,-d} \tag{2.4}$$
$$+ \sum_{d\le e+1} (-1)^{\tau+1}\beta_{\tau+1,d}\gamma_{\le\tau,-d} + \sum_{d>e+1} (-1)^{\tau+1}\beta_{\tau+1,d}\gamma_{\le\tau-1,-d}$$
$$+ \sum_{\substack{i>\tau+1, \\ d\in\mathbb{Z}}} (-1)^i \beta_{i,d}\gamma_{\le i-2,-d}.$$

When $e = f_\tau$ and γ is the cohomology table of the supernatural bundle of Proposition 2.15, this reduces to the expression given there. If $F_\bullet$ is a resolution and $\mathcal{F}$ is a coherent sheaf on $\mathbb{P}^{n-1}$ we let $\gamma(\mathcal{F})$ be its cohomology table and define

$$\langle F_\bullet, \mathcal{F}\rangle_{e,\tau} = \langle \beta(F_\bullet), \gamma(\mathcal{F})\rangle_{e,\tau}.$$

That this pairing is the natural one is established by the following which is the key result of the paper [12], extended somewhat in [15].

Theorem 2.16. *For any minimal free resolution $F_\bullet$ of length $\le c$ and coherent sheaf $\mathcal{F}$ on $\mathbb{P}^{c-1}$ the pairing*

$$\langle F_\bullet, \mathcal{F}\rangle_{e,\tau} \ge 0.$$

The proofs of this uses the spectral sequence of a double complex. It is not long but somewhat technical so we do not reproduce it here, but refer the reader to Theorem 4.1 of [12] and Theorem 4.1 of [15], or the latest Theorem 3.3 of [14]. It is essential that $F_\bullet$ is a *minimal* free resolution. Using the above results we are now in a position to prove Theorem 1.11 or equivalently Theorem 1.16 (ii).

Proof. If $\mathbf{facet}(f, \tau)$ is a facet of type (iii) of the fan $\Sigma(\mathbf{a}, \mathbf{b})$, the upper hyperplane equation is $\langle -, \gamma(\mathcal{E})\rangle_{e,\tau} = 0$ where $e = f_\tau$ and $\mathcal{E}$ given in Proposition 2.15. For facets of type (i) or (ii) the hyperplane equations are $\beta_{ij} = 0$ for suitable i, j.

Each exterior facet determines a nonnegative half plane H^+. Since the forms above are nonnegative on all Betti diagrams $\beta(M)$ in $\mathbb{D}(\mathbf{a}, \mathbf{b})$ by Theorem 2.16, the cone $B(\mathbf{a}, \mathbf{b})$ is contained in the intersection of all the half planes H^+ which again is contained in the fan $\Sigma(\mathbf{a}, \mathbf{b})$. □

3 The Existence of Pure Free Resolutions and of Vector Bundles with Supernatural Cohomology

There are three main constructions of pure free resolutions. The first appeared on the arXiv.org in September 2007 [11]. This construction works in char $\mathbb{k} = 0$ and is the $\mathrm{GL}(n)$-equivariant resolution. Then in December 2007 appeared the simpler but rougher construction of [12] which works in all characteristics. In the paper [11] there also appeared another construction, resolutions of modules supported on determinantal loci. This construction is somewhat less celebrated but certainly deserves more attention for its naturality and beauty. It is a comprehensive generalization of the Eagon–Northcott complexes and Buchsbaum–Rim complexes in a generic setting.

Quite parallel to the first two constructions of pure free resolutions, there are analogous constructions of vector bundles on $\mathbb{P}^{n-1}$ with supernatural cohomology. These constructions are actually simpler than the constructions of pure free resolutions, and were to some extent known before the term supernatural cohomology was coined in [12]. In the following we let V be a finite dimensional vector space and let S be the symmetric algebra $S(V)$ with unique graded maximal ideal $\mathfrak{m}$.

3.1 The Equivariant Pure Free Resolution

We shall first give the construction of the $\mathrm{GL}(V)$-equivariant pure resolution of type $(1, 1, \ldots, 1)$ and more generally of type $(r, 1, \ldots, 1)$ for $r \geq 1$. These cases are known classically, and provide the hint for how to search for equivariant pure resolutions of any type $\mathbf{d}$.

Pure resolutions of type $(1, 1, \ldots, 1)$. In this case the resolution is the Koszul complex

$$S \leftarrow S \otimes_{\mathbb{k}} V \leftarrow S \otimes_{\mathbb{k}} \textstyle\bigwedge^2 V \leftarrow \cdots \leftarrow S \otimes_{\mathbb{k}} \bigwedge^n V$$

which is a resolution of the module $\mathbb{k} = S/\mathfrak{m}$. (We consider V to have degree 1, and $\bigwedge^p V$ to have degree p.) The general linear group $\mathrm{GL}(V)$ acts on each term $S \otimes_{\mathbb{k}} \bigwedge^p V$ since it acts on S and $\bigwedge^p V$. And the differentials respect this action so they are maps of $\mathrm{GL}(V)$-modules. We say the resolution is $\mathrm{GL}(V)$-equivariant. To define the differentials note that there are $\mathrm{GL}(V)$-equivariant maps $\bigwedge^{p+1} V \xrightarrow{\rho}$

$V \otimes_{\Bbbk} \bigwedge^p V$. Also let $\mu : S \otimes_{\Bbbk} V \to S$ be the multiplication map. The differential in the Koszul complex is then:

$$S \otimes_{\Bbbk} \bigwedge^{p+1} V \xrightarrow{1_S \otimes \rho} S \otimes_{\Bbbk} V \otimes_{\Bbbk} \bigwedge^p V \xrightarrow{\mu \otimes 1} S \otimes_{\Bbbk} \bigwedge^p V.$$

Pure resolutions of type $(r, 1, \dots, 1)$. Let us consider resolutions of type (3, 1, 1).

Example 3.1. In $S = \Bbbk[x_1, x_2, x_3]$ the ideal $\mathfrak{m}^3 = (x_1, x_2, x_3)^3$ has 3-linear resolution. The resolution of the quotient ring (a Cohen–Macaulay module of codimension three) is :

$$S \xleftarrow{d_0} S(-3)^{10} \xleftarrow{d_1} S(-4)^{15} \xleftarrow{d_2} S(-5)^6.$$

Looking at this complex in degree 3, the exponent 10 is the third symmetric power $S_3(V)$ of $V = \langle x_1, x_2, x_3 \rangle$. Looking at the complex in degree 4, we see that 15 is the dimension of the kernel of the multiplication $V \otimes_{\Bbbk} S_3(V) \to S_4(V)$. This map is $\mathrm{GL}(V)$-equivariant and the kernel is the representation $S_{3,1}(V)$ which has dimension 15 (see below for references explaining this representation). The inclusion

$$S_{3,1}(V) \to V \otimes_{\Bbbk} S_3(V)$$

induces a composition

$$V \otimes_{\Bbbk} S_{3,1}(V) \to V \otimes_{\Bbbk} V \otimes_{\Bbbk} S_3(V) \to S_2(V) \otimes_{\Bbbk} S_3(V).$$

This is the map d_1 in degree 5 and the kernel of this map is the representation $S_{3,1,1}(V)$ whose dimension is 6, accounting for the last term in the resolution above.

In general it is classically known that the resolution of $S/\mathfrak{m}^r$ is

$$S \leftarrow S \otimes_{\Bbbk} S_r(V) \leftarrow S \otimes_{\Bbbk} S_{r,1}(V) \leftarrow \cdots \leftarrow S \otimes_{\Bbbk} S_{r,1^{n-1}}(V). \tag{3.1}$$

This is a pure resolution of type $(r, 1, 1, \dots, 1)$.

Representations of $\mathbf{GL}(V)$. Let us pause to give a brief explanation of the terms $S_{r,1^{n-1}}(V)$. The irreducible representations of $\mathrm{GL}(V)$ where $n = \dim_{\Bbbk} V$ are classified by partitions of integers

$$\lambda : \lambda_1 \geq \lambda_2 \geq \cdots \geq \lambda_n.$$

For each such partition there is a representation denoted by $S_\lambda(V)$. The details of the construction are easily found in various textbooks like [23], [8] or [29].

Such a partition may be displayed in a Young diagram if $\lambda_n \geq 0$. With row index going downwards, put λ_i boxes in row i, and align the rows to the left. (Call this Horizontal display. Another convention is to display λ_i boxes in column i and top align them. Call this Vertical display.)

If

$$\lambda' : \lambda_1 + a \geq \lambda_2 + a \geq \cdots \geq \lambda_n + a,$$

then $S_{\lambda'}(V) = (\bigwedge^n V)^a \otimes_{\Bbbk} S_\lambda(V)$ where $\bigwedge^n V$ is the one-dimensional determinant representation. In Example 3.1 we needed to consider tensor products $S_\lambda(V) \otimes_{\Bbbk} S_\mu(V)$. In char $\Bbbk = 0$ this decomposes into a direct sum of irreducible representations. In general this is complicated, but in one case there is a simple rule which is of central use for us in the construction of equivariant resolutions.

For two partitions μ and λ with $\mu_i \geq \lambda_i$ for each i, say that $\mu \backslash \lambda$ is a *horizontal strip* (*vertical strip* with vertical display) if $\mu_i \leq \lambda_{i-1}$ for all i. Thus when removing the diagram of λ from that of μ, no two boxes are in the same column.

Pieri's rule.

$$S_r(V) \otimes_{\Bbbk} S_\lambda(V) = \bigoplus_{\substack{\mu\backslash\lambda \\ \text{is a horizontal strip} \\ \text{with } r \text{ boxes}}} S_\mu(V).$$

Resolutions of length two. Let us examine one more case where equivariant pure free resolutions are easily constructed: The case when $S = \Bbbk[x, y]$.

Example 3.2.

$$S^2 \xleftarrow{\begin{bmatrix} 4x^3 & 3x^2y & 2xy^2 & y^3 & 0 \\ 0 & x^3 & 2x^2y & 3xy^2 & 4y^3 \end{bmatrix}} S(-3)^5 \xleftarrow{\begin{bmatrix} y^2 & 0 & 0 \\ -2xy & y^2 & 0 \\ x^2 & -2xy & y^2 \\ 0 & x^2 & -2xy \\ 0 & 0 & x^2 \end{bmatrix}} S(-5)^3.$$

The matrices here are chosen so that the complex is GL(2)-equivariant, but we are really free to vary the coefficients of the first matrix as we like, in an open set, and there will be a suitable match for the second matrix.

In general one may construct a GL(2)-equivariant complex

$$S \otimes_{\Bbbk} S_{a-1,0} \leftarrow S \otimes_{\Bbbk} S_{a+b-1,0} \leftarrow S \otimes_{\Bbbk} S_{a+b-1,a}. \tag{3.2}$$

This is a resolution of type $(a-1, a+b-1, 2a+b-1)$. By twisting it with $a-1$ it becomes of type $(0, b, a+b)$.

Resolutions of length three. Now suppose we want to construct pure resolutions of type (d_0, d_1, d_2, d_3). Since we may twist the complex by $-r$ to get a pure resolution of type $(d_0 + r, d_1 + r, d_2 + r, d_3 + r)$, what really matters are the differences $e_1 = d_1 - d_0, e_2 = d_2 - d_1$, and $e_3 = d_3 - d_2$. Looking at the complexes (3.1) and (3.2) it seems that one should try to construct a complex as follows

$$S \otimes_{\Bbbk} S_{\lambda_1,\lambda_2,\lambda_3} \xleftarrow{d_1} S \otimes_{\Bbbk} S_{\lambda_1+e_1,\lambda_2,\lambda_3} \tag{3.3}$$
$$\xleftarrow{d_2} S \otimes_{\Bbbk} S_{\lambda_1+e_1,\lambda_2+e_2,\lambda_3} \xleftarrow{d_3} S \otimes_{\Bbbk} S_{\lambda_1+e_1,\lambda_2+e_2,\lambda_3+e_3}.$$

After looking at the numerics, i.e. the dimensions of the representations and the Herzog–Kühl equations, there is one choice that fits exactly. This is taking

$$\lambda_3 = 0, \quad \lambda_2 = e_3 - 1, \quad \lambda_1 = (e_2 - 1) + (e_3 - 1).$$

We must then construct these complexes. To construct d_1 one must chose a map

$$S_{\lambda_1+e_1,\lambda_2,\lambda_3} \to S_{e_1} \otimes_{\Bbbk} S_{\lambda_1,\lambda_2,\lambda_3}.$$

But by Pieri's rule the first module occurs exactly once as a component in the second tensor product. Hence there is a nonzero map as above, unique up to a nonzero constant. Similarly d_2 is given by

$$S_{\lambda_1+e_1,\lambda_2+e_2,\lambda_3} \to S_{e_2} \otimes_{\Bbbk} S_{\lambda_1+e_1,\lambda_2,\lambda_3}$$

and again by Pieri's rule there is a nonzero such map unique up to a nonzero constant. Similarly for d_3. Hence up to multiplying the differentials with constants there is a unique possible such complex (3.3) with nonzero differentials. What must be demonstrated is that this is a resolution, i.e. the only homology is the cokernel of d_1. And in fact this is the challenging part.

General construction of equivariant resolutions. To construct a pure resolution of type $(d_0, d_1, \ldots, d_n)$ in general one lets $e_i = d_i - d_{i-1}$. Let $\lambda_i = \sum_{j>i}(e_j - 1)$ and define the partition

$$\alpha(\mathbf{e}, i): \ \lambda_1 + e_1, \ldots, \lambda_i + e_i, \lambda_{i+1}, \ldots, \lambda_n.$$

Theorem 3.3 ([11]). *There is a* $\mathrm{GL}(n)$*-equivariant resolution*

$$E(\mathbf{e}): S \otimes_{\Bbbk} S_{\alpha(\mathbf{e},0)} \leftarrow S \otimes_{\Bbbk} S_{\alpha(\mathbf{e},1)} \leftarrow \cdots \leftarrow S \otimes_{\Bbbk} S_{\alpha(\mathbf{e},n)}.$$

This complex is uniquely defined up to multiplying the differentials by nonzero constants.

These equivariant complexes have a canonical position as follows. Since the complex above is equivariant for $\mathrm{GL}(n)$ it is equivariant for the diagonal matrices in $\mathrm{GL}(n)$. Hence it is a $\mathbb{Z}^n$-graded complex. Fix a sequence of differences $(e_1, \ldots, e_n)$. Consider $\mathbb{Z}^n$-graded resolutions

$$F_0 \leftarrow F_1 \leftarrow \cdots \leftarrow F_n \tag{3.4}$$

of Artinian $\mathbb{Z}^n$-graded modules which (i) become pure when taking total degrees, i.e. when making a new grading by the map $\mathbb{Z}^n \to \mathbb{Z}$ given by $(a_1, \ldots, a_n) \to \sum_1^n a_i$, and (ii) such that the differences of these total degrees are the fixed numbers $e_1, \ldots, e_n$. Each $F_i = \bigoplus_{\mathbf{j} \in \mathbb{Z}^n} S(-\mathbf{j})^{\beta_{i\mathbf{j}}}$. We may encode the information of all the multigraded Betti numbers $\beta_{i\mathbf{j}}$ as an element in the Laurent polynomial ring $T = \mathbb{Z}[t_1, t_1^{-1}, \ldots, t_n, t_n^{-1}]$. Namely for $i = 0, \ldots, n$ let $B_i = \bigoplus_{\mathbf{j} \in \mathbb{Z}^n} \beta_{i\mathbf{j}} t^{\mathbf{j}}$, which we call the Betti polynomials. Consider the lattice ($\mathbb{Z}$-submodule) L of T^{n+1} generated by all tuples of Betti polynomials $(B_0, \ldots, B_n)$ derived from resolutions (3.4). This is in fact a T-submodule of T^{n+1}.

The Betti polynomial of the module $S \otimes_{\mathbb{k}} S_\lambda(V)$ is the Schur polynomial s_λ and so the tuple of the equivariant resolutions $E(\mathbf{e})$ is

$$s(\mathbf{e}) = (s_{\alpha(\mathbf{e},0)}, s_{\alpha(\mathbf{e},1)}, \ldots, s_{\alpha(\mathbf{e},n)}).$$

Fløystad shows that this tuple has a distinguished status among tuples of Betti polynomials of $\mathbb{Z}^n$-graded resolutions of Artinian modules.

Theorem 3.4 (Theorem 1.2, [21]). *Let* char $\mathbb{k} = 0$ *and assume the greatest common divisor of* $e_1, \ldots, e_n$ *is* 1. *The T-submodule L of T^{n+1} is a free T-module of rank one. The tuple $s(\mathbf{e})$ is, up to a unit in T (which is $\pm t^{\mathbf{a}}$, where $t^{\mathbf{a}}$ is a Laurent monomial), the unique generator of this T-module.*

Generalizations

The diagram $\alpha(\mathbf{e}, 1) \backslash \alpha(\mathbf{e}, 0)$ is a horizontal strip living only in the first row. In [28], S. Sam and J. Weyman consider partitions β and α such that $\beta \backslash \alpha$ is a horizontal strip (vertical strip in Vertical display). They give explicitly the minimal free resolutions [28, Theorem 2.8], of the cokernel of the map (char $\mathbb{k} = 0$):

$$S \otimes_{\mathbb{k}} S_\beta(V) \to S \otimes_{\mathbb{k}} S_\alpha(V).$$

This cokernel may no longer be a Cohen–Macaulay module. It may even have positive rank. In the case that $\beta \backslash \alpha$ contains boxes only in the ith and nth row for some i, they show that the resolution is pure, [28, Corollary 2.11]. The methods used in this paper have the advantage that they are more direct and explicit than the inductive arguments given in [11].

More generally they give the (not necessarily minimal) free resolution of the cokernel of

$$\bigoplus_i S \otimes_{\mathbb{k}} S_{\beta^i}(V) \to S \otimes_{\mathbb{k}} S_\alpha(V)$$

where $\beta^i \backslash \alpha$ are horizontal strips, and if for each i the horizontal strip lives in one row, the resolution is minimal.

Sam and Weyman in [28, Section 3] also generalize the construction of GL(V)-equivariant pure resolutions to resolutions equivariant for the symplectic and orthogonal groups.

3.2 Equivariant Supernatural Bundles

The equivariant resolution has an analog in the construction of bundles with supernatural cohomology. Given any ring R and an R-module F, one may for any partition λ in a functorial way construct the Schur module $S_\lambda F$, see [8], [23] or [29]. In the case when $R = \mathbb{k}$ and F is a vector space V in char $\mathbb{k} = 0$, with $\mathrm{GL}(V)$ acting, the Schur modules $S_\lambda V$ give the irreducible representations of $\mathrm{GL}(V)$. The construction of $S_\lambda F$ respects localization and so for a locally free sheaf $\mathcal{E}$, an algebraic vector bundle on a scheme, we get Schur bundles $S_\lambda \mathcal{E}$. In particular consider the sheaf of differentials on $\mathbb{P}^n$, the kernel of the natural map *ev*:

$$0 \leftarrow \mathcal{O}_{\mathbb{P}^n} \overset{ev}{\longleftarrow} H^0\mathcal{O}_{\mathbb{P}^n}(1) \otimes_{\mathbb{k}} \mathcal{O}_{\mathbb{P}^n}(-1) \leftarrow \Omega_{\mathbb{P}^n} \leftarrow 0,$$

we may construct Schur bundles $S_\lambda(\Omega_{\mathbb{P}^n}(1))$.

Example 3.5. The cohomology of the bundle $\Omega_{\mathbb{P}^n}(1)$ is well known and it has supernatural cohomology. It is easily computed by the long exact cohomology sequence associated to the short exact sequence above.

- In the range $1 \leq p \leq n-1$ the only nonvanishing cohomology $H^p\Omega_{\mathbb{P}^n}(i)$ is when $p = 1$ and $i = 0$: $H^1\Omega_{\mathbb{P}^n} \cong \mathbb{k}$.
- $H^0\Omega_{\mathbb{P}^n}(i)$ vanishes for $i \leq 1$ and is nonvanishing for $i \geq 2$.
- $H^n\Omega_{\mathbb{P}^n}(i)$ vanishes for $i \geq -(n-1)$ and is nonvanishing for $i \leq -n$.

The root sequence of $\Omega_{\mathbb{P}^n}(1)$ is $0, -2, -3, \ldots, -n$ and its Hilbert polynomial is

$$\frac{1}{(n-1)!} \cdot z \cdot \prod_{i=2}^{n} (z+i).$$

In general the bundle $S_\lambda(\Omega_{\mathbb{P}^n}(1))$ has supernatural cohomology. It is standard to compute its cohomology by the Borel–Bott–Weil formula in the theory of linear algebraic groups, [27] or [29]. The computation of its cohomology is done explicitly in [20, Section 4], or in [16, Theorem 5.6] for the dual bundle. In fact the nonzero cohomology modules $H^p S_\lambda(\Omega_{\mathbb{P}^n}(i))$ are all irreducible representations $S_\mu V$, where μ depends on λ, i and p.

Theorem 3.6. *The Schur bundle $S_\lambda(\Omega_{\mathbb{P}^n}(1))$ has supernatural cohomology with root sequence*

$$\lambda_1 - 1, \lambda_2 - 2, \lambda_3 - 3, \ldots, \lambda_n - n.$$

3.3 Characteristic Free Supernatural Bundles

It is a general fact that if $\mathcal{F}$ is a coherent sheaf on $\mathbb{P}^N$ and $\mathbb{P}^N \overset{\pi}{\dashrightarrow} \mathbb{P}^n$ is a projection whose center of projection is disjoint from the support of $\mathcal{F}$, then $\pi_*(\mathcal{F})$ and $\mathcal{F}$ have the same cohomology

$$H^i(\mathbb{P}^n, (\pi_*\mathcal{F})(p)) \cong H^i(\mathbb{P}^N, \mathcal{F}(p))$$

for all i and p. (By the projection formula [24, page 124], and since one can use flasque resolutions to compute cohomology,[24, Chapter III.2].)

Example 3.7. The Segre embedding embeds $\mathbb{P}^1 \times \mathbb{P}^{n-1}$ as a variety of degree n into $\mathbb{P}^{2n-1}$. If we take a general projection $\mathbb{P}^{2n-1} \dashrightarrow \mathbb{P}^n$, the line bundle $\mathcal{O}_{\mathbb{P}^1}(-2) \otimes \mathcal{O}_{\mathbb{P}^{n-1}}$ projects down to a vector bundle of rank n on $\mathbb{P}^n$ which is the sheaf of differentials $\Omega_{\mathbb{P}^n}$. In fact the cohomology of the line bundle $\mathcal{O}_{\mathbb{P}^1}(-2) \otimes \mathcal{O}_{\mathbb{P}^{n-1}}$ and its successive twists by $\mathcal{O}_{\mathbb{P}^1}(1) \otimes \mathcal{O}_{\mathbb{P}^{n-1}}(1)$ is readily computed by the Künneth formula and it is a *sheaf* with supernatural cohomology. It has the same cohomology as $\Omega_{\mathbb{P}^n}$, and it is not difficult to argue that the projection is actually this bundle.

This example may be generalized as follows. The Segre embedding embeds $\mathbb{P}^a \times \mathbb{P}^b$ into $\mathbb{P}^{ab+a+b}$ as a variety of degree $\binom{a+b}{a}$. Consider the line bundle $\mathcal{O}_{\mathbb{P}^a}(-a-1) \otimes \mathcal{O}_{\mathbb{P}^b}$ on $\mathbb{P}^a \times \mathbb{P}^b$. The line bundle of the hyperplane divisor on $\mathbb{P}^{ab+a+b}$ pulls back to $\mathcal{O}_{\mathbb{P}^a}(1) \otimes \mathcal{O}_{\mathbb{P}^b}(1)$ and by twisting with this line bundle, the above line bundle is a sheaf with supernatural cohomology. Taking a general projection of $\mathbb{P}^{ab+a+b}$ to $\mathbb{P}^{a+b}$ this line bundle projects down to the bundle $\bigwedge^a \Omega_{\mathbb{P}^{a+b}}$ of rank $\binom{a+b}{a}$, as may be argued using Tate resolutions, [22, Proposition 3.4]. The root sequence of this bundle is $a, a-1, \ldots, 1, -1, -2, \ldots, -b$. It is natural to generalize this by looking at Segre embeddings $\mathbb{P}^{a_1} \times \cdots \times \mathbb{P}^{a_r} \hookrightarrow \mathbb{P}^N$ composed with a general projection $\mathbb{P}^N \dashrightarrow \mathbb{P}^n$ where $n = \sum_i a_i$.

Theorem 3.8. *Let a root sequence be the union of sets of consecutive integers*

$$\bigcup_{i=1}^{r} \{z_i, z_i - 1, \ldots, z_i - (a_i - 1)\}$$

where $z_i \geq z_{i+1} + a_i$, *and let* $n = a_1 + a_2 + \cdots + a_r$. *The line bundle*

$$p_1^* \mathcal{O}_{\mathbb{P}^{a_1}}(-z_1 - 1) \otimes \cdots \otimes p_{r-1}^* \mathcal{O}_{\mathbb{P}^{a_{r-1}}}(-z_{r-1} - 1) \otimes p_r^* \mathcal{O}_{\mathbb{P}^{a_r}}(-z_r - 1)$$

considered on the Segre embedding of $\mathbb{P}^{a_1} \times \cdots \times \mathbb{P}^{a_r}$ *has supernatural cohomology. By a general projection down to* $\mathbb{P}^n$ *it projects down to a vector bundle with supernatural cohomology of rank* $\binom{n}{a_1 a_2 \cdots a_r}$ *and root sequence given above.*

Remark 3.9. Although in the case $r = 2$ the projection is a Schur bundle $\bigwedge^a \Omega_{\mathbb{P}^n}$, it is no longer true for $r > 2$ that one gets twists of Schur bundles $S_\lambda(\Omega_{\mathbb{P}^n})(p)$, as may be seen from the ranks.

3.4 The Characteristic Free Pure Resolutions

In this construction of [12, Section 5] one starts with a complex of locally free sheaves on a product of projective spaces $\mathbb{P}^{m_0} \times \mathbb{P}^{m_1} \times \cdots \times \mathbb{P}^{m_r}$, whose terms are direct sums of line bundles $\mathcal{O}(t_0, \dots, t_r)$. The complex is linear in each coordinate twist and is exact except at the start, so is a locally free resolution. Then we successively push this complex forward by omitting one factor in the product of projective spaces at a time. Each time some linear part of the complex "collapses", so that at each step we get "multitwisted" pure resolutions. In the end we will have a singly twisted pure resolution on $\mathbb{P}^{m_0}$ of the type we desire.

The main ingredient in the construction is the following.

Proposition 3.10 (Proposition 5.3, [12]). *Let $\mathcal{F}$ be a sheaf on $X \times \mathbb{P}^m$, and denote by p_1 and p_2 the projections onto the factors of this product. Suppose $\mathcal{F}$ has a resolution of the form*

$$p_1^*\mathcal{G}_0 \otimes p_2^*\mathcal{O}_{\mathbb{P}^m}(-e_0) \leftarrow \cdots \leftarrow p_1^*\mathcal{G}_N \otimes p_2^*\mathcal{O}_{\mathbb{P}^m}(-e_N) \leftarrow 0.$$

where $e_0 < e_1 < \cdots < e_N$. Suppose for some $k \geq 0$ the subsequence $(e_{k+1}, \dots, e_{k+m})$ is equal to $(1, 2 \dots, m)$. Then $R^l p_{1*}\mathcal{F} = 0$ for $l > 0$ and $p_{1*}\mathcal{F}$ has a resolution on X of the form

$$\begin{aligned} &\mathcal{G}_0 \otimes H^0\mathcal{O}_{\mathbb{P}^m}(-e_0) \leftarrow \mathcal{G}_1 \otimes H^0\mathcal{O}_{\mathbb{P}^m}(-e_1) \leftarrow \cdots \\ &\leftarrow \mathcal{G}_k \otimes H^0\mathcal{O}_{\mathbb{P}^m}(-e_k) \leftarrow \mathcal{G}_{k+m+1} \otimes H^m\mathcal{O}_{\mathbb{P}^m}(-e_{k+m+1}) \leftarrow \cdots \\ &\leftarrow \mathcal{G}_N \otimes H^m\mathcal{O}_{\mathbb{P}^m}(-e_N). \end{aligned}$$

The proof of this is quite short and uses the hypercohomology spectral sequence.

Example 3.11. Let Y be the complete intersection of m forms of type $(1,1)$ on $\mathbb{P}^a \times \mathbb{P}^b$. Assume $m \geq b + d$ where d is a nonnegative integer. Let $\mathcal{F}$ be the twisted structure sheaf $\mathcal{O}_Y(0, d)$. The resolution of $\mathcal{O}_Y(0, d)$ is

$$\begin{aligned} &\mathcal{O}(0, d)^{\alpha_0} \leftarrow \mathcal{O}(-1, d-1)^{\alpha_1} \leftarrow \cdots \leftarrow \mathcal{O}(-d, 0)^{\alpha_d} \\ &\leftarrow \mathcal{O}(-d-1, -1)^{\alpha_{d+1}} \leftarrow \cdots \leftarrow \mathcal{O}(-d-b, -b)^{\alpha_{d+b}} \\ &\leftarrow \mathcal{O}(-d-b-1, -b-1)^{\alpha_{d+b+1}} \leftarrow \cdots \leftarrow \mathcal{O}(-m, d-m)^{\alpha_m}. \end{aligned}$$

The first coordinate twist is the one we are interested in. If we push this complex forward to $\mathbb{P}^a$, the above Proposition 3.10, shows that $p_*\mathcal{O}_Y(0, d)$ has a resolution

$$\begin{aligned} &\mathcal{O}(0)^{\alpha'_0} \leftarrow \mathcal{O}(-1)^{\alpha'_1} \leftarrow \cdots \leftarrow \mathcal{O}(-d)^{\alpha'_d} \\ &\qquad \leftarrow \mathcal{O}(-d-b-1)^{\alpha'_{d+b+1}} \leftarrow \cdots \leftarrow \mathcal{O}(-m)^{\alpha'_m}. \end{aligned}$$

We see that we adjusted the second coordinate twist so that we got a collapse in the first coordinate twist resulting in a gap from d to $d + b + 1$. We have a complex which is pure but no longer linear.

For the general construction, suppose we want a pure resolution of a sheaf on $\mathbb{P}^n$ by sums of line bundles,

$$\mathcal{O}_{\mathbb{P}^n}(-d_0)^{\alpha_0} \leftarrow \mathcal{O}_{\mathbb{P}^n}(-d_1)^{\alpha_1} \leftarrow \cdots \leftarrow \mathcal{O}_{\mathbb{P}^n}(-d_n)^{\alpha_n}. \tag{3.5}$$

We see in the example above that we get a leap in twist $\mathcal{O}(-d) \leftarrow \mathcal{O}(-d-b-1)$ by pushing down omitting a factor $\mathbb{P}^b$. The leaps in the pure resolution we want are $d_i - d_{i-1}$, so we consider a projective space $\mathbb{P}^{m_i}$ where $m_i = d_i - d_{i-1} - 1$. On the product $\mathbb{P}^n \times \mathbb{P}^{m_1} \times \cdots \times \mathbb{P}^{m_n}$ we let Y be the complete intersection of $M = d_n$ forms of type $(1, 1, \ldots, 1)$, and let $\mathcal{F}$ be $\mathcal{O}_Y(0, d_0, d_1, \ldots, d_{n-1})$. Its resolution is

$$\begin{aligned}\mathcal{O}(0, d_0, \ldots, d_{n-1}) &\leftarrow \mathcal{O}(-1, d_0 - 1, \ldots, d_{n-1} - 1) \leftarrow \cdots \\ &\leftarrow \mathcal{O}(-d_{n-1}, d_0 - d_{n-1}, \ldots, 0)^{\alpha'} \leftarrow \cdots \\ &\leftarrow \mathcal{O}(-d_{n-1} - m_n, d_0 - d_{n-1} - m_n, \ldots, -m_n)^{\alpha''} \\ &\leftarrow \mathcal{O}(-d_n, \ldots, -m_n - 1)^{\alpha'''}\end{aligned}$$

Note that when the coordinate twist corresponding to $\mathbb{P}^{m_i}$ varies through $0, -1, \ldots$ $-m_i, -m_i - 1$ (displayed when $i = n$ in the second and third line above), the first coordinate twist varies through

$$\begin{aligned}-d_{i-1}, -d_{i-1} - 1, \ldots, -d_{i-1} - m_i, -d_{i-1} - m_i - 1 \\ (= -d_i + 1), (= -d_i).\end{aligned}$$

Hence after the projection omitting $\mathbb{P}^{m_i}$, only the first twist $-d_{i-1}$ and the last $-d_i$ survive in the first coordinate. After all the projections we get a pure resolution consisting of sums of line bundles (3.5) on $\mathbb{P}^n$. Taking global sections of all twists of this complex, we get a complex

$$S(-d_0)^{\alpha_0} \leftarrow \cdots \leftarrow S(-d_n)^{\alpha_n}.$$

That this is a resolution follows by the Acyclicity Lemma [11, Lemma 20.11] or may be verified by breaking (3.5) into short exact sequences

$$0 \leftarrow \mathcal{K}_{i-1} \leftarrow \mathcal{O}(-d_i)^{\alpha_i} \leftarrow \mathcal{K}_i \leftarrow 0.$$

By descending induction on i starting from $i = n$ one easily checks that there are exact sequences of graded modules

$$0 \leftarrow \Gamma_* \mathcal{K}_{i-1} \leftarrow S(-d_i)^{\alpha_i} \leftarrow \Gamma_* \mathcal{K}_i \leftarrow 0.$$

Generalizations

In [3] this method of collapsing part of the complex by suitable projections is generalized considerably. They construct wide classes of multilinear complexes from tensors ϕ in $R^a \otimes R^{b_1} \otimes \cdots \otimes R^{b_n}$ where R is a commutative ring, and weights w in $\mathbb{Z}^{n+1}$ (the twist $(0, d_0, d_1, \ldots, d_{n-1})$ above is such a weight). In a generic setting these are resolutions generalizing many known complexes, like for instance Eagon–Northcott and Buchsbaum–Rim complexes which arise from a 2-tensor (a matrix).

In particular, Theorem 1.9 of [3] provides infinitely many new families of pure resolutions of type **d** for any degree sequence **d**. The essential idea is that given a degree sequence, say $(0, 4, 7)$, the integers in the complement

$$\ldots, -3, -2, -1, 1, 2, 3, 5, 6, 8, 9, 10, \ldots$$

may be partitioned in many ways into sequences of successive integers, and by cleverly adjusting the construction above, all twists in such sequences may be collapsed.

The paper also give explicit constructions of the differentials of the complexes, and in particular of those in the resolutions constructed above by Eisenbud and Schreyer.

3.5 Pure Resolutions Constructed from Generic Matrices

We now describe the second construction of pure resolutions in [11], which also requires that char $\Bbbk = 0$. It gives a comprehensive generalization of the Eagon–Northcott and Buchsbaum–Rim complexes in a generic setting.

Resolutions of length two. For a map $R^{r-1} \xrightarrow{\phi} R^r$ of free modules over a commutative ring R, let m_i be the minor we get by deleting row i in the map ϕ_i. The well-known Hilbert–Burch theorem says that if the ideal $I = (m_1, \ldots, m_r)$ has depth ≥ 2 then there is a resolution of R/I

$$R \xleftarrow{[m_1, -m_2, \ldots, (-1)^r m_r]} R^r \xleftarrow{\phi} R^{r-1}.$$

We get a generic situation if we let F and G be vector spaces with bases $f_1, \ldots, f_{r-1}$ and $g_1, \ldots, g_r$ respectively, and set $S = \mathrm{Symm}(G^* \otimes F)$. Then $G^* \otimes F$ has basis $e_{ij} = g_i^* \otimes f_j$, where the g_i^* are a dual basis for G^*. We have a generic map

$$G \otimes_{\Bbbk} S \overset{[e_{ij}]}{\leftarrow} F \otimes_{\Bbbk} S(-1).$$

The Hilbert–Burch theorem gives a resolution

$$S \xleftarrow{[m_1, -m_2, \ldots, (-1)^{r-1} m_r]} G \otimes_{\Bbbk} S(-r+1) \xleftarrow{[e_{ij}]} F \otimes_{\Bbbk} S(-r).$$

The construction of [11, Section 4] generalizes this to pure resolutions of type $(0, s, r)$ for all $0 < s < r$. This is a resolution

$$\bigwedge^{r-1} F \otimes_{\Bbbk} \bigwedge^{s} F^* \otimes_{\Bbbk} S \longleftarrow \bigwedge^{r-1} F \otimes_{\Bbbk} \bigwedge^{s} G^* \otimes_{\Bbbk} S(-s)$$
$$\longleftarrow \bigwedge^{r-1} F \otimes_{\Bbbk} \bigwedge^{r-s} F \otimes_{\Bbbk} \bigwedge^{r} G^* \otimes_{\Bbbk} S(-r).$$

Note that the right map here identifies as the natural map

$$\bigwedge^{r-1} F \otimes_{\Bbbk} \bigwedge^{r} G^* \otimes_{\Bbbk} \bigwedge^{r-s} G \otimes_{\Bbbk} S(-s) \longleftarrow \bigwedge^{r-1} F \otimes_{\Bbbk} \bigwedge^{r} G^* \otimes_{\Bbbk} \bigwedge^{r-s} F \otimes_{\Bbbk} S(-r).$$

Also note that the ranks of the modules here are different from the ranks of the modules in the equivariant case. For instance the rank of the middle term in this construction is $\binom{r}{s}$ while in the equivariant construction this rank is simply r.

Example 3.12. When $r = 5$ we derive in the case $s = 4$ the Hilbert–Burch complex

$$S \xleftarrow{[m_1, -m_2, m_3, -m_4, m_5]} S(-4)^5 \xleftarrow{[e_{ij}]} S(-5)^4.$$

When $s = 3$ we get a complex

$$S^{\binom{4}{3}} \xleftarrow{\alpha} S(-3)^{\binom{5}{3} = \binom{5}{2}} \xleftarrow{\beta} S(-5)^{\binom{4}{2}}.$$

There is a natural basis of the first term $S^{\binom{4}{3}}$ consisting of three-sets of basis elements $\{f_{i_1}^*, f_{i_2}^*, f_{i_3}^*\}$ of F^* and a basis for $S(-3)^{\binom{5}{3}}$ consisting of three-sets of basis elements $\{g_{j_1}^*, g_{j_2}^*, g_{j_3}^*\}$ of G^*. The entries of α are the 3×3 -minors corresponding to the columns i_1, i_2, i_3 and rows j_1, j_2, j_3 of the 5×4 matrix $[e_{ij}]$. Similarly the entries of β consists of 2×2-minors of the matrix $[e_{ij}]$.

When $s = 2$ we get a complex

$$S^{\binom{4}{2}} \xleftarrow{\alpha} S(-2)^{\binom{5}{2} = \binom{5}{3}} \xleftarrow{\beta} S(-5)^{\binom{4}{3}},$$

and when $s = 1$ we get a complex

$$S^{\binom{4}{1}} \xleftarrow{\alpha} S(-1)^{\binom{5}{1} = \binom{5}{4}} \xleftarrow{\beta} S(-5)^{\binom{4}{4}}.$$

Resolutions of length three and longer. To construct resolutions of length three one must start with a vector space F of rank $r - 2$ and a vector space G of rank r. The resolution of pure type $(0, a, a + b, a + b + c = r)$ has the following form

$$S_{2^{c-1}, 1^{b-1}} F \otimes_{\Bbbk} S \xleftarrow{\alpha_1} \bigwedge^{r-2} F \otimes_{\Bbbk} \bigwedge^{c-1} F \otimes_{\Bbbk} \bigwedge^{a} G^* \otimes_{\Bbbk} S(-a)$$
$$\xleftarrow{\alpha_2} \bigwedge^{r-2} F \otimes_{\Bbbk} \bigwedge^{b+c-1} F \otimes_{\Bbbk} \bigwedge^{a+b} G^* \otimes_{\Bbbk} S(-a-b)$$
$$\xleftarrow{\alpha_3} \bigwedge^{r-2} F \otimes_{\Bbbk} S_{2^c, 1^{b-1}} F \otimes_{\Bbbk} \bigwedge^{r} G^* \otimes_{\Bbbk} S(-r).$$

When $b = c = 1$ this is the Eagon–Northcott complex associated to a generic map. When $a = c = 1$ it is the Buchsbaum–Rim complex, and when $a = b = 1$ we get the third complex occurring naturally in this family as given in [8, Appendix A.3].

In general for a degree sequence $\mathbf{d} = (d_0, \ldots, d_c)$, denote $e_i = d_i - d_{i-1}$. One chooses G of rank $r = \sum_1^c e_i$ and F of rank $r - c + 1$. Let $\gamma(\mathbf{e}, i)$ be the partition

$$((c-1)^{e_c - 1}, (c-2)^{e_{c-1}-1}, \ldots, i^{e_{i+1}-1}, i^{e_i}, (i-1)^{e_{i-1}-1}, \ldots, 1^{e_1 - 1}).$$

This is the dual of the partition $\alpha(\mathbf{e}, i)$ defined in the equivariant case. The terms in our complex will be

$$H(\mathbf{d}, i) = S_{\gamma(\mathbf{e},i)} F \otimes_{\Bbbk} \bigwedge^{d_i} G^* \otimes_{\Bbbk} S(-d_i).$$

The differentials $H(\mathbf{d}, i) \xrightarrow{\gamma_i} H(\mathbf{d}, i-1)$ in the complex are given by

$$\begin{gathered} S_{\gamma(\mathbf{e},i)} F \otimes_{\Bbbk} \bigwedge^{d_i} G^* \otimes_{\Bbbk} S(-d_i) \\ \downarrow \\ S_{\gamma(\mathbf{e},i-1)} F \otimes_{\Bbbk} \bigwedge^{d_i - d_{i-1}} F \otimes_{\Bbbk} \bigwedge^{d_i - d_{i-1}} G^* \otimes_{\Bbbk} \bigwedge^{d_{i-1}} G^* \otimes_{\Bbbk} S(-d_i) \\ \downarrow \\ S_{\gamma(\mathbf{e},i-1)} F \otimes_{\Bbbk} \bigwedge^{d_{i-1}} G^* \otimes_{\Bbbk} S(-d_{i-1}). \end{gathered}$$

The last map is due to $\bigwedge^{d_i - d_{i-1}} F \otimes_{\Bbbk} \bigwedge^{d_i - d_{i-1}} G^*$ being a summand of $\mathrm{Symm}(F \otimes G^*)_{d_i - d_{i-1}}$.

Theorem 3.13 (Theorem 0.2, [11]). *The complex $H(\mathbf{d}, \cdot)$ is a $\mathrm{GL}(F) \times \mathrm{GL}(G)$ equivariant pure resolution of type $\mathbf{d}$.*

4 Cohomology of Vector Bundles on Projective Spaces

In their paper [12], Eisenbud and Schreyer also achieved a complete classification of cohomology tables of vector bundles on projective spaces up to a rational multiple. This runs fairly analogous to the classification of Betti diagrams of Cohen–Macaulay modules up to rational multiple. First we introduce cohomology tables of coherent sheaves and vector bundles, and notation related to these.

4.1 Cohomology Tables

For a coherent sheaf $\mathcal{F}$ on the projective space $\mathbb{P}^m$ our interest shall be the cohomological dimensions

$$\gamma_{i,d}(\mathcal{F}) = \dim_{\Bbbk} H^i \mathcal{F}(d).$$

The indexed set $(\gamma_{i,d})_{i=0,\dots,m,d\in\mathbb{Z}}$ is the *cohomology table* of $\mathcal{F}$, which lives in the vector space $\mathbb{T} = \mathbb{D}^* = \Pi_{d\in\mathbb{Z}}\mathbb{Q}^{m+1}$ with the $\gamma_{i,d}$ as coordinate functions. An element in this vector space will be called a *table*.

We shall normally display a table as follows.

$$\begin{array}{cccccc|c}
\cdots & \gamma_{n,-n-1} & \gamma_{n,-n} & \gamma_{n,-n+1} & \cdots & & n \\
 & \vdots & \vdots & \vdots & & & \\
\cdots & \gamma_{1,-2} & \gamma_{1,-1} & \gamma_{1,0} & \cdots & & 1 \\
\cdots & \gamma_{0,-1} & \gamma_{0,0} & \gamma_{0,1} & \cdots & & 0 \\
\hline
\cdots & -1 & 0 & 1 & \cdots & & d\backslash i
\end{array}$$

Compared to the natural way of displaying $\gamma_{i,d}$ in row i and column d, we have shifted row i to the right i steps. With the above way of displaying the cohomology table, the columns correspond to the terms in the Tate resolution (see Subsection 6.5) of the coherent sheaf $\mathcal{F}$. We write $H^i_*\mathcal{F} = \bigoplus_{n\in\mathbb{Z}} H^i\mathcal{F}(n)$. This is an S-module, the ith cohomology module of $\mathcal{F}$.

Example 4.1. The cohomology table of the ideal sheaf of two points in $\mathbb{P}^2$ is

$$\begin{array}{ccccccccc}
\cdots & 6 & 3 & 1 & \llcorner 0 & 0 & 0 & 0 & \cdots \\
\cdots & 2 & 2 & 2 & 2 & 1 & \llcorner 0 & 0 & \cdots \\
\cdots & 0 & 0 & 0 & 0 & 1 & 3 & 8 & \cdots \\
\hline
\cdots & -3 & -2 & -1 & 0 & 1 & 2 & 3 & \cdots
\end{array}$$

In this table there are in the two upper rows two distinguished corners with 0, indicated with a $\llcorner$, such that the quadrant determined by it only consists of zeroes. The 0 in the H^1-row is in the column labelled by 2 so it is in degree $z_1 = 2 - 1$. The 0 in the H^2-row is in the column labelled by 0 so its degree is $z_2 = 0 - 2$. The sequence z_1, z_2 is called the root sequence of the cohomology table.

Recall that the classical Castelnuovo–Mumford regularity of a coherent sheaf $\mathcal{F}$ is defined by

$$r = \inf\{m \mid H^i\mathcal{F}(m-i) = 0 \text{ for } i \geq 1\}.$$

Definition 4.2. For $p \geq 1$ the *p-regularity* of a coherent sheaf is defined to be

$$r_p = \inf\{d \mid H^i\mathcal{F}(m-i) = 0 \text{ for } i \geq p, m \geq d\}.$$

(Is is not difficult to show that the numbers r_1 and r are the same.) The *root sequence* of $\mathcal{F}$ is $z_p = r_p - p$ for $p \geq 1$.

(Eisenbud and Schreyer call in [13] the z_p the regularity sequence, but by private communication from Schreyer the notions of root sequence and regularity sequence were mixed up in that paper.)

Example 4.3. Let $\mathcal{E}$ be the vector bundle on $\mathbb{P}^3$ which is the cohomology of the complex

$$\mathcal{O}_{\mathbb{P}^3} \xleftarrow{[x_0,x_1,x_2^2,x_3^2]} \mathcal{O}_{\mathbb{P}^3}(-1)^2 \oplus \mathcal{O}_{\mathbb{P}^3}(-2)^2 \xleftarrow{[-x_2^2,-x_3^2,x_0,x_1]^t} \mathcal{O}_{\mathbb{P}^3}(-3).$$

The cohomology table of this is

...	21	7	1	0	0	0	0	...	3
...	0⌉	1	2	1	0	0	0	...	2
...	0	0	0⌉	1	2	1	0	...	1
...	0	0	0	0⌉	1	7	21	...	0
...	-2	-1	0	1	2	3	4	...	$d\backslash i$

For a vector bundle $\mathcal{E}$ all the intermediate cohomology modules $H^i_*\mathcal{E}$ have finite length for $i = 1, \ldots, m-1$. Also $H^0\mathcal{E}(d)$ vanishes for $d \ll 0$. Hence in this case it is for rows $0, 1, \ldots, m-1$ also meaningful to speak of the corners with 0, extending downwards and to the left, indicated by ⌉ in the diagram above.

4.2 The Fan of Cohomology Tables of Vector Bundles

We want to consider vector bundles whose cohomology tables live in a finite dimensional subspace of $\mathbb{T}$. Let $\mathbb{Z}^m_{\text{root}}$ be the set of strictly decreasing integer sequences $(a_1, \ldots, a_m)$. Such sequences are called *root sequences.* This is a partially ordered set with $\mathbf{a} \le \mathbf{b}$ if $a_i \le b_i$ for $i = 1, \ldots, m$. The interval $[\mathbf{a}, \mathbf{b}]_{\text{root}}$ is the set of all root sequences $\mathbf{z}$ such that $\mathbf{a} \le \mathbf{z} \le \mathbf{b}$. We will consider vector bundles $\mathcal{E}$ such that for each $i = 1, \ldots, m$ we have $H^i\mathcal{E}(p) = 0$ for $p \ge b_i$. (This is the same as the root sequence of $\mathcal{E}$ being $\le \mathbf{b}$.)

As shown in Example 4.3, for a vector bundle we may also bound below the ranges of the cohomology modules $H^i_*\mathcal{E}$ for $i = 0, \ldots, m-1$, and we assume that $H^i\mathcal{E}(p) = 0$ for $p \le a_{i+1}$ for $i = 0, \ldots, m-1$. In particular note that $b_i - 1$ bounds above the supporting range of $H^i_*\mathcal{E}$ and $a_{i+1} + 1$ bounds below the supporting range of $H^i_*\mathcal{E}$. If $\mathcal{E}$ has supernatural cohomology, the conditions means that its root sequence is in the interval $[\mathbf{a}, \mathbf{b}]_{\text{root}}$.

Definition 4.4. $\mathbb{T}(\mathbf{a}, \mathbf{b})$ is the subspace of $\mathbb{T}$ consisting of all tables such that

- $\gamma_{i,d} = 0$ for $i = 1, \ldots, m$ and $d \ge b_i$.
- $\gamma_{i,d} = 0$ for $i = 0, \ldots, m-1$ and $d \le a_{i+1}$.
- The alternating sum $\gamma_{0,d} - \gamma_{1,d} + \cdots + (-1)^m \gamma_{m,d}$ is a polynomial in d of degree $\le m$ for $d \ge b_1$ and for $d \le a_m$.

The space $\mathbb{T}(\mathbf{a}, \mathbf{b})$ is a finite dimensional vector space as is easily verified, since the values of a polynomial of degree $\le m$ is determined by any of $m+1$ successive values.

The last condition for $\mathbb{T}(\mathbf{a}, \mathbf{b})$ is not really canonical. The conditions are just to get a simply defined finite-dimensional space containing the cohomology tables of vector bundles with supernatural cohomology and root sequences in the interval $[\mathbf{a}, \mathbf{b}]_{\text{root}}$. Note that set of all positive rational multiples of cohomology tables of vector bundles whose tables are in $\mathbb{T}(\mathbf{a}, \mathbf{b})$, forms a positive cone which we denote by $C(\mathbf{a}, \mathbf{b})$.

For a root sequence $\mathbf{z} : z_1 > z_2 > \cdots > z_m$ we associate a table $\gamma^{\mathbf{z}}$ given by

$$\gamma^{\mathbf{z}}_{i,d} = \begin{cases} \frac{1}{m!}\Pi_{i=1}^{m}|d - z_i|, & z_i > d > z_{i+1}, \\ 0, & \text{otherwise}. \end{cases}$$

This is the *supernatural table* associated to this root sequence.

Lemma 4.5. *If* $Z : \mathbf{z}^1 > \mathbf{z}^2 > \cdots > \mathbf{z}^r$ *is a chain of root sequences, then* $\gamma^{\mathbf{z}^1}, \gamma^{\mathbf{z}^2}, \ldots,$ $\gamma^{\mathbf{z}^r}$ *are linearly independent.*

Hence these supernatural tables span a simplicial cone $\sigma(Z)$ in $\mathbb{T}$.

Proposition 4.6. *The set of simplicial cones* $\sigma(Z)$ *where* Z *ranges over the chains* $Z : \mathbf{z}^1 < \mathbf{z}^2 < \cdots < \mathbf{z}^r$ *in* $[\mathbf{a}, \mathbf{b}]_{\text{root}}$, *form a simplicial fan in* $\mathbb{T}(\mathbf{a}, \mathbf{b})$ *which we denote as* $\Gamma(\mathbf{a}, \mathbf{b})$.

Here is the analog of Theorem 1.16.

Theorem 4.7. (i) *The realization of the fan* $\Gamma(\mathbf{a}, \mathbf{b})$ *is contained in the positive cone* $C(\mathbf{a}, \mathbf{b})$.

(ii) *The positive cone* $C(\mathbf{a}, \mathbf{b})$ *is contained in the realization of the fan* $\Gamma(\mathbf{a}, \mathbf{b})$.

In conclusion the realization of $\Gamma(\mathbf{a}, \mathbf{b})$ *and the positive cone* $C(\mathbf{a}, \mathbf{b})$ *are equal.*

Part (i) is a consequence of the existence of vector bundles with supernatural cohomology which is treated in Subsections 3.2 and 3.3. The proof of part (ii) is analogous to the proof of Theorem 1.16 (ii), which we developed in Section 2. We outline this in the next subsection and the essential part is again to find the facet equations of $\Gamma(\mathbf{a}, \mathbf{b})$.

4.3 Facet Equations

Example 4.8. Let $\mathbf{a} = (0, -4, -5)$ and $\mathbf{b} = (0, -2, -4)$. Considering the interval $[\mathbf{a}, \mathbf{b}]_{\text{root}}$ as a partially ordered set, its Hasse diagram is:

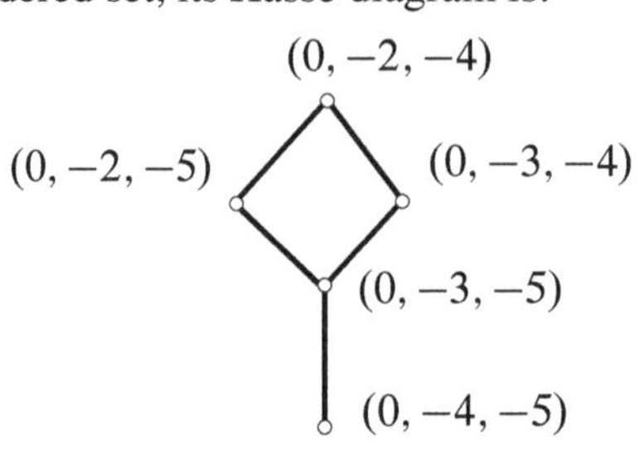

There are two maximal chains in this diagram

$$Z: \ (0,-4,-5) < (0,-3,-5) < (0,-3,-4) < (0,-2,-4)$$
$$Y: \ (0,-4,-5) < (0,-3,-5) < (0,-2,-5) < (0,-2,-4)$$

so the realization of the Boij–Söderberg fan consists of the union of two simplicial cones of dimension four. Cutting it with a hyperplane, we get two tetrahedra. (The vertices are labelled by the pure diagrams on its rays.)

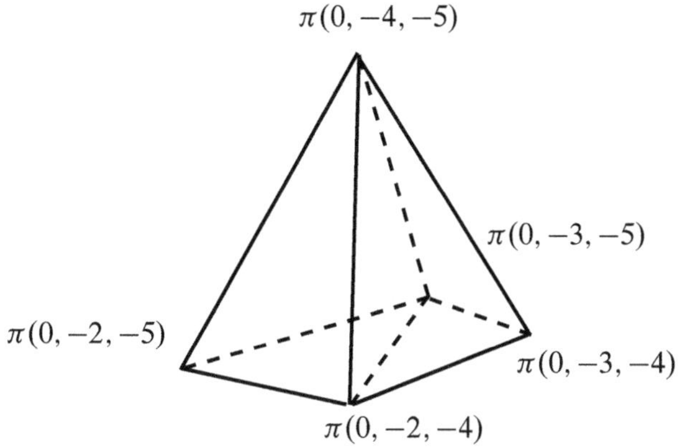

There is one interior facet of the fan, while all other facets are exterior. The exterior facets are of three types. We give an example of each case by giving the chain.

(i) $Z\backslash\{(0,-2,-4)\}$. Here we omit the maximal element **b**. Clearly this can only be completed to a maximal chain in one way so this gives an exterior facet. The nonzero values of the table $\gamma^{(0,-2,-4)}$ is

...	*	*	*	·	·	·	·	...
...	·	·	·	*	·	·	·	...
...	·	·	·	·	*	·	·	...
...	·	·	·	·	·	*	*	...
...	−4	−3	−2	−1	0	1	2	...

The second coordinate is changing from **b** to its predecessor. Hence the facet equation is $\gamma_{2,-3} = 0$, since $\gamma_{2,-3}$ is nonzero on $\gamma^{(0,-2,-4)}$ but vanishes on the other elements in Z.

(ii) $Z\backslash\{(0,-3,-5)\}$. This chain contains $(0,-3,-4)$ and $(0,-4,-5)$. Clearly the only way to complete this to a maximal chain is by including $(0,-3,-5)$, so this gives an exterior facet. The tables associated to these root sequences has nonzero positions as follows

$$
\begin{array}{ccccccccc}
\cdots & * & * & + & \cdot & \cdot & \cdot & \cdot & \cdots \\
\cdots & \cdot & \cdot & \sim & \cdot & \cdot & \cdot & \cdot & \cdots \\
\cdots & \cdot & \cdot & - & * & * & \cdot & \cdot & \cdots \\
\cdots & \cdot & \cdot & \cdot & \cdot & \cdot & * & * & \cdots \\
\hline
\cdots & -4 & -3 & -2 & -1 & 0 & 1 & 2 & \cdots
\end{array}
$$

In column -2 each of $\gamma^{(0,-3,-4)}$, $\gamma^{(0,-3,-5)}$ and $\gamma^{(0,-4,-5)}$ has only one nonzero value, indicated by a $+, \sim$ and $-$ respectively. We see that $\gamma_{2,-4}$ is nonzero on $\gamma^{(0,-3,-5)}$ but vanishes on the other elements in the chain, giving the facet equation $\gamma_{2,-4} = 0$.

(iii) $Y \backslash \{(0,-3,-5)\}$. This chain contains $(0,-2,-5)$ and $(0,-4,-5)$. Clearly the only way to complete this to a maximal chain is by including $(0,-3,-5)$, so this gives an exterior facet. The nonzero cohomology groups of $\gamma^{(0,-2,-5)}$ are indicated by $*$ and $+$ in the following diagram, those of $\gamma^{(0,-4,-5)}$ are indicated by $*$ and $-$, while those of the element omitted, $\gamma^{(0,-3,-5)}$, are indicated by $*$'s, the first $+$ and the second $-$.

$$
\begin{array}{cccccccccc}
\cdots & * & * & * & \cdot & \cdot & \cdot & \cdot & \cdot & \cdots \\
\cdots & \cdot & \cdot & \cdot & + & + & \cdot & \cdot & \cdot & \cdots \\
\cdots & \cdot & \cdot & \cdot & - & - & * & \cdot & \cdot & \cdots \\
\cdots & \cdot & \cdot & \cdot & \cdot & \cdot & \cdot & * & * & \cdots
\end{array}
$$

The diagram is divided into two parts. The upper part consists of all positions above the $*$ and $-$ positions, and the lower part below the $*$ and $+$ positions. There will be an upper and a lower facet equation. Working it out in a way analogous to Example 2.6, the lower facet equation is given by the following table

⋯	0^*	0^*	0^*	0	0	0	0	0	⋯
⋯	0	0	4	0^+	0^+	0	0	0	⋯
⋯	0	−4	15	−20	10	0^*	0	0	⋯
⋯	4	−15	20	−10	0	1	0^*	0^*	⋯

The meaning of the numbers turns out to be as follows. Taking the negative of the union of the degree sequences of z^+, z and z^- we get $\mathbf{d} = (0,2,3,4,5)$. The pure resolution of this type has exactly the absolute values of the nonzero numbers in the bottom row as Betti numbers:

$$S \leftarrow S(-2)^{10} \leftarrow S(-3)^{20} \leftarrow S(-4)^{15} \leftarrow S(-5)^4.$$

Proposition 4.9. *Let Z be a maximal chain in $[\mathbf{a}, \mathbf{b}]_{\text{root}}$ and $z \in Z$. Then $\sigma(Z \backslash \{z\})$ is an exterior facet of $\Gamma(\mathbf{a}, \mathbf{b})$ if either of the following holds.*

(i) *z is either* **a** *or* **b**. *The facet equation is $\gamma_{i,d} = 0$ for appropriate i and d.*

(ii) *The root sequences of z^- and z^+ immediately before and after z in Z differ in exactly one position. So for some r we have*

$$z^- = (\dots, -(r+1), \dots), z = (\dots, -r, \dots), z^+ = (\dots, -(r-1), \dots).$$

(iii) *The root sequences of z^- and z^+ immediately before and after z in Z differ in two consecutive positions such that for some r we have*

$$z^- = (\dots, -r, -(r+1), \dots), z = (\dots, -(r-1), -(r+1), \dots),$$
$$z^+ = (\dots, -(r-1), -r, \dots).$$

Letting i be the position of $-(r-1)$, the facet equation is $\gamma_{i,-r} = 0$.

For facets of type (ii) the description of the facet equations are as follows.

Theorem 4.10. *Let Z be a chain giving an exterior facet of type* (ii), *and let z^-, z and z^+ be successive elements in this chain which differ only in the ith position. Let f be the degree sequence which is the union of z^+, z and z^- and let $F_\bullet$ be a pure resolution corresponding to the degree sequence f. The facet equation of this exterior facet is then*

$$\langle \beta(F_\bullet), \gamma \rangle_{e,i} = 0$$

where $e = -z_i - 1$.

We may now prove Theorem 4.7 (ii).

Proof. Consider a facet of type (ii) of the fan $\Gamma(\mathbf{a}, \mathbf{b})$ associated to the root sequence z and position i. The upper equation is $\langle \beta(F_\bullet), - \rangle_{e,i} = 0$ where $e = -z_i - 1$ and $F_\bullet$ is given in Theorem 4.10. For facets of type (i) or (iii) the hyperplane equations are $\gamma_{i,d} = 0$ for suitable i, d.

Each exterior facet determines a nonnegative half plane H^+. Since the forms above are nonnegative on all cohomology tables $\gamma(\mathcal{E})$ in $\mathbb{T}(\mathbf{a}, \mathbf{b})$ by Theorem 2.16, the cone $C(\mathbf{a}, \mathbf{b})$ is contained in the intersection of all the half planes H^+ which again is contained in the fan $\Gamma(\mathbf{a}, \mathbf{b})$. □

5 Extensions to Non-Cohen–Macaulay Modules and to Coherent Sheaves

We have in Sections 1 and 2 considered Betti diagrams of Cohen–Macaulay modules over $S = \Bbbk[x_1, \dots, x_n]$ of a given codimension. Shortly after Eisenbud and

Schreyer proved the Boij–Söderberg conjectures, Boij and Söderberg, [6], extended the theorems to the case of arbitrary (finitely generated and graded) modules over this polynomial ring. The description here is just as complete as in the Cohen–Macaulay case.

In [13] Eisenbud and Schreyer extended the decomposition algorithm for vector bundles to a decomposition algorithm for coherent sheaves. This cannot however be seen as a final achievement since it does not give a way to determine if a table is the cohomology table, up to rational multiple, of a coherent sheaf on a projective space.

5.1 Betti Diagrams of Graded Modules in General

The modifications needed to extend the Boij–Söderberg conjectures (theorems actually) to graded modules in general are not great. Let $\mathbb{Z}_{\text{deg}}^{\leq n+1}$ be the set of increasing sequences of integers $\mathbf{d} = (d_0, \ldots, d_s)$ with $s \leq n$ and consider a partial order on this by letting

$$(d_0, \ldots, d_s) \geq (e_0, \ldots, e_t)$$

if $s \leq t$ and $d_i \geq e_i$ when i ranges from $0, \ldots, s$. Note that if we identify the sequence $\mathbf{d}$ with the sequence $(d_0, \ldots, d_n)$ where $d_{s+1}, \ldots, d_n$ are all equal to $+\infty$, then this is completely natural.

Associated to $\mathbf{d}$, we have a pure diagram $\pi(\mathbf{d})$ by Subsection 1.4, such that any Cohen–Macaulay module of codimension s with pure resolution of type $\mathbf{d}$, will have a Betti diagram which is a multiple of $\pi(\mathbf{d})$. Boij–Söderberg prove the following variation of Theorem 1.11 for an arbitrary module.

Theorem 5.1. *Let $\beta(M)$ be the Betti diagram of a graded S-module M. Then there exists positive rational numbers c_i and a chain of sequences $\mathbf{d}^1 < \mathbf{d}^2 < \cdots < \mathbf{d}^p$ in $\mathbb{Z}_{\text{deg}}^{\leq n+1}$ such that*

$$\beta(M) = c_1\pi(\mathbf{d}^1) + \cdots + c_p\pi(\mathbf{d}^p).$$

The algorithm for this decomposition goes exactly as the algorithm in Subsection 1.7.

Example 5.2. Let $M = \Bbbk[x, y, z]/(x^2, xy, xz^2)$. This is a module with Betti diagram

$$\begin{matrix} 0 \\ 1 \\ 2 \end{matrix} \begin{bmatrix} 1 & 0 & 0 & 0 \\ 0 & 2 & 1 & 0 \\ 0 & 1 & 2 & 1 \end{bmatrix}$$

which can be decomposed as

$$\frac{1}{5} \cdot \begin{bmatrix} 1 & 0 & 0 & 0 \\ 0 & 5 & 5 & 0 \\ 0 & 0 & 0 & 1 \end{bmatrix} + \frac{1}{10} \cdot \begin{bmatrix} 3 & 0 & 0 & 0 \\ 0 & 10 & 0 & 0 \\ 0 & 0 & 15 & 8 \end{bmatrix} + \frac{1}{6} \cdot \begin{bmatrix} 1 & 0 & 0 & 0 \\ 0 & 0 & 0 & 0 \\ 0 & 4 & 3 & 0 \end{bmatrix} + \frac{1}{3} \cdot \begin{bmatrix} 1 & 0 & 0 & 0 \\ 0 & 0 & 0 & 0 \\ 0 & 1 & 0 & 0 \end{bmatrix}.$$

The cone cut out by the facet equations. Let $\mathbf{a}, \mathbf{b}$ in $\mathbb{Z}^{n+1}_{\deg}$ be degree sequences of length $(n+1)$. In the linear space $L^{\mathrm{HK}}(\mathbf{a}, \mathbf{b})$ we know that the positive cone cut out by the functionals β_{ij} and $\langle -, \mathcal{E} \rangle_{e,\tau}$, where e is an integer, $0 \le \tau \le n-1$, and $\mathcal{E}$ is a vector bundle on $\mathbb{P}^{n-1}$, is the positive cone $B(\mathbf{a}, \mathbf{b})$. We may then ask what is the positive cone $B_{\mathrm{eq}}(\mathbf{a}, \mathbf{b})$ cut out by these functionals in the space of *all* diagrams in the window $\mathbb{D}(\mathbf{a}, \mathbf{b})$.

Let $B_{\mathrm{mod}}(\mathbf{a}, \mathbf{b})$ be the positive cone consisting of all rational multiples of Betti diagrams of graded modules whose diagram is in the window $\mathbb{D}(\mathbf{a}, \mathbf{b})$. Since the functionals are nonnegative on all Betti diagrams of modules, by Proposition 2.16, it is clear that $B_{\mathrm{mod}}(\mathbf{a}, \mathbf{b}) \subseteq B_{\mathrm{eq}}(\mathbf{a}, \mathbf{b})$. In [5] they describe the facet equations of $B_{\mathrm{mod}}(\mathbf{a}, \mathbf{b})$. They are limits of facet equations of the type $\langle -, \mathcal{E} \rangle_{e,\tau}$ where elements in the root sequence of $\mathcal{E}$ tend to infinity. This shows that also $B_{\mathrm{mod}}(\mathbf{a}, \mathbf{b}) \supseteq B_{\mathrm{eq}}(\mathbf{a}, \mathbf{b})$. Hence the cone $B_{\mathrm{eq}}(\mathbf{a}, \mathbf{b})$ in $\mathbb{D}(\mathbf{a}, \mathbf{b})$ cut out by the functionals is simply $B_{\mathrm{mod}}(\mathbf{a}, \mathbf{b})$, the positive cone generated by all Betti diagrams of graded modules with support in in the window $\mathbb{D}(\mathbf{a}, \mathbf{b})$.

When $c = n$, the exterior facets of type (i) (when removing a minimal element), (ii), and (iii) in Proposition 2.2 are on unique exterior facets of the full-dimensional cone $B_{\mathrm{mod}}(\mathbf{a}, \mathbf{b}) = B_{\mathrm{eq}}(\mathbf{a}, \mathbf{b})$ in $\mathbb{D}(\mathbf{a}, \mathbf{b})$. The unique hyperplane equation (up to scalar) of these latter facets are given by the β_{ij} and the upper equation respectively, testifying to the naturality of these choices in Section 2.

5.2 Cohomology of Coherent Sheaves

In contrast to the case of vector bundles the decomposition algorithm for coherent sheaves on projective space is not of a finite number of steps.

In order to extend the algorithm we need to define sheaves with supernatural cohomology. Let

$$\mathbf{z} : z_1 > z_2 > \cdots > z_s$$

be a sequence of integers. It will be convenient to let $z_0 = \infty$ and $z_{s+1} = z_{s+2} = \cdots = -\infty$. A coherent sheaf $\mathcal{F}$ on $\mathbb{P}^m$ has supernatural cohomology if:

(i) The Hilbert polynomial is $\chi\mathcal{F}(d) = \frac{d^0}{s!} \cdot \Pi_{i=1}^{s}(d - z_i)$ for a constant d^0 (which must be the degree of $\mathcal{F}$).

(ii) For each d let i be such that $z_i > d > z_{i+1}$. Then

$$H^i\mathcal{F}(d) = \begin{cases} \frac{d^0}{s!} \cdot \Pi_{i=1}^{s}|d - z_i|, & z_i > d > z_{i+1}, \\ 0, & \text{otherwise.} \end{cases}$$

In particular we see that for each d there is at most one nonvanishing cohomology group.

The typical example of such a sheaf is a vector bundle with supernatural cohomology living on a linear subspace $\mathbb{P}^s \subseteq \mathbb{P}^m$. Let $\gamma^{\mathbf{z}}$ be the cohomology table of the sheaf with supernatural cohomology with root sequence $\mathbf{z}$.

We need to define one more notion derived from a cohomology table of a coherent sheaf.

Example 5.3. Consider the cohomology table :

$\cdots$	23	11	5	$\lfloor 1$	0	0	0	$\cdots$	3
$\cdots$	6	5	4	3	2	$\lfloor 1$	0	$\cdots$	2
$\cdots$	0	0	1	1	1	1	0	$\cdots$	1
$\cdots$	0	0	0	0	1	3	8	$\cdots$	0
$\cdots$	-2	-1	0	1	2	3	4	$\cdots$	$d\backslash i.$

In rows 3 and 2 there are two distinguished corners with nonzero values, marked with a $\lfloor$ such that in the first quadrant determined by them, these are the only nonzero values. In this case the root sequence is $z_1 = 4-1 = 3$, $z_2 = 4-2 = 2$ and $z_3 = 2-3 = -1$. We see that there is no corner position in row 1 because $z_2 = z_1 - 1$.

Definition 5.4. Given a root sequence $z_1 > \cdots > z_s$. The position $(i, d) = (i, z_i - 1)$ is a *corner position* if $z_{i+1} < z_i - 1$.

We may verify that $\gamma^{\mathbf{z}}$ has nonzero values at each corner position. Assume $\mathbf{z}$ is the root sequence of the cohomology table γ of a coherent sheaf. Let $\alpha_r, \alpha_{r-1}, \ldots, \alpha_0$ be the values of the corner positions of γ, and let $a_r, a_{r-1}, \ldots, a_0$ be the values of the corresponding corner positions in $\gamma^{\mathbf{z}}$.

Define
$$q_{\mathbf{z}} = \min\left\{\frac{\alpha_0}{a_0}, \ldots, \frac{\alpha_r}{a_r}\right\}.$$

Eisenbud and Schreyer [13] show the following.

- The table $\gamma - q_{\mathbf{z}}\gamma^{\mathbf{z}}$ has nonnegative entries.
- The root sequence $\mathbf{z}'$ of this new table is $<$ than the root sequence $\mathbf{z}$.

The algorithm of Eisenbud and Schreyer is now to continue this process. For a table γ, let $\dim \gamma$ be the largest i such that row i is nonzero.

0. Let $s = \dim \gamma$ and $\gamma_0 = \gamma$.
1. $\gamma_1 = \gamma_0 - q_{\mathbf{z}_0}\gamma^{\mathbf{z}^0}$ where $\mathbf{z}^0$ is the root sequence of γ_0.
2. $\gamma_2 = \gamma_1 - q_{\mathbf{z}_1}\gamma^{\mathbf{z}^1}$ where $\mathbf{z}^1$ is the root sequence of γ_1.

$\vdots$

In the case of vector bundles in $\mathbb{T}(\mathbf{a}, \mathbf{b})$ we are guaranteed that this process stops at latest when $\mathbf{z}^i = \mathbf{a}$, and we get the decomposition derived from the simplicial fan structure of $C(\mathbf{a}, \mathbf{b})$, Theorem 4.7. For coherent sheaves this process gives a strictly decreasing chain of root sequences

$$\mathbf{z}^0 > \mathbf{z}^1 > \mathbf{z}^2 > \cdots$$

and may continue an infinite number of steps. Clearly the top value z_s^i must tend to $-\infty$ as i tends to infinity. In the end we get a table γ_∞ where row s is zero so $\dim \gamma_\infty < s$. Note that we are not guaranteed that the entries of γ_∞ are rational numbers.

We may repeat this process with $\gamma' = \gamma_\infty$, which has dimension strictly smaller than that of γ. Eisenbud and Schreyer [13] show the following.

Theorem 5.5. *Let $\gamma(\mathcal{F})$ be a cohomology table of a coherent sheaf $\mathcal{F}$ on $\mathbb{P}^m$. There is a chain of root sequences Z and positive real numbers $q_{\mathbf{z}}$ for $\mathbf{z} \in Z$ such that*

$$\gamma(\mathcal{F}) = \sum q_{\mathbf{z}} \gamma^{\mathbf{z}}.$$

Both Z and the numbers $q_{\mathbf{z}}$ are uniquely determined by these conditions. The $q_{\mathbf{z}}$ are rational numbers if $\dim \gamma^{\mathbf{z}} = \dim \gamma$.

The way $q_{\mathbf{z}}$ is defined we are only sure that the corner values of $\gamma - q_{\mathbf{z}}\gamma^{\mathbf{z}}$ stays nonnegative. The essential ingredient in the proof is to show that not only the corner values stay nonnegative but that every entry in the table stays nonnegative. In order to prove the theorem, Eisenbud and Schreyer show that certain linear functionals are nonnegative when applied to the cohomology table of a coherent sheaf.

6 Further Topics

6.1 The Semigroup of Betti Diagrams of Modules

Theorem 1.9 gives a complete description of the positive rational cone $B(\mathbf{a}, \mathbf{b})$ generated by Betti diagrams of Cohen–Macaulay modules in the window $\mathbb{D}(\mathbf{a}, \mathbf{b})$. Of course a more ultimate goal is to describe precisely what the possible Betti diagrams of modules really are.

This is a much harder problem and the results so far may mostly be described as families of examples. Investigations into this has been done mainly by D. Erman in [17] and by Eisenbud, Erman and Schreyer in [9].

Denote by $B_{\text{int}} = B(\mathbf{a}, \mathbf{b})_{\text{int}}$ the semigroup of integer diagrams in $B(\mathbf{a}, \mathbf{b})$, which we call the semigroup of *virtual Betti diagrams*, and let $B_{\text{mod}} = B(\mathbf{a}, \mathbf{b})_{\text{mod}}$ be the semigroup of diagrams in $B(\mathbf{a}, \mathbf{b})$ which are actual Betti diagrams of modules of codimension n.

As a general result Erman shows:

Theorem 6.1 ([17]). *The semigroups B_{int} and B_{mod} are finitely generated.*

Not every virtual Betti diagram may be an actual Betti diagram of a module.

Example 6.2. The pure diagram $\pi = \pi(0,1,3,4)$ is

$$\begin{bmatrix} 1 & 2 & - & - \\ - & - & 2 & 1 \end{bmatrix}.$$

If this were the Betti diagram of a module, this module would have resolution

$$S \xleftarrow{[l_1,\ l_2]} S(-1)^2 \leftarrow S(-3)^2 \leftarrow S(-4).$$

But this is not possible since writing $S(-1)^2 = Se_1 \oplus Se_2$ with $e_i \mapsto l_i$, there would be a syzygy $l_2e_1 - l_1e_2$ of degree 2.

However 2π is an actual Betti diagram. Take a sufficiently general map $S^2 \xleftarrow{d} S(-1)^4$, for instance

$$d = \begin{pmatrix} x_1 & x_2 & x_3 & x_4 \\ x_1 & 2x_2 & 3x_3 & 4x_4 \end{pmatrix}.$$

The resolution of the cokernel of d is then

$$S^2 \xleftarrow{d} S(-1)^4 \leftarrow S(-3)^4 \leftarrow S(-4)^2.$$

Also, the equivariant resolution $E(1,2,1)$ (recall that $1,2,1$ are the differences of $0,1,3,4$) is given by

$$S^3 \leftarrow S(-1)^6 \leftarrow S(-3)^6 \leftarrow S(-4)^3.$$

So we see that on the ray determined by π the integer diagrams

$$\begin{bmatrix} m & 2m & - & - \\ - & - & 2m & m \end{bmatrix}$$

are actual Betti diagrams for $m \geq 2$ but not for $m = 1$.

Recall that if $\mathbf{d}$ is a degree sequence, then $\pi(\mathbf{d})$ is the smallest integer diagram on the ray $t\pi(\mathbf{d}), t > 0$. The integer diagrams are then $m\pi(\mathbf{d}), m \in \mathbb{N}$.

Conjecture 6.3 ([11, Conjecture 6.1]). For every degree sequence $\mathbf{d}$ there is an integer m_0 such that for $m \geq m_0$ the diagram $m\pi(\mathbf{d})$ is the Betti diagram of a module.

For rays in the positive cone $B(\mathbf{a}, \mathbf{b})$ which are not extremal, things are more refined. The following examples are due to Erman [17].

Example 6.4. The diagram

$$\beta = \begin{bmatrix} 2 & 3 & 2 & - \\ - & 5 & 7 & 3 \end{bmatrix}$$

is a virtual Betti diagram in B_{int}, and

$$\pi = \pi(0,2,3,4) = \begin{bmatrix} 1 & - & - & - \\ - & 6 & 8 & 3 \end{bmatrix}$$

is the Betti diagram of the module $S/(x_1,x_2,x_3)^2$. Erman shows that $\beta + m\pi$ is not in B_{mod} for any integer $m \geq 0$. In particular $B_{\text{int}} \backslash B_{\text{mod}}$ may not be finite.

In the following example we let $S = \Bbbk[x_1,\ldots,x_{p+1}]$ where p is a prime. It generalizes Example 6.2 above which is the case $p = 2$.

Example 6.5. Erman calculates that the diagram

$$\pi = \pi(0,1,p+1,\ldots,2p) = \begin{bmatrix} 1 & 2 & - & - & \cdots & - \\ \vdots & - & \vdots & & & \vdots \\ - & - & * & * & \cdots & * \end{bmatrix}.$$

If $m\pi$ is the Betti diagram of a CM-module then its resolution starts

$$S^m \xleftarrow{d} S(-1)^{2m} \leftarrow \cdots.$$

An $a \times (a+m)$ matrix degenerates in codimension $\leq m+1$. Since we are considering CM-modules, their codimension is one less than the length of the degree sequence defining π, which is $p+1$. So $m\pi$ is in B_{mod} only if $m \geq p$.

Example 6.6. The diagram

$$\beta = \begin{bmatrix} 2 & 4 & 3 & - \\ - & 3 & 4 & 2 \end{bmatrix}$$

is a virtual Betti diagram. Erman shows that β and 3β are not in B_{mod}, but 2β is in B_{mod}. In particular the points on B_{mod} on a ray in B_{int} may contain nonconsecutive lattice points.

In [18] Erman is able to apply Boij–Söderberg theory to prove the Buchsbaum–Eisenbud–Horrocks conjecture in some special cases.

Module theoretic interpretations of the decomposition. When decomposing the Betti diagram of a module M into a linear combination of pure diagrams associated to a chain of degree sequences

$$\beta(M) = \sum_{i=1}^{t} c_i \pi(\mathbf{d}^i) \tag{6.1}$$

where $\mathbf{d}^1 < \cdots < \mathbf{d}^t$, one may ask if the decomposition reflects some decomposition of the the minimal free resolution of M.

Eisenbud, Erman and Schreyer, [9], ask if there exists a filtration of

$$M = M_t \supset M_{t-1} \supset M_{t-2} \supset \cdots \supset M_1 \supset M_0 = 0 \tag{6.2}$$

such that the M_i/M_{i-1} have pure resolutions $c_i\pi(\mathbf{d}^i)$. Of course in general this cannot be so since the coefficient c_i in the decomposition of $\beta(M)$ may not be integers. However even in the case of integer c_i, examples show there may not exist such a filtration of M or even of $M^{\oplus r}$ for $r \geq 1$, [28, Example 4.5]. There is the question though if some deformation or specialization of M or $M^{\oplus r}$ for $r \geq 1$ could have such a filtration.

In [9] they give sufficient conditions on chains of degree sequences such that if M has a decomposition (6.1), the c_i are integers and there is a filtration (6.2). As a particular striking application they give the following.

Example 6.7. Let $S = \Bbbk[x_1, x_2, x_3]$ and $p = 2k + 1$ be an odd prime. Consider the pure diagrams

$$\pi(0,1,2,p) = \begin{bmatrix} \binom{p-1}{2} & p^2-2p & \binom{p}{2} & - \\ \vdots & & & \vdots \\ - & - & - & 1 \end{bmatrix},$$

$$\pi(0,p-2,p-1,p) = \begin{bmatrix} 1 & - & - & - \\ \vdots & & & \vdots \\ - & \binom{p}{2} & p^2-2p & \binom{p-1}{2} \end{bmatrix}$$

and

$$\pi(0,k,k+1,p) = \begin{bmatrix} 1 & - & - & - \\ & \vdots & & \\ - & p & p & - \\ & & \vdots & \\ - & - & - & 1 \end{bmatrix}.$$

If $\alpha + 1 + \binom{p-1}{2} \equiv 0 \pmod p$ the diagram

$$\beta = \frac{1}{p}\pi(0,1,2,p) + \frac{\alpha}{p}\pi(0,k,k+1,p) + \frac{1}{p}\pi(0,p-2,p-1,p)$$

is an integer diagram and the integer diagrams on the ray of β are $m\beta$ where m is a positive integer. In [9] they show that if M is a module with Betti diagram $m\beta$, then it has a filtration (6.2). Thus the coefficients in the decomposition of $m\beta$ are integers and so m must be divisible by p. Hence we have a ray where only $\frac{1}{p}$ of the lattice points are actual Betti diagrams of modules.

6.2 Variations on the Grading

The Boij–Söderberg conjectures concerns modules over the standard graded polynomial ring $\mathbb{k}[x_1, \ldots, x_r]$ where each $\deg x_i = 1$. Since the conjectures have been settled, it is natural to consider variations on the grading or finer gradings on the polynomial ring and module categories. Specifically each degree of x_i may be an element of $\mathbb{N}_0^r \backslash \{\mathbf{0}\}$ and the modules $\mathbb{Z}^r$-graded. When the module category we consider is closed under direct sums, the positive rational multiples of their Betti diagrams form a cone. The topic one has been most interested in concerning other gradings, is what are the extremal rays in this cone. This is the analog of the rays of pure resolutions. More precisely one is interested in finding the Betti diagrams $\beta(M)$ such that if one has a positive linear combination

$$\beta(M) = q_1\beta(M_1) + q_2\beta(M_2),$$

where M_1 and M_2 are other modules in the category, then each $\beta(M_i)$ is a multiple of $\beta(M)$.

In the case $S = \mathbb{k}[x, y]$ and $\deg x = 1, \deg y = 2$, this is investigated in the note [1] by B. Barwick et al. where they give candidates for what the extremal rays are, although no proofs.

Also with the polynomial ring in two variables, Boij and Fløystad, [4], consider the case when $\deg x = (1, 0), \deg y = (0, 1)$. They fix a degree sequence $(0, p, p + q)$ and consider bigraded Artinian modules whose resolution becomes pure of this type when taking total degrees by the map $\mathbb{Z}^2 \to \mathbb{Z}$ given by $(d_1, d_2) \mapsto d_1 + d_2$. Let $P(p, q)$ be the positive rational cone generated by such modules.

Theorem 6.8 ([5]). *When p and q are relatively prime the extremal rays in the cone $P(p,q)$ are parametrized by pairs (a, I) where a is an integer and I is an order ideal (down set) in the partially ordered set $\mathbb{N}^2$, contained in the region $px + qy < (p - 1)(q - 1)$.*

In particular there is a maximal order ideal in this region; it corresponds to the equivariant resolution. And there is a minimal order ideal, the empty set; it corresponds to a resolution of a quotient of monomial ideals given in the original [5, Remark 3.2].

For the polynomial ring in any number r of variables, Fløystad [21] lets $\deg x_i$ be the ith unit vector e_i. He considers $\mathbb{Z}^r$-graded Artinian modules whose resolutions becomes pure of a given type $(d_0, d_1, \ldots, d_c)$ when taking total degrees. He gives a complete description of the linear space generated by their multigraded Betti diagrams, see Theorem 3.4 in this survey.

Instead of Betti diagrams one may consider cohomology tables arising from other gradings. Eisenbud and Schreyer in [12] consider vector bundles $\mathcal{F}$ on $\mathbb{P}^1 \times \mathbb{P}^1$. The cohomology groups

$$H^i\mathcal{F}(a,b), \quad i = 0, 1, 2, \quad (a,b) \in \mathbb{Z}^2 \tag{6.3}$$

give a cohomology table in $\bigoplus_{(a,b)\in\mathbb{Z}^2}\mathbb{Q}^3$. One gets a positive rational cone of bigraded cohomology tables and one may ask what are the extremal rays of this cone. If for each (a,b) the cohomology groups (6.3) are nonvanishing for at most one i, $\mathcal{F}$ is said to have natural cohomology. In [12] they give sufficient conditions for a vector bundle with natural cohomology to be on an extremal ray.

Let us end the subsection with a quote by F.-O. Schreyer [14]: "Very little is known for the extension of this theory to the multi-graded setting. I believe that there will be beautiful results ahead in this direction."

6.3 Poset Structures

In the unique decomposition of a Betti diagram

$$\beta(M) = \sum_{i=1}^{s} c_i \pi(\mathbf{d}^i)$$

that we consider, we require that the degree sequences form a chain $\mathbf{d}^1 < \mathbf{d}^2 < \cdots < \mathbf{d}^r$.

C. Berkesch et al. [2] show that this order condition is reflected on modules with pure resolutions.

Theorem 6.9 ([2]). *Let* $\mathbf{d}$ *and* $\mathbf{d}'$ *be degree sequences. Then* $\mathbf{d} \leq \mathbf{d}'$ *if and only if there exists Cohen–Macaulay modules* M *and* M' *with pure resolutions of types* $\mathbf{d}$ *and* $\mathbf{d}'$ *with a nonzero morphism* $M' \to M$ *of degree* ≤ 0.

They also show the analog of this for vector bundles. This point of view may be fruitful when trying to understand decomposition algorithms of Betti diagrams under variations on the gradings.

D. Cook, [7], investigates the posets $[\mathbf{a},\mathbf{b}]_{\deg}$ and shows that they are vertex-decomposable, Cohen–Macaulay and square-free glicci.

6.4 Computer Packages

Macaulay 2 has the package "BoijSoederberg". We mention the most important routines in this package.

- **decompose**: Decomposes a Betti diagram B as a positive linear combination of pure diagrams.
- **pureBettiDiagram**: Lists the smallest positive integral Betti diagram of a pure resolution of a given type.
- **pureCohomologyTable**: Gives the smallest positive integral cohomology table for a given root sequence.
- **facetEquation**: Computes the upper facet equation of a given facet of type (iii).

- Routines to compute the Betti numbers for all three pure resolutions constructed in Section 3.
  - The equivariant resolution.
  - The characteristic free resolution.
  - The resolutions associated to generic matrices.

The package "PieriMaps" contains the routine **PureFree** to compute the equivariant resolutions constructed in Subsection 3.1, and the routine **pieriMaps** to compute the more general resolutions of [28], see the end of Subsection 3.1.

6.5 Three Basic Problems

The notes [19] is a collection of open questions and problems related to Boij–Söderberg theory. We mention here three problems, which we consider to be fundamental. (They are not explicitly in the notes.)

In [12] Eisenbud and Schreyer give a decomposition of the cohomology table of a coherent sheaf on $\mathbb{P}^n$ involving an infinite number of data. It does not seem possible from this to determine the possible cohomology tables of coherent sheaves up to rational multiple.

Problem 1. Determine the possible cohomology tables of coherent sheaves on $\mathbb{P}^n$ up to rational multiple. Can it by done by essentially a finite number of data? (At least if you fix a suitable "window".)

Let $E = \bigoplus_{i=0}^{\dim V} \bigwedge^i V$ be the exterior algebra. A *Tate resolution* is an acyclic complex unbounded in each direction

$$\cdots \to G^{i-1} \to G^i \to G^{i+1} \to \cdots$$

where each G^i is a free graded E-module $\bigoplus_{j\in\mathbb{Z}} E(j)^{\gamma_{i,j}}$. To any coherent sheaf $\mathcal{F}$ is associated a Tate resolution $T(\mathcal{F})$, see [10]. Tate resolutions associated to coherent sheaves constitute the class of Tate resolutions which are eventually linear i.e. such that $G^i = E(i - i_0)$ for $i \gg 0$ and some integer i_0. Hence the following is a generalization of the above Problem 1.

Problem 2. Determine the tables $(\gamma_{i,j})$ of Tate resolutions, up to rational multiple.

A complex $F_\bullet$ of free S-modules comes with three natural sets of invariants: The graded Betti numbers B, the Hilbert functions H of its homology modules, and the Hilbert functions C of the homology modules of the dualized complex $D(K)$, where $D = \mathrm{Hom}(-, \omega_S)$ is the standard duality.

When H and C each live in only one homological degree, $F_\bullet$ is a resolution of a Cohen–Macaulay module and Boij–Söderberg theory describes the positive rational cone of Betti diagrams B and, since H and C are determined by B, the set of the triples (B, H, C). If H only lives in one homological degree, i.e. $F_\bullet$ is a resolution, we saw in Subsection 5.1 that Boij and Söderberg, [6], gave a description of the possible B which are projections onto the first coordinate of such triples, up to rational multiple.

Problem 3. Describe all triples (B, H, C) that can occur for a complex of free S-modules $F_\bullet$, up to rational multiple. Also describe all such triples under various natural conditions on B, H and C.

Acknowledgments. We thank the referee for several corrections and useful suggestions for improving the presentation.

Bibliography

[1] Barwick, B., Biermann, J., Cook II, D., Moore, W. F., Raicu, C., Stamate, D., Boij–Söderberg theory for non-standard graded rings; 2010. http://math.berkeley.edu/~claudiu/mrc/nonStdBetti.pdf.

[2] Berkesch, C., Erman, D., Kummini, M., Sam. S. V., Poset structures in Boij–Söderberg theory. Preprint. arXiv:10102663. 2010.

[3] Berkesch, C., Erman, D., Kummini, M., Sam, S. V., Tensor complexes: Multilinear free resolutions constructed from higher tensors. Preprint. arXiv:11014604. 2011.

[4] Boij, M., Fløystad, G., The positive cone of Betti diagrams of bigraded artinian modules of codimension two. In: Fløystad, G., Johnsen, T., Knutsen, A. L., editors. Combinatorial Aspects of Commutative Algebra and Algebraic Geometry. Abel Symposia 6. pp. 1–16. Springer; 2011.

[5] Boij, M., Söderberg, J., Graded Betti numbers of Cohen–Macaulay modules and the multiplicity conjecture. J Lond Math Soc (2). 2008;78(1):78–101.

[6] Boij, M., Söderberg, J., Betti numbers of graded modules and the Multiplicity Conjecture in the non-Cohen–Macaulay case. Preprint. arXiv:803. 2008.

[7] Cook II, D., The structure of the Boij–Söderberg posets. Proc Amer Math Soc. 2011; 139(6):2009–2015.

[8] Eisenbud, D., Commutative algebra with a view toward algebraic geometry. vol. 150 of Graduate texts in mathematics. Springer; 1995.

[9] Eisenbud, D., Erman, D., Schreyer, F. O., Beyond Numerics: The Existence of Pure Filtrations. Preprint. arXiv:10010585. 2010.

[10] Eisenbud, D., Fløystad, G., Schreyer, F. O., Sheaf cohomology and free resolutions over exterior algebras. Trans Amer Math Soc. 2003;355(11):4397–4426.

[11] Eisenbud, D., Fløystad, G., Weyman, J., The existence of equivariant pure free resolutions. Ann Inst Fourier (Grenoble). 2011;61(3):905–926.

[12] Eisenbud, D., Schreyer, F. O., Betti numbers of graded modules and cohomology of vector bundles. J Amer Math Soc. 2009;22(3):859–888.

[13] Eisenbud, D., Schreyer, F. O., Cohomology of coherent sheaves and series of supernatural bundles. J Eur Math Soc. 2010;12(3):703–722.

[14] Eisenbud, D., Schreyer, F. O., Betti Numbers of Syzygies and Cohomology of Coherent Sheaves. Preprint. arXiv:11023559. 2011.

[15] Eisenbud, D., Schreyer, F. O., Boij–Söderberg theory. In: Fløystad, G., Johnsen, T., Knutsen, A. L., editors. Combinatorial Aspects of Commutative Algeba and Algebraic Geometry. Proceedings, Abel Symposia 6. pp. 35–48. Springer-Verlag; 2011.

[16] Eisenbud, D., Schreyer, F. O., Weyman, J., Resultants and Chow forms via exterior syzygies. J Amer Math Soc. 2003;16(3):537–579.

[17] Erman, D., The semigroup of Betti diagrams. Algebra Number Theory. 2009;3(3):341–365.

[18] Erman, D., A special case of the Buchsbaum–Eisenbud–Horrocks rank conjecture. Math Res Lett. 2010;17(6):1079–1089.

[19] Erman, D., Moore, F., Nilsson, J., Open Questions Related to the Boij–Söderberg Theorems. Available from: http://math.stanford.edu/ derman/Papers/questions.pdf.

[20] Fløystad, G., Exterior algebra resolutions arising from homogeneous bundles. Math Scand. 2004;94(2):191–201.

[21] Fløystad, G., The linear space of Betti diagrams of multigraded artinian modules. Math Res Lett. 2010;17(05):943–958.

[22] Fløystad, G., Vatne, J. E., (Bi-)Cohen–Macaulay Simplicial Complexes and Their Associated Coherent Sheaves. Comm Algebra. 2005;33(9):3121–3136.

[23] Fulton, W., Harris, J., Representation theory: A first course. vol. 129 of Graduate texts in mathematics. Springer; 1991.

[24] Hartshorne, R., Algebraic geometry. vol. 52 of Graduate texts in mathematics. Springer Verlag; 1977.

[25] Herzog, J., Srinivasan, H., Bounds for multiplicities. Trans Amer Math Soc. 1998; 350(7):2879–2902.

[26] Huneke, C., Miller, M., A note on the multiplicity of Cohen–Macaulay algebras with pure resolutions. Canad J Math. 1985;37(6):1149–1162.

[27] Jantzen, J. C., Representations of algebraic groups. vol. 107 of Mathematical surveys and monographs. American Mathematical Society; 2003.

[28] Sam, S. V., Weyman, J., Pieri resolutions for classical groups. J Algebra. 2011;329(1): 222–259.

[29] Weyman, J., Cohomology of vector bundles and syzygies. vol. 149 of Cambridge tracts in mathematics. Cambridge University Press; 2003.

Author Information

Gunnar Fløystad, Matematisk Institutt, Universitetet i Bergen, Bergen, Norway.
E-mail: gunnar@mi.uib.no

Progress in Commutative Algebra 1, 55–83

Hilbert Functions of Fat Point Subschemes of the Plane: the Two-fold Way

Anthony V. Geramita, Brian Harbourne, and Juan C. Migliore

Abstract. Two approaches for determining Hilbert functions of fat point subschemes of $\mathbb{P}^2$ are demonstrated. A complete determination of the Hilbert functions which occur for 9 double points is given using the first approach, extending results obtained in a previous paper using the second approach. In addition the second approach is used to obtain a complete determination of the Hilbert functions for $n \geq 9$ m-multiple points for every m if the points are smooth points of an irreducible plane cubic curve. Additional results are obtained using the first approach for $n \geq 9$ double points when the points lie on an irreducible cubic (but now are not assumed to be smooth points of the cubic).

Keywords. Fat Points, Hilbert Functions, Linkage, Blow Ups, Projective Plane.

2010 Mathematics Subject Classification. Primary: 13D40, 14J26, 14M07; Secondary: 14M06, 14N05.

1 Introduction

If X is a reduced set of n points in $\mathbb{P}^2$, the *fat point subscheme* $Z = mX \subset \mathbb{P}^2$ is the $(m-1)$-st infinitesimal neighborhood of X. Thus mX is the subscheme defined by the symbolic power $I(X)^{(m)} \subset R = k[\mathbb{P}^2]$ (that is, by the saturation of the ideal $I(X)^m$ with respect to the ideal generated by the coordinate variables in the ring $k[\mathbb{P}^2]$). The question motivating this paper is: What are the Hilbert functions of such subschemes of $\mathbb{P}^2$? There have been two main approaches to this question, and one goal of this paper is to demonstrate them in various situations.

The two approaches are exemplified by the papers [9] and [7]. The approach of [9] is to identify constraints that Hilbert functions must satisfy and then for each function satisfying those constraints to try to find a specific subscheme having that function as its Hilbert function. A complete classification of all Hilbert functions of reduced 0-dimensional subschemes of projective space was given in [8] using essentially this approach. The paper [9] then uses [8] as the starting point for classifying Hilbert functions for subschemes of the form $Z = 2X \subset \mathbb{P}^2$ with X reduced and 0-dimensional.

We thank the NSF, whose support for the MAGIC05 conference at the University of Notre Dame gave us an opportunity to begin work on this paper. Geramita also thanks the NSERC for research support, and Harbourne and Migliore thank the NSA for its partial support of their research (under grants H98230-11-1-0139 and H98230-07-1-0066 for Harbourne and under grants H98230-07-1-0036 and H98230-09-1-0031 for Migliore).

This approach is most effective when the class of possible functions is fairly limited, hence the restriction in [9] to the case $m = 2$. This approach has the advantage of providing explicit results often without needing detailed information about the disposition of the points, but it has the disadvantage of not providing a complete dictionary of which point sets give which Hilbert function. The approach of [7] is to use the geometry of the surface Y obtained by blowing up the points of the support of Z to obtain information about the Hilbert function of Z. This approach is most effective when the geometry of Y is well-understood, hence the restriction in [7] to the case $n \leq 8$. Given points p_i and non-negative integers m_i, the subscheme defined by the ideal $\bigcap_{i=1}^{n}(I(p_i)^{m_i})$ is also called a fat point subscheme, and is denoted by $m_1 p_1 + \cdots + m_n p_n$. The advantage of the second approach, as implemented in [7], is that it provided complete results for all fat point subschemes $Z = m_1 p_1 + \cdots + m_n p_n$ with $n \leq 8$, together with a complete determination of which Z give the same Hilbert function, but it had the cost of needing a lengthy analysis of the geometry of Y, and gives only recursive determinations of the Hilbert functions. However, for $n \leq 8$ and $k = 2$ there are only finitely many cases, so a complete list of the Hilbert functions which occur can be given. See [7] for this list.

The first case left open by [7] is $n = 9$ points of $\mathbb{P}^2$. It should, in principle, be possible to carry out the necessary analysis to obtain a complete recursive classification of Hilbert functions and corresponding points sets for $n = 9$, but whereas for $n \leq 8$ there are only finitely many classes of sets of n points, there will certainly be infinitely many when $n = 9$ (related to the fact that there can be infinitely many prime divisors on Y of negative self-intersection, and to the fact that effective nef divisors F can occur with $h^1(Y, \mathcal{O}_Y(F)) > 0$). Thus a complete classification in this case using the methods of [7] will be a substantial effort, which we leave for future research (not necessarily by us).

Instead, in this paper we will focus on some special cases. We devote Section 2 to demonstrating the first approach by obtaining a complete answer in the case of $n = 9$ and $m = 2$. This also shows how one could recover the result for $n = 8$ and $m = 2$ obtained in [7] using the methods of [9].

The rest of the paper is devoted to demonstrating both methods for the case of n points of multiplicity m on cubics, under somewhat different hypotheses chosen to play to the strengths of each method. The Philosophy of the First Way is to use known facts about Hilbert functions to say things about what Hilbert functions are possible. The Philosophy of the Second Way is to use known facts about cohomology of blown up surfaces to say things about what dimensions of linear systems are possible. Sections 3 (using the First Way) and 4 (using the Second Way) illustrate how we can attack the same problem and obtain overlapping and sometimes complementary results, but using dramatically different ways to do so.

So, given points on a plane cubic, for the First Way we will assume the cubic is irreducible, that $m = 2$ and, in some cases, that n is not too small. Our main results here are Theorem 3.4 and Theorem 3.7. For the Second Way we will make no restrictions on m nor assume the cubic is irreducible but we will assume the points are smooth points of the cubic and we will assume that the points are evenly distributed (meaning essentially that no component contains too many of the points). Under these two assumptions we give a complete determination of all possible Hilbert functions in Theorem 4.2. Using the same techniques we will, in Remark 4.5, also recover the Hilbert functions for X and $2X$ when X is a reduced set of points contained in a reduced, irreducible singular cubic curve in case the singular point of the curve is one of the points of X.

We now discuss both methods in somewhat more detail. For the first approach we will follow [7] and [9] and sometimes work with the first difference, Δh_{2X}, of the Hilbert function h_{2X} rather than with h_{2X} directly, since for our purposes Δh_{2X} is easier to work with. (For clarity we will refer to it as the Hilbert difference function, but we regard Δh_{2X} as just an equivalent formulation of the Hilbert function, so whenever our goal is to determine a Hilbert function, we will regard specifying the Hilbert difference function as having achieved the goal.) The first approach can be summarized as follows. We start by listing all Hilbert difference functions Δh_X for reduced sets X of $n = 9$ points, using [8], and then we analyze each case in turn using h_X to constrain the behavior of h_{2X}. For example, in some extreme cases the form of Δh_X forces many of the points of X to lie on a line; knowing this can be very useful in determining h_{2X}.

Our analysis uses the following tools: (i) a crude bound on the regularity of $I(2X)$, giving an upper bound for the last degree in which Δh_{2X} can be non-zero; (ii) Bézout considerations giving the values of Δh_{2X} in most degrees; (iii) the fact that the sum of the values of Δh_{2X} is 27; and (iv) a theorem of Davis [4] giving geometric consequences for certain behavior of the function Δh_{2X}. The idea is that we know the value of the Hilbert function for most degrees by (i), (ii) and (iii), and we can exhaustively list the possibilities for the remaining degrees. Then we use (iv) to rule out many of these. Finally, for the cases that remain, we try to construct examples of them (and in the situations studied in this paper, we succeed).

For the second approach we study h_Z for an arbitrary fat point subscheme $Z = m_1p_1 + \cdots + m_np_n \subset \mathbb{P}^2$ using the geometry of the surface Y, where $\pi : Y \to \mathbb{P}^2$ is the morphism obtained by blowing up the points p_i. This depends on the well known fact that $\dim I(Z)_t = h^0(Y, \mathcal{O}_Y(F))$ where $F = tL - m_1E_1 - \cdots - m_nE_n$, $\mathcal{O}_Y(L) = \pi^*\mathcal{O}_{\mathbb{P}^2}(1)$ and $E_i = \pi^{-1}(p_i)$. The fundamental fact here is the theorem of Riemann–Roch:

$$\begin{aligned} &h^0(Y, \mathcal{O}_Y(F)) - h^1(Y, \mathcal{O}_Y(F)) + h^2(Y, \mathcal{O}_Y(F)) \\ &\qquad = \frac{F^2 - K_Y \cdot F}{2} + 1 = \binom{t+2}{2} - \sum_i \binom{m_i+1}{2}. \end{aligned} \tag{1.1}$$

To see the relevance of (1.1), note that $K_Y = -3L + E_1 + \cdots + E_n$, so we have by duality that $h^2(Y, \mathcal{O}_Y(F)) = h^0(Y, \mathcal{O}_Y(K_Y - F))$ and thus $h^2(Y, \mathcal{O}_Y(F)) = 0$ if

$t > -3$. Now, since we are interested in the values of Hilbert functions when $t \geq 0$, we have $h_Z(t) = \dim(R_t) - \dim(I(Z)_t) = \binom{t+2}{2} - h^0(Y, \mathcal{O}_Y(F))$ which using (1.1) becomes

$$h_Z(t) = \sum_i \binom{m_i+1}{2} - h^1(Y, \mathcal{O}_Y(F)). \tag{1.2}$$

This second approach, as applied in [7], depended on knowing two things: the set $\mathrm{Neg}(Y)$ of all prime divisors C on Y with $C^2 < 0$ and on knowing $h^0(Y, \mathcal{O}_Y(F))$ for every divisor F for which we have $F \cdot C \geq 0$ for all $C \in \mathrm{Neg}(Y)$. Given $\mathrm{Neg}(Y)$, one can in principle reduce the problem of computing $h^0(Y, \mathcal{O}_Y(F))$ for an arbitrary divisor F to the case that $F \cdot C \geq 0$ for all $C \in \mathrm{Neg}(Y)$. If $n \geq 2$ and $F \cdot C \geq 0$ for all $C \in \mathrm{Neg}(Y)$, then $h^2(Y, \mathcal{O}_Y(F)) = 0$, so from Riemann–Roch we have only $h^0(Y, \mathcal{O}_Y(F)) \geq 1 + (F^2 - K_Y \cdot F)/2$.

When $n \leq 8$ or the points p_i lie on a conic (possibly singular), this inequality is always an equality, but for $n \geq 9$ points not contained in a conic it need not be, so more information in general is needed. Similarly, in case $n \leq 8$ or the points p_i lie on a conic (possibly singular), it turns out, in fact, that $\mathrm{Neg}(Y)$ is a finite set, but this also can fail for $n \geq 9$ points not contained in a conic. As a consequence, given $\mathrm{Neg}(Y)$ one can determine h_Z for any fat point subscheme $Z = m_1 p_1 + \cdots + m_n p_n \subset \mathbb{P}^2$ if either $n \leq 8$ or the points p_i lie on a conic. This raises the question of what sets $\mathrm{Neg}(Y)$ occur under these assumptions. We answered this question in [7]. There are only finitely many possibilities and [7] gives a complete list.

When $n \geq 9$ and the points p_i do not lie on a conic then not only can $\mathrm{Neg}(Y)$ fail to be finite but $h^1(Y, \mathcal{O}_Y(F))$ need not vanish, even if $F \cdot C \geq 0$ for all $C \in \mathrm{Neg}(Y)$ and even if F is effective. Assuming that the points p_i lie on a cubic curve does not eliminate either difficulty, but it does mean that $-K_Y$ is effective (whether the cubic is irreducible or not), and thus the results of [12] can be applied to the problem of computing $h^0(Y, \mathcal{O}_Y(F))$. In case $-K_Y$ is effective, it is known what kinds of classes can be elements of $\mathrm{Neg}(Y)$, but no one has yet classified precisely which sets $\mathrm{Neg}(Y)$ arise for $n \geq 9$ (doing this for $n = 7, 8$ was the new contribution in [7]). On the other hand, even without this complete classification, partial results can still be obtained using the second approach, as we will show here using information about the geometry of Y developed in [12].

2 Approach I: Nine Double Points

It is natural to ask what can be said for fat point schemes Z supported at $n > 8$ points. As observed in [7, Remark 2.2], there are infinitely many configuration types of $n > 8$ points, so we will restrict our attention to subschemes $2Z = 2(p_1 + \cdots + p_n)$ of $\mathbb{P}^2$. Since we are now restricting the multiplicities of the points to be at most 2, it is not necessary to make an exhaustive list of the configuration types – indeed, we will point out situations where different configurations exist but nevertheless do not give different

Hilbert functions. Instead, in this situation we can bring to bear the methods developed in [9], and to demonstrate additional methods which can be used. We will determine all Hilbert functions that occur for double point subschemes $2Z = 2(p_1 + \cdots + p_9)$ of $\mathbb{P}^2$, for every Hilbert function occurring as the Hilbert function of a simple point subscheme $Z = p_1 + \cdots + p_9$.

Definition 2.1. Let Z be a zero-dimensional subscheme of $\mathbb{P}^N$ with Hilbert function h_Z. The *difference function* of Z is the first difference of the Hilbert function of Z, $\Delta h_Z(t) = h_Z(t) - h_Z(t-1)$. (This is sometimes also called the *h-vector* of Z, and sometimes the *Castelnuovo function* of Z.)

The Hilbert function and its difference function clearly give equivalent information and it is primarily because of the simpler bookkeeping allowed by the first difference that we use it. Notice that Δh_Z is the Hilbert function of any Artinian reduction of R/I_Z by a linear form.

One problem raised in [9] is the existence and determination of maximal and minimal Hilbert functions. In the current context, this means that we fix an underlying Hilbert function $\underline{h}$ that exists for some set of 9 points in $\mathbb{P}^2$, and letting X move in the irreducible flat family of all sets of points with Hilbert function $\underline{h}$, we ask whether there is a maximal and a minimal Hilbert function for the corresponding schemes $Z = 2X$. It was shown in [9] that there *does* exist a maximal such Hilbert function, denoted $\underline{h}^{\max}$ (for any number of points). The proof in [9] is nonconstructive, and [9] determines $\underline{h}^{\max}$ in only a few special cases. The paper [9] also raises the question of whether $\underline{h}^{\min}$ always exists; i.e., whether there exists an X' such that h_{2X} is at least as big in every degree as $h_{2X'}$ for every X with $h_X = h_{X'}$. This question remains open.

A useful tool is the following lemma. This lemma, and generalizations of it, are well-known. For a very short proof of the statement given here see [9, Lemma 2.18].

Lemma 2.2. *Let X be a reduced set of points in $\mathbb{P}^2$ with regularity $r+1$. Then the regularity of I_{2X} is bounded by* $\operatorname{reg}(I_{2X}) \leq 2 \cdot \operatorname{reg}(I_X) = 2r+2$.

We will also use the following result of Davis [4]. It is a special case of a more general phenomenon [1] related to maximal growth of the first difference of the Hilbert function.

Theorem 2.3. *Let $X \subset \mathbb{P}^2$ be a zero-dimensional subscheme, and assume that $\Delta h_X(t) = \Delta h_X(t+1) = d$ for some t, d. Then the degree t and the degree $t+1$ components of I_X have a GCD, F, of degree d. Furthermore, the subscheme W_1 of X lying on the curve defined by F (i.e. I_{W_1} is the saturation of the ideal (I_X, F)) has Hilbert function whose first difference is given by the truncation*

$$\Delta h_{W_1}(s) = \min\{\Delta h_X(s), d\}.$$

Furthermore, the Hilbert function of the points W_2 not on F (defined by $I_{W_2} = I_X : (F)$) has first difference given by the (shifted) part above *the truncation:*

$$\Delta h_{W_2}(s) = \max\{\Delta h_X(s+d) - d, 0\}.$$

We will see precisely the possibilities that occur for the first infinitesimal neighborhood of nine points, and we will see that there is in each case a maximum and minimum Hilbert function. All together, there occur eight Hilbert functions for schemes $X = p_1 + \cdots + p_9$. We give their difference functions, and the possible Hilbert functions that occur for double point schemes $2X$, in the following theorem.

Theorem 2.4. *The following table lists all possibilities for the Hilbert difference function for nine double points, in terms of the Hilbert difference function of the underlying nine points. In particular, for each Hilbert function $\underline{h}$, both $\underline{h}^{\max}$ and $\underline{h}^{\min}$ exist, and we indicate by "max" or "min" the function that achieves $\underline{h}^{\max}$ or $\underline{h}^{\min}$, respectively, for each $\underline{h}$. Of course when we have "max = min," the Hilbert function of $2X$ is uniquely determined by that of X.*

difference function of X	possible difference functions of $2X$	max/min
1 1 1 1 1 1 1 1 1	1 2 2 2 2 2 2 2 2 2 1 1 1 1 1 1 1 1	max = min
1 2 1 1 1 1 1 1	1 2 3 4 2 2 2 2 2 1 1 1 1 1 1 1	max = min
1 2 2 1 1 1 1	1 2 3 4 4 3 2 2 1 1 1 1 1 1	max = min
1 2 2 2 1 1	1 2 3 4 4 4 3 2 1 1 1 1	max = min
1 2 2 2 2	1 2 3 4 4 4 4 2 2 1	max
	1 2 3 4 4 4 3 2 2 2	min
1 2 3 1 1 1	1 2 3 4 5 5 2 1 1 1 1 1	max = min
1 2 3 2 1	1 2 3 4 5 6 4 2	max
	1 2 3 4 5 6 3 2 1	
	1 2 3 4 5 6 3 1 1 1	
	1 2 3 4 5 6 2 2 1 1	min
1 2 3 3	1 2 3 4 5 6 6	max
	1 2 3 4 5 6 5 1	
	1 2 3 4 5 6 4 2	
	1 2 3 4 5 6 3 3	
	1 2 3 4 5 5 4 3	min

(2.1)

Proof. One has to "integrate" the difference functions in order to verify the claims about $\underline{h}^{\max}$ or $\underline{h}^{\min}$. We leave this to the reader. The fact that the eight Hilbert functions specified above for X give a complete list is standard, and we omit the proof.

Case 1: `1 1 1 1 1 1 1 1 1`. If X has this difference function then X must be a set of 9 collinear points in $\mathbb{P}^2$. Such a set of points is necessarily a complete intersection, so it is easy to check that the difference function for $2X$ is the one claimed. (Even the minimal free resolution is well known; see for instance [2, 3, 10, 11].)

Case 2: `1 2 1 1 1 1 1 1`. If X has this difference function then X must consist of 8 points on a line and one point off the line (it follows from Theorem 2.3). It is not hard to check, using Bézout arguments, that then $2X$ has the claimed difference function.

Case 3: `1 2 2 1 1 1 1`. If X has this difference function then X must consist of seven points on a line, say λ_1, and two points off the line (again using Theorem 2.3). Let Q_1, Q_2 be these latter points. We will see that the Hilbert function is independent of whether Q_1 and Q_2 are collinear with one of the seven other points or not. Note first that $2X$ contains a subscheme of degree 14 lying on a line. Hence the regularity is ≥ 14, so the difference function ends in degree ≥ 13.

Let L_1 be a linear form defining λ_1 and let L_2 be a linear form defining the line joining Q_1 and Q_2. Using Bézout's theorem, it is clear that there is no form of degree ≤ 3 vanishing on $2X$. Furthermore, $L_1^2 L_2^2$ is the only form (up to scalar multiples) of degree 4 vanishing on $2X$. Now, in degrees 5, 6 and 7 we have that L_1^2 is a common factor for all forms in the ideal of $2X$. Hence $(I_{2Q_1+2Q_2})_{t-2} \cong (I_{2X})_t$ for $5 \leq t \leq 7$, where the isomorphism is obtained by multiplying by L_1^2. But $2Q_1 + 2Q_2$ imposes independent conditions on forms of degree 3, so we can compute that $\dim(I_{2X})_t = 4, 9, 15$ for $t = 5, 6, 7$ respectively.

The calculations above give the claimed difference function up to degree 7, namely $(1, 2, 3, 4, 4, 3, 2, 2)$. But the sum of the terms of the difference function has to equal 27 ($= \deg 2X$), and the terms past degree 7 must be non-increasing, and positive through degree 13. This is enough to force the claimed difference function.

Case 4: `1 2 2 2 1 1`. By Theorem 2.3, X must consist of six points, X_1, on a line, λ_1, and three collinear points, X_2, on another line, λ_2. The intersection of λ_1 and λ_2 may or may not be a point of X_1; it is not a point of X_2. We will see, as in Case 3, that this combinatorial distinction does not affect the Hilbert function of $2X$. Pictorially we have the following two possibilities:

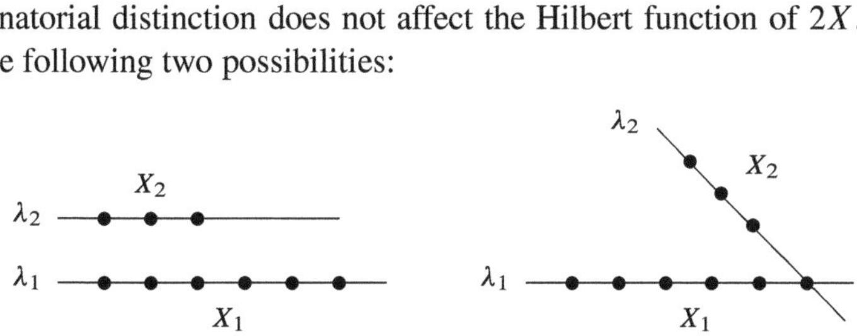

Combining Lemma 2.2 with the fact that $2X$ contains a subscheme of degree 12 on a line, we get that the difference function of $2X$ ends in degree exactly 11. Using

Bézout it is not hard to check that

$$
\begin{aligned}
5lll h^0(\mathcal{I}_{2X}) &= h^0(\mathcal{I}_{2X}(1)) = h^0(\mathcal{I}_{2X}(2)) = h^0(\mathcal{I}_{2X}(3)) = 0 \\
h^0(\mathcal{I}_{2X}(4)) &= 1 \\
h^0(\mathcal{I}_{2X}(5)) &= h^0(\mathcal{I}_{2X_2}(3)) = h^0(\mathcal{I}_{X_2}(2)) = 3 \\
h^0(\mathcal{I}_{2X}(6)) &= h^0(\mathcal{I}_{2X_2}(4)) = h^0(\mathcal{I}_{X_2}(3)) = 7.
\end{aligned}
$$

This means that the difference function of $2X$ begins `1 2 3 4 4 4 3` ... and arguing as in Case 3 gives the result.

Case 5: `1 2 2 2 2`. This case corresponds to nine points on a reduced conic curve. There are three possibilities. If the conic is smooth then the nine points are arbitrary. If the conic consists of two lines then this case takes the form of five points on one line and four points on the other line. Here we can have (i) none of the nine points is the point of intersection of the two lines, or (ii) one of the five points is the point of intersection. All of these cases have been studied in [7], and we omit the details.

Case 6: `1 2 3 1 1 1`. Now X consists of six point on a line plus three non-collinear points off the line. It is easy to check, using the same methods, that there is only one possibility for the Hilbert function of $2X$, independent of whether the line through two of the non-collinear points meets one of the six collinear points or not. We omit the details.

Case 7: `1 2 3 2 1`. By Lemma 2.2, the difference function for $2X$ ends in degree ≤ 9 and the entries again add up to 27. Furthermore, it is not hard to see that X has at most 5 points on a line, and X has at most one set of 5 collinear points.

The first main step in the proof is the following assertion:

Claim 2.6. $h^0(\mathcal{I}_{2X}(5)) = 0$.

Note that this implies that $h^0(\mathcal{I}_{2X}(t)) = 0$ for $t \leq 5$. Suppose that there is a curve F, of degree 5 containing $2X$. There are several possibilities. By abuse of notation we will denote by F also a form defining this curve.

- F is reduced. Then F has to contain 9 singular points, which form the points of X (and hence have the difference function `1 2 3 2 1`). This can happen in one of two ways:
  - F consists of the union of five lines, and X consists of nine of the resulting ten double points. But from Bézout we note that the 10 double points do not lie on a cubic curve (since each of the five lines would have to be a component of the cubic), so the ten points have difference function `1 2 3 4`, and hence X cannot have difference function `1 2 3 2 1`.

- F consists of the union of three lines and a smooth conic, and X consists of all nine resulting double points. Now the three lines have to be components of any cubic containing X, so there is a unique such cubic, and again X does not have difference function `1 2 3 2 1`.

- F has a double conic. Then all the singular points of F must lie on this conic. But, X does not lie on a conic, so this is impossible.
- F has a double line, i.e. $F = L^2G$, $\deg G = 3$. Then G contains at most 3 singular points of F. This forces the remaining 6 singular points to be on the line, contradicting the fact that at most 5 points of X can lie on a line.

This concludes the proof of Claim 2.6.

Thanks to Claim 2.6, we now know that the difference function for $2X$ has the form

`1 2 3 4 5 6 □ □ □ □`

where the last four spaces correspond to entries that are ≥ 0 and add up to $27-21 = 6$. Now notice that there is an irreducible flat family of subschemes of degree 9 with difference function `1 2 3 2 1` [6], and the general such is a complete intersection of two cubics. The difference function for the corresponding scheme $2X$ is easily checked to be `1 2 3 4 5 6 4 2`. It follows that not only does this difference function exist, but in fact it corresponds to $\underline{h}^{\max}$. (See also [9, Remark 7.4].) In particular, `1 2 3 4 5 6 6` and `1 2 3 4 5 6 5 1` do not occur. The following, then, are the remaining possibilities for the difference function of $2X$:

(i) `1 2 3 4 5 6 4 2`

(ii) `1 2 3 4 5 6 4 1 1`

(iii) `1 2 3 4 5 6 3 3`

(iv) `1 2 3 4 5 6 3 2 1`

(v) `1 2 3 4 5 6 3 1 1 1`

(vi) `1 2 3 4 5 6 2 2 2`

(vii) `1 2 3 4 5 6 2 2 1 1`

For each of these we will either give a specific example (that the reader can verify directly, either by hand or on a computer program) or a proof of non-existence.

(i) `1 2 3 4 5 6 4 2`. As we saw above, this occurs when X is the complete intersection of two cubics, and this corresponds to $\underline{h}^{\max}$.

(ii) `1 2 3 4 5 6 4 1 1`. This does not exist. Indeed, this difference function forces the existence of a line λ that contains a subscheme of $2X$ of degree 9, which is impossible. (Any such subscheme must have even degree.)

(iii) `1 2 3 4 5 6 3 3`. This does not exist in our context. Note that it *does* exist when X has difference function `1 2 3 3`, as we will verify below. To see that this does not exist, note that by Theorem 2.3, the `3 3` at the end forces the existence of a cubic curve C that cuts out from $2X$ a subscheme W of degree 21 with difference function `1 2 3 3 3 3 3 3`. Observe that if P is a point of X which is a smooth point of C, then C cuts out a non-reduced point of degree 2 at P. If P is a point of X which is a singular point of C, then C contains the fat point $2P$ (which has degree 3). Note also that our h-vector does not permit the existence of a subscheme of degree more than 8 on a line.

Suppose first that C is reduced. Since we only have the nine points of X to work with, it is not hard to check, using the above observation, that the only way that C can cut out from $2X$ a subscheme of degree 21 is if X has the following configuration:

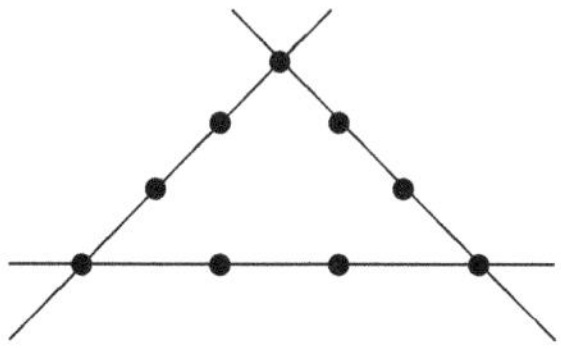

But this uses all nine points, and its support lies on a *unique* cubic, contradicting the fact that X has difference function `1 2 3 2 1`. This configuration provides one of the correct difference functions for `1 2 3 3` below.

Now suppose that C is not reduced. Without loss of generality, C has a double line. The difference function for X would, in principle, allow five points of X to lie on a line, but because the hypothetical difference function for the subscheme W ends in degree 7, in fact at most four points of X can lie on a line. So the double line contains at most four fat points of $2X$, which have degree 12. In order for C to cut out a subscheme of degree 21, then, we must have a reduced line that cuts out an additional subscheme of degree at least 9. This forces at least five points of X to be collinear, which again is impossible.

(iv) `1 2 3 4 5 6 3 2 1`. This difference function does exist. It occurs when X is the union of one point and the complete intersection of a conic and a general quartic curve.

(v) `1 2 3 4 5 6 3 1 1 1`. This difference function does exist. It occurs when X is the union of five general points on a line, three general points on a second line, and one additional general point off both lines.

(vi) `1 2 3 4 5 6 2 2 2`. This difference function does not exist. Indeed, suppose that it did exist. Because of the `2 2 2`, there must be a curve C of degree 2 that cuts out on $2X$ a subscheme W of degree 17 having difference function `1 2 2 2 2 2 2 2 2`.

First note that X cannot contain five points on a line (and hence a subscheme of W of degree at least 10) since the hypothetical difference function ends in degree 8. Now consider cases.

(a) C is smooth: then it cannot cut out a subscheme of odd degree.

(b) C is reduced and reducible: then we cannot obtain the desired subscheme W of degree 17 unless X contains 5 points on a line, in which case W contains a subscheme of degree at least 10 on that line.

(c) C non-reduced: then we cannot have a subscheme of degree 17 supported on that line.

(vii) `1 2 3 4 5 6 2 2 1 1`. This difference function does exist. It occurs when X has the following configuration:

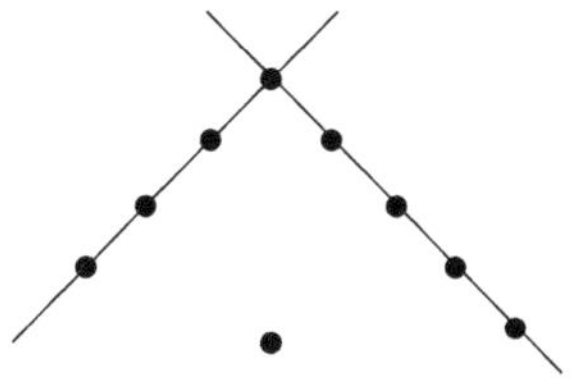

Case 8: `1 2 3 3`. This is the difference function for a general set of nine points in $\mathbb{P}^2$. We know (from [12], for example) that the "generic" difference function for nine general double points is `1 2 3 4 5 6 6`. Hence this occurs and corresponds to the maximum possible Hilbert function. Clearly all other possibilities will end in degree ≥ 7. On the other hand, Lemma 2.2 guarantees that all other examples end in degree ≤ 7. Note that again, X can have at most four points on a line.

Claim 8.1. $h^0(\mathcal{I}_{2X}(5)) \leq 1$.

Notice that as a consequence of this claim we also obtain $h^0(\mathcal{I}_{2X}(4)) = 0$. Keeping in mind that it is possible that $h^0(\mathcal{I}_{2X}(5)) = 0$ (e.g. the generic case), we will assume that $h^0(\mathcal{I}_{2X}(5)) \neq 0$ and deduce that then it must be $= 1$. So let C be a curve of degree 5 containing the scheme $2X$. As before (Claim 2.6) there are a few possibilities.

- If C is reduced then since it must have nine double points, it must consist of either the union of five lines, no three through a point, or the union of three lines and a smooth conic, with no three components meeting in a point. By Bézout, each component of C is then a fixed component of the linear system $|(I_{2X})_5|$, so the claim follows.
- If C contains a double line then at most four (fat) points of $2X$ lie on this line, so we must have a cubic curve that contains the remaining five double points. Consider the support, X_1, of these five double points. The points of X_1 are not collinear, and they do not have four collinear points since X lies on only one

cubic. With these restrictions, clearly there is no cubic curve double at such a set of five points.

- If C contains a double conic (smooth or not), this conic contains at most seven points of X, because of the Hilbert function of X. Hence C must have a line that contains two double points, which is impossible.

This concludes the proof of Claim 8.1.

It follows that the possibilities for the difference function of $2X$ are the following:

(i) 1 2 3 4 5 6 6

(ii) 1 2 3 4 5 6 5 1

(iii) 1 2 3 4 5 6 4 2

(iv) 1 2 3 4 5 6 3 3

(v) 1 2 3 4 5 5 5 2

(vi) 1 2 3 4 5 5 4 3

As before, we examine these each in turn.

(i) 1 2 3 4 5 6 6. We have seen that this occurs generically.

(ii) 1 2 3 4 5 6 5 1. This exists, for instance from the following configuration:

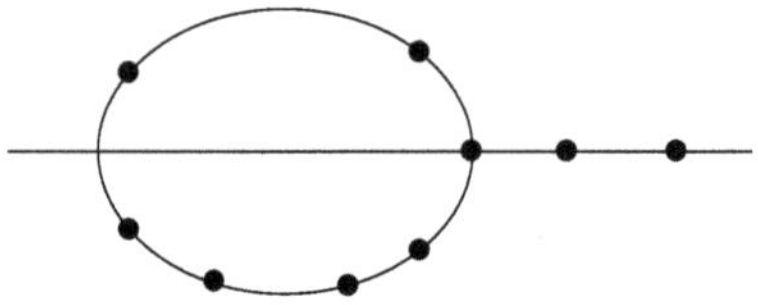

(That is, seven points on a conic, three points on a line, with one point in common.)

(iii) 1 2 3 4 5 6 4 2. This exists, for instance from the following configuration:

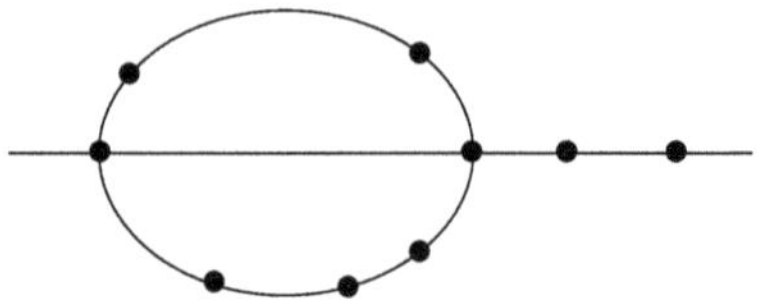

(iv) `1 2 3 4 5 6 3 3`. This exists, for instance from the configuration mentioned earlier:

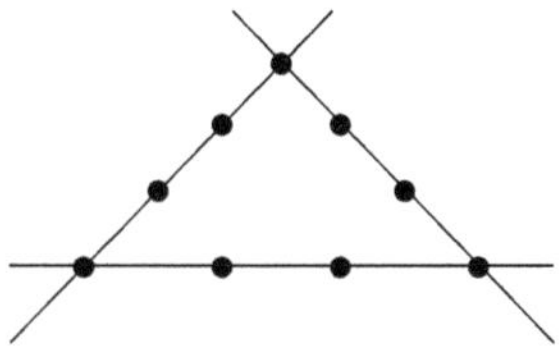

(v) `1 2 3 4 5 5 5 2`. We claim that this does not exist. The key is that such a double point scheme, $2X$, would have to lie on a unique quintic curve, say C. To see that this is impossible, the argument is very similar to that of Claim 2.6, but with a small difference. One checks as before that C must consist either of five lines or the union of three lines and a conic, and in both cases we must have that no three components share a common point. In the first case, X consists of nine of the ten double points of C (it does not matter which nine), and in the second case X consists of all nine double points of C. But in both of these cases one can check geometrically or on a computer that $h^0(\mathcal{I}_{2X}(6)) = 4$, while the hypothetical difference function would require this dimension to be 3.

(vi) `1 2 3 4 5 5 4 3`. This exists, and can be achieved by the configuration mentioned above: it is supported on nine of the ten intersection points of five general lines in $\mathbb{P}^2$. □

3 Approach I: Points on Cubics

For this section we will always let $C \subset \mathbb{P}^2$ be an irreducible cubic curve defined by a polynomial F of degree 3. Let X be a reduced set of $n = 3t + \delta$ points on C, where $0 \leq \delta \leq 2$. Let $Z = 2X$ be the double point scheme in $\mathbb{P}^2$ supported on X. The object of this section is to describe the possible Hilbert functions of X and of the corresponding Z. In some instances we assume that t is "big enough" (with mild bounds), and in one instance (Theorem 3.4 (ii)) we assume that the points are not too special and that C is smooth.

Proposition 3.1. *Assume that $\delta = 0$, $t \geq 3$, and the Hilbert function of X has first difference*

deg	0	1	2	...	$t-1$	t	$t+1$	$t+2$
Δh_X	1	2	3	...	3	2	1	0

(3.1)

(where the values between 2 and $t-1$, if any, are all 3). Then X is a complete intersection with ideal (F, G), where $\deg F = 3$ *and* $\deg G = t$*. Furthermore, if C*

is singular then the singular point is not a point of X. Assume that $t > 3$, so that $t + 3 > 6$ and $2t > t + 3$. Then we have the first difference of the Hilbert function of Z is

($t = 3$) 1 2 3 4 5 6 4 2 0;

($t = 4$) 1 2 3 4 5 6 6 5 3 1 0;

($t = 5$) 1 2 3 4 5 6 6 6 5 4 2 1 0;

($t \geq 6$)

deg	0	1	2	3	4	5	6	...	$t+2$	$t+3$
Δh_Z	1	2	3	4	5	6	6	...	6	5

deg	$t+4$	$t+5$	...	$2t-1$	$2t$	$2t+1$	$2t+2$
Δh_Z	4	3	...	3	2	1	0

Proof. We first show that X must be a complete intersection. From the Hilbert difference function (3.1), it is clear that F is a factor of every form in I_X up to degree $t-1$, and that in fact it generates the ideal up to this point. In degree t there is exactly one new form, G, in the ideal, and since F is irreducible, F and G have no common factor. But (F, G) is a saturated ideal that is contained in I_X and defines a zero-dimensional scheme of the same degree as X, hence $I_X = (F, G)$.

Since X is a complete intersection, if C is singular and $P \in X$ is the singular point of C, then X must be non-reduced at P, contradicting our assumption.

Now, it is a simple (and standard) argument that $I_Z = (F^2, FG, G^2)$, and one can verify the claimed Hilbert function of R/I_Z, for instance by using the fact that (F, G) is directly linked to the ideal of Z by the complete intersection (F^2, G^2), and using the formula for the behavior of Hilbert functions under linkage [5] (see also [15]). We omit the details. See also [3, pages 176, 177], and [2, Remark 4.9]. □

Because the form F of least degree is irreducible, the Hilbert function of X has first difference that is strictly decreasing from the first degree where it has value < 3 until it reaches 0. Having proved Proposition 3.1, we can now assume without loss of generality that the Hilbert function of X has first difference

deg	0	1	2	3	...	t	$t+1$	$t+2$
Δh_X	1	2	3	3	...	3	δ	0

(3.2)

where $0 \leq \delta \leq 2$.

Theorem 3.4. *Assume that either C is smooth, or else that no point of X is the singular point of C. Assume further that $t > 5 - \delta$. Then the Hilbert difference function of the double point scheme Z supported on X is*

deg	0	1	2	3	4	5	...	$t+3$	$t+4$	$t+5$	...	$2t+\delta-1$	$2t+\delta$
Δh_Z	1	2	3	4	5	6	...	6	$3+\delta$	3	...	3	??

For the behavior in degree $\geq 2t + \delta$*, we have the following conclusions.*

(i) *If* $\delta = 1$ *or* $\delta = 2$ *then* $\Delta h_Z(2t + \delta) = 3 - \delta$ *and* $\Delta h_Z(k) = 0$ *for* $k > 2t + \delta$.

(ii) *If* $\delta = 0$*, there are two possible Hilbert functions, these being determined by*

(a) $\Delta h_Z(2t) = 3$ *and* $\Delta h_Z(k) = 0$ *for* $k > 2t$*, and*

(b) $\Delta h_Z(2t) = 2, \Delta h_Z(2t + 1) = 1, \Delta h_Z(2t + 2) = 0$.

Moreover, if the points p_i *are sufficiently general and* C *is smooth, then the Hilbert function is given by the first of these two.*

Proof. A complete analysis of all cases with $\delta = 0$, where C is a reduced cubic and the points p_i either are arbitrary smooth points of C or they are completely arbitrary and C is also irreducible, is given in the next section using the Second Way. The interested reader can complete the current proof to those cases using the techniques of this section, as a further comparison of the methods.

First note that the condition $t > 5 - \delta$ implies $2t + \delta > t + 5$. We proceed via a number of claims.

Claim 1. For $\ell < 2t + \delta$, $(I_Z)_\ell$ has the cubic form F as a common factor (i.e. C is part of the base locus).

Suppose that $G \in (I_Z)_\ell$ does not have F as a factor. Then at each point of X, the intersection multiplicity of F and G is at least 2 since G is double at each point. Hence by Bézout's theorem, $3\ell \geq 2n = 2(3t + \delta) = 6t + 2\delta$. Hence $\ell \geq 2t + \frac{2}{3}\delta$, and the claim follows.

Claim 2. For $\ell \leq t + 3$, $(I_Z)_\ell$ has F^2 as a common factor.

By Claim 1, since F is not double at any point of X, for $\ell < 2t + \delta$ we have an isomorphism

$$(I_X)_{\ell-3} \cong (I_Z)_\ell \tag{3.3}$$

where the isomorphism is given by multiplication by F. But from (3.2), we see that F is a common factor for $(I_X)_k$ for all $k \leq t$. Hence $(I_Z)_\ell$ has F^2 as a factor whenever $\ell - 3 \leq t$, as claimed.

This verifies the claimed first difference of the Hilbert function up to degree $t + 3$. Note that the Hilbert function, in degree $t + 3$, has value equal to

$$1 + 2 + 3 + 4 + 5 + 6 \cdot [(t + 3) - 4] = 6t + 9.$$

We now compute the value in degree $t + 4 < 2t$. Using the isomorphism (3.3), we have

$$\begin{aligned} h_Z(t+4) &= \binom{t+6}{2} - h^0(\mathcal{I}_Z(t+4)) \\ &= \binom{t+6}{2} - h^0(\mathcal{I}_X(t+1)) \\ &= \binom{t+6}{2} - \left[\binom{t+3}{2} - h_X(t+1)\right] \\ &= \binom{t+6}{2} - \left[\binom{t+3}{2} - (3t+\delta)\right] \\ &= 6t + 12 + \delta. \end{aligned}$$

Then we easily see that $\Delta h_Z(t+4) = 3 + \delta$ as claimed.

Next we compute the value in degree $t + 5$. We have $2t + \delta > t + 5$, so we can use Claim 1. Then a similar computation gives

$$h_Z(t+5) = 6t + 15 + \delta.$$

From this we immediately confirm $\Delta h_Z(t+5) = 3$.

Since F is a common factor in all components $< 2t + \delta$, and since Δh_Z takes the value 3 already in degree $t + 5$, it repeats this value until F is no longer a common factor. In particular, it takes the value 3 up to degree $2t + \delta - 1$.

We now have to see what happens past degree $2t + \delta - 1$. Note that using our above calculations, it follows that

$$\begin{aligned} h_Z(2t+\delta-1) &= 6t + 12 + \delta + 3[2t + \delta - 1 - (t+4)] \\ &= 3(3t+\delta) - 3 + \delta. \end{aligned}$$

Since $\deg Z = 3(3t+\delta)$, we have reached the multiplicity minus $(3-\delta)$. We consider these cases separately. When $\delta = 1$ or $\delta = 2$, we are adding only 2 or 1, respectively, and since the first difference of the Hilbert function cannot be flat at this point, Δh_Z must be as claimed in (a). This completes (a). Since the sum of the values of Δh_Z up to degree $2t - 1$ is $9t - 3$, this observation that Δh_Z cannot be flat at this point also proves that the possibilities listed in (b) are the only ones possible.

If $\delta = 0$, though, Δh_Z can either end $\ldots 3, 3, 0$ or $\ldots 3, 2, 1$. We now consider these two possibilities. The former means that also in degree $2t + \delta = 2t$, all forms in I_Z have F as a factor. The latter means that there is a form, G, of degree $2t + \delta = 2t$ in I_Z that does not have F as a factor, and hence (F, G) is a regular sequence (since F is irreducible).

Suppose that the latter holds. Note that the complete intersection defined by (F, G) has degree $3 \cdot 2t = 6t = 2n$. As in Claim 1, G cuts out on C a divisor of degree at least $2n$, so in fact G cuts out exactly the divisor $2X_C$ on C (where by $2X_C$ we mean the subscheme $2X \cap C$, which is a divisor on C). So X itself is not a complete intersection (since it has the Hilbert difference function given by (3.2)), but the divisor $2X_C$ (as a subscheme of $\mathbb{P}^2$) is a complete intersection, namely of type $(3, 2t)$. Note that $2X_C$, which is curvilinear, is not the same as Z.

Now suppose that C is smooth. We know that then two effective divisors of the same degree are linearly equivalent if and only if they have the same sum in the group of C. The condition described in the previous paragraph implies that the divisor $X_C - nQ$, where Q is a point of inflection for C, is a 2-torsion element in the Picard group of C but is not zero. Since there are at most three 2-torsion elements in the Picard group of C, for general choices we have a contradiction, and so such a G cannot exist (in general), and we have proved the assertion about the general choice of the points.

Finally, we show that the Hilbert difference function (b) of (ii) also occurs. We begin with four general lines, $\lambda_1, \lambda_2, \lambda_3, \lambda_4 \subset \mathbb{P}^2$ and let $P_1, P_2, P_3, P_4, P_5, P_6$ be the six points of pairwise intersection of these lines. Let G_1 be the form defining the union of these four lines. Let $X_1 = \bigcup_{1 \le i \le 6} P_i$. Notice that X_1 does not lie on any conic, since by Bézout any conic containing X_1 has to contain all four lines $\lambda_1, \dots, \lambda_4$, hence must have G_1 as a factor. Thus the Hilbert function of X_1 has first difference $(1, 2, 3)$, and X_1 is not a complete intersection.

Let C be a general cubic curve containing X_1 (hence C is smooth), and let F be the defining polynomial of C. Notice that the degree of the complete intersection of F and G_1 is 12, and this complete intersection is at least double at each P_i, so in fact it is exactly double at each P_i. In particular, there is no additional multiplicity at any of the P_i coming from tangency. As a divisor on C, note that X_1 is not cut out by any conic, since it is not a complete intersection. However, the divisor $2X_1$ is cut out by a quartic, namely G_1.

Now let X be the union of X_1 with a general hypersurface section, W_1, of C cut out by a curve of degree $t - 2$. Note that W_1 is a complete intersection defined by (F, H) for some form H of degree $t - 2$. We first claim that X is not a complete intersection. Indeed, suppose that X were a complete intersection defined by (F, H') for some H' of degree t. Then I_X links W_1 to X_1. But W_1 and X are both complete intersections sharing a generator, so by liaison theory the residual is also a complete intersection. But we have seen that X_1 is not a complete intersection. Contradiction. In particular, Δh_X is given by (3.2).

Now let Z be the fat point scheme supported on X, and consider the form $G_1 H^2$. This has degree $2t$, and cuts out the divisor $2X$ on C. Even more, $G_1 H^2$ is an element of I_Z in degree $2t$ that does not have F as a factor. As we saw above, this gives the values $\Delta h_Z(2t) = 2$ and $\Delta h_Z(2t + 1) = 1$ as desired. This completes the proof of Theorem 3.4. □

Now we wish to explore the possibilities when C is singular and one point, P, of X is the singular point of C. The arguments are very similar, and we will primarily highlight the differences. The main observation is that C is already double at P so we have to focus on the remaining $n-1$ points.

Lemma 3.6. *Assume that C is singular, that $P \in X \subset C$ is the singular point of C, and that $n \geq 5$. Then X is not a complete intersection.*

Proof. More precisely, we will show that if $P \in X \subset C$ with X a complete intersection, and if P is the singular point of C, then X has one of the following types: $CI(1,1)$, $CI(1,2)$, $CI(2,2)$.

First note that if X is a complete intersection defined by forms (F,G), where F is the defining polynomial for C, then X has multiplicity ≥ 2 at P, so X is not reduced. Hence we have to determine all the possibilities for reduced complete intersections on C that do not use F as a minimal generator. The listed possibilities are clear: one point, two points, four points, and these all exist even including P as one of the points. Using the irreducibility of F, it is not hard to show that these are the only possibilities, and we omit the details. □

Theorem 3.7. *Assume that C is an irreducible singular cubic with singular point P, and assume that $P \in X$, where X is a reduced set of $3t+\delta$ points of C, with $0 \leq \delta \leq 2$. Assume further that $t > 3$. Then the Hilbert difference function of the double point scheme Z supported on X is as follows.*

(i) *If $\delta = 0$ then*

deg	0	1	2	3	4	5	...	$t+2$	$t+3$	$t+4$	$t+5$	...	$2t$	$2t+1$	$2t+2$
Δh_Z	1	2	3	4	5	6	...	6	5	3	3	...	3	1	0

(ii) *If $\delta = 1$ then either*

deg	0	1	2	3	4	5	...	$t+2$	$t+3$	$t+4$	$t+5$	...	$2t$	$2t+1$	$2t+2$
Δh_Z	1	2	3	4	5	6	...	6	6	3	3	...	3	3	0

or

deg	0	1	2	3	4	5	...	$t+2$	$t+3$	$t+4$	$t+5$	...	$2t$	$2t+1$	$2t+2$
Δh_Z	1	2	3	4	5	6	...	6	5	4	3	...	3	3	0

(iii) *If $\delta = 2$ then*

deg	0	1	2	3	4	5	...	$t+2$	$t+3$	$t+4$	$t+5$	...	$2t$	$2t+1$	$2t+2$	$2t+3$
Δh_Z	1	2	3	4	5	6	...	6	6	4	3	...	3	3	2	0

Proof. The bound $t > 3$ is simply to ensure that in each case, some value of the Hilbert difference function Δh_Z takes the value 3. For instance, in the case $\delta = 0$, we have $2t > t + 3$. As a consequence of Lemma 3.6, when $n = 3t + \delta \geq 5$ the Hilbert function of X must have first difference

deg	0	1	2	3	...	t	$t+1$	$t+2$
Δh_X	1	2	3	3	...	3	δ	0

(3.4)

In analogy with Theorem 3.4, we first have

Claim 1. Assume that

$$\ell \leq \begin{cases} 2t & \text{if } \delta = 0 \\ 2t+1 & \text{if } \delta = 1, 2 \end{cases}$$

Then F is a common factor of $(I_Z)_\ell$.

The proof is the same as that of Claim 1 in Theorem 3.4, except that the intersection multiplicity of F and G at P is now at least 4.

Claim 2. For $\ell \leq t + 2$, $(I_Z)_\ell$ has F^2 as a common factor. Furthermore,

- If $\delta = 0$ then F^2 is *not* a common factor of $(I_Z)_{t+3}$.
- If $\delta = 2$ then F^2 is a common factor of $(I_Z)_{t+3}$.
- If $\delta = 1$ then F^2 may or may not be a common factor of $(I_Z)_{t+3}$ (examples exist for either option).

The proof of Claim 2 hinges on the possible Hilbert functions for $X - \{P\}$. In particular, we show that $(I_{X-\{P\}})_{t-1}$ always has F as a common factor, and the differences in the three cases rest with the possibilities for $(I_{X-\{P\}})_t$, which we get by comparing to those for I_X, obtained using Lemma 3.6.

- If $\delta = 0$ then X has Hilbert function with first difference

deg	0	1	2	3	...	t	$t+1$
Δh_X	1	2	3	3	...	3	0

so clearly the only possibility for $\Delta h_{X-\{P\}}$ is

deg	0	1	2	3	...	$t-1$	t	$t+1$
$\Delta h_{X-\{P\}}$	1	2	3	3	...	3	2	0

Hence there is a form G of degree t vanishing on $X - \{P\}$ but not containing F as a factor, so $FG \in (I_Z)_{t+3}$ does not have F^2 as a factor.

- If $\delta = 2$ then X has Hilbert function with first difference

deg	0	1	2	3	...	$t-1$	t	$t+1$	$t+2$
$\Delta h_{X-\{P\}}$	1	2	3	3	...	3	3	2	0

so $\Delta h_{X-\{P\}}$ is

deg	0	1	2	3	...	$t-1$	t	$t+1$	$t+2$
$\Delta h_{X-\{P\}}$	1	2	3	3	...	3	3	1	0

We know that $(I_Z)_\ell \cong (I_{X-\{P\}})_{\ell-3}$ for ℓ satisfying the bounds of Claim 1, and as a result of the above observations we know when $(I_{X-\{P\}})_{\ell-3}$ is forced to have F as a common factor, so the claim follows.

If $\delta = 1$ then X has Hilbert function with first difference

deg	0	1	2	3	...	t	$t+1$	$t+2$
Δh_X	1	2	3	3	...	3	1	0

so $\Delta h_{X-\{P\}}$ is either

deg	0	1	2	3	...	$t-1$	t	$t+1$
$\Delta h_{X-\{P\}}$	1	2	3	3	...	3	3	0

or

deg	0	1	2	3	...	$t-1$	t	$t+1$	$t+2$
$\Delta h_{X-\{P\}}$	1	2	3	3	...	3	2	1	0

Since we have removed P, the remaining points could be a complete intersection, so F^2 is a common factor of $(I_Z)_{t+3}$ if and only if the points of $X - \{P\}$ are not a complete intersection of a curve of degree t with F. This completes the proof of Claim 2.

The rest of the proof is very similar to that of Theorem 3.4 and we omit the details. □

4 Approach II: Points on Cubics

Let $Z = m_1p_1 + \cdots + m_np_n \subset \mathbb{P}^2$, where the points $p_1, \ldots, p_n$ are distinct and arbitrary. When $n < 9$, a complete determination of h_Z is given in [7], but the case of $n \geq 9$ remains of interest. Giving a complete determination of h_Z for all $n \geq 9$ arbitrary distinct points $p_1, \ldots, p_n$ would involve solving some extremely hard open problems. For example, it is even an open problem to determine h_Z for $n > 9$ when the points $p_1, \ldots, p_n$ are general and $m_1 = \cdots = m_n$. So here, as in Section 3, we consider the case of $n \geq 9$ points p_i in special cases. These cases include those considered in Section 3. We recover and in some cases extend the results of Section 3, but the methods we use here are different. To start, let $p_1, \ldots, p_n$ be $n \geq 9$ distinct points on a reduced plane cubic C. If C is not irreducible, we assume further that all the points are smooth points of C. If D is a component of C, let n_D be the number of these points on D. We will say that the points are *evenly distributed* if $n_D = n(\deg(D))/3$ for every reduced irreducible component D of C. Note that for n points to be evenly distributed, it is necessary either that 3 divide n or that C be irreducible.

We will use some facts about surfaces obtained by blowing up points in the plane, in particular we'll make use of the intersection form on such surfaces, which we now briefly recall. Given distinct points $p_1, \ldots, p_n \in \mathbb{P}^2$, let $\pi : Y \to \mathbf{P}^2$ be the morphism obtained by blowing up the points p_i. The divisor class group $\mathrm{Cl}(Y)$ of divisors modulo linear equivalence is a free Abelian group with basis $[L], [E_1], \ldots, [E_n]$, where L is the pullback to Y of a general line, and $E_i = \pi^{-1}(p_i)$. There is a bilinear form, called the *intersection form*, defined on the group of divisors, which descends to $\mathrm{Cl}(Y)$. It is uniquely determined by the fact that $L, E_1, \cdots, E_n$ are orthogonal with respect to the intersection form, with $L \cdot L = L^2 = 1$ and $E_i^2 = -1$ for $i = 1, \ldots, n$. For two distinct, reduced, irreducible curves C_1 and C_2 on Y, $C_1 \cdot C_2$ is just the number of points of intersection of the two curves, counted with multiplicity. We recall that a divisor F is *nef* if $F \cdot C \geq 0$ for every effective divisor C. A useful criterion for nefness is that if F is an effective divisor such that $F \cdot C \geq 0$ for every component C of F, then F is nef.

In preparation for stating Theorem 4.2, our main result in this section, we set some additional notation. Let $Z = m(p_1 + \cdots + p_n)$. When t satisfies $3t = mn$, the value of $h_Z(t)$ is influenced by torsion in the group $\mathrm{Pic}(C)$. Our formula for h_Z as given in Theorem 4.2 accounts for this influence via an integer-valued function we will denote by s. In fact, s depends on the points p_i, on m and on t, but for a fixed set of points p_i it is convenient to mostly suppress the dependence on the points and denote s as $s(t,n,m)$, where the parameter n is a reminder of the dependence on the n points. To define $s(t,n,m)$, let L be a general line in the plane and fix evenly distributed smooth points $p_1, \ldots, p_n$ of a reduced cubic C. Since Theorem 4.2 applies only for $n \geq 9$ and we need $s(t,n,m)$ only when $t \geq nm/3$, we define $s(t,n,m)$ only for $n \geq 9$ when $t \geq nm/3$:

(i) If $t > nm/3$, we set $s(t,n,m) = 0$.

(ii) If $n = 9$ and $t = 3m$, let λ be the order (possibly infinite) of $\mathcal{O}_C(3L) \otimes \mathcal{O}_C(-p_1 - \cdots - p_9)$ in $\mathrm{Pic}(C)$. We then set $s(t,n,m) = \lfloor m/\lambda \rfloor$.

(iii) If $n > 9$ and $t = nm/3$, we set $s(t,n,m) = 1$ if $\mathcal{O}_C(tL) \otimes \mathcal{O}_C(-mp_1 - \cdots - mp_n) = \mathcal{O}_C$ in $\mathrm{Pic}(C)$, and we set $s(t,n,m) = 0$ otherwise.

The value of $s(t,n,m)$ depends on whether $\mathcal{O}_C(tL) \otimes \mathcal{O}_C(-mp_1 - \cdots - mp_n)$ is trivial. Note that triviality of this line bundle is equivalent to the divisor $mp_1 + \cdots + mp_n$ on C being the intersection of C with a curve H, necessarily of degree $t = mn/3$. Of course it can happen that $\mathcal{O}_C(tL) \otimes \mathcal{O}_C(-p_1 - \cdots - p_n)$ is non-trivial even though $\mathcal{O}_C(tmL) \otimes \mathcal{O}_C(-mp_1 - \cdots - mp_n)$ is trivial. For example, if p_1, p_2 and p_3 are flexes on C but not collinear, then $\mathcal{O}_C(L) \otimes \mathcal{O}_C(-p_1 - p_2 - p_3)$ is not trivial, but $\mathcal{O}_C(3L) \otimes \mathcal{O}_C(-3p_1 - 3p_2 - 3p_3)$ is trivial, and H in this case is the union of the lines tangent to C at the points p_1, p_2 and p_3. When C is a smooth cubic curve, triviality of $\mathcal{O}_C(tL) \otimes \mathcal{O}_C(-mp_1 - \cdots - mp_n)$ is equivalent to the sum $mp_1 + \cdots + mp_n$ being trivial in the group law on the cubic (with respect to a flex being taken as the identity element). (The divisor X_1 given in the proof of part (ii) of Theorem 3.4 gives another example, and shows that this issue arose also with the first approach.)

Remark 4.1. When $n = 9$, the values of λ that can occur depend on the torsion in $\mathrm{Pic}(C)$, and this depends on C and on the characteristic of the ground field; see Remark 4.4. Thus knowing something about C tells us something about what Hilbert functions can occur for points on C, but the Hilbert functions themselves depend only on λ, and already for a smooth irreducible non-supersingular cubic C, there is torsion of all orders.

Theorem 4.2. *Let $X = p_1 + \cdots + p_n$ be a set of $n \geq 9$ evenly distributed smooth points on a reduced plane cubic C. Let $Z = mX$. The value $h_Z(t) = \dim(k[\mathbf{P}^2]/(I(Z)))_t$ of the Hilbert function in degree t is:*

(i) $\binom{t+2}{2}$ *if $t < 3m$;*

(ii) $n\binom{m+1}{2} - s(t, n, m)$ *if $t \geq nm/3$; and*

(iii) $\binom{t+2}{2} - \binom{t-3r+2}{2} + n\binom{m-r+1}{2} - s(t-3r, n, m-r)$ *if $n > 9$ and $3m \leq t < mn/3$, where $r = \lceil (mn - 3t)/(n - 9) \rceil$.*

Proof. This result is a corollary of the main result of [12]. Let $F = tL - mE_1 - \cdots - mE_n$ with respect to the morphism $\pi : Y \to \mathbf{P}^2$ obtained by blowing up the points p_i.

Let $C' \subset Y$ be the proper transform of C with respect to π. Since the points p_i blown up are smooth points on C, we see $[C'] = [3L - E_1 - \cdots - E_n]$ (and hence C' is an anticanonical divisor). Moreover, each component of C' is the proper transform D' of a component D of C, and each of the components of C (and hence of C') is reduced. (To see this note that $n_D > 0$ for each component D of C since the points p_i are evenly distributed, but the number of points p_i which lie on D is n_D and all of the points p_i are smooth points of C, so each component of C has a smooth point and hence must be reduced.)

In addition, the following statements are equivalent:

(a) $F \cdot D' \geq 0$ for every irreducible component D' of C';

(b) $F \cdot C' \geq 0$; and

(c) $F \cdot D' \geq 0$ for some irreducible component D' of C'.

Clearly, (a) implies (b), and (b) implies (c). We now show that (c) implies (a). If C' has only one component, then (c) and (a) are trivially equivalent, so suppose D'_1 and D'_2 are distinct components of C'. In order to show that $F \cdot D'_1 \geq 0$ implies

$F \cdot D_2' \geq 0$, we will use the assumption that the points p_i are evenly distributed smooth points of C. Let $D_j = \pi(D_j')$, so D_j' is the proper transform of D_j. Because the points are evenly distributed, we have $n_{D_j} = n(\deg(D_j))/3$. Thus n_{D_j} of the n points p_i lie on D_j. Because the points are smooth points of C, we have $[D_j'] = [\deg(D_j)L - \sum_{p_i \in D_j} E_i]$, where the sum involves n_{D_j} terms. Thus $F \cdot D_j' \geq 0$ can be rewritten as $t \deg(D_j) - mn_{D_j} \geq 0$. Substituting $n(\deg(D_j))/3$ for n_{D_j} gives $t \deg(D_j) - mn(\deg(D_j))/3 \geq 0$ which is equivalent to $3t - mn \geq 0$, which is itself just $F \cdot C' \geq 0$. Thus $F \cdot D_1' \geq 0$ and $F \cdot D_2' \geq 0$ are both equivalent to $F \cdot C' \geq 0$, and hence $F \cdot D_1' \geq 0$ if and only if $F \cdot D_2' \geq 0$. This shows (c) implies (a).

We now show that $h^0(Y, \mathcal{O}_Y(F)) = 0$ if and only if $t < 3m$. For $t \geq 3m$, we have $\mathcal{O}_Y(F) = \mathcal{O}_Y((t-3m)L + mC')$, and hence $h^0(Y, \mathcal{O}_Y(F)) > 0$. If, however, $t < 3m$, then $3t < 9m \leq nm$ so $F \cdot C' < 0$, and hence, as we saw above, $F \cdot D' < 0$ for each component D' of C', in which case each component D' of C' is a fixed component of $|F|$ so $h^0(Y, \mathcal{O}_Y(F)) = h^0(Y, \mathcal{O}_Y(F - C')) = h^0(Y, \mathcal{O}_Y((t-3)L - (m-1)E_1 - \cdots - (m-1)E_n))$. But $t - 3 < 3(m-1)$, so, by the same argument, we can again subtract off C' without changing h^0. Continuing in this way we eventually obtain $h^0(Y, \mathcal{O}_Y(F)) = h^0(Y, \mathcal{O}_Y((t-3m)L)) = h^0(\mathbb{P}^2, \mathcal{O}_{\mathbb{P}^2}(t-3m))$, but $h^0(\mathbb{P}^2, \mathcal{O}_{\mathbb{P}^2}(t-3m)) = 0$ since $t - 3m < 0$. Thus $h_Z(t) = \binom{t+2}{2}$ for $t < 3m$, which proves (i).

Next consider (ii). If $t \geq nm/3$, i.e., if $F \cdot C' \geq 0$, then as we saw above $F \cdot D' \geq 0$ for every component D' of C'. But as we also saw above, $(t-3m)L + mC' \in |F|$, hence F is nef. If $t > nm/3$ (in which case $s(t,n,m) = 0$), then $F \cdot C' > 0$, so by [12, Theorem III.1 (a,b)], $h^1(Y, \mathcal{O}_Y(F)) = 0$. Thus (1.2) gives $h_Z(t) = n\binom{m+1}{2} = n\binom{m+1}{2} - s(t,n,m)$ as claimed. We are left with the case that $t = nm/3$.

Suppose $t = nm/3$ and $n = 9$. Thus $F = mC'$ and $F \cdot C' = 0$ (because $n = 9$ and $t = 3m$), so $(C')^2 = 0$. By duality we have $h^2(Y, \mathcal{O}_Y(mC')) = h^0(Y, \mathcal{O}_Y(-(m+1)C')) = 0$, so by Riemann–Roch we have $h^0(Y, \mathcal{O}_Y(F)) - h^1(Y, \mathcal{O}_Y(F)) = 1 + (F^2 + C' \cdot F)/2 = 1$. Since F is nef, so is iC' for all $i \geq 0$. Since $F \cdot C' = 0$, either $|F|$ has an element disjoint from C' or F and C' share a common component.

If $|F|$ has an element disjoint from C', then $\mathcal{O}_{C'}(F)$ is trivial, so $h^0(C', \mathcal{O}_{C'}(F)) = 1$ since C (and hence C') is connected and reduced. Suppose F and C' share a common component. Then C' is in the base locus of $|F|$ by [12, Corollary III.2], and hence $h^0(Y, \mathcal{O}_Y(F)) = h^0(Y, \mathcal{O}_Y(F - C'))$. Let ϕ be the least $i > 0$ (possibly infinite) such that C' is not in the base locus of $|iC'|$. Then we have that $h^0(Y, \mathcal{O}_Y(jC')) = h^0(Y, \mathcal{O}_Y((j+1)C'))$ for $0 \leq j < \phi - 1$, so by induction (using the base case $h^0(Y, \mathcal{O}_Y) = 1$ and the fact $h^0(Y, \mathcal{O}_Y(jC')) - h^1(Y, \mathcal{O}_Y(jC')) = 1$) we have $h^0(Y, \mathcal{O}_Y(jC')) = 1$ and $h^1(Y, \mathcal{O}_Y(jC')) = 0$ for all $0 \leq j < \phi$. It follows that

$$0 \to \mathcal{O}_Y((s-1)C') \to \mathcal{O}_Y(sC') \to \mathcal{O}_{C'}(sC') \to 0 \qquad (\star)$$

is exact on global sections for $1 \leq s \leq \phi$, and that $h^0(C', \mathcal{O}_{C'}(sC')) = 0$ and $h^1(C', \mathcal{O}_{C'}(sC')) = 0$ for $0 < s < \phi$. Thus $\mathcal{O}_{C'}(sC')$ is nontrivial for $0 < s < \phi$. Since for all m, $|mC'|$ either has an element disjoint from C' or C' is in the base locus of $|mC'|$, we see that $\mathcal{O}_{C'}(\phi C')$ is trivial, and hence ϕ is the order of $\mathcal{O}_{C'}(C')$ in

$\mathrm{Pic}(C')$. But since the points p_i blown up are smooth points of C, the morphism $\pi: Y \to \mathbb{P}^2$ induces an isomorphism $C \to C'$, and under this isomorphism, $\mathcal{O}_C(3L) \otimes \mathcal{O}_C(-p_1 - \cdots - p_n)$ corresponds to $\mathcal{O}_{C'}(C')$, so we see $\phi = \lambda$. It follows that $h^0(C', \mathcal{O}_{C'}(sC')) = h^1(C', \mathcal{O}_{C'}(sC'))$ for all $s \geq 0$, and these are both 1 if $\mathcal{O}_{C'}(sC')$ is trivial (i.e., if s is a multiple of λ) and they are 0 otherwise.

We now claim that $(\star)$ is exact on global sections for all $s \geq 1$. It is enough to show this when s is a multiple of λ, because otherwise, as we noted above, $h^0(C', \mathcal{O}_{C'}(sC'))$ equals 0 and hence $(\star)$ is automatically exact on global sections. But $|\lambda C'|$ (and hence also $|i\lambda C'|$ for all $i \geq 1$) has an element disjoint from C', so $H^0(Y, \mathcal{O}_Y(i\lambda C')) \to H^0(C', \mathcal{O}_{C'}(i\lambda C'))$ is onto, which shows that $(\star)$ is exact on global sections when s is a multiple of λ.

It follows that $(\star)$ is also exact on h^1's (since as above $h^2(Y, \mathcal{O}_Y(iC')) = 0$ for all $i \geq 0$), and hence that $h^1(Y, \mathcal{O}_Y(mC')) = h^1(Y, \mathcal{O}_Y) + \sum_{1 \leq i \leq m} h^1(C', \mathcal{O}_{C'}(iC'))$. Now $h^1(Y, \mathcal{O}_Y) = h^1(\mathbb{P}^2, \mathcal{O}_{\mathbb{P}^2}) = 0$ and $h^1(C', \mathcal{O}_{C'}(iC'))$ is 1 if and only if i is a multiple of λ and it is 0 otherwise. Thus $h^1(Y, \mathcal{O}_Y(mC'))$ is the number of summands $h^1(C', \mathcal{O}_{C'}(iC'))$ for which i is a multiple of λ; i.e., $h^1(Y, \mathcal{O}_Y(mC')) = \lfloor m/\lambda \rfloor$, which is just $s(t,n,m)$. This implies that $h_Z(t) = n\binom{m+1}{2} - s(t,n,m)$, as claimed.

If $t = nm/3$ but $n > 9$, then $F^2 > 0$ so by [12, Theorem III.1 (c)] either $\mathcal{O}_{C'}(F)$ is trivial (in which case $s(t,n,m) = 1$) and $h^1(Y, \mathcal{O}_Y(F)) = 1$ (and hence $h_Z(t) = n\binom{m+1}{2} - 1 = n\binom{m+1}{2} - s(t,n,m)$), or C' is in the base locus of $|F|$. If C' is in the base locus, then by [12, Theorem III.1 (d)] and the fact that $F^2 > 0$ we have $\mathcal{O}_{C'}(F)$ is not trivial (in which case $s(t,n,m) = 0$) and $h^1(Y, \mathcal{O}_Y(F)) = 0$, and hence $h_Z(t) = n\binom{m+1}{2} - s(t,n,m)$, as claimed.

Now consider case (iii); i.e., $3m \leq t < nm/3$ and $n > 9$. Then $F \cdot D' < 0$ for each component D' of C' (since the points are evenly distributed), in which case $h^0(Y, \mathcal{O}_Y(F)) = h^0(Y, \mathcal{O}_Y(F - C')) = h^0(Y, \mathcal{O}_Y((t-3)L - (m-1)E_1 - \cdots - (m-1)E_n))$. If $t - 3 < n(m-1)/3$, we can subtract C' off again. This continues until we have subtracted C' off $r = \lceil (mn - 3t)/(n-9) \rceil$ times, at which point we have that $F - rC'$ is nef (and hence $t - 3r \geq n(m-r)/3$) and effective and $h^0(Y, \mathcal{O}_Y(F)) = h^0(Y, \mathcal{O}_Y(F - rC'))$. Applying (ii) to $F - rC'$ gives $\binom{t-3r+2}{2} - h^0(Y, \mathcal{O}_Y(F - rC')) = h_{(m-r)X}(t - 3r) = n\binom{m-r+1}{2} - s(t - 3r, n, m - r)$ or $h^0(Y, \mathcal{O}_Y(F - rC')) = \binom{t-3r+2}{2} - \left(n\binom{m-r+1}{2} - s(t-3r, n, m-r)\right)$. Substituting this in for $h^0(Y, \mathcal{O}_Y(F))$ in $h_Z(t) = \binom{t+2}{2} - h^0(Y, \mathcal{O}_Y(F))$ gives (iii). □

Remark 4.3. We can now write down all possible Hilbert functions for $n \geq 9$ points of multiplicity m for each possible choice of Hilbert function for the reduced scheme given by the points, if the points are smooth points of a reduced cubic curve and evenly distributed. Suppose $X = p_1 + \cdots + p_n$ and $m = 1$. If 3 does not divide n, or it does but $s(n/3, n, 1) = 0$, then the difference function for the Hilbert function of X is the same as given in (3.2), but if 3 divides n and $s(n/3, n, 1) = 1$, then X is a complete intersection and the difference function for the Hilbert function of X is the same as given in (3.1).

We now compare our results for $Z = 2X = 2(p_1 + \cdots + p_n)$ with those obtained in Proposition 3.1 and Theorem 3.4, and we explicitly list those cases skipped there (because there we assumed $n = 3t$ with $t \geq 3$ in Proposition 3.1 and $n = 3t + \delta$ with $t > 5 - \delta$ in Theorem 3.4).

Say $n \equiv 1 \mod 3$. Then the difference function for the Hilbert function is:

$n = 10$: 1 2 3 4 5 6 6 3 0

$n = 13$: 1 2 3 4 5 6 6 6 4 2 0, and for

$n = 10 + 3x$ for $x > 1$: the result is the same as given in Theorem 3.4(i).

Next, say $n \equiv 2 \mod 3$. Then the difference function for the Hilbert function is:

$n = 11$: 1 2 3 4 5 6 6 5 1 0, and for

$n = 11 + 3x$ for $x > 0$: the result is the same as given in Theorem 3.4(i).

If $n = 3x$, there are two possibilities. If $s(2x, n, 2) = 0$ for the given points (i.e., the divisor $2p_1 + \cdots + 2p_n$ on C is not cut out by a curve of degree $2x$, or equivalently $\mathcal{O}_C(2xL - 2E_1 - \cdots - 2E_n)$ is not trivial), then the difference function for the Hilbert function is:

$n = 9$: 1 2 3 4 5 6 6 0, and

$n = 3x$: 1 2 3 4 5 6 ... 6 3 ... 3 0 for $x \geq 4$, where the number of 6's is $x - 1$ and the number of trailing 3's is $x - 3$. For $x > 5$, this is the same as the result given in Theorem 3.4 (ii).

If $s(2x, n, 2) = 1$ for the given points (i.e., the divisor $2p_1 + \cdots + 2p_n$ on C is cut out by a curve of degree $2x$, or equivalently $\mathcal{O}_C(2xL - 2E_1 - \cdots - 2E_n)$ is trivial), but $s(n/3, n, 1) = 0$ (so $p_1 + \cdots + p_n$ is not cut out by a curve of degree x, which is equivalent to saying that $\mathcal{O}_C(xL - E_1 - \cdots - E_n)$ is not trivial), then the difference function for the Hilbert function is:

$n = 9$: 1 2 3 4 5 6 5 1 0,

$n = 12$: 1 2 3 4 5 6 6 6 2 1 0, and

$n = 3x$: 1 2 3 4 5 6 ... 6 3 ... 3 2 1 0 for $x > 4$, where the number of 6's is $x - 1$ and the number of trailing 3's is $x - 4$. For $x > 5$, this is the same as the result given in Theorem 3.4 (ii).

Now say $n = 3x$ and $s(x, n, 1) = 1$. In this case, X is the complete intersection of C and a form of degree t, and the difference function for the Hilbert function of $2X$ is:

$n = 9$: 1 2 3 4 5 6 4 2 0

$n = 12$: 1 2 3 4 5 6 6 5 3 1 0 and

$n = 3x$ for $x > 4$: the result is the same as given in Proposition 3.1.

Remark 4.4. The possible values of the Hilbert functions as given in Theorem 4.2 depend partly on what torsion occurs in $\mathrm{Pic}(C)$, and this in turn is affected by the characteristic of k. When C is smooth, see [14, Example IV.4.8.1] for a discussion of the torsion. When C is reduced but not smooth, the torsion is easy to understand since it is all contained in the identity component $\mathrm{Pic}^0(C)$ of $\mathrm{Pic}(C)$, whose group structure is isomorphic either to the additive or multiplicative groups of the ground field. (See for example [13, Proposition 5.2], which states a result for curves of so-called canonical type. But for any reduced cubic C, one can always find a set of 9 evenly distributed smooth points of C, and the proper transform C' with respect to blowing those points up is a *curve of canonical type*, meaning that $C' \cdot D = K_X \cdot D = 0$ for every component D of C'. Since the points blown up are smooth on C, C and C' are isomorphic and thus so are $\mathrm{Pic}(C)$ and $\mathrm{Pic}(C')$, hence the conclusion of [13, Proposition 5.2] applies to C, even though C is not itself of canonical type.) When C is reduced and irreducible but singular, for example, the result is that $\mathrm{Pic}^0(C)$ is the additive group of the ground field when C is cuspidal and it is the multiplicative group of the field when C is nodal [14, Exercise II.6.9]. In particular, if C is an irreducible cuspidal cubic curve over a field of characteristic zero, then $\mathrm{Pic}^0(C)$ is torsion free, so h_{2X} cannot be $(1, 2, 3, 4, 5, 6, 6, 6, 2, 1)$; indeed, this follows, after a simple calculation, because if $\mathcal{O}_C(2xL - 2p_1 - \cdots - 2p_n)$ is trivial, then so is $\mathcal{O}_C(xL - p_1 - \cdots - p_n)$. On the other hand, $\mathcal{O}_C(xL - p_1 - \cdots - p_n)$ can be nontrivial even if $\mathcal{O}_C(2xL - 2p_1 - \cdots - 2p_n)$ is trivial if the characteristic is 2 or if the singular point is a node but the characteristic is not 2, since in those cases $\mathrm{Pic}(C)$ has elements of order 2.

Remark 4.5. We can also use the method of proof of Theorem 4.2 to recover the result of Theorem 3.7 for the Hilbert function of $mX = m(p_1 + \cdots + p_n)$ for $n \geq 9$ points on a reduced, irreducible cubic curve C where p_1, say, is the singular point and m is 1 or 2. As is now clear, the approach of Theorem 4.2 is to determine $h^0(Y, \mathcal{O}_Y(tL - mE_1 - \cdots - mE_n))$ for all t, and then translate this into the Hilbert function or the difference function for mX.

This translation is purely mechanical and the resulting Hilbert functions in the case that $n \geq 12$ are already given in Theorem 3.7 (we leave writing down the Hilbert functions for $9 \leq n \leq 11$ using the results that follow as an exercise for the reader). Thus it is the calculation of $h^0(Y, \mathcal{O}_Y(tL - mE_1 - \cdots - mE_n))$ that is of most interest, and it is on this that we now focus.

Let Y be the blow up of the points, let C' be the proper transform of C, and let $F_t = tL - E_1 - \cdots - E_n$ and $G_t = tL - 2(E_1 + \cdots + E_n)$, where we denote by L both a general line in the plane and its pullback to Y. Up to linear equivalence, note that $C' = 3L - 2E_1 - E_2 - \cdots - E_n$.

The goal here is to compute the values of $h^0(Y, \mathcal{O}_Y(F_t))$ and $h^0(Y, \mathcal{O}_Y(G_t))$. For $t < 3$, Bézout tells us that $h^0(X, \mathcal{O}_Y(F_t)) = 0$, since $F_t \cdot C' < 0$ (hence $h^0(Y, \mathcal{O}_Y(F_t))$ equals $h^0(Y, \mathcal{O}_Y(F_t - C'))$ and $(F_t - C') \cdot L < 0$ (hence $h^0(Y, \mathcal{O}_Y(F_t - C')) = 0$). If $t \geq 3$, then certainly $h^0(Y, \mathcal{O}_Y(F_t)) > 0$, since $F_t = (t-3)L + C' + E_1$. We consider three cases, according to whether $F_t \cdot C' < 0$, $F_t \cdot C' > 0$ or $F_t \cdot C' = 0$.

If $0 > F_t \cdot C' = 3t - 2 - (n-1)$ (i.e., if $3 \leq t < (n+1)/3$), then $h^0(Y, \mathcal{O}_Y(F_t)) = h^0(Y, \mathcal{O}_Y(F_t - C')) = h^0(Y, \mathcal{O}_Y((t-3)L + E_1)) = h^0(Y, \mathcal{O}_Y((t-3)L)) = h^0(\mathbb{P}^2, \mathcal{O}_{\mathbb{P}^2}((t-3)L)) = \binom{t-3+2}{2}$, since $F_t - C' = (t-3)L + E_1$. If $F_t \cdot C' > 0$ (i.e., $t > (n+1)/3$), then $h^0(Y, \mathcal{O}_Y(F_t)) = \binom{t+2}{2} - n$ (since F_t, meeting both components of $-K_Y = C' + E_1$ positively, is nef and hence $h^1(Y, \mathcal{O}_Y(F_t)) = 0$ by [12, Theorem III.1 (a,b)]).

This leaves the case that $t = (n+1)/3$. This means that $\mathcal{O}_{C'}(F_t)$ has degree 0. Consider the exact sequence

$$0 \to \mathcal{O}_Y((t-3)L + E_1) \to \mathcal{O}_Y(F_t) \to \mathcal{O}_{C'}(F_t) \to 0.$$

By an analogous argument to the one used to show $h^1(Y, \mathcal{O}_Y(F_t)) = 0$ when $t > (n+1)/3$, we obtain that $h^1(\mathcal{O}_Y((t-3)L + E_1)) = 0$. But C' is a smooth rational curve, so also the third sheaf in the sequence has vanishing first cohomology. Thus we obtain $h^1(Y, O_Y(F_t)) = 0$, hence the points impose independent conditions. It follows that $h^0(Y, O_Y(F_t)) = \binom{t+2}{2} - n$ also for $t = (n+1)/3$.

We thus have: $h^0(Y, \mathcal{O}_Y(F_t)) = 0$ for $0 \le t < 3$; $h^0(Y, \mathcal{O}_Y(F_t)) = \binom{t-1}{2}$ for $3 \le t < (n+1)/3$; and $h^0(Y, \mathcal{O}_Y(F_t)) = \binom{t+2}{2} - n$ for $t \ge (n+1)/3$.

A similar analysis works for $2X$. There are now four ranges of degrees. The first range is $t < 6$, in which case $h^0(Y, \mathcal{O}_Y(G_t)) = 0$ by Bézout, arguing as above. For $t \ge 6$, we have $h^0(Y, \mathcal{O}_Y(G_t)) > 0$, since up to linear equivalence we have $G_t = (t-6)L + 2(C' + E_1)$. The second range is now $6 \le t < (n+8)/3$; in this case $2C'$ is, by Bézout, a fixed component of $|G_t|$, so $h^0(Y, \mathcal{O}_Y(G_t)) = h^0(Y, \mathcal{O}_Y((t-6)L + 2E_1)) = \binom{t-4}{2}$. The third range is $(n+8)/3 \le t < (2/3)(n+1)$, for which C' is a fixed component of $|G_t|$ (and $G_t - C' = (t-6)L + C' + 2E_1$ is nef) so $h^0(Y, \mathcal{O}_Y(G_t)) = h^0(Y, \mathcal{O}_Y(G_t - C'))$ and we know $h^0(Y, \mathcal{O}_Y(G_t - C'))$ by Theorem 4.2 (ii) if $n > 9$, while $h^1(Y, \mathcal{O}_Y(G_t - C')) = 0$ by [12, Theorem III.1 (a,b)]) if $n = 9$, so again we know $h^0(Y, \mathcal{O}_Y(G_t - C'))$. The last range is $t \ge (2/3)(n+1)$, in which case G_t is nef. If $t > (2/3)(n+1)$, then G_t meets $-K_Y$ positively, so $h^1(Y, \mathcal{O}_Y(G_t)) = 0$ [12, Theorem III.1 (a,b)]), and $h^0(Y, \mathcal{O}_Y(G_t)) = \binom{t+2}{2} - 3n$. We are left with the case that $t = (2/3)(n+1)$. Consider the exact sequence

$$0 \to \mathcal{O}_Y((t-3)L - E_2 - \cdots - E_n) \to \mathcal{O}_Y(G_t) \to \mathcal{O}_{C'}(G_t) \to 0.$$

Since $G_t \cdot C' \ge 0$ and C' is smooth and rational, we have $h^1(C', \mathcal{O}_{C'}(G_t)) = 0$, and since $\mathcal{O}_Y((t-3)L - E_2 - \cdots - E_n) = \mathcal{O}_Y((t-6)L + C' + 2E_1)$ and $(t-6)L + C' + 2E_1$ is nef (as observed above) with $(G_t - C') \cdot C' > 0$, we have $h^1(Y, \mathcal{O}_Y(G_t - C')) = 0$ [12, Theorem III.1 (a,b)]) and hence $h^1(Y, \mathcal{O}_Y(G_t)) = 0$, so in fact $h^0(Y, \mathcal{O}_Y(G_t)) = \binom{t+2}{2} - 3n$.

Remark 4.6. Here we comment on what is left to do if one wants to recover the results of Section 2 using the methods of Section 4. So consider $n = 9$ points on a given cubic C (but note that there may be more than one cubic through the points), either all of multiplicity 1 or all of multiplicity 2. The case that the points are evenly distributed smooth points of C is done above, as is the case that the curve C is reduced and irreducible. The case that the points all lie on a conic follows from the known result for configuration types of points on a conic [7]. What is left is that the points do not all lie on any given conic (and hence C is reduced) and either: one or more of the points is not a smooth point of C and C is not irreducible, or the points are not distributed evenly (and hence again C is not irreducible). The four reducible cubics that arise are: a conic and a line tangent to the conic; a conic and a transverse line; three lines passing through a point; and three lines with no point common to all three. Each of these cases leads to a number of cases depending on how the points are placed (such as how many are on each component and whether one or more is a singular point of the cubic, but also depending on the group law of the cubic). Analyzing these cases would give a complete result of the Hilbert functions of the form h_X and h_{2X} for a reduced scheme X consisting of 9 distinct points of the plane.

Acknowledgments. We thank Z. Teitler for helpful comments and the referee for a careful reading of the paper.

Bibliography

[1] Bigatti, A., Geramita, A. V., Migliore, J., Geometric consequences of extremal behavior in a theorem of Macaulay. Trans Amer Math Soc. 1994;346:203–235.

[2] Buckles, M., Guardo, E., Van Tuyl, A., Fat points on a generic almost complete intersection. Matematiche (Catania). 2000;55:191–202.

[3] Buckles, M., Guardo, E., Van Tuyl, A., Fat points on a grid in $\mathbb{P}^2$. Matematiche (Catania). 2000;55:169–189.

[4] Davis, E., 0-dimensional subschemes of $\mathbb{P}^2$: new application of Castelnuovo's function. Ann Univ Ferrara Sez VII Sci Mat. 1986;XXXII:93–107.

[5] Davis, E., Geramita, A. V., Orecchia, F., Gorenstein algebras and the Cayley–Bacharach theorem. Proc Amer Math Soc. 1985;93:593–597.

[6] Ellingsrud, G., Sur le schéma de Hilbert des variétés de codimension 2 dans $\mathbb{P}^e$ à cône de Cohen–Macaulay. Ann Sci Éc Norm Supér (4) 1975;8:423–431.

[7] Geramita, A. V., Harbourne, B., Migliore, J., Classifying Hilbert functions of fat point subschemes in $\mathbb{P}^2$. Collect Math. 2009;60(2):159–192.

[8] Geramita, A. V., Maroscia, P., Roberts, L., The Hilbert function of a reduced k-algebra. J Lond Math Soc (2). 1983;28:443–452.

[9] Geramita, A. V., Migliore, J., Sabourin, L., On the first infinitesimal neighborhood of a linear configuration of points in $\mathbb{P}^2$. J Algebra. 2006;298(2):563–611.

[10] Guardo, E., Van Tuyl, A., Some results on fat points whose support is a complete intersection minus a point. In: Projective varieties with unexpected properties. De Gruyter; 2005. p. 257–266.

[11] Guardo, E., Van Tuyl, A., Powers of complete intersections: graded Betti numbers and applications. Illinois J Math. 2005;49:265–279.

[12] Harbourne, B., Anticanonical rational surfaces. Trans Amer Math Soc. 1997;349:1191–1208.

[13] Harbourne, B., Lang, W. E., Multiple fibres on rational elliptic surfaces. Trans Amer Math Soc. 1988;307:205–223.

[14] Hartshorne, R., Algebraic Geometry. vol. 52 of Graduate texts in mathematics. New York: Springer-Verlag; 1977.

[15] Migliore, J., Introduction to liaison theory and deficiency modules. vol. 165 of Progress in mathematics. Birkhäuser; 1998.

Author Information

Anthony V. Geramita, Dipartimento di Matematica, Università di Genova, Genova, Italia and Department of Mathematics, Queen's University, Kingston, ON, Canada.
E-mail: anthony.geramita@gmail.com

Brian Harbourne, Department of Mathematics, University of Nebraska, Lincoln, NE, USA.
E-mail: bharbour@math.unl.edu

Juan C. Migliore, Department of Mathematics, University of Notre Dame, South Bend, IN, USA.
E-mail: migliore.1@nd.edu

Progress in Commutative Algebra 1, 85–126

Edge Ideals: Algebraic and Combinatorial Properties

Susan Morey and Rafael H. Villarreal

Abstract. Let $\mathcal{C}$ be a clutter and let $I(\mathcal{C}) \subset R$ be its edge ideal. This is a survey paper on the algebraic and combinatorial properties of $R/I(\mathcal{C})$ and $\mathcal{C}$, respectively. We give a criterion to estimate the regularity of $R/I(\mathcal{C})$ and apply this criterion to give new proofs of some formulas for the regularity. If $R/I(\mathcal{C})$ is sequentially Cohen–Macaulay, we present a formula for the regularity of the ideal of vertex covers of $\mathcal{C}$ and give a formula for the projective dimension of $R/I(\mathcal{C})$. We also examine the associated primes of powers of edge ideals, and show that for a graph with a leaf, these sets form an ascending chain.

Keywords. Edge Ideal, Regularity, Associated Prime, Sequentially Cohen–Macaulay.

2010 Mathematics Subject Classification. Primary: 13-02, 13F55; Secondary: 05C10, 05C25, 05C65, 05E40.

1 Introduction

A *clutter* $\mathcal{C}$ is a finite ground set X together with a family E of subsets of X such that if $f_1, f_2 \in E$, then $f_1 \not\subset f_2$. The ground set X is called the *vertex set* of $\mathcal{C}$ and E is called the *edge set* of $\mathcal{C}$, denoted by $V(\mathcal{C})$ and $E(\mathcal{C})$ respectively. Clutters are simple hypergraphs and are sometimes called *Sperner families* in the literature. We can also think of a clutter as the maximal faces of a simplicial complex over a ground set. One example of a clutter is a graph with the vertices and edges defined in the usual way.

Let $\mathcal{C}$ be a clutter with vertex set $X = \{x_1, \ldots, x_n\}$ and with edge set $E(\mathcal{C})$. Permitting an abuse of notation, we will also denote by x_i the ith variable in the polynomial ring $R = K[x_1, \ldots, x_n]$ over a field K. The *edge ideal* of $\mathcal{C}$, denoted by $I(\mathcal{C})$, is the ideal of R generated by all monomials $x_e = \prod_{x_i \in e} x_i$ such that $e \in E(\mathcal{C})$. Edge ideals of graphs and clutters were introduced in [109] and [39, 47, 53], respectively. The assignment $\mathcal{C} \mapsto I(\mathcal{C})$ establishes a natural one-to-one correspondence between the family of clutters and the family of square-free monomial ideals. Edge ideals of clutters are also called *facet ideals* [39].

This is a survey paper on edge ideals, which includes some new proofs of known results and some new results. The study of algebraic and combinatorial properties of edge ideals and clutters (e.g., Cohen–Macaulayness, unmixedness, normality, normally torsion-freeness, shellability, vertex decomposability, stability of associated primes) is of current interest, see [22, 24, 25, 32, 39, 41, 42, 45, 51, 61, 62, 89, 116] and

The second author was partially supported by CONACyT grant 49251-F and SNI.

the references there. In this paper we will focus on the following algebraic properties: the sequentially Cohen–Macaulay property, the stability of associated primes, and the connection between torsion-freeness and combinatorial problems.

The numerical invariants of edge ideals have attracted a great deal of interest [1,46, 53,77,87,90,91,106,110,117,118]. In this paper we focus on the following invariants: projective dimension, regularity, depth and Krull dimension.

We present a few new results on edge ideals. We give a criterion to estimate the regularity of edge ideals (see Theorem 3.14). We apply this criterion to give new proofs of some formulas for the regularity of edge ideals (see Corollary 3.15). If $\mathcal{C}$ is a clutter and $R/I(\mathcal{C})$ is sequentially Cohen–Macaulay, we present a formula for the regularity of the ideal of vertex covers of $\mathcal{C}$ (see Theorem 3.31) and give a formula for the projective dimension of $R/I(\mathcal{C})$ (see Corollary 3.33). We also give a new class of monomial ideals for which the sets of associate primes of powers are known to form ascending chains (Proposition 4.23).

For undefined terminology on commutative algebra, edge ideals, graph theory, and the theory of clutters and hypergraphs we refer to [10, 28, 102], [38, 110], [7, 54], [17,97], respectively.

2 Algebraic and Combinatorial Properties of Edge Ideals

Let $\mathcal{C}$ be a clutter with vertex set $X = \{x_1, \ldots, x_n\}$ and let $I = I(\mathcal{C}) \subset R$ be its edge ideal. A subset F of X is called *independent* or *stable* if $e \not\subset F$ for any $e \in E(\mathcal{C})$. The dual concept of a stable vertex set is a *vertex cover*, i.e., a subset C of X is a vertex cover of $\mathcal{C}$ if and only if $X \setminus C$ is a stable vertex set. A first hint of the rich interaction between the combinatorics of $\mathcal{C}$ and the algebra of $I(\mathcal{C})$ is that the number of vertices in a minimum vertex cover of $\mathcal{C}$ (the *covering number* $\alpha_0(\mathcal{C})$ of $\mathcal{C}$) coincides with the *height* ht $I(\mathcal{C})$ of the ideal $I(\mathcal{C})$. The number of vertices in a maximum stable set (the *stability number* of $\mathcal{C}$) is denoted by $\beta_0(\mathcal{C})$. Notice that $n = \alpha_0(\mathcal{C}) + \beta_0(\mathcal{C})$.

A less immediate interaction between the two fields comes from passing to a simplicial complex and relating combinatorial properties of the complex to algebraic properties of the ideal. The Stanley–Reisner complex of $I(\mathcal{C})$, denoted by $\Delta_{\mathcal{C}}$, is the simplicial complex whose faces are the independent vertex sets of $\mathcal{C}$. The complex $\Delta_{\mathcal{C}}$ is also called the *independence complex* of $\mathcal{C}$. Recall that $\Delta_{\mathcal{C}}$ is called *pure* if all maximal independent vertex sets of $\mathcal{C}$, with respect to inclusion, have the same number of elements. If $\Delta_{\mathcal{C}}$ is pure (resp. Cohen–Macaulay, shellable, vertex decomposable), we say that $\mathcal{C}$ is *unmixed* (resp. Cohen–Macaulay, shellable, vertex decomposable). Since minor variations of the definition of shellability exist in the literature, we state here the definition used throughout this article.

Definition 2.1. A simplicial complex Δ is *shellable* if the facets (maximal faces) of Δ can be ordered $F_1, \ldots, F_s$ such that for all $1 \leq i < j \leq s$, there exists some $v \in F_j \setminus F_i$ and some $\ell \in \{1, \ldots, j-1\}$ with $F_j \setminus F_\ell = \{v\}$.

We are interested in determining which families of clutters have the property that $\Delta_{\mathcal{C}}$ is pure, Cohen–Macaulay, or shellable. These properties have been extensively studied, see [10, 89, 92–94, 100, 102, 110, 111] and the references there.

The above definition of shellable is due to Björner and Wachs [6] and is usually referred to as *nonpure shellable*, although here we will drop the adjective "nonpure". Originally, the definition of shellable also required that the simplicial complex be pure, that is, all facets have the same dimension. We will say Δ is *pure shellable* if it also satisfies this hypothesis. These properties are related to other important properties [10, 102, 110]:

$$\text{pure shellable} \implies \text{constructible} \implies \text{Cohen–Macaulay} \impliedby \text{Gorenstein.}$$

If a shellable complex is not pure, an implication similar to that above holds when Cohen–Macaulay is replaced by sequentially Cohen–Macaulay.

Definition 2.2. Let $R = K[x_1, \ldots, x_n]$. A graded R-module M is called *sequentially Cohen–Macaulay* (over K) if there exists a finite filtration of graded R-modules

$$(0) = M_0 \subset M_1 \subset \cdots \subset M_r = M$$

such that each M_i/M_{i-1} is Cohen–Macaulay, and the Krull dimensions of the quotients are increasing:

$$\dim(M_1/M_0) < \dim(M_2/M_1) < \cdots < \dim(M_r/M_{r-1}).$$

We call a clutter $\mathcal{C}$ *sequentially Cohen–Macaulay* if $R/I(\mathcal{C})$ is sequentially Cohen–Macaulay. As first shown by Stanley [102], shellable implies sequentially Cohen–Macaulay.

A related notion for a simplicial complex is that of *vertex decomposability* [5]. If Δ is a simplicial complex and v is a vertex of Δ, then the subcomplex formed by deleting v is the simplicial complex consisting of the faces of Δ that do not contain v, and the *link* of v is

$$lk(v) = \{F \in \Delta | v \notin F \text{ and } F \cup \{v\} \in \Delta\}.$$

Suppose Δ is a (not necessarily pure) simplicial complex. We say that Δ is vertex-decomposable if either Δ is a simplex, or Δ contains a vertex v such that both the link of v and the subcomplex formed by deleting v are vertex-decomposable, and such that every facet of the deletion is a facet of Δ. If $\mathcal{C}$ is vertex decomposable, i.e., $\Delta_{\mathcal{C}}$ is vertex decomposable, then $\mathcal{C}$ is shellable and sequentially Cohen–Macaulay [5, 116]. Thus, we have:

$$\text{vertex decomposable} \implies \text{shellable} \implies \text{sequentially Cohen–Macaulay.}$$

Two additional properties related to the properties above are also of interest in this area. One is the unmixed property, which is implied by the Cohen–Macaulay property.

The other is balanced. To define balanced, it is useful to have a matrix that encodes the edges of a graph or clutter.

Definition 2.3. Let $f_1, \dots, f_q$ be the edges of a clutter $\mathcal{C}$. The *incidence matrix* or *clutter matrix* of $\mathcal{C}$ is the $n \times q$ matrix $A = (a_{ij})$ given by $a_{ij} = 1$ if $x_i \in f_j$ and $a_{ij} = 0$ otherwise. We say that $\mathcal{C}$ is a *totally balanced* clutter (resp. *balanced* clutter) if A has no square submatrix of order at least 3 (resp. of odd order) with exactly two 1's in each row and column.

If G is a graph, then G is balanced if and only if G is bipartite and G is totally balanced if and only if G is a forest [96, 97].

While the implications between the properties mentioned above are interesting in their own right, it is useful to identify classes of ideals that satisfy the various properties. We begin with the Cohen–Macaulay and unmixed properties. There are classifications of the following families in terms of combinatorial properties of the graph or clutter:

(c_1) [94, 111] unmixed bipartite graphs,

(c_2) [36, 60] Cohen–Macaulay bipartite graphs,

(c_3) [109] Cohen–Macaulay trees,

(c_4) [38] totally balanced unmixed clutters,

(c_5) [89] unmixed clutters with the König property without cycles of length 3 or 4,

(c_6) [89] unmixed balanced clutters.

We now focus on the sequentially Cohen–Macaulay property.

Proposition 2.4 ([43]). *The only sequentially Cohen–Macaulay cycles are C_3 and C_5.*

The next theorem generalizes a result of [36] (see (c_2) above) which shows that a bipartite graph G is Cohen–Macaulay if and only if Δ_G has a pure shelling.

Theorem 2.5 ([107]). *Let G be a bipartite graph. Then G is shellable if and only if G is sequentially Cohen–Macaulay.*

Recently Van Tuyl [106] has shown that Theorem 2.5 remains valid if we replace shellable by vertex decomposable.

Additional examples of sequentially Cohen–Macaulay ideals depend on the chordal structure of the graph. A graph G is said to be *chordal* if every cycle of G of length at least 4 has a chord. A *chord* of a cycle is an edge joining two non-adjacent vertices of the cycle. Chordal graphs have been extensively studied, and they can be constructed according to a result of G. A. Dirac, see [21, 63, 104]. A chordal graph is called *strongly chordal* if every cycle C of even length at least six has a chord that divides C into two odd length paths. A *clique* of a graph is a set of mutually adjacent vertices. Totally balanced clutters are precisely the clutters of maximal cliques of strongly chordal graphs by a result of Farber [37]. Faridi [39] introduced the notion of a simplicial forest. In [62, Theorem 3.2] it is shown that $\mathcal{C}$ is the clutter of the facets of a simplicial forest if and only if $\mathcal{C}$ is a totally balanced clutter. Additionally, a clutter $\mathcal{C}$ is called *d-uniform* if all its edges have size d.

Theorem 2.6. *Any of the following clutters is sequentially Cohen–Macaulay:*

(i) [116] *graphs with no chordless cycles of length other than* 3 *or* 5,

(ii) [43] *chordal graphs,*

(iii) [63] *clutters whose ideal of covers has linear quotients (see Definitions* 2.7 *and* 3.1*),*

(iv) [55] *clutters of paths of length t of directed rooted trees,*

(v) [39] *simplicial forests, i.e., totally balanced clutters,*

(vi) [52] *uniform admissible clutters whose covering number is* 3.

The clutters of parts (i)–(v) are in fact shellable, and the clutters of parts (i)–(ii) are in fact vertex decomposable, see [22, 63, 106, 107, 115, 116]. The family of graphs in part (ii) is contained in the family of graphs of part (i) because the only induced cycles of a chordal graph are 3-cycles.

A useful tool in examining invariants related to resolutions comes from a carefully chosen ordering of the generators.

Definition 2.7. A monomial ideal I has *linear quotients* if the monomials that generate I can be ordered $g_1, \ldots, g_q$ such that for all $1 \leq i \leq q-1$, $((g_1, \ldots, g_i) : g_{i+1})$ is generated by linear forms.

If an edge ideal I is generated in a single degree and I has linear quotients, then I has a linear resolution (cf. [39, Lemma 5.2]). If I is the edge ideal of a graph, then I has linear quotients if and only if I has a linear resolution if and only if each power of I has a linear resolution [64].

Let G be a graph. Given a subset $A \subset V(G)$, by $G \setminus A$, we mean the graph formed from G by deleting all the vertices in A, and all edges incident to a vertex in A. A graph G is called *vertex-critical* if $\alpha_0(G \setminus \{v\}) < \alpha_0(G)$ for all $v \in V(G)$. An *edge critical* graph is defined similarly. The final property introduced in this section is a combinatorial decomposition of the vertex set of a graph.

Definition 2.8 ([2]). A graph G without isolated vertices is called a *B-graph* if there is a family $\mathcal{G}$ consisting of independent sets of G such that $V(G) = \bigcup_{C \in \mathcal{G}} C$ and $|C| = \beta_0(G)$ for all $C \in \mathcal{G}$.

The notion of a *B-graph* is at the center of several interesting families of graphs. One has the following implications for any graph G without isolated vertices [2, 110]:

$$\begin{array}{ccccccc} & & \text{edge-critical} & \Longrightarrow & & & \\ & & & & B\text{-graph} & \Longrightarrow & \text{vertex-critical} \\ \text{Cohen–Macaulay} & \Longrightarrow & \text{unmixed} & \Longrightarrow & & & \end{array}$$

In [2] the integer $\alpha_0(G)$ is called the *transversal number* of G.

Theorem 2.9 ([34, 46]). *If G is a B-graph, then $\beta_0(G) \leq \alpha_0(G)$.*

3 Invariants of Edge Ideals: Regularity, Projective Dimension, Depth

Let $\mathcal{C}$ be a clutter and let $I = I(\mathcal{C})$ be its edge ideal. In this section we study the regularity, depth, projective dimension, and Krull dimension of $R/I(\mathcal{C})$. There are several well-known results relating these invariants that will prove useful. We collect some of them here for ease of reference.

The first result is a basic relation between the dimension and the depth (see for example [28, Proposition 18.2]):

$$\operatorname{depth} R/I(\mathcal{C}) \leq \dim R/I(\mathcal{C}). \tag{3.1}$$

The deviation from equality in the above relationship can be quantified using the projective dimension, as is seen in a formula discovered by Auslander and Buchsbaum (see [28, Theorem 19.9]):

$$\operatorname{pd}_R(R/I(\mathcal{C})) + \operatorname{depth} R/I(\mathcal{C}) = \operatorname{depth}(R). \tag{3.2}$$

Notice that since in the setting of this article R is a polynomial ring in n variables, $\operatorname{depth}(R) = n$.

Another invariant of interest also follows from a closer inspection of a minimal projective resolution of R/I. Consider the minimal graded free resolution of $M = R/I$ as an R-module:

$$\mathbb{F}_\star : \quad 0 \to \bigoplus_j R(-j)^{b_{gj}} \to \cdots \to \bigoplus_j R(-j)^{b_{1j}} \to R \to R/I \to 0.$$

The *Castelnuovo–Mumford regularity* or simply the *regularity* of M is defined as

$$\operatorname{reg}(M) = \max\{j - i \mid b_{ij} \neq 0\}.$$

An excellent reference for the regularity is the book of Eisenbud [29]. There are methods to compute the regularity of R/I avoiding the construction of a minimal graded free resolution, see [3] and [50, page 614]. These methods work for any homogeneous ideal over an arbitrary field.

We are interested in finding good bounds for the regularity. Of particular interest is to be able to express $\mathrm{reg}(R/I(\mathcal{C}))$ in terms of the combinatorics of $\mathcal{C}$, at least for some special families of clutters. Several authors have studied the regularity of edge ideals of graphs and clutters [18, 20, 53, 74, 75, 77, 81, 90, 103, 106, 114]. The main results are general bounds for the regularity and combinatorial formulas for the regularity of special families of clutters. The estimates for the regularity are in terms of matching numbers and the number of cliques needed to cover the vertex set. Covers will play a particularly important role since they form the basis for a duality.

Definition 3.1. The *ideal of covers* of $I(\mathcal{C})$, denoted by $I_c(\mathcal{C})$, is the ideal of R generated by all the monomials $x_{i_1}\cdots x_{i_k}$ such that $\{x_{i_1},\dots,x_{i_k}\}$ is a vertex cover of $\mathcal{C}$. The ideal $I_c(\mathcal{C})$ is also called the *Alexander dual* of $I(\mathcal{C})$ and is also denoted by $I(\mathcal{C})^\vee$. The clutter of minimal vertex covers of $\mathcal{C}$, denoted by $\mathcal{C}^\vee$, is called the *Alexander dual clutter* or *blocker* of $\mathcal{C}$.

To better understand the Alexander dual, let $e \in E(\mathcal{C})$ and consider the monomial prime ideal $(e) = (\{x_i | x_i \in e\})$. Then the duality is given by:

$$\begin{array}{ccc} I(\mathcal{C}) = & (x_{e_1}, x_{e_2}, \dots, x_{e_q}) & = \mathfrak{p}_1 \cap \mathfrak{p}_2 \cap \cdots \cap \mathfrak{p}_s \\ & \updownarrow & \updownarrow \\ I_c(\mathcal{C}) = (e_1) \cap (e_2) \cap \cdots \cap (e_q) & = & (x_{\mathfrak{p}_1}, x_{\mathfrak{p}_2}, \dots, x_{\mathfrak{p}_s}), \end{array} \tag{3.3}$$

where $\mathfrak{p}_1,\dots,\mathfrak{p}_s$ are the associated primes of $I(\mathcal{C})$ and $x_{\mathfrak{p}_k} = \prod_{x_i \in \mathfrak{p}_k} x_i$ for $1 \le k \le s$. Notice the equality $I_c(\mathcal{C}) = I(\mathcal{C}^\vee)$. Since $(\mathcal{C}^\vee)^\vee = \mathcal{C}$, we have $I_c(\mathcal{C}^\vee) = I(\mathcal{C})$. In many cases $I(\mathcal{C})$ reflects properties of $I_c(\mathcal{C})$ and viceversa [27, 58, 86]. The following result illustrates this interaction.

Theorem 3.2 ([103]). *Let $\mathcal{C}$ be a clutter. If* $\mathrm{ht}(I(\mathcal{C})) \ge 2$, *then*

$$\mathrm{reg}\, I(\mathcal{C}) = 1 + \mathrm{reg}\, R/I(\mathcal{C}) = \mathrm{pd}\, R/I_c(\mathcal{C}),$$

where $I_c(\mathcal{C})$ is the ideal of minimal vertex covers of $\mathcal{C}$.

If $|e| \ge 2$ for all $e \in E(\mathcal{C})$, then this formula says that the regularity of $R/I(\mathcal{C})$ equals 1 if and only if $I_c(\mathcal{C})$ is a Cohen–Macaulay ideal of height 2. This formula will be used to show that regularity behaves well when working with edge ideals with disjoint sets of variables (see Proposition 3.4). This formula also holds for edge ideals of height one [61, Proposition 8.1.10].

Corollary 3.3. *If* $\mathrm{ht}(I(\mathcal{C})) = 1$, *then* $\mathrm{reg}\, R/I(\mathcal{C}) = \mathrm{pd}\, R/I_c(\mathcal{C}) - 1$.

Proof. We set $I = I(\mathcal{C})$. The formula clearly holds if $I = (x_1 \cdots x_r)$ is a principal ideal. Assume that I is not principal. Consider the primary decomposition of I

$$I = (x_1) \cap \cdots \cap (x_r) \cap \mathfrak{p}_1 \cap \cdots \cap \mathfrak{p}_m,$$

where $L = \mathfrak{p}_1 \cap \cdots \cap \mathfrak{p}_m$ is an edge ideal of height at least 2. Notice that $I = fL$, where $f = x_1 \cdots x_r$. Then the Alexander dual of I is

$$I^\vee = (x_1, \ldots, x_r, x_{\mathfrak{p}_1}, x_{\mathfrak{p}_2}, \ldots, x_{\mathfrak{p}_m}) = (x_1, \ldots, x_r) + L^\vee.$$

The multiplication map $L[-r] \xrightarrow{f} fL$ induces an isomorphism of graded R-modules. Thus $\operatorname{reg}(L[-r]) = r + \operatorname{reg}(L) = \operatorname{reg}(I)$. By the Auslander–Buchsbaum formula, one has the equality $\operatorname{pd}(R/I^\vee) = r + \operatorname{pd}(R/L^\vee)$. Therefore, using Theorem 3.2, we get

$$\operatorname{reg}(R/I) = \operatorname{reg}(R/L) + r \overset{3.2}{=} (\operatorname{pd}(R/L^\vee) - 1) + r = \operatorname{pd}(R/I^\vee) - 1.$$

Thus $\operatorname{reg}(R/I) = \operatorname{pd}(R/I^\vee) - 1$, as required. □

Next we show some basic properties of regularity. The first such property is that regularity behaves well when working with the edge ideal of a graph with multiple disjoint components or with isolated vertices, as can be seen by the following proposition.

Proposition 3.4 ([117, Lemma 7]). *Let $R_1 = K[\mathbf{x}]$ and $R_2 = K[\mathbf{y}]$ be two polynomial rings over a field K and let $R = K[\mathbf{x}, \mathbf{y}]$. If I_1 and I_2 are edge ideals of R_1 and R_2 respectively, then*

$$\operatorname{reg} R/(I_1R + I_2R) = \operatorname{reg}(R_1/I_1) + \operatorname{reg}(R_2/I_2).$$

Proof. By abuse of notation, we will write I_i in place of I_iR for $i = 1, 2$ when it is clear from context that we are using the generators of I_i but extending to an ideal of the larger ring. Let $\mathbf{x} = \{x_1, \ldots, x_n\}$ and $\mathbf{y} = \{y_1, \ldots, y_m\}$ be two disjoint sets of variables. Notice that $(I_1 + I_2)^\vee = I_1^\vee I_2^\vee = I_1^\vee \cap I_2^\vee$ where $I_i^\vee$ is the Alexander dual of I_i (see Definition 3.1). Hence by Theorem 3.2 and using the Auslander–Buchsbaum formula, we get

$$\operatorname{reg}(R/(I_1 + I_2)) = n + m - \operatorname{depth}(R/(I_1^\vee \cap I_2^\vee)) - 1,$$
$$\operatorname{reg}(R_1/I_1) + \operatorname{reg}(R_2/I_2) = n - \operatorname{depth}(R_1/I_1^\vee) - 1 + m - \operatorname{depth}(R_2/I_2^\vee) - 1.$$

Therefore we need only show the equality

$$\operatorname{depth}(R/(I_1^\vee \cap I_2^\vee)) = \operatorname{depth}(R_1/I_1^\vee) + \operatorname{depth}(R_2/I_2^\vee) + 1.$$

Since $\operatorname{depth}(R/(I_1^\vee + I_2^\vee)) = \operatorname{depth}(R_1/I_1^\vee) + \operatorname{depth}(R_2/I_2^\vee)$, the proof reduces to showing the equality

$$\operatorname{depth}(R/(I_1^\vee \cap I_2^\vee)) = \operatorname{depth}(R/(I_1^\vee + I_2^\vee)) + 1. \tag{3.4}$$

We may assume that $\operatorname{depth}(R/I_1^\vee) \geq \operatorname{depth}(R/I_2^\vee)$. There is an exact sequence of graded R-modules:

$$0 \longrightarrow R/(I_1^\vee \cap I_2^\vee) \xrightarrow{\varphi} R/I_1^\vee \oplus R/I_2^\vee \xrightarrow{\phi} R/(I_1^\vee + I_2^\vee) \longrightarrow 0, \tag{3.5}$$

where $\varphi(\overline{r}) = (\overline{r}, -\overline{r})$ and $\phi(\overline{r}_1, \overline{r}_2) = \overline{r_1 + r_2}$. From the inequality

$$\begin{aligned}\operatorname{depth}(R/I_1^\vee \oplus R/I_2^\vee) &= \max\{\operatorname{depth}(R/I_i^\vee)\}_{i=1}^2 = \operatorname{depth}(R/I_1^\vee)\\ &= \operatorname{depth}(R_1/I_1^\vee) + m\\ &> \operatorname{depth}(R_1/I_1^\vee) + \operatorname{depth}(R_2/I_2^\vee) = \operatorname{depth}(R/(I_1^\vee + I_2^\vee))\end{aligned}$$

and applying the depth lemma (see [10, Proposition 1.2.9] for example) to (3.5), we obtain (3.4). □

Another useful property of regularity is that one can delete isolated vertices of a graph without changing the regularity of the edge ideal. The following lemma shows that this can be done without significant changes to the projective dimension as well.

Lemma 3.5. *Let $R = K[x_1, \ldots, x_n]$ and I be an ideal of R. If $I \subset (x_1, \ldots, x_{n-1})$, and $R' = R/(x_n) \cong K[x_1, \ldots, x_{n-1}]$, then $\operatorname{reg}(R/I) = \operatorname{reg}(R'/I)$ and $\operatorname{pd}_R(R/I) = \operatorname{pd}_{R'}(R'/I)$. Similarly, if $x_n \in I$ and $I' = I/(x_n)$, then $\operatorname{reg}(R/I) = \operatorname{reg}(R'/I')$ and $\operatorname{pd}_R(R/I) = \operatorname{pd}_{R'}(R'/I') + 1$.*

Proof. The first projective dimension result follows from the Auslander–Buchsbaum formula since $\operatorname{depth}(R/I) = \operatorname{depth}(R'/I) + 1$ and $\operatorname{depth}(R) = \operatorname{depth}(R') + 1$. Since $\operatorname{depth}(R/I) = \operatorname{depth}(R'/I')$ the second result for projective dimension holds as well. The results for regularity follow from Proposition 3.4 by noting that the regularity of a polynomial ring $K[x]$ is 0, as is the regularity of the field $K = K[x]/(x)$. □

While adding variables to the ring will preserve the regularity, other changes to the base ring, such as changing the characteristic, will affect this invariant. The following example shows that, even for graphs, a purely combinatorial description of the regularity might not be possible. Results regarding the role the characteristic of the field plays in the resolution of the ideal appear in [19, 74].

Example 3.6. Consider the edge ideal $I \subset R = K[x_1, \ldots, x_{10}]$ generated by the monomials

$$\begin{array}{lllllllll} x_1x_3, & x_1x_4, & x_1x_7, & x_1x_{10}, & x_1x_{11}, & x_2x_4, & x_2x_5, & x_2x_8, & x_2x_{10},\\ x_2x_{11}, & x_3x_5, & x_3x_6, & x_3x_8, & x_3x_{11}, & x_4x_6, & x_4x_9, & x_4x_{11}, & x_5x_7,\\ x_5x_9, & x_5x_{11}, & x_6x_8, & x_6x_9, & x_7x_9, & x_7x_{10}, & x_8x_{10}. & & \end{array}$$

Using *Macaulay*2 [49] we get that $\operatorname{reg}(R/I) = 3$ if $\operatorname{char}(K) = 2$, and $\operatorname{reg}(R/I) = 2$ if $\operatorname{char}(K) = 3$.

As mentioned in Theorem 2.6 (ii), chordal graphs provide a key example of a class of clutters whose edge ideals are sequentially Cohen–Macaulay. Much work has been done toward finding hypergraph generalizations of chordal graphs, typically by looking at cycles of edges or at tree hypergraphs [32, 53, 113]. The papers [53, 113, 115] contribute to the algebraic approach that is largely motivated by finding hypergraph generalizations that have edge ideals with linear resolutions.

It is useful to consider the homogeneous components of the ideals when using linear resolutions. Let (I_d) denote the ideal generated by all degree d elements of a homogeneous ideal I. Then I is called *componentwise linear* if (I_d) has a linear resolution for all d. If I is the edge ideal of a clutter, we write $I_{[d]}$ for the ideal generated by all the squarefree monomials of degree d in I.

Theorem 3.7. *Let K be a field and $\mathcal{C}$ be a clutter. Then*

(i) [27] *$R/I(\mathcal{C})$ is Cohen–Macaulay if and only if $I_c(\mathcal{C})$ has a linear resolution.*

(ii) [58] *$R/I(\mathcal{C})$ is sequentially Cohen–Macaulay if and only if $I_c(\mathcal{C})$ is componentwise linear.*

(iii) [58] *$I(\mathcal{C})$ is componentwise linear if and only if $I(\mathcal{C})_{[d]}$ has a linear resolution for $d \geq 0$.*

(iv) [44] *If G is a graph, then $I(G)$ has a linear resolution if and only if G^c is chordal.*

(v) [103] *If G is a graph, then* $\operatorname{reg}(R/I(G)) = 1$ *if and only if $I_c(G)$ is Cohen–Macaulay.*

A graph whose complement is chordal is called *co-chordal*. A consequence of this result and Theorem 3.2 is that an edge ideal $I(\mathcal{C})$ has regularity 2 if and only if $\Delta_{\mathcal{C}}$ is the independence complex of a co-chordal graph. In this case the complex $\Delta_{\mathcal{C}}$ turns out to be a quasi-forest in the sense of Zheng [118]. In [61, Theorem 9.2.12] it is shown that a complex Δ is a quasi-forest if and only if Δ is the clique complex of a chordal graph.

Information about the regularity of a clutter can also be found by examining smaller, closely related clutters. Let S be a set of vertices of a clutter $\mathcal{C}$. The *induced subclutter* on S, denoted by $\mathcal{C}[S]$, is the maximal subclutter of $\mathcal{C}$ with vertex set S. Thus the vertex set of $\mathcal{C}[S]$ is S and the edges of $\mathcal{C}[S]$ are exactly the edges of $\mathcal{C}$ contained in S. Notice that $\mathcal{C}[S]$ may have isolated vertices, i.e., vertices that do not belong to any edge of $\mathcal{C}[S]$. If $\mathcal{C}$ is a discrete clutter, i.e., all the vertices of $\mathcal{C}$ are isolated, we set $I(\mathcal{C}) = 0$ and $\alpha_0(\mathcal{C}) = 0$. A clutter of the form $\mathcal{C}[S]$ for some $S \subset V(\mathcal{C})$ is called an *induced subclutter* of $\mathcal{C}$.

Proposition 3.8. *Suppose $\mathcal{D}$ is an induced subclutter of $\mathcal{C}$. Then* $\operatorname{reg}(R/I(\mathcal{D})) \leq \operatorname{reg}(R/I(\mathcal{C}))$.

Proof. There is $S \subset V(\mathcal{C})$ such that $\mathcal{D} = \mathcal{C}[S]$. Let $\mathfrak{p}$ be the prime ideal of R generated by the variables in S. By duality, we have

$$I_c(\mathcal{C}) = \bigcap_{e \in E(\mathcal{C})} (e) \Longrightarrow I_c(\mathcal{C})_{\mathfrak{p}} = \bigcap_{e \in E(\mathcal{C})} (e)_{\mathfrak{p}} = \bigcap_{e \in E(\mathcal{D})} (e)_{\mathfrak{p}} = I_c(\mathcal{D})_{\mathfrak{p}}.$$

Therefore, using Theorem 3.2 and Lemma 3.5, we get

$$\begin{aligned}\operatorname{reg}(R/I(\mathcal{C})) &= \operatorname{pd}(R/I_c(\mathcal{C})) - 1 \\ &\geq \operatorname{pd}(R_{\mathfrak{p}}/I_c(\mathcal{C})_{\mathfrak{p}}) - 1 = \operatorname{pd}(R_{\mathfrak{p}}/I_c(\mathcal{D})_{\mathfrak{p}}) - 1 \\ &= \operatorname{pd}(R'/I_c(\mathcal{D})) - 1 = \operatorname{pd}(R/I_c(\mathcal{D})) - 1 = \operatorname{reg}(R/I(\mathcal{D})),\end{aligned}$$

where R' is the polynomial ring $K[S]$. Thus, $\operatorname{reg}(R/I(\mathcal{C})) \geq \operatorname{reg}(R/I(\mathcal{D}))$. □

Several combinatorially defined invariants that bound the regularity or other invariants of a clutter are given in terms of subsets of the edge set of the clutter. An *induced matching* in a clutter $\mathcal{C}$ is a set of pairwise disjoint edges $f_1, \ldots, f_r$ such that the only edges of $\mathcal{C}$ contained in $\bigcup_{i=1}^r f_i$ are $f_1, \ldots, f_r$. We let $\operatorname{im}(\mathcal{C})$ be the number of edges in the largest induced matching.

The next result was shown in [53, Theorem 6.5] for the family of uniform properly-connected hypergraphs.

Corollary 3.9. *Let $\mathcal{C}$ be a clutter and let $f_1, \ldots, f_r$ be an induced matching of $\mathcal{C}$ with $d_i = |f_i|$ for $i = 1, \ldots, r$. Then*

(i) $(\sum_{i=1}^r d_i) - r \leq \operatorname{reg}(R/I(\mathcal{C}))$.

(ii) [74, *Lemma* 2.2] $\operatorname{im}(G) \leq \operatorname{reg}(R/I(G))$ *for any graph G.*

Proof. Let $\mathcal{D} = \mathcal{C}[\cup_{i=1}^r f_i]$. Notice that $I(\mathcal{D}) = (x_{f_1}, \ldots, x_{f_r})$. Thus $I(\mathcal{D})$ is a complete intersection and the regularity of $R/I(\mathcal{D})$ is the degree of its h-polynomial. The Hilbert series of $R/I(\mathcal{D})$ is given by

$$HS_{\mathcal{D}}(t) = \frac{\prod_{i=1}^r (1 + t + \cdots + t^{d_i - 1})}{(1-t)^{n-r}}.$$

Thus, the degree of the h-polynomial equals $(\sum_{i=1}^r d_i) - r$. Therefore, part (i) follows from Proposition 3.8. Part (ii) follows from part (i). □

Corollary 3.10. *If $\mathcal{C}$ is a clutter and $R/I_c(\mathcal{C})$ is Cohen–Macaulay, then* $\operatorname{im}(\mathcal{C}) = 1$.

Proof. Let r be the induced matching number of $\mathcal{C}$ and let d be the cardinality of any edge of $\mathcal{C}$. Using Theorem 3.2 and Corollary 3.9, we obtain $d - 1 \geq r(d-1)$. Thus $r = 1$, as required. □

The following example shows that the inequality obtained in Corollary 3.9 (ii) can be strict.

Example 3.11. Let G be the complement of a cycle $C_6 = \{x_1, \ldots, x_6\}$ of length six. The edge ideal of G is

$$I(G) = (x_1x_3, x_1x_5, x_1x_4, x_2x_6, x_2x_4, x_2x_5, x_3x_5, x_3x_6, x_4x_6).$$

Using *Macaulay*2 [49], we get $\operatorname{reg}(R/I(G)) = 2$ and $\operatorname{im}(G) = 1$.

Lemma 3.12 ([28, Corollary 20.19]). *If* $0 \to N \to M \to L \to 0$ *is a short exact sequence of graded finitely generated* R*-modules, then*

(i) $\operatorname{reg}(N) \leq \max(\operatorname{reg}(M), \operatorname{reg}(L) + 1)$.

(ii) $\operatorname{reg}(M) \leq \max(\operatorname{reg}(N), \operatorname{reg}(L))$.

(iii) $\operatorname{reg}(L) \leq \max(\operatorname{reg}(N) - 1, \operatorname{reg}(M))$.

Definition 3.13. If x is a vertex of a graph G, then its *neighbor set*, denoted by $N_G(x)$, is the set of vertices of G adjacent to x.

The following theorem gives a precise sense in which passing to induced subgraphs can be used to bound the regularity. Recall that a discrete graph is one in which all the vertices are isolated.

Theorem 3.14. *Let* $\mathcal{F}$ *be a family of graphs containing any discrete graph and let* $\beta\colon \mathcal{F} \to \mathbb{N}$ *be a function satisfying that* $\beta(G) = 0$ *for any discrete graph G, and such that given* $G \in \mathcal{F}$, *with* $E(G) \neq \emptyset$, *there is* $x \in V(G)$ *such that the following two conditions hold:*

(i) $G \setminus \{x\}$ *and* $G \setminus (\{x\} \cup N_G(x))$ *are in* $\mathcal{F}$.

(ii) $\beta(G \setminus (\{x\} \cup N_G(x))) < \beta(G)$ *and* $\beta(G \setminus \{x\}) \leq \beta(G)$.

Then $\operatorname{reg}(R/I(G)) \leq \beta(G)$ *for any* $G \in \mathcal{F}$.

Proof. The proof is by induction on the number of vertices. Let G be a graph in $\mathcal{F}$. If G is a discrete graph, then $I(G) = (0)$ and $\operatorname{reg}(R) = \beta(G) = 0$. Assume that G has at least one edge. There is a vertex $x \in V(G)$ such that the induced subgraphs $G_1 = G \setminus \{x\}$ and $G_2 = G \setminus (\{x\} \cup N_G(x))$ satisfy (i) and (ii). There is an exact sequence of graded R-modules

$$0 \longrightarrow R/(I(G)\colon x)[-1] \stackrel{x}{\longrightarrow} R/I(G) \longrightarrow R/(x, I(G)) \longrightarrow 0.$$

Notice that $(I(G)\colon x) = (N_G(x), I(G_2))$ and $(x, I(G)) = (x, I(G_1))$. The graphs G_1 and G_2 have fewer vertices than G. It follows directly from the definition of regularity that $\operatorname{reg}(M[-1]) = 1 + \operatorname{reg}(M)$ for any graded R-module M. Therefore applying the induction hypothesis to G_1 and G_2, and using conditions (i) and (ii) and Lemma 3.5, we get

$$\begin{aligned}\operatorname{reg}(R/(I(G)\colon x)[-1]) &= \operatorname{reg}(R/(I(G)\colon x)) + 1 = \operatorname{reg}(R'/I(G_2)) + 1\\ &\leq \beta(G_2) + 1 \leq \beta(G),\\ \operatorname{reg}(R/(x, I(G))) &\leq \beta(G_1) \leq \beta(G)\end{aligned}$$

where R' is the ring in the variables $V(G_2)$. Therefore from Lemma 3.12, the regularity of $R/I(G)$ is bounded by the maximum of the regularities of $R/(I(G)\colon x)[-1]$ and $R/(x, I(G))$. Thus $\operatorname{reg}(R/I(G)) \leq \beta(G)$, as required. □

As an example of how Theorem 3.14 can be applied to obtain combinatorial bounds for the regularity, we provide new proofs for several previously known results. Let G be a graph. We let $\beta'(G)$ be the cardinality of any smallest maximal matching of G. Hà and Van Tuyl proved that the regularity of $R/I(G)$ is bounded from above by the matching number of G and Woodroofe improved this result showing that $\beta'(G)$ is an upper bound for the regularity.

Corollary 3.15. *Let G be a graph and let $R = K[V(G)]$. Then*

(i) [53, *Corollary* 6.9] $\operatorname{reg}(R/I(G)) = \operatorname{im}(G)$ *for any chordal graph G.*

(ii) [53, *Theorem* 6.7; 117] $\operatorname{reg}(R/I(G)) \leq \beta'(G)$.

(iii) [106, *Theorem* 3.3] $\operatorname{reg}(R/I(G)) = \operatorname{im}(G)$ *if G is bipartite and $R/I(G)$ is sequentially Cohen–Macaulay.*

Proof. (i) Let $\mathcal{F}$ be the family of chordal graphs and let G be a chordal graph with $E(G) \neq \emptyset$. By Corollary 3.9 and Theorem 3.14 it suffices to prove that there is $x \in V(G)$ such that $\operatorname{im}(G_1) \leq \operatorname{im}(G)$ and $\operatorname{im}(G_2) < \operatorname{im}(G)$, where G_1 and G_2 are the subgraphs $G \setminus \{x\}$ and $G \setminus (\{x\} \cup N_G(x))$, respectively. The inequality $\operatorname{im}(G_1) \leq \operatorname{im}(G)$ is clear because any induced matching of G_1 is an induced matching of G. We now show the other inequality. By [104, Theorem 8.3], there is $y \in V(G)$ such that $G[N_G(y) \cup \{y\}]$ is a complete subgraph. Pick $x \in N_G(y)$ and set $f_0 = \{x, y\}$. Consider an induced matching $f_1, \ldots, f_r$ of G_2 with $r = \operatorname{im}(G_2)$. We claim that $f_0, f_1, \ldots, f_r$ is an induced matching of G. Let e be an edge of G contained in $\bigcup_{i=0}^r f_i$. We may assume that $e \cap f_0 \neq \emptyset$ and $e \cap f_i \neq \emptyset$ for some $i \geq 1$, otherwise $e = f_0$ or $e = f_i$ for some $i \geq 1$. Then $e = \{y, z\}$ or $e = \{x, z\}$ for some $z \in f_i$. If $e = \{y, z\}$, then $z \in N_G(y)$ and $x \in N_G(y)$. Hence $\{z, x\} \in E(G)$ and $z \in N_G(x)$, a contradiction because the vertex set of G_2 is disjoint from $N_G(x) \cup \{x\}$. If $e = \{x, z\}$, then $z \in N_G(x)$, a contradiction. This completes the proof of the claim. Hence $\operatorname{im}(G_2) < \operatorname{im}(G)$.

(ii) Let $\mathcal{F}$ be the family of all graphs and let G be a graph with $E(G) \neq \emptyset$. By Theorem 3.14 it suffices to prove that there is $x \in V(G)$ such that $\beta'(G_1) \leq \beta'(G)$ and

$\beta'(G_2) < \beta'(G)$, where G_1 and G_2 are the subgraphs $G \setminus \{x\}$ and $G \setminus (\{x\} \cup N_G(x))$, respectively.

Let $f_1, \ldots, f_r$ be a maximal matching of G with $r = \beta'(G)$ and let x, y be the vertices of f_1. Clearly $f_2, \ldots, f_r$ is a matching of G_1. Thus we can extend it to a maximal matching $f_2, \ldots, f_r, h_1, \ldots, h_s$ of G_1. Notice that $s \leq 1$. Indeed if $s \geq 2$, then $h_i \cap f_1 = \emptyset$ for some $i \in \{1, 2\}$ (otherwise $y \in h_1 \cap h_2$, which is impossible). Hence $f_1, \ldots, f_r, h_i$ is a matching of G, a contradiction because $f_1, \ldots, f_r$ is maximal. Therefore $\beta'(G_1) \leq r - 1 + s \leq \beta'(G)$.

The set $f_2, \ldots, f_r$ contains a matching of G_2, namely those edges f_i that do not degenerate. Reorder the edges so that $f_2, \ldots, f_m$ are the edges that do not degenerate. Then this set can be extended to a maximal matching $f_2, \ldots, f_m, f'_{m+1}, \ldots, f'_k$ of G_2. Now consider f'_{m+1}. Since $f_1, \ldots, f_r$ is a maximal matching of G, f'_{m+1} has a nontrivial intersection with f_i for some i. Note that $i \neq 1$ since f'_{m+1} is an edge of G_2, and $i \geq m + 1$ since $f_2, \ldots, f_m$ and f'_{m+1} are all part of a matching of G_2. Reorder so that $i = m + 1$. Repeat the process with f'_{m+2}. As before, f'_{m+2} has a nontrivial intersection with f_i for some $i \geq m + 1$. If $i = m + 1$, then since f'_{m+1} and f'_{m+2} are disjoint, each must share a different vertex with f_{m+1}. But f_{m+1} degenerated when passing to G_2, and G_2 is induced on the remaining vertices, so this is a contradiction. Thus we may reorder so that f'_{m+2} nontrivially intersects f_{m+2}. Repeating this process, we see that f'_j nontrivially intersects f_j for all $m+1 \leq j \leq k$. Thus $k \leq r$. Now $\beta'(G_2) \leq k - 1 \leq r - 1 < \beta'(G)$.

(iii) Let $\mathcal{F}$ be the family of all bipartite graphs G such that $R/I(G)$ is sequentially Cohen–Macaulay, and let $\beta\colon \mathcal{F} \to \mathbb{N}$ be the function $\beta(G) = \operatorname{im}(G)$. Let G be a graph in $\mathcal{F}$ with $E(G) \neq \emptyset$. By Corollary 3.9 and Theorem 3.14 it suffices to observe that, according to [107, Corollary 2.10], there are adjacent vertices x and y with $\deg(y) = 1$ such that the bipartite graphs $G \setminus (\{x\} \cup N_G(x))$ and $G \setminus (\{y\} \cup N_G(y))$ are sequentially Cohen–Macaulay. Thus conditions (i) and (ii) of Theorem 3.14 are satisfied. □

Corollary 3.15 shows that the regularity of $R/I(G)$ equals $\operatorname{im}(G)$ for any forest G, which was first proved by Zheng [118]. If G is an unmixed graph, Kummini [77] showed that $\operatorname{reg}(R/I(G))$ equals the induced matching number of G. If G is claw-free and its complement has no induced 4-cycles, then $\operatorname{reg}(R/I(G)) \leq 2$ with equality if its complement is not chordal [90] (note that in this case $\operatorname{reg}(R/I(G)) = \operatorname{im}(G) + 1$). Formulas for the regularity of ideals of mixed products are given in [71]. The regularity and depth of lex segment edge ideals are computed in [33]. The regularity and other algebraic properties of the edge ideal $I(G)$ associated to a Ferrers graph G are studied in detail in [18]. If $R/I(G)$ is Cohen–Macaulay, then a formula for $\operatorname{reg}(R/I(G))$ follows from [95, Corollary 4.2].

The following result about regularity was shown by Kalai and Meshulam for squarefree monomial ideals and by Herzog for arbitrary monomial ideals. Similar inequalities hold for the projective dimension.

Proposition 3.16 ([57,73]). *Let I_1 and I_2 be monomial ideals of R. Then*

(i) $\operatorname{reg} R/(I_1 + I_2) \leq \operatorname{reg}(R/I_1) + \operatorname{reg}(R/I_2)$,

(ii) $\operatorname{reg} R/(I_1 \cap I_2) \leq \operatorname{reg}(R/I_1) + \operatorname{reg}(R/I_2) + 1$.

Corollary 3.17. *If $\mathcal{C}_1, \ldots, \mathcal{C}_s$ are clutters on the vertex set X, then*

$$\operatorname{reg}\Big(R/I\Big(\bigcup_{i=1}^{s} \mathcal{C}_i\Big)\Big) \leq \operatorname{reg}(R/I(\mathcal{C}_1)) + \cdots + \operatorname{reg}(R/I(\mathcal{C}_s)).$$

Proof. The set of edges of $\mathcal{C} = \bigcup_{i=1}^{s} \mathcal{C}_i$ equals $\bigcup_{i=1}^{s} E(\mathcal{C}_i)$. By Proposition 3.16, it suffices to notice the equality $I(\bigcup_{i=1}^{s} \mathcal{C}_i) = \sum_{i=1}^{s} I(\mathcal{C}_i)$. □

A clutter $\mathcal{C}$ is called *co-CM* if $I_c(\mathcal{C})$ is Cohen–Macaulay. A co-CM clutter is uniform because Cohen–Macaulay clutters are unmixed.

Corollary 3.18. *If $\mathcal{C}_1, \ldots, \mathcal{C}_s$ are co-CM clutters on the vertex set X, then*

$$\operatorname{reg}(R/I(\cup_{i=1}^{s} \mathcal{C}_i)) \leq (d_1 - 1) + \cdots + (d_s - 1),$$

where d_i is the number of elements in any edge of $\mathcal{C}_i$.

Proof. By Theorem 3.2, we get that $\operatorname{reg} R/I(\mathcal{C}_i) = d_i - 1$ for all i. Thus the result follows from Corollary 3.17. □

This result is especially useful for graphs. A graph G is *weakly chordal* if every induced cycle in both G and G^c has length at most 4. It was pointed out in [117] that a weakly chordal graph G can be covered by $\operatorname{im}(G)$ co-CM graphs (this fact was shown in [13]). Thus we have:

Theorem 3.19 ([117]). *If G is a weakly chordal graph, then* $\operatorname{reg}(R/I(G)) = \operatorname{im}(G)$.

There are bounds for the regularity of R/I in terms of some other algebraic invariants of R/I. Recall that the a-invariant of R/I, denoted by $a(R/I)$, is the degree (as a rational function) of the Hilbert series of R/I. Also recall that the independence complex of $I(\mathcal{C})$, denoted by $\Delta_{\mathcal{C}}$, is the simplicial complex whose faces are the independent vertex sets of $\mathcal{C}$. The *arithmetic degree* of $I = I(\mathcal{C})$, denoted by arith-deg(I), is the number of *facets* (maximal faces with respect to inclusion) of $\Delta_{\mathcal{C}}$. The *arithmetical rank* of I, denoted by $\operatorname{ara}(I)$, is the least number of elements of R which generate the ideal I up to radical.

Theorem 3.20 ([108, Corollary B.4.1]). *$a(R/I) \leq \operatorname{reg}(R/I) - \operatorname{depth}(R/I)$, with equality if R/I is Cohen–Macaulay.*

Theorem 3.21 ([79; 80, Proposition 3]). $\operatorname{reg}(R/I^{\vee}) = \operatorname{pd}(R/I) - 1 \leq \operatorname{ara}(I) - 1$.

The equality $\operatorname{reg}(R/I^\vee) = \operatorname{pd}(R/I) - 1$ was pointed out earlier in Theorem 3.2. There are many instances where the equality $\operatorname{pd}(R/I) = \operatorname{ara}(I)$ holds, see [1,33,76] and the references there. For example, for paths, one has $\operatorname{pd}(R/I) = \operatorname{ara}(I)$ [1]. Barile [1] has conjectured that the equality holds for edge ideals of forests. We also have that $n - \min_i\{\operatorname{depth}(R/I^{(i)})\}$ is an upper bound for $\operatorname{ara}(I)$, see [80]. This upper bound tends to be very loose. If I is the edge ideal of a tree, then I is normally torsion free (see Section 4 together with Theorems 4.34 and 4.8). Then $\min_i\{\operatorname{depth}(R/I^{(i)})\} = 1$ by [87, Lemma 2.6]. But when I is the edge ideal of a path with 8 vertices, then the actual value of $\operatorname{ara}(I)$ is 5.

Theorem 3.22 ([103, Theorem 3.1]). *If* $\operatorname{ht}(I) \geq 2$, *then* $\operatorname{reg}(I) \leq \operatorname{arith-deg}(I)$.

The next open problem is known as the Eisenbud–Goto regularity conjecture [30].

Conjecture 3.23. If $\mathfrak{p} \subset (x_1, \ldots, x_n)^2$ is a prime graded ideal, then

$$\operatorname{reg}(R/\mathfrak{p}) \leq \deg(R/\mathfrak{p}) - \operatorname{codim}(R/\mathfrak{p}).$$

A pure d-dimensional complex Δ is called *connected in codimension* 1 if each pair of facets F, G can be connected by a sequence of facets $F = F_0, F_1, \ldots, F_s = G$, such that $\dim(F_{i-1} \cap F_i) = d - 1$ for $1 \leq i \leq s$. According to [5, Proposition 11.7], every Cohen–Macaulay complex is connected in codimension 1.

The following gives a partial answer to the monomial version of the Eisenbud–Goto regularity conjecture.

Theorem 3.24 ([103]). *Let* $I = I(\mathcal{C})$ *be an edge ideal. If* $\Delta_{\mathcal{C}}$ *is connected in codimension* 1, *then*

$$\operatorname{reg}(R/I) \leq \deg(R/I) - \operatorname{codim}(R/I).$$

The dual notion to the independence complex of $I(\mathcal{C})$ is to start with a complex Δ and associate to it an ideal whose independence complex is Δ.

Definition 3.25. Given a simplicial complex Δ with vertex set $X = \{x_1, \ldots, x_n\}$, the *Stanley–Reisner ideal* of Δ is defined as

$$I_\Delta = (\{x_{i_1} \cdots x_{i_r} \mid i_1 < \cdots < i_r,\ \{x_{i_1}, \ldots, x_{i_r}\} \notin \Delta\}),$$

and its *Stanley–Reisner ring* $K[\Delta]$ is defined as the quotient ring R/I_Δ.

A simple proof the next result is given in [44].

Theorem 3.26 ([101]). *Let* $\mathcal{C}$ *be a clutter and let* $\Delta = \Delta_{\mathcal{C}}$ *be its independence complex. Then*

$$\operatorname{depth} R/I(\mathcal{C}) = 1 + \max\{i \mid K[\Delta^i] \text{ is Cohen–Macaulay}\},$$

where $\Delta^i = \{F \in \Delta \mid \dim(F) \leq i\}$ *is the* i*-skeleton of* Δ *and* $-1 \leq i \leq \dim(\Delta)$.

A variation on the concept of the i-skeleton will facilitate an extension of the result above to the sequentially Cohen–Macaulay case.

Definition 3.27. Let Δ be a simplicial complex. The *pure i-skeleton* of Δ is defined as:

$$\Delta^{[i]} = \langle\{F \in \Delta |\ \dim(F) = i\}\rangle; \quad -1 \leq i \leq \dim(\Delta),$$

where $\langle \mathcal{F} \rangle$ denotes the subcomplex generated by $\mathcal{F}$.

Note that $\Delta^{[i]}$ is always pure of dimension i. We say that a simplicial complex Δ is *sequentially Cohen–Macaulay* if its Stanley–Reisner ring has this property. The following results link the sequentially Cohen–Macaulay property to the Cohen–Macaulay property and to the regularity and projective dimension. The first is an interesting result of Duval.

Theorem 3.28 ([26, Theorem 3.3]). *Let Δ be a simplicial complex. Then Δ is sequentially Cohen–Macaulay if and only if the pure i-skeleton $\Delta^{[i]}$ is Cohen–Macaulay for $-1 \leq i \leq \dim(\Delta)$.*

Corollary 3.29. *$R/I(\mathcal{C})$ is Cohen–Macaulay if and only if $R/I(\mathcal{C})$ is sequentially Cohen–Macaulay and $\mathcal{C}$ is unmixed.*

Lemma 3.30. *Let $\mathcal{C}$ be a clutter and let $\Delta = \Delta_{\mathcal{C}}$ be its independence complex. If $\beta_0'(\mathcal{C})$ is the cardinality of a smallest maximal independent set of $\mathcal{C}$, then $\Delta^{[i]} = \Delta^i$ for $i \leq \beta_0'(\mathcal{C}) - 1$.*

Proof. First we prove the inclusion $\Delta^{[i]} \subset \Delta^i$. Let F be a face of $\Delta^{[i]}$. Then F is contained in a face of Δ of dimension i, and so F is in Δ^i. Conversely, let F be a face of Δ^i. Then

$$\dim(F) \leq i \leq \beta_0'(\mathcal{C}) - 1 \Longrightarrow |F| \leq i + 1 \leq \beta_0'(\mathcal{C}).$$

Since $\beta_0'(\mathcal{C})$ is the cardinality of any smallest maximal independent set of $\mathcal{C}$, we can extend F to an independent set of $\mathcal{C}$ with $i + 1$ vertices. Thus F is in $\Delta^{[i]}$. □

While β_0' regulates the equality of the i-skeleton and the pure i-skeleton of the independence complex, its complement provides a lower bound for the regularity of the ideal of covers.

Theorem 3.31. *Let $\mathcal{C}$ be a clutter, let $I_c(\mathcal{C})$ be its ideal of vertex covers, and let $\alpha_0'(\mathcal{C})$ be the cardinality of a largest minimal vertex cover of $\mathcal{C}$. Then*

$$\operatorname{reg} R/I_c(\mathcal{C}) \geq \alpha_0'(\mathcal{C}) - 1,$$

with equality if $R/I(\mathcal{C})$ is sequentially Cohen–Macaulay.

Proof. We set $\beta_0'(\mathcal{C}) = n - \alpha_0'(\mathcal{C})$. Using Theorem 3.2 and the Auslander–Buchsbaum formula (see (3.2)), the proof reduces to showing: depth $R/I(\mathcal{C}) \leq \beta_0'(\mathcal{C})$, with equality if $R/I(\mathcal{C})$ is sequentially Cohen–Macaulay.

First we show that depth $R/I(\mathcal{C}) \leq \beta_0'(\mathcal{C})$. Assume Δ^i is Cohen–Macaulay for some $-1 \leq i \leq \dim(\Delta)$, where Δ is the independence complex of $\mathcal{C}$. According to Theorem 3.26, it suffices to prove that $1 + i \leq \beta_0'(\mathcal{C})$. Notice that $\beta_0'(\mathcal{C})$ is the cardinality of any smallest maximal independent set of $\mathcal{C}$. Thus, we can pick a maximal independent set F of $\mathcal{C}$ with $\beta_0'(\mathcal{C})$ vertices. Since Δ^i is Cohen–Macaulay, the complex Δ^i is pure, that is, all maximal faces of Δ have dimension i. If $1 + i > \beta_0'(\mathcal{C})$, then F is a maximal face of Δ^i of dimension $\beta_0'(\mathcal{C}) - 1$, a contradiction to the purity of Δ^i.

Assume that $R/I(\mathcal{C})$ is sequentially Cohen–Macaulay. By Lemma 3.30 $\Delta^{[i]} = \Delta^i$ for $i \leq \beta_0'(\mathcal{C}) - 1$. Then by Theorem 3.28, the ring $K[\Delta^i]$ is Cohen–Macaulay for $i \leq \beta_0'(\mathcal{C}) - 1$. Therefore, applying Theorem 3.26, we get that the depth of $R/I(\mathcal{C})$ is at least $\beta_0'(\mathcal{C})$. Consequently, in this case one has the equality depth $R/I(\mathcal{C}) = \beta_0'(\mathcal{C})$. □

The inequality in Theorem 3.31 also follows directly from the definition of regularity because $\operatorname{reg}(I_c(\mathcal{C}))$ is an upper bound for the largest degree of a minimal generator of $I_c(\mathcal{C})$.

Remark 3.32. $\alpha_0'(\mathcal{C})$ is $\max\{|e| : e \in E(\mathcal{C}^\vee)\}$ and $\alpha_0'(\mathcal{C}^\vee)$ is $\max\{|e| : e \in E(\mathcal{C})\}$. This follows by Alexander duality, see (3.3).

Corollary 3.33. *If $I(\mathcal{C})$ is an edge ideal, then* $\operatorname{pd}_R(R/I(\mathcal{C})) \geq \alpha_0'(\mathcal{C})$, *with equality if $R/I(\mathcal{C})$ is sequentially Cohen–Macaulay.*

Proof. It follows from the proof of Theorem 3.31. □

There are many interesting classes of sequentially Cohen–Macaulay clutters where this formula for the projective dimension applies (see Theorem 2.6). The projective dimension of edge ideals of forests was studied in [22, 53], where some recursive formulas are presented. Explicit formulas for the projective dimension for some path ideals of directed rooted trees can be found in [55, Theorem 1.2]. Path ideals of directed graphs were introduced by Conca and De Negri [15]. Fix an integer $t \geq 2$, and suppose that $\mathcal{D}$ is a directed graph, i.e., each edge has been assigned a direction. A sequence of t vertices $x_{i_1}, \ldots, x_{i_t}$ is said to be a *path* of length t if there are $t - 1$ distinct edges $e_1, \ldots, e_{t-1}$ such that $e_j = (x_{i_j}, x_{i_{j+1}})$ is a directed edge from x_{i_j} to $x_{i_{j+1}}$. The *path ideal* of $\mathcal{D}$ of length t, denoted by $I_t(\mathcal{D})$, is the ideal generated by all monomials $x_{i_1} \cdots x_{i_t}$ such that $x_{i_1}, \ldots, x_{i_t}$ is a path of length t in $\mathcal{D}$. Note that when $t = 2$, then $I_2(\mathcal{D})$ is simply the edge ideal of $\mathcal{D}$.

Example 3.34. Let K be any field and let G be the following chordal graph

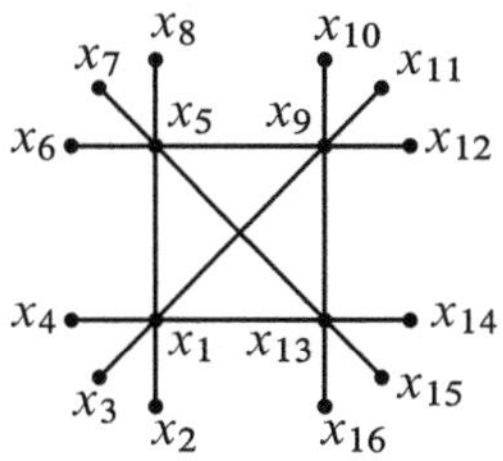

Then, by Theorem 2.6 and Corollary 3.31, $\mathrm{pd}_R(R/I(G)) = 6$ and depth $R/I(G) = 10$.

Corollary 3.35. *Let $\mathcal{C}$ be a clutter. If $I(\mathcal{C})$ has linear quotients, then*

$$\operatorname{reg} R/I(\mathcal{C}) = \max\{|e|: e \in E(\mathcal{C})\} - 1.$$

Proof. The ideal of covers $I_c(\mathcal{C})$ is sequentially Cohen–Macaulay by Theorem 2.6 (iv). Hence, using Theorem 3.31, we get $\operatorname{reg}(R/I(\mathcal{C})) = \alpha_0'(\mathcal{C}^\vee) - 1$. To complete the proof notice that $\alpha_0'(\mathcal{C}^\vee) = \max\{|e|: e \in E(\mathcal{C})\}$ (see Remark 3.32). □

The converse of Theorem 3.33 is not true.

Example 3.36. Let C_6 be a cycle of length 6. Then $R/I(C_6)$ is not sequentially Cohen–Macaulay by Proposition 2.4. Using *Macaulay*2, we get $\mathrm{pd}(R/I(C_6)) = \alpha_0'(C_6) = 4$.

When R/I is not known to be Cohen–Macaulay, it can prove useful to have effective bounds on the depth of R/I.

Theorem 3.37. *Let G be a bipartite graph without isolated vertices. If G has n vertices, then*

$$\operatorname{depth} R/I(G) \leq \left\lfloor \frac{n}{2} \right\rfloor.$$

Proof. Let (V_1, V_2) be a bipartition of G with $|V_1| \leq |V_2|$. Note $2|V_1| \leq n$ because $|V_1| + |V_2| = n$. Since V_1 is a maximal independent set of vertices one has $\beta_0'(G) \leq |V_1| \leq n/2$. Therefore, using Corollary 3.33 and the Auslander–Buchsbaum formula, we get depth $R/I(G) \leq n/2$. □

Corollary 3.38. *If G is a B-graph with n vertices, then*

$$\operatorname{depth} R/I(G) \leq \dim R/I(G) \leq \left\lfloor \frac{n}{2} \right\rfloor.$$

Proof. Recall that $n = \alpha_0(G) + \beta_0(G)$. By Theorem 2.9, $\beta_0(G) \leq \alpha_0(G)$, and so $\beta_0(G) \leq \lfloor \frac{n}{2} \rfloor$. The result now follows because $\beta_0(G) = \dim R/I(\mathcal{C})$. □

Lower bounds are given in [87] for the depths of $R/I(G)^t$ for $t \geq 1$ when $I(G)$ is the edge ideal of a tree or forest. Upper bounds for the depth of $R/I(G)$ are given in [46, Corollary 4.15] when G is any graph without isolated vertices. The depth and the Cohen–Macaulay property of ideals of mixed products is studied in [71].

We close this section with an upper bound for the multiplicity of edge rings. Let $\mathcal{C}$ be a clutter. The *multiplicity* of the edge-ring $R/I(\mathcal{C})$, denoted by $e(R/I(\mathcal{C}))$, equals the number of faces of maximum dimension of the independence complex $\Delta_{\mathcal{C}}$, i.e., the multiplicity of $R/I(\mathcal{C})$ equals the number of independent sets of $\mathcal{C}$ with $\beta_0(\mathcal{C})$ vertices. A related invariant that was considered earlier is arith-deg$(I(\mathcal{C}))$, the number of maximal independent sets of $\mathcal{C}$.

Proposition 3.39 ([46]). *If $\mathcal{C}$ is a d-uniform clutter and $I = I(\mathcal{C})$, then $e(R/I) \leq d^{\alpha_0(\mathcal{C})}$.*

4 Stability of Associated Primes

One method of gathering information about an ideal is through its associated primes. Let I be an ideal of a ring R. In this section, we will examine the sets of associated primes of powers of I, that is, the sets

$$\text{Ass}(R/I^t) = \{\mathfrak{p} \subset R \mid \mathfrak{p} \text{ is prime and } \mathfrak{p} = (I^t : c) \text{ for some } c \in R\}.$$

When I is a monomial ideal of a polynomial ring $R = K[x_1, \dots, x_n]$, the associated primes will be monomial ideals, that is, prime ideals which are generated by a subset of the variables. When I is a square-free monomial ideal, the minimal primes of I, Min(R/I), correspond to minimal vertex covers of the clutter $\mathcal{C}$ associated to I. In general Min$(R/I) \subset$ Ass(R/I^t) for all positive integers t. For a square-free monomial ideal, in the case where equality holds for all t, the ideal I is said to be *normally torsion-free*. More generally, an ideal $I \subset R$ is called *normally torsion-free* if Ass(R/I^i) is contained in Ass(R/I) for all $i \geq 1$ and $I \neq R$.

In [9], Brodmann showed that when R is a Noetherian ring and I is an ideal of R, the sets Ass(R/I^t) stabilize for large t. That is, there exists a positive integer N such that Ass$(R/I^t) =$ Ass(R/I^N) for all $t \geq N$. We will refer to a minimal such N as the *index of stability* of I. There are two natural questions following from this result. In this article, we will focus on the monomial versions of the questions.

Question 4.1. Given a monomial ideal I, what is an effective upper bound on the index of stability for a given class of monomial ideals?

Question 4.2. Given a monomial ideal I, which primes are in Ass(R/I^t) for all sufficiently large t?

An interesting variation on Questions 4.1 and 4.2 was posed in [99].

Question 4.3. Suppose that N is the index of stability of an ideal I. Given a prime $\mathfrak{p} \in \operatorname{Ass}(R/I^N)$, can you find an integer $N_{\mathfrak{p}}$ for which $\mathfrak{p} \in \operatorname{Ass}(R/I^t)$ for $t \geq N_{\mathfrak{p}}$?

Brodmann also showed that the sets $\operatorname{Ass}(I^{t-1}/I^t)$ stabilize. Thus in the general setting, one could ask similar questions about these sets. However, for monomial ideals the following lemma shows that in order to find information about $\operatorname{Ass}(R/I^t)$, one may instead study $\operatorname{Ass}(I^{t-1}/I^t)$.

Lemma 4.4. *Let I be a monomial ideal. Then* $\operatorname{Ass}(I^{t-1}/I^t) = \operatorname{Ass}(R/I^t)$.

Proof. Suppose that $\mathfrak{p} \in \operatorname{Ass}(R/I^t)$. Then $\mathfrak{p} = (I^t : c)$ for some monomial $c \in R$. But since $\mathfrak{p}$ is necessarily a monomial prime, generated by a subset of the variables, then if $xc \in I^t$ for a variable $x \in \mathfrak{p}$, then $c \in I^{t-1}$ and so $\mathfrak{p} \in \operatorname{Ass}(I^{t-1}/I^t)$. The other inclusion is automatic. □

Note that this method was used in [105] to show that the corresponding equality also holds for the integral closures of the powers of I.

For special classes of ideals, there have been some results that use properties of the ideals to find bounds on N. For example, if I is generated by a regular sequence, then by [67], I is normally torsion-free, or $\operatorname{Ass}(R/I^t) = \operatorname{Min}(R/I)$ for all t, and thus $N = 1$. If instead I is generated by a d-sequence and is strongly Cohen–Macaulay, then it was shown in [88] that N is bounded above by the dimension of the ring. In particular, $N \leq n-g+1$ where n is the dimension of the ring and g is the height of the ideal. We are particularly interested in finding similar bounds for classes of monomial ideals.

In [65], Hoa used integer programming techniques to give an upper bound on N for general monomial ideals. Let n be the number of variables, s the number of generators of I, and d the maximal degree of a generator.

Theorem 4.5 ([65, Theorem 2.12]). *If I is a monomial ideal, then the index of stability is bounded above by*

$$\max\{d(ns+s+d)(\sqrt{n})^{n+1}(\sqrt{2}d)^{(n+1)(s-1)}, s(s+n)^4 s^{n+2} d^2 (2d^2)^{s^2-s+1}\}.$$

Notice that this bound can be extremely large. For general monomial ideals, examples are given in [65] to show that the bound should depend on d and n. However, if we restrict to special classes of monomial ideals, much smaller bounds can be found. For example, an alternate bound has been shown to hold for integral closures of powers of monomial ideals.

Theorem 4.6 ([105, Theorem 16]). *If I is a monomial ideal, and $N_0 = n2^{n-1}d^{n-2}$, then* $\operatorname{Ass}(R/\overline{I^t}) = \operatorname{Ass}(R/\overline{I^{N_0}})$ *for $t \geq N_0$ when $n \geq 2$.*

Here again n is the number of variables and d is the maximal degree of a generator. For the class of normal monomial ideals, this bound on the index of stability can be significantly lower than the general bound given above. When $n = 2$, the index of stability of the integral closures is lower still.

Lemma 4.7 ([85]). *If $n \leq 2$, then $\mathrm{Ass}(R/\overline{I^t}) = \mathrm{Ass}(R/\overline{I})$ for all $t \geq 1$.*

Note that this result is of interest for general monomial ideals; however, when $n = 2$ a square-free monomial ideal will be a complete intersection. Of particular interest for this article are results that use combinatorial and graph-theoretic properties to yield insights into the associated primes and index of stability of monomial ideals. One pivotal result in this area establishes a classification of all graphs for which $N = 1$.

Theorem 4.8 ([100, Theorem 5.9]). *Let G be a graph and I its edge ideal. Then G is bipartite if and only if I is normally torsion-free.*

The result above shows that $N = 1$ for the edge ideal of a graph if and only if the graph is bipartite. Since minimal primes correspond to minimal vertex covers, this completely answers Questions 4.1 and 4.2 for bipartite graphs. In addition if I is the edge ideal of a balanced clutter, then $N = 1$ [45].

Suppose now that G is a graph that is not bipartite. Then G contains at least one odd cycle. For such graphs, a method of describing embedded associated primes, and a bound on where the stability occurs, were given in [14]. The method of building embedded primes centered around the odd cycles, so we first give an alternate proof of the description of the associated primes for this base case.

Lemma 4.9. *Suppose G is a cycle of length $n = 2k + 1$ and I is the edge ideal of G. Then $\mathrm{Ass}(R/I^t) = \mathrm{Min}(R/I)$ if $t \leq k$ and $\mathrm{Ass}(R/I^t) = \mathrm{Min}(R/I) \cup \{\mathfrak{m}\}$ if $t \geq k + 1$. Moreover, when $t \geq k + 1$, $\mathfrak{m} = (I^t : c)$ for a monomial c of degree $2t - 1$.*

Proof. If $\mathfrak{p} \neq \mathfrak{m}$ is a prime ideal, then $I_{\mathfrak{p}}$ is the edge ideal of a bipartite graph, and thus by Theorem 4.8 $\mathfrak{p} \in \mathrm{Ass}(R/I^t)$ (for any $t \geq 1$) if and only if $\mathfrak{p}$ is a minimal prime of I. Notice also that the deletion of any vertex x_i (which corresponds to passing to the quotient ring $R/(x_i)$) results in a bipartite graph as well. Thus by [51, Corollary 3.6], $\mathfrak{m} \notin \mathrm{Ass}(R/I^t)$ for $t \leq k$ since a maximal matching has k edges. For $t \geq k + 1$, define $b = \left(\prod_{i=1}^{n} x_i\right)$ and $c = b(x_1x_2)^{t-k-1}$ where x_1x_2 is any edge of G. Then since c has degree $2t - 1$, $c \notin I^t$, but G is a cycle, $x_ib \in I^{k+1}$ and $x_ic \in I^t$. Thus $\mathfrak{m} = (I^t : c)$ and so $\mathfrak{m} \in \mathrm{Ass}(R/I^t)$ for $t \geq k + 1$. □

Corollary 4.10. *Suppose G is a connected graph containing an odd cycle of length $2k + 1$ and suppose that every vertex of G that is not in the cycle is a leaf. Then $\mathrm{Ass}(R/I^t) = \mathrm{Min}(R/I) \cup \{\mathfrak{m}\}$ if $t \geq k + 1$. Moreover, when $t \geq k + 1$, $\mathfrak{m} = (I^t : c)$ for a monomial c of degree $2t - 1$.*

Proof. Let b and c be defined as in the proof of Lemma 4.9. Notice that if x is a leaf, then x is connected to a unique vertex in the cycle and that $xb \in I^{k+1}$. The remainder of the proof follows as in Lemma 4.9. □

If G is a more general graph, the embedded associated primes of $I = I(G)$ are formed by working outward from the odd cycles. This was done in [14], including a detailed explanation of how to work outward from multiple odd cycles. Before providing more concise proofs of the process, we first give an informal, but illustrative, description. Suppose C is a cycle with $2k+1$ vertices $x_1, \dots, x_{2k+1}$. Color the vertices of C red and color any noncolored vertex that is adjacent to a red vertex blue. The set of colored vertices, together with a minimal vertex cover of the set of edges neither of whose vertices is colored, will be an embedded associated prime of I^t for all $t \geq k+1$. To find additional embedded primes of higher powers, select any blue vertex to turn red and turn any uncolored neighbors of this vertex blue. The set of colored vertices, together with a minimal vertex cover of the noncolored edges, will be an embedded associated prime of I^t for all $t \geq k+2$. This process continues until all vertices are colored red or blue.

The method of building new associated primes for a power of I from primes associated to lower powers relies on localization. Since localization will generally cause the graph (or clutter) to become disconnected, we first need the following lemma.

Lemma 4.11 ([51, Lemma 3.4], see also [14, Lemma 2.1]). *Suppose I is a square-free monomial ideal in $S = K[x_1, \dots, x_r, y_1, \dots, y_s]$ such that $I = I_1S + I_2S$, where $I_1 \subset S_1 = K[x_1, \dots, x_r]$ and $I_2 \subset S_2 = K[y_1, \dots, y_s]$. Then $\mathfrak{p} \in \operatorname{Ass}(S/I^t)$ if and only if $\mathfrak{p} = \mathfrak{p}_1S + \mathfrak{p}_2S$, where $\mathfrak{p}_1 \in \operatorname{Ass}(S_1/I_1^{t_1})$ and $\mathfrak{p}_2 \in \operatorname{Ass}(S_2/I_2^{t_2})$ with $(t_1-1)+(t_2-1) = t-1$.*

Note that this lemma easily generalizes to an ideal $I = (I_1, I_2, \dots, I_s)$ where the I_i are edge ideals of disjoint clutters. Then $\mathfrak{p} \in \operatorname{Ass}(R/I^t)$ if and only if $\mathfrak{p} = (\mathfrak{p}_1, \dots, \mathfrak{p}_s)$ with $\mathfrak{p}_i \in \operatorname{Ass}(R/I_i^{t_i})$ where $(t_1-1)+(t_2-1)+\cdots+(t_s-1) = (t-1)$.

We now fix a notation to show how to build embedded associated primes. Consider $\mathfrak{p} \in \operatorname{Ass}(R/I^t)$ for I the edge ideal of a graph G. Without loss of generality, Lemma 4.11 allows us to assume G does not have isolated vertices. If $\mathfrak{p} \neq \mathfrak{m}$, then since $\mathfrak{p} \in \operatorname{Ass}(R/I^t) \Leftrightarrow \mathfrak{p}R_{\mathfrak{p}} \in \operatorname{Ass}(R_{\mathfrak{p}}/(I_{\mathfrak{p}})^t)$, consider $I_{\mathfrak{p}}$. Write $I_{\mathfrak{p}} = (I_a, I_b)$ where I_a is generated by all generators of $I_{\mathfrak{p}}$ of degree two and I_b is the prime ideal generated by the degree one generators of $I_{\mathfrak{p}}$, which correspond to the isolated vertices of the graph associated to $I_{\mathfrak{p}}$. Note that the graph corresponding to I_a need not be connected. If $I_a = (0)$, then $\mathfrak{p}$ is a minimal prime of I, so assume $I_a \neq (0)$. Define $\mathfrak{p}_a$ to be the monomial prime generated by variables of I_a. Define $N_1 = \bigcup_{x \in \mathfrak{p}_a} N(x)$, where $N(x)$ is the neighbor set of x in G, and let $\mathfrak{p}_1 = \mathfrak{p}_a \cup N_1$. Notice that if $x \in \mathfrak{p}_a$, then x is not isolated in $I_{\mathfrak{p}}$, so $N_1 \subset \mathfrak{p}$ and thus $N_1 \subset \mathfrak{p}_a \cup I_b$. Define $\mathfrak{p}_2 = \mathfrak{p} \backslash \mathfrak{p}_1 = I_b \backslash N_1$, and $N_2 = \bigcup_{x \in \mathfrak{p}_1} N(x) \backslash \mathfrak{p}$. If G_1 is the induced subgraph of G on the vertices in $\mathfrak{p}_1 \cup N_2$, G_2 is the induced subgraph of G on vertices in $V \backslash \mathfrak{p}_1$, and

$I_i = I(G_i)$ for $i = 1, 2$, then $I_\mathfrak{p} = ((I_1)_{\mathfrak{p}_1}, (I_2)_{\mathfrak{p}_2})$ and $\mathfrak{p}_2$ is a minimal vertex cover of I_2. By design, any vertex appearing in both G_1 and G_2 is not in $\mathfrak{p}$, and thus $(I_1)_{\mathfrak{p}_1}$ and $(I_2)_{\mathfrak{p}_2}$ do not share a vertex. Thus by Lemma 4.11 and the fact that associated primes localize, $\mathfrak{p} \in \mathrm{Ass}(R_\mathfrak{p}/(I_\mathfrak{p})^t)$ if and only if $\mathfrak{p}_1 \in \mathrm{Ass}(R_1/(I_1)^t)$ and $\mathfrak{p}_1$ is the maximal ideal of $R_1 = K[x \mid x \in \mathfrak{p}_1]$. For convenience, define $R_a = K[x \mid x \in \mathfrak{p}_a]$.

Proposition 4.12. *Let $\mathfrak{p} \in \mathrm{Ass}(R/I^t)$. Using the notation from above, assume $\mathfrak{p}_1 = (I_1^t : c)$ for some monomial $c \in R_a$ of degree at most $2t-1$. Let $x \in \mathfrak{p}_1$. Let $\mathfrak{p}_1' = \mathfrak{p}_1 \cup N(x)$, and let $\mathfrak{p}_2'$ be any minimal vertex cover of the edges of G_2' where G_2' is the induced subgraph of G on the vertices $V\backslash\mathfrak{p}_1'$. Let $N_2' = \bigcup_{x\in\mathfrak{p}_1'} N(x)\backslash(\mathfrak{p}_1' \cup \mathfrak{p}_2')$. Let G_1' be the induced subgraph of G with vertices in $\mathfrak{p}_1' \cup N_2'$. Then $\mathfrak{p}' = (\mathfrak{p}_1', \mathfrak{p}_2') \in \mathrm{Ass}(R/I^{t+1})$.*

Proof. If v is an isolated vertex of G_2' then $N(v) \subset \mathfrak{p}_1'$ and thus every edge of G containing v is covered by $\mathfrak{p}_1'$. Hence $\mathfrak{p}'$ is a vertex cover of G. Since $x \in \mathfrak{p}_1$, there is an edge $xy \in G_1$ with $y \in \mathfrak{p}_a$. Consider $c' = cxy$. Then the degree of c' is at most $2t+1$, so $c' \notin (I_1)^{t+1}$. If $I_1' = I(G_1')$, then $c' \notin (I_1')^{t+1}$ as well. If $z \in \mathfrak{p}_1$, then $z(cxy) = (zc)(xy) \in (I_1')^{t+1}$. If $z \in N(x)$, then $z(cxy) = (cy)(zx) \in (I_1')^{t+1}$ since $y \in \mathfrak{p}_1$. Thus $\mathfrak{p}_1' \subset ((I_1')^{t+1} : c')$. Suppose $z \notin \mathfrak{p}_1'$ is a vertex of G_1'. Then $z \in N_2'$. Then $z \notin N(x)$ and $z \notin N(y)$ since $y \in \mathfrak{p}_a$, so zx and zy are not edges of G_1'. Also $z \notin \mathfrak{p}_1$ and $c \in R_a$, so $zc \notin (I_1')^{t+1}$. Thus the inclusion must be an equality.

Since $\mathfrak{p}_2'$ is a minimal vertex cover of the edges of G_2', then $\mathfrak{p}_2' \in \mathrm{Ass}(R/I_2')$ where I_2' is the edge ideal of G_2' (where isolated vertices of G_2' are not included in I_2'). Note that $I_{\mathfrak{p}'} = ((I_1')_{\mathfrak{p}_1'}, (I_2')_{\mathfrak{p}_2'})$ and so the result follows from Lemma 4.11. □

Note that if G contains an odd cycle of length $2k+1$, then embedded associated primes satisfying the hypotheses of Proposition 4.12 exist for $t \geq k+1$ by Lemma 4.9 and Corollary 4.10. Starting with an induced odd cycle C one can now recover all the primes described in [14, Theorem 3.3]. In addition, combining Corollary 4.10 with Lemma 4.11 as a starting place for Proposition 4.12 recovers the result from [14, Theorem 3.7] as well. Define $\mathrm{Ass}(R/I^t)^*$ to be the set of embedded associated primes of I^t produced in Proposition 4.12 by starting from any odd cycle, or collection of odd cycles, of the graph. Then $\mathrm{Ass}(R/I^t)^* \subset \mathrm{Ass}(R/I^s)$ for all $s \geq t$. To see this, recall that if $\mathfrak{p}$ is not a minimal prime, then there is a vertex x such that $x \cup N(x) \subset \mathfrak{p}$. Choosing such an x results in $\mathfrak{p}_1 = \mathfrak{p}_1'$ and the process shows that $\mathfrak{p} \in \mathrm{Ass}(R/I^{t+1})$. Notice also that the sets $\mathrm{Ass}(R/I^t)^*$ stabilize. In particular, $\mathrm{Ass}(R/I^t)^* = \mathrm{Ass}(R/I^n)^*$ for all $t \geq n$ where n is the number of variables. Notice that choosing $x \in N_1$ each time will eventually result in $\mathfrak{m} \in \mathrm{Ass}(R/I^t)^*$ for some t. Counting the maximal number of steps this could take provides a bound on the index of stability. Following the process above for a particular graph can often yield a significantly lower power M for which $\mathrm{Ass}(R/I^t)^*$ stabilize. These results are collected below.

Theorem 4.13. *Let I be the edge ideal of a connected graph G that is not bipartite. Suppose G has n vertices and s leaves, and N is the index of stability of I.*

(i) [14, *Theorem* 4.1] *The process used in Proposition* 4.12 *produces all embedded associated primes in the stable set. That is,* $\operatorname{Ass}(R/I^N) = \operatorname{Min}(R/I) \cup \operatorname{Ass}(R/I^N)^*$.

(ii) [14, *Corollary* 4.3], *(Proposition* 4.12*) If the smallest odd cycle of G has length $2k+1$, then $N \leq n-k-s$.*

(iii) [14, *Theorem* 5.6, *Corollary* 5.7] *If G has a unique odd cycle, then* $\operatorname{Ass}(R/I^t) = \operatorname{Min}(R/I) \cup \operatorname{Ass}(R/I^t)^*$ *for all t. Moreover, the sets* $\operatorname{Ass}(R/I^t)$ *form an ascending chain.*

(iv) *Suppose $\mathfrak{p} \in \operatorname{Ass}(R/I^N)$, and N_0 is the smallest positive integer for which $\mathfrak{p} \in \operatorname{Ass}(R/I^{N_0})^*$. Then $\mathfrak{p} \in \operatorname{Ass}(R/I^t)$ for all $t \geq N_0$.*

To interpret Theorem 4.13 in light of our earlier questions, notice that (i) answers Question 4.2, (ii) answers Question 4.1, and (iv) provides a good upper bound for $N_\mathfrak{p}$ in Question 4.3. The significance of (iii) is to answer a fourth question of interest. Before presenting that question, we first discuss some extensions of the above results to graphs containing loops.

Corollary 4.14. *Let I be a monomial ideal, not necessarily square-free, such that the generators of I have degree at most two. Define* $\operatorname{Ass}(R/I^t)^*$ *to be the set of embedded associate primes of I^t produced in Proposition* 4.12 *by starting from any odd cycle, or collection of odd cycles where generators of I that are not square-free are considered to be cycles of length one. Then the results of Theorem* 4.13 *hold for I.*

Proof. If I has generators of degree one, then write $I = (I_1, I_2)$ where I_2 is generated in degree one. Then I_2 is a complete intersection, so by using Lemma 4.11 we may replace I by I_1 and assume I is generated in degree two. If I is not square-free, consider a generator $x^2 \in I$. This generator can be represented as a loop (cycle of length one) in the graph. Define $\mathfrak{p}_a = (x)$ and $N_1 = N(x)$. Note that $\mathfrak{p}_1 = \mathfrak{p}_a \cup N_1 = (I_1 : c)$ where I_1 is the induced graph on $x \cup N(x)$ and $c = x$. Then $\mathfrak{p}_1$ satisfies the hypotheses of Proposition 4.12. The results now follow from the proof of Proposition 4.12. □

Notice that ideals that are not square-free will generally have embedded primes starting with $t = 1$ since the smallest odd cycle has length $1 = 2(0)+1$, so $k+1 = 1$. The above corollary can be extended to allow for any pure powers of variables to be generators of the ideal I.

Corollary 4.15. *Let $I = (I_1, I_2)$ where I_2 is the edge ideal of a graph G and $I_1 = (x_{i_1}^{s_1}, \ldots, x_{i_r}^{s_r})$ for any powers $s_j \geq 1$. Then the results of Theorem* 4.13 *hold for I with* $\operatorname{Ass}(R/I^t)^*$ *defined as in Corollary* 4.14.

Proof. As before, we may assume $x_j \geq 2$ for all j. Let $K = (x_{i_1}^2, \ldots, x_{i_r}^2)$ and let $J = (K, I_2)$. Then J satisfies the hypotheses of Corollary 4.14. Let $\mathfrak{p} \in \text{Ass}(R/J^t)^*$ be formed by starting with $\mathfrak{p}_a = (x_{i_1})$ and let $\mathfrak{p}_1 = (J_1^t : c)$ where J_1 and c are defined as in Corollary 4.14. Suppose $x_{i_1}, \ldots, x_{i_v} \in \mathfrak{p}_1$. Let $q_j \geq 0$ be the least integers such that $c' = x_{i_1}^{q_1} \cdots x_{i_v}^{q_v} \cdot c \notin I_1^t$. Then it is straightforward to check that $\mathfrak{p}_1 = (I_1^t : c')$ and so $\mathfrak{p} \in \text{Ass}(R/I^t)^*$. Thus higher powers of variables can also be treated as loops and the results of Theorem 4.13 hold. □

We now return to Theorem 4.13 (iii). In general, the sets $\text{Ass}(R/I^t)^*$ form an ascending chain. Theorem 4.13 (iii) gives a class of graphs for which $\text{Ass}(R/I^t)^*$ describe every embedded prime of a power of I^t optimally. Thus $\text{Ass}(R/I^t)$ will form a chain. This happens for many classes of monomial ideals, and leads to the fourth question.

Question 4.16. If I is a square-free monomial ideal, is $\text{Ass}(R/I^t) \subset \text{Ass}(R/I^{t+1})$ for all t?

For monomial ideals, Question 4.16 is of interest for low powers of I. For sufficiently large powers, the sets of associated primes are known to form an ascending chain, and a bound beyond which the sets $\text{Ass}(I^t/I^{t+1})$ form a chain has been shown by multiple authors (see [85, 99]). This bound depends on two graded algebras which encode information on the powers of I, and which will prove useful in other results. The first is the *Rees algebra* $R[It]$ of I, which is defined by

$$R[It] = R \oplus It \oplus I^2t^2 \oplus I^3t^3 \oplus \cdots$$

and the second is the *associated graded ring* of I,

$$\text{gr}_I(R) = R/I \oplus I/I^2 \oplus I^2/I^3 \oplus \cdots .$$

Notice that while the result is for $\text{Ass}(I^t/I^{t+1})$, for monomial ideals $\text{Ass}(R/I^{t+1})$ will also form a chain.

Theorem 4.17 ([85, 99]). $\text{Ass}(I^t/I^{t+1})$ *is increasing for* $t > a^0_{R[It]_+}(\text{gr}_I(R))$.

Note that square-free is essential in Question 4.16. Examples of monomial ideals for which the associated primes do not form an ascending chain have been given in [59, 65]. Those examples were designed for other purposes and so are more complex than what is needed here. A simple example can be found by taking the product of consecutive edges of an odd cycle.

Example 4.18. Let $I = (x_1x_2^2x_3, x_2x_3^2x_4, x_3x_4^2x_5, x_4x_5^2x_1, x_5x_1^2x_2)$, and let $\mathfrak{m} = (x_1, x_2, x_3, x_4, x_5)$. Then $\mathfrak{m} \in \text{Ass}(R/I^t)$ for $t = 1, 4$, but $\mathfrak{m} \notin \text{Ass}(R/I^t)$ for $t = 2, 3$.

The ideal in Example 4.18 can be viewed as multiplying adjacent edges in a 5-cycle to form generators of I, and so has a simple combinatorial realization. A similar result holds for longer odd cycles, where the maximal ideal is not the only associate prime to appear and disappear. However, if instead

$$I = (x_1x_2x_3, x_2x_3x_4, x_3x_4x_5, x_4x_5x_1, x_5x_1x_2)$$

is the path ideal of the pentagon, then $\mathrm{Ass}(R/I^t) = \mathrm{Min}(R/I) \cup \{\mathfrak{m}\}$ for $t \geq 2$ and thus $\mathrm{Ass}(R/I^t)$ form an ascending chain (see [51, Example 3.14]).

There are some interesting cases where associated primes are known to form ascending chains. The first listed is quite general, but has applications to square-free monomial ideals.

Theorem 4.19 ([84, Proposition 3.9], see also [56, Proposition 16.3]). *If R is a Noetherian ring, then* $\mathrm{Ass}(R/\overline{I^t})$ *form an ascending chain.*

In order to present the next class of ideals for which the associated primes are known to form ascending chains, we first need some some background definitions.

Definition 4.20. Let G be a graph. A *colouring* of the vertices of G is an assignment of colours to the vertices of G such that adjacent vertices have distinct colours. The *chromatic number* of G is the minimal number of colours in a colouring of G. A graph is called *perfect* if for every induced subgraph H, the chromatic number of H equals the size of the largest complete subgraph of H.

An excellent reference for the theory of perfect graphs is the book of Golumbic [48]. Using perfect graphs, we now give an example to show how Theorem 4.19 can be applied to classes of square-free monomial ideals. An alternate proof appears in [42, Corollary 5.11].

Example 4.21. If I is the ideal of minimal vertex covers of a perfect graph, then $\mathrm{Ass}(R/I^t)$ form an ascending chain.

Proof. By [112, Theorem 2.10], $R[It]$ is normal. Thus $I^t = \overline{I^t}$ for all t, so by Theorem 4.19, $\mathrm{Ass}(R/I^t)$ form an ascending chain. □

Similar results hold for other classes of monomial ideals for which $R[It]$ is known to be normal. For example, in [15, Corollary 4.2] it is shown that a path ideal of a rooted tree has a normal Rees algebra. Thus by Theorem 4.19, $\mathrm{Ass}(R/I^t)$ form an ascending chain for such ideals. Note that a path ideal can be viewed as the edge ideal of a carefully chosen uniform clutter.

It is interesting to compare the result of Example 4.21 to [42, Theorem 5.9], where it is shown that if I is the ideal of minimal vertex covers of a perfect graph, then the set of primes associated to any fixed power has the saturated chain property. Here $\mathrm{Ass}(R/I^t)$ has the saturated chain property if for every $\mathfrak{p} \in \mathrm{Ass}(R/I^t)$, either $\mathfrak{p}$ is minimal or there is a $Q \subsetneq \mathfrak{p}$ with $Q \in \mathrm{Ass}(R/I^t)$ and height $Q =$ height $\mathfrak{p} - 1$.

Notice that Theorem 4.13 shows that in the case of a graph with a unique odd cycle, Question 4.16 has an affirmative answer. This result can be generalized to any graph containing a leaf. First we need a slight variation of a previously known result.

Lemma 4.22 ([88, Lemma 2.3]). *Suppose $I = I(G)$ is the edge ideal of a graph and $a \in I/I^2$ is a regular element of the associated graded ring $\mathrm{gr}_I(R)$. Then the sets $\mathrm{Ass}(R/I^t)$ form an ascending chain. Moreover, $\mathrm{Ass}(R/I^t) = \mathrm{Ass}(I^{t-1}/I^t)$ for all $t \geq 1$.*

Proof. Let $a \in I/I^2$ be a regular element of $\mathrm{gr}_I(R)$. Assume $\mathfrak{p} \in \mathrm{Ass}(I^t/I^{t+1})$. Then there is a $c \in I^t/I^{t+1}$ with $\mathfrak{p} = (0 :_{R/I} c)$. But then $\mathfrak{p} = (0 :_{R/I} ac)$, and a lives in degree one, so $\mathfrak{p} \in \mathrm{Ass}(I^{t+1}/I^{t+2})$. So these sets form an ascending chain. Now the standard short exact sequence

$$0 \to I^t/I^{t+1} \to R/I^{t+1} \to R/I^t \to 0$$

gives

$$\mathrm{Ass}(I^t/I^{t+1}) \subset \mathrm{Ass}(R/I^{t+1}) \subset \mathrm{Ass}(R/I^t) \cup \mathrm{Ass}(I^t/I^{t+1})$$

and the result follows by induction. □

Proposition 4.23. *Let G be a graph containing a leaf x and let $I = I(G)$ be its edge ideal. Then $\mathrm{Ass}(R/I^t) \subset \mathrm{Ass}(R/I^{t+1})$ for all t. That is, the sets of associated primes of the powers of I form an ascending chain.*

Proof. Since x is a leaf of G, there is a unique generator $e = xy \in I$ divisible by x. Let a denote the image of e in I/I^2. We claim a is a regular element of $\mathrm{gr}_I(R)$. To see this, it suffices to show that if $fe \in I^{t+1}$ for some t, then $f \in I^t$. Since I is a monomial ideal and e is a monomial, $fe \in I^{t+1}$ if and only if every term of fe is in I^{t+1}. Thus we may assume f is a monomial and $fxy = e_1 e_2 \cdots e_{t+1} h$ for some edges e_i of G and some monomial h. Suppose $f \notin I^t$. Then x divides e_i for some i, say $i = t + 1$. Since x is a leaf, $e_i = xy$ and by cancellation $f = e_1 \cdots e_t h \in I^t$. Thus a is a regular element of $\mathrm{gr}_I(R)$ and by Lemma 4.22 the result follows. □

When extending the above results to more general square-free monomial ideals, one needs to pass from graphs to clutters. An obstruction to extending the results is the lack of an analog to Theorem 4.8 [100, Theorem 5.9]. One possible analog appears as a conjecture of Conforti and Cornuéjols, see [17, Conjecture 1.6], which we discuss later in this section. This conjecture is stated in the language of combinatorial optimization. It says that a clutter $\mathcal{C}$ has the *max-flow min-cut* (MFMC, see Definition 4.30) property if and only if $\mathcal{C}$ has the *packing* property. These criterion for clutters have been shown in recent years to have algebraic translations [45] which will be discussed in greater detail later in the section. An ideal satisfies the packing property if the monomial grade of I (see Definition 4.29) is equal to the height of I and this same equality holds for every minor of I [17,45]. Here a minor is formed by either localizing at a collection of variables, passing to the image of I in a quotient ring $R/(x_{i_1}, x_{i_2}, \dots, x_{i_s})$, or a combination of the two. In [47, Corollary 3.14] and [62, Corollary 1.6], it was shown that $\mathcal{C}$ satisfies MFMC if and only if the corresponding edge ideal $I(\mathcal{C})$ is normally torsion-free. This allows the conjecture to be restated (cf. [45, Conjecture 4.18]) as: if $\mathcal{C}$ has the packing property, then $I(\mathcal{C})$ is normally torsion-free.

Since a proof of this conjecture does not yet exist, the techniques used to describe the embedded associated primes, the stable set of associated primes, and the index of stability for graphs are difficult to extend. However some partial results are known. The first gives some conditions under which it is known that the maximal ideal is, or is not, an associated prime. In special cases, this can provide a seed for additional embedded associated primes using techniques such as those in Proposition 4.12.

Theorem 4.24. *If I is a square-free monomial ideal, every proper minor of I is normally torsion-free, and β_1 is the monomial grade of I, then*

(i) [51, *Corollary* 3.6] $\mathfrak{m} \notin \operatorname{Ass}(R/I^t)$ *for* $t \leq \beta_1$.

(ii) [51, *Theorem* 4.6] *If I fails the packing property, then* $\mathfrak{m} \in \operatorname{Ass}(R/I^{\beta_1+1})$.

(iii) [51, *Proposition* 3.9] *If I is unmixed and satisfies the packing property, then I is normally torsion-free.*

Other recent results have taken a different approach. Instead of working directly with the edge ideal of a clutter $\mathcal{C}$, one can work with its Alexander dual, which is again the edge ideal of a clutter. Using this approach, the embedded associated primes of the Alexander dual have been linked to colorings of a clutter. Recall that $\chi(\mathcal{C})$ is the minimal number d for which there is a partition $X_1, \dots, X_d$ of the vertices of $\mathcal{C}$ for which for all edges f of $\mathcal{C}$, $f \not\subseteq X_i$ for every i. A clutter is *critically d-chromatic* if $\chi(\mathcal{C}) = d$ but $\chi(\mathcal{C}\backslash\{x\}) < d$ for every vertex x.

Theorem 4.25. (i) [42, Corollary 4.6] *If I is the ideal of covers of a clutter $\mathcal{C}$, and if the induced subclutter $\mathcal{C}_\mathfrak{p}$ on the vertices in $\mathfrak{p}$ is critically $(d+1)$-chromatic, then* $\mathfrak{p} \in \operatorname{Ass}(R/I^d)$ *but* $\mathfrak{p} \notin \operatorname{Ass}(R/I^t)$ *for any* $t \leq d-1$.

(ii) [42, *Theorem* 5.9] *If* I *is the ideal of covers of a perfect graph* G*, then* $\mathfrak{p} \in \mathrm{Ass}(R/I^t)$ *if and only if the induced graph on the vertices in* $\mathfrak{p}$ *is a clique of size at most* $t+1$.

If one restricts to a particular power, then additional results on embedded associate primes are known. For example, in [41, Corollary 3.4] it is shown that if I is the edge ideal of the Alexander dual of a graph G, then embedded primes of R/I^2 are in one-to-one correspondence with induced odd cycles of G. More precisely, $\mathfrak{p} \in \mathrm{Ass}(R/I^2)$ is an embedded prime if and only if the induced subgraph of G on the vertices in $\mathfrak{p}$ is an induced odd cycle of G.

An interesting class of ideals, which is in a sense dual to the edge ideals of graphs, is unmixed square-free monomial ideals of height two. These are the Alexander duals of edge ideals of graphs, which can be viewed as edge ideals of clutters where, instead of requiring that each edge has two vertices, it is instead required that each minimal vertex cover has two vertices. For such ideals it has been shown in [40, Theorem 1.2] that an affirmative answer to a conjecture on graph colorings, [40, Conjecture 1.1], would imply an affirmative answer to Question 4.16. In [40, Corollary 3.11] it is shown that this conjecture holds for cliques, odd holes, and odd antiholes. Thus the Alexander duals of these special classes of graphs provide additional examples where Question 4.16 has an affirmative answer.

We now provide a more detailed discussion of the Conforti–Cornuéjols conjecture, followed by a collection of results which provide families of clutters where the conjecture is known to be true (such a family was already given in Theorem 4.24 (iii)). We also discuss some algebraic versions of this conjecture and how it relates to the depth of powers of edge ideals and to normality and torsion-freeness.

Having defined the notion of a *minor* for edge ideals, using the correpondence between clutters and square-free monomias ideals, we also have the notion of a *minor* of a clutter. We say that $\mathcal{C}$ has the *packing property* if $I(\mathcal{C})$ has this property.

Definition 4.26. Let A be the incidence matrix of a clutter $\mathcal{C}$. The *set covering polyhedron* is the rational polyhedron:

$$Q(A) = \{x \in \mathbb{R}^n \mid x \geq \mathbf{0},\ xA \geq \mathbf{1}\},$$

where $\mathbf{0}$ and $\mathbf{1}$ are vectors whose entries are equal to 0 and 1 respectively. Often we denote the vectors $\mathbf{0}$, $\mathbf{1}$ simply by 0, 1. We say that $Q(A)$ is *integral* if it has only integral vertices.

Theorem 4.27 (A. Lehman [78; 17, Theorem 1.8]). *If a clutter* $\mathcal{C}$ *has the packing property, then* $Q(A)$ *is integral.*

The converse is not true. A famous example is the clutter $\mathcal{Q}_6$, given below. It does not pack and $Q(A)$ is integral.

Example 4.28. Let $I = (x_1x_2x_5, x_1x_3x_4, x_2x_3x_6, x_4x_5x_6)$. The figure:

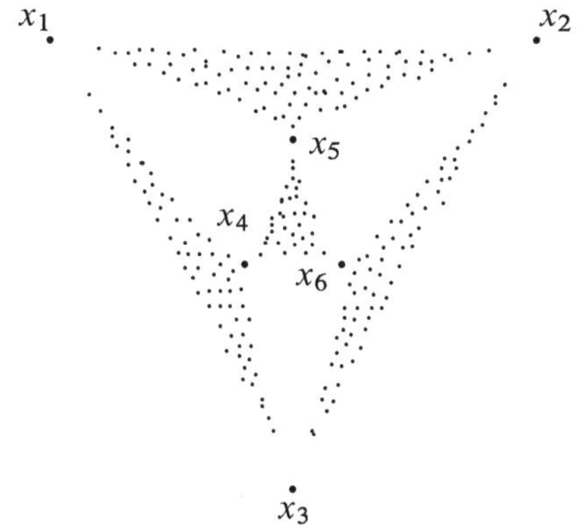

corresponds to the clutter associated to I. This clutter will be denoted by $\mathcal{Q}_6$. Using *Normaliz* [11] we obtain that $\overline{R[It]} = R[It][x_1 \cdots x_6t^2]$. Thus $R[It]$ is not normal. An interesting property of this example is that $\mathrm{Ass}(R/\overline{I^i}) = \mathrm{Ass}(R/I)$ for all i (see [45]).

Definition 4.29. A set of edges of a clutter $\mathcal{C}$ is *independent* if no two of them have a common vertex. We denote the maximum number of independent edges of $\mathcal{C}$ by $\beta_1(\mathcal{C})$. We call $\beta_1(\mathcal{C})$ the *edge independence number* of $\mathcal{C}$ or the *monomial grade* of I.

Let A be the incidence matrix of $\mathcal{C}$. The edge independence number and the covering number are related to min-max problems because they satisfy:

$$\begin{aligned}\alpha_0(\mathcal{C}) &\geq \min\{\langle 1, x\rangle \mid x \geq 0; xA \geq 1\}\\ &= \max\{\langle y, 1\rangle \mid y \geq 0; Ay \leq 1\} \geq \beta_1(\mathcal{C}).\end{aligned}$$

Notice that $\alpha_0(\mathcal{C}) = \beta_1(\mathcal{C})$ if and only if both sides of the equality have integral optimum solutions.

Definition 4.30. A clutter $\mathcal{C}$, with incidence matrix A, satisfies the *max-flow min-cut* (MFMC) property if both sides of the LP-duality equation

$$\min\{\langle \alpha, x\rangle \mid x \geq 0; xA \geq 1\} = \max\{\langle y, 1\rangle \mid y \geq 0; Ay \leq \alpha\} \tag{4.1}$$

have integral optimum solutions x and y for each non-negative integral vector α. The system $x \geq 0; xA \geq 1$ is called *totally dual integral* (TDI) if the maximum in (4.1) has an integral optimum solution y for each integral vector α with finite maximum.

Definition 4.31. If $\alpha_0(\mathcal{C}) = \beta_1(\mathcal{C})$ we say that the clutter $\mathcal{C}$ (or the ideal I) has the *König property*.

Note that $\mathcal{C}$ has the packing property if and only if every minor of $\mathcal{C}$ satisfies the König property. This leads to the following well-known result.

Corollary 4.32 ([17]). *If a clutter $\mathcal{C}$ has the max-flow min-cut property, then $\mathcal{C}$ has the packing property.*

Proof. Assume that the clutter $\mathcal{C}$ has the max-flow min-cut property. This property is closed under taking minors. Thus it suffices to prove that $\mathcal{C}$ has the König property. We denote the incidence matrix of $\mathcal{C}$ by A. By hypothesis the LP-duality equation

$$\min\{\langle 1, x\rangle \mid x \geq 0; xA \geq 1\} = \max\{\langle y, 1\rangle \mid y \geq 0; Ay \leq 1\}$$

has optimum integral solutions x, y. To complete the proof notice that the left hand side of this equality is $\alpha_0(\mathcal{C})$ and the right-hand side is $\beta_1(\mathcal{C})$. □

Conforti and Cornuéjols [16] conjecture that the converse is also true.

Conjecture 4.33 (Conforti–Cornuéjols). If a clutter $\mathcal{C}$ has the packing property, then $\mathcal{C}$ has the max-flow min-cut property.

An algebraic description of the packing property has already been given. In order to use algebraic techniques to attack this combinatorial conjecture, an algebraic translation is needed for the max-flow min-cut property. There are several equivalent algebraic descriptions of the max-flow min-cut property, as seen in the following result.

Theorem 4.34 ([35,47,69]). *Let $\mathcal{C}$ be a clutter and let I be its edge ideal. The following conditions are equivalent*:

(i) $\mathrm{gr}_I(R)$ *is reduced.*

(ii) $R[It]$ *is normal and* $Q(A)$ *is an integral polyhedron.*

(iii) $x \geq 0$; $xA \geq 1$ *is a* TDI system.

(iv) $\mathcal{C}$ *has the max-flow min-cut property.*

(v) $I^i = I^{(i)}$ *for* $i \geq 1$, *where* $I^{(i)}$ *is the* i*th symbolic power.*

(vi) I *is normally torsion-free.*

By Theorems 4.34 and 4.27, Conjecture 4.33 reduces to:

Conjecture 4.35 ([45]). If I has the packing property, then $R[It]$ is normal.

Several variations of condition (ii) of Theorem 4.34 are possible. In particular, there are combinatorial conditions on the clutter that can be used to replace the normality of the Rees algebra. One such condition is defined below.

Definition 4.36. Let $\mathcal{C}^\vee$ be the clutter of minimal vertex covers of $\mathcal{C}$. The clutter $\mathcal{C}$ is called *diadic* if $|e \cap e'| \leq 2$ for $e \in E(\mathcal{C})$ and $e' \in E(\mathcal{C}^\vee)$

Proposition 4.37 ([45]). *If $Q(A)$ is integral and $\mathcal{C}$ is diadic, then I is normally torsion-free.*

Theorem 4.34 can be used to exhibit families of normally torsion-free ideals. Recall that a matrix A is called *totally unimodular* if each $i \times i$ subdeterminant of A is 0 or ± 1 for all $i \geq 1$.

Corollary 4.38. *If A is totally unimodular, then I and $I^\vee$ are normally torsion-free.*

Proof. By [97] the linear system $x \geq 0$; $xA \geq 1$ is TDI. Hence I is normally torsion-free by Theorem 4.34. Let $\mathcal{C}^\vee$ be the blocker (or Alexander dual) of $\mathcal{C}$. By [97, Corollary 83.1a (v), page 1441], we get that $\mathcal{C}^\vee$ satisfies the max-flow min-cut property. Hence $I(\mathcal{C}^\vee)$ is normally torsion-free by Theorem 4.34. Thus $I^\vee$ is normally torsion-free because $I(\mathcal{C}^\vee) = I^\vee$. □

In particular if I is the edge ideal of a bipartite graph, then I and $I^\vee$ are normally torsion-free.

Theorem 4.34 shows that the Rees algebra and the associated graded ring play an important role in the study of the max-flow min-cut property. An invariant related to the blowup algebras will also be useful. The *analytic spread* of an edge ideal I is given by $\ell(I) = \dim R[It]/\mathfrak{m}R[It]$. If $\mathcal{C}$ is uniform, the analytic spread of I is the rank of the incidence matrix of $\mathcal{C}$. The analytic spread of a monomial ideal can be computed in terms of the Newton polyhedron of I, see [4]. The next result follows directly from [83, Theorem 3].

Proposition 4.39. *If $Q(A)$ is integral, then $\ell(I) < n = \dim(R)$.*

To relate this result on $\ell(I)$ to Conjecture 4.33 (or equivalently to Conjecture 4.35) we first need to recall the following bound on the depths of the powers of an ideal I.

Theorem 4.40. $\inf_i\{\mathrm{depth}(R/I^i)\} \leq \dim(R) - \ell(I)$. *If $\mathrm{gr}_I(R)$ is Cohen–Macaulay, then the equality holds.*

This inequality is due to Burch [12] (cf. [70, Theorem 5.4.7]), while the equality comes from [31]. By a result of Brodmann [8], depth R/I^k is constant for $k \gg 0$. Broadmann improved Burch's inequality by showing that the constant value is bounded by $\dim(R) - \ell(I)$. For a study of the initial and limit behaviour of the numerical function $f(k) = \mathrm{depth}\, R/I^k$ see [59].

Theorem 4.41 ([68]). *Let R be a Cohen–Macaulay ring and let I be an ideal of R containing regular elements. If $R[It]$ is Cohen–Macaulay, then $\mathrm{gr}_I(R)$ is Cohen–Macaulay.*

Proposition 4.42. *Let $\mathcal{C}$ be a clutter and let I be its edge ideal. Let J_i be the ideal obtained from I by making $x_i = 1$. If $Q(A)$ is integral, then I is normal if and only if J_i is normal for all i and* $\text{depth}(R/I^k) \geq 1$ *for all $k \geq 1$.*

Proof. Assume that I is normal. The normality of an edge ideal is closed under taking minors [35], hence J_i is normal for all i. By hypothesis the Rees algebra $R[It]$ is normal. Then $R[It]$ is Cohen–Macaulay by a theorem of Hochster [66]. Then the ring $\text{gr}_I(R)$ is Cohen–Macaulay by Theorem 4.41. Hence using Theorem 4.40 and Proposition 4.39 we get that $\text{depth}(R/I^i) \geq 1$ for all i. The converse follows readily adapting the arguments given in the proof of the normality criterion presented in [35]. □

By Proposition 4.42 and Theorem 4.27, we get that Conjecture 4.33 also reduces to:

Conjecture 4.43. If I has the packing property, then $\text{depth}(R/I^i) \geq 1$ for all $i \geq 1$.

We conclude this section with a collection of results giving conditions under which Conjecture 4.33, or its equivalent statements mentioned above, is known to hold. For uniform clutters it suffices to prove Conjecture 4.33 for Cohen–Macaulay clutters [23].

Proposition 4.44 ([45]). *Let $\mathcal{C}$ be the collection of bases of a matroid. If $\mathcal{C}$ satisfies the packing property, then $\mathcal{C}$ satisfies the max-flow min-cut property.*

When G is a graph, integrality of $Q(A)$ is sufficient in condition (ii) of Theorem 4.34, and the packing property is sufficient to imply the max-flow min-cut property, thus providing another class of examples for which Conjecture 4.33 holds.

Proposition 4.45 ([17, 45]). *If G is a graph and $I = I(G)$, then the following are equivalent:*

(i) $\text{gr}_I(R)$ *is reduced.*

(ii) *G is bipartite.*

(iii) *$Q(A)$ is integral.*

(iv) *G has the packing property.*

(v) *G has the max-flow min-cut property.*

(vi) $\overline{I^i} = I^{(i)}$ *for $i \geq 1$.*

Definition 4.46. A clutter is *binary* if its edges and its minimal vertex covers intersect in an odd number of vertices.

Theorem 4.47 ([98]). *A binary clutter $\mathcal{C}$ has the max-flow min-cut property if and only if $\mathcal{Q}_6$ is not a minor of $\mathcal{C}$.*

Corollary 4.48. *If $\mathcal{C}$ is a binary clutter with the packing property, then $\mathcal{C}$ has the max-flow min-cut property.*

Proposition 4.49 ([112]). *Let $\mathcal{C}$ be a uniform clutter and let A be its incidence matrix. If the polyhedra*

$$P(A) = \{x \,|\, x \geq 0;\ xA \leq 1\} \quad \text{and} \quad Q(A) = \{x \,|\, x \geq 0;\ xA \geq 1\}$$

are integral, then $\mathcal{C}$ has the max-flow min-cut property.

In light of Theorem 4.27, this result implies that if $P(A)$ is integral and $\mathcal{C}$ has the packing property, then $\mathcal{C}$ has the max-flow min-cut property. An open problem is to show that this result holds for non-uniform clutters (see [82, Conjecture 1.1]).

A *Meyniel graph* is a simple graph in which every odd cycle of length at least five has at least two chords. The following gives some support to [82, Conjecture 1.1] because Meyniel graphs are perfect [97, Theorem 66.6].

Theorem 4.50 ([82]). *Let $\mathcal{C}$ be the clutter of maximal cliques of a Meyniel graph. If $\mathcal{C}$ has the packing property, then $\mathcal{C}$ has the max-flow min-cut property.*

Let $P = (X, \prec)$ be a *partially ordered set* (*poset* for short) on the finite vertex set X and let G be its *comparability graph.* Recall that the vertex set of G is X and the edge set of G is the set of all unordered pairs $\{x_i, x_j\}$ such that x_i and x_j are comparable.

Theorem 4.51 ([25]). *If G is a comparability graph and $\mathcal{C}$ is the clutter of maximal cliques of G, then the edge ideal $I(\mathcal{C})$ is normally torsion free.*

Theorem 4.52 ([24]). *Let $\mathcal{C}$ be a uniform clutter with a perfect matching such that $\mathcal{C}$ has the packing property and $\alpha_0(\mathcal{C}) = 2$. If the columns of the incidence matrix of $\mathcal{C}$ are linearly independent, then $\mathcal{C}$ has the max-flow min-cut property.*

Acknowledgments. The software package *Macaulay2* [49] was used to compute many of the examples in this paper, including Example 4.18, which was originally discovered by three undegraduate students (Mike Alwill, Katherine Pavelek, and Jennifer von Reis) in an unpublished project in 2002. The authors would like to thank an anonymous referee for providing us with useful comments and suggestions.

Bibliography

[1] Barile, M., On the arithmetical rank of the edge ideals of forests. Comm Algebra. 2008; 36(12):4678–4703.

[2] Berge, C., Some common properties for regularizable graphs, edge-critical graphs and B-graphs. In: Rosa, A., Sabidussi, G., Turgeon, J., editors. Theory and practice of combinatorics. vol. 60 of North-Holland mathematics studies. Amsterdam: North-Holland; 1982. p. 31–44.

[3] Bermejo, I., Gimenez, P., Saturation and Castelnuovo–Mumford regularity. J. Algebra. 2006;303(2):592–617.

[4] Bivià-Ausina, C., The analytic spread of monomial ideals. Comm Algebra. 2003;31(7): 3487–3496.

[5] Björner, A., Topological methods. In: Handbook of combinatorics, Vol. 1, 2. Amsterdam: Elsevier; 1819–1872, 1995.

[6] Björner, A., Wachs, M., Shellable nonpure complexes and posets I, Trans Amer Math Soc. 1996;348:1299–1327.

[7] Bollobás, B., Modern graph theory. vol. 184 ofGraduate texts in mathematics 184. New York: Springer-Verlag; 1998.

[8] Brodmann, M., The asymptotic nature of the analytic spread. Math Proc Cambridge Philos Soc. 1979;86:35–39.

[9] Brodmann, M., Asymptotic stability of Ass$(M/I^n M)$. Proc Amer Math Soc. 1979;74: 16–18.

[10] Bruns, W., Herzog, J., Cohen–Macaulay rings. Revised ed. Cambridge: Cambridge University Press; 1997.

[11] Bruns, W., Ichim, B., NORMALIZ 2.0, Computing normalizations of affine semigroups. 2008. Available from: http://www.math.uos.de/normaliz.

[12] Burch, L., Codimension and analytic spread. Proc Cambridge Philos Soc. 1972;72:369–373.

[13] Busch, A. H., Dragan, F. F., Sritharan, R., New min-max theorems for weakly chordal and dually chordal graphs. vol. 6509 of Lecture notes in computer science. Springer-Verlag; 2010. p. 207–218.

[14] Chen, J., Morey, S., Sung, A., The stable set of associated primes of the ideal of a graph. Rocky Mountain J Math. 2002;32:71–89.

[15] Conca, A., De Negri, E., M-sequences, graph ideals, and ladder ideals of linear type. J Algebra. 1999;211:599–624.

[16] Conforti, M., Cornuéjols, G., Clutters that pack and the Max-Flow Min-Cut property: a conjecture. In: Pulleyblank, W. R., Shepherd, F. B., editors. The Fourth Bellairs Workshop on Combinatorial Optimization. 1993.

[17] Cornuéjols, G., Combinatorial optimization: packing and covering. vol. 74 of CBMS-NSF regional conference series in applied mathematics. SIAM; 2001.

[18] Corso, A., Nagel, U., Monomial and toric ideals associated to ferrers graphs. Trans Amer Math Soc. 2009;361(3):1371–1395.

[19] Dalili, K., Kummini, M., Dependence of betti numbers on characteristic. Preprint. arXiv:1009.4243v1. 2010.

[20] Dalili, K., Faridi, S., Traves, W., The reconstruction conjecture and edge ideals. Discrete Math. 2008;308:2002–2010.

[21] Dirac, G. A., On rigid circuit graphs. Abh Math Semin Univ Hambg. 1961;38:71–76.

[22] Dochtermann, A., Engstrom, A., Algebraic properties of edge ideals via combinatorial topology. Electron J Combin. 2009;16(2):R2.

[23] Dupont, L. A., Reyes, E., Villarreal, R. H., Cohen–Macaulay clutters with combinatorial optimization properties and parallelizations of normal edge ideals. São Paulo J Math Sci. 2009;3(1):61–75.

[24] Dupont, L. A., Villarreal, R. H., Algebraic and combinatorial properties of ideals and algebras of uniform clutters of TDI systems. J Comb Optim. 2011;21(3):269–292.

[25] Dupont, L. A., Villarreal, R. H., Edge ideals of clique clutters of comparability graphs and the normality of monomial ideals. Math Scand. 2010;106(1):88–98.

[26] Duval, A. M., Algebraic shifting and sequentially Cohen–Macaulay simplicial complexes. Electron J Combin. 1996;3(1):Research paper 21.

[27] Eagon, J. A., Reiner, V., Resolutions of Stanley–Reisner rings and Alexander duality. J Pure Appl Algebra. 1998;130:265–275.

[28] Eisenbud, D., Commutative algebra with a view toward algebraic geometry. vol. 150 of Graduate texts in mathematics. Springer-Verlag; 1995.

[29] Eisenbud, D., The geometry of syzygies: a second course in commutative algebra and algebraic geometry. vol. 229 of Graduate texts in mathematics. New York: Springer; 2005.

[30] Eisenbud, D., Goto, S., Linear free resolutions and minimal multiplicity. J Algebra. 1984;88(1):189–133.

[31] Eisenbud, D., Huneke, C., Cohen–Macaulay Rees algebras and their specializations. J Algebra. 1983;81:202–224.

[32] Emtander, E., A class of hypergraphs that generalizes chordal graphs. Math Scand. 2010;106(1):50–66.

[33] Ene, V., Olteanu, O., Terai, N., Arithmetical rank of lexsegment edge ideals. Bull Math Soc Sci Math Roumanie (N S). 2010;53(4):315–327.

[34] Erdős, P., Gallai, T., On the minimal number of vertices representing the edges of a graph. Magyar Tud Akad Mat Kutató Int Közl. 1961;6:181–203.

[35] Escobar, C., Villarreal, R. H., Yoshino, Y., Torsion freeness and normality of blowup rings of monomial ideals. In: Commutative algebra. vol. 244 of Lecture notes in pure and applied mathematics. Boca Raton, FL: Chapman & Hall/CRC; 2006. p. 69–84.

[36] Estrada, M., Villarreal, R. H., Cohen–Macaulay bipartite graphs. Arch Math. 1997;68: 124–128.

[37] Farber, M., Characterizations of strongly chordal graphs. Discrete Math. 1983;43(2–3): 173–189.

[38] Faridi, S., Cohen–Macaulay properties of square-free monomial ideals. J Combin Theory Ser A. 2005;109(2):299–329.

[39] Faridi, S., Simplicial trees are sequentially Cohen–Macaulay. J Pure Appl Algebra. 2004;190:121–136.

[40] Francisco, C., Hà, H. T., Van Tuyl, A., A conjecture on critical graphs and connections to the persistence of associated primes. Discrete Math. 2010;310:2176–2182.

[41] Francisco, C., Hà, H. T., Van Tuyl, A., Associated primes of monomial ideals and odd holes in graphs. J Algebraic Combin. 2010;32:287–301.

[42] Francisco, C., Hà, H. T., Van Tuyl, A., Colorings of hypergraphs, perfect graphs, and associated primes of powers of monomial ideals. J Algebra. 2011;331(1):224–242.

[43] Francisco, C. A., Van Tuyl, A., Sequentially Cohen–Macaulay edge ideals. Proc Amer Math Soc. 2007;135:2327–2337.

[44] Fröberg, R., On Stanley–Reisner rings. In: Balcerzyk, S., Józefiak, T., Krempa, J., Simson, D., Vogel, W., editors. Topics in algebra, Part 2. Polish Scientific Publishers; 1990. p. 57–70.

[45] Gitler, I., Reyes, E., Villarreal, R. H., Blowup algebras of square-free monomial ideals and some links to combinatorial optimization problems. Rocky Mountain J Math. 2009; 39(1):71–102.

[46] Gitler, I., Valencia, C., Bounds for invariants of edge-rings. Comm Algebra. 2005;33: 1603–1616.

[47] Gitler, I., Valencia, C., Villarreal, R. H., A note on Rees algebras and the MFMC property. Beiträge Algebra Geom. 2007;48(1):141–150.

[48] Golumbic, M. C., Algorithmic graph theory and perfect graphs, second edition. vol. 57 of Annals of discrete mathematics. Amsterdam: Elsevier Science B.V.; 2004.

[49] Grayson, D. R., Stillman, M. E., Macaulay2, a software system for research in algebraic geometry. 1996. Available from: http://www.math.uiuc.edu/Macaulay2/.

[50] Greuel, G. M., Pfister, G., A singular introduction to commutative algebra. 2nd extended ed. Berlin: Springer; 2008.

[51] Hà, H. T., Morey, S., Embedded associated primes of powers of square-free monomial ideals. J Pure Appl Algebra. 2010;214(4):301–308.

[52] Hà, H. T., Morey, S., Villarreal, R. H., Cohen–Macaulay admissible clutters. J Commut Algebra. 2009;1(3):463–480.

[53] Hà, H. T., Van Tuyl, A., Monomial ideals, edge ideals of hypergraphs, and their graded Betti numbers. J Algebraic Combin. 2008;27:215–245.

[54] Harary, F., Graph theory. Reading, MA: Addison-Wesley; 1972.

[55] He, J., Van Tuyl, A., Algebraic properties of the path ideal of a tree. Comm Algebra. 2010;38(5):1725–1742.

[56] Herrmann, M., Ikeda, S., Orbanz, U., Equimultiplicity and blowing up. An algebraic study. Berlin: Springer-Verlag; 1988.

[57] Herzog, J., A Generalization of the Taylor complex construction. Comm Algebra. 2007; 35(5):1747–1756.

[58] Herzog, J., Hibi, T., Componentwise linear ideals. Nagoya Math J. 1999;153:141–153.

[59] Herzog, J., Hibi, T., The depth of powers of an ideal. J Algebra. 2005;291:534–550.

[60] Herzog, J., Hibi, T., Distributive lattices, bipartite graphs and Alexander duality. J Algebraic Combin. 2005;22(3):289–302.

[61] Herzog, J., Hibi, T., Monomial ideals. vol. 260 of Graduate texts in mathematics. Springer-Verlag; 2011.

[62] Herzog, J., Hibi, T., Trung, N. V., Zheng, X., Standard graded vertex cover algebras, cycles and leaves. Trans Amer Math Soc. 2008;360:6231–6249.

[63] Herzog, J., Hibi, T., Zheng, X., Dirac's theorem on chordal graphs and Alexander duality. European J Combin. 2004;25(7):949–960.

[64] Herzog, J., Hibi, T., Zheng, X., Monomial ideals whose powers have a linear resolution. Math Scand. 2004;95:23–32.

[65] Hoa, L. T., Stability of associated primes of monomial ideals. Vietnam J Math. 2006; 34(4):473–487.

[66] Hochster, M., Rings of invariants of tori, Cohen–Macaulay rings generated by monomials, and polytopes. Ann of Math. 1972;96:318–337.

[67] Hochster, M., Criteria for the equality of ordinary and symbolic powers of primes. Math Z. 1973;133:53–65.

[68] Huneke, C., On the associated graded ring of an ideal. Illinois J Math. 1982;26:121–137.

[69] Huneke, C., Simis, A., Vasconcelos, W. V., Reduced normal cones are domains. Contemp Math. 1989;88:95–101.

[70] Huneke, C., Swanson, I., Integral closure of ideals rings, and modules. vol. 336 of London Mathematical Society lecture note series. Cambridge: Cambridge University Press; 2006.

[71] Ionescu, C., Rinaldo, G., Some algebraic invariants related to mixed product ideals. Arch Math. 2008;91:20–30.

[72] Jacques, S., Betti numbers of graph ideals [PhD thesis]. University of Sheffield; 2004. arXiv:math/0410107v1.

[73] Kalai, G., Meshulam, R., Unions and intersections of Leray complexes. J Combin Theory Ser A. 2006;113:1586–1592.

[74] Katzman, M., Characteristic-independence of Betti numbers of graph ideals. J Combin Theory Ser A. 2006;113(3):435–454.

[75] Kiani, D., Moradi, S., Bounds for the regularity of edge ideal of vertex decomposable and shellable graphs. Preprint. arXiv:1007.4056v1. 2010.

[76] Kimura, K., Terai, N., Yoshida, K., Arithmetical rank of squarefree monomial ideals of small arithmetic degree. J Algebraic Combin. 2009;29:389–404.

[77] Kummini, M., Regularity, depth and arithmetic rank of bipartite edge ideals. J Algebraic Combin. 2009;30(4):429–445.

[78] Lehman, A., On the width-length inequality and degenerate projective planes. In: Cook, W., Seymour, P., editors. Polyhedral Combinatorics. vol. 1 of DIMACS series in discrete mathematics and theoretical computer science. American Mathematical Society; 1990, p. 101–105.

[79] Lyubeznik, G., On the local cohomology modules $H^i_{\mathcal{A}}(R)$ for ideals $\mathcal{A}$ generated by monomials in an R-sequence. In: Greco, S., Strano, R., editors. Complete intersections. Lectures given at the 1st 1983 Session of the Centro Internazionale Matematico Estivo (C.I.M.E.); 1983, Jun 13–21; Acireale, Italy. Berlin-Heidelberg: Springer. p. 214–220.

[80] Lyubeznik, G., On the arithmetical rank of monomial ideals. J Algebra. 1988;112:86–89.

[81] Mahmoudi, M., Mousivand, A., Crupi, M., Rinaldo, G., Terai, N., Yassemi, S., Vertex decomposability and regularity of very well-covered graphs. J Pure Appl Algebra. 2011; 215(10):2473–2480.

[82] Martínez-Bernal, J., O'Shea, E., Villarreal, R. H., Ehrhart clutters: regularity and max-flow min-cut. Electron J Combin. 2010;17(1):R52.

[83] McAdam, S., Asymptotic prime divisors and analytic spreads. Proc Amer Math Soc. 1980;80:555–559.

[84] McAdam, S., Asymptotic prime divisors. vol. 103 of Lecture notes in mathematics. New York: Springer-Verlag; 1983.

[85] McAdam, S., Eakin, P., The asymptotic Ass. J Algebra. 1979;61(1):71–81.

[86] Miller, E., Sturmfels, B., Combinatorial commutative algebra. vol. 227 of Graduate texts in mathematics. Springer; 2004.

[87] Morey, S., Depths of powers of the edge ideal of a tree. Comm Algebra. 2010;38(11): 4042–4055.

[88] Morey, S., Stability of associated primes and equality of ordinary and symbolic powers of ideals. Comm Algebra. 1999;27:3221–3231.

[89] Morey, S., Reyes, E., Villarreal, R. H., Cohen–Macaulay, shellable and unmixed clutters with a perfect matching of König type. J Pure Appl Algebra. 2008;212(7):1770–1786.

[90] Nevo, E., Regularity of edge ideals of C_4-free graphs via the topology of the lcm-lattice. J Combin Theory Ser A. 2011;118(2):491–501.

[91] Peeva, I., Stillman, M., Open problems on syzygies and Hilbert functions. J Commut Algebra. 2009;1(1):159–195.

[92] Plummer, M. D., Some covering concepts in graphs. J Combin Theory. 1970;8:91–98.

[93] Plummer, M. D., Well-covered graphs: a survey. Quaest Math. 1993;16(3):253–287.

[94] Ravindra, G., Well-covered graphs. J Comb Inf Syst Sci. 1977;2(1):20–21.

[95] Renteln, P., The Hilbert series of the face ring of a flag complex. Graphs Combin. 2002; 18(3):605–619.

[96] Schrijver, A., Theory of linear and integer programming. New York: John Wiley & Sons; 1986.

[97] Schrijver, A., Combinatorial optimization. vol. 24 of Algorithms and combinatorics. Berlin: Springer-Verlag; 2003.

[98] Seymour, P. D., The matroids with the max-flow min-cut property. J Combin Theory Ser B. 1977;23:189–222.

[99] Sharp, R., Convergence of sequences of sets of associated primes. Proc Amer Math Soc. 2003;131:3009–3017.

[100] Simis, A., Vasconcelos, W. V., Villarreal, R. H., On the ideal theory of graphs. J Algebra. 1994;167:389–416.

[101] Smith, D. E., On the Cohen–Macaulay property in commutative algebra and simplicial topology. Pacific J Math. 1990;141:165–196.

[102] Stanley, R. P., Combinatorics and commutative algebra. 2nd ed. vol. 41 of Progress in mathematics. Boston, MA: Birkhäuser Boston, Inc., 1996.

[103] Terai, N., Alexander duality theorem and Stanley–Reisner rings. Sūrikaisekikenkyūsho Kōkyūroku 1999;1078(Free resolutions of coordinate rings of projective varieties and related topics; 1998; Kyoto):174–184.

[104] Toft, B., Colouring, stable sets and perfect graphs. In: Graham, R. L., Grötschel, M., Lovász, L., editors. Handbook of combinatorics I. Elsevier; 1995. p. 233–288.

[105] Trung, T. N., Stability of associated primes of integral closures of monomial ideals. J Combin Theory Ser A 2009;116:44–54.

[106] Van Tuyl, A., Sequentially Cohen–Macaulay bipartite graphs: vertex decomposability and regularity. Arch Math. 2009;93(5):451–459.

[107] Van Tuyl, A., Villarreal, R. H., Shellable graphs and sequentially Cohen–Macaulay bipartite graphs. J Combin Theory Ser A. 2008;115(5):799–814.

[108] Vasconcelos, W. V., Computational methods in commutative algebra and algebraic geometry. Springer-Verlag: 1998.

[109] Villarreal, R. H., Cohen–Macaulay graphs. Manuscripta Math. 1990;66:277–293.

[110] Villarreal, R. H., Monomial algebras. vol. 238 of Monographs and textbooks in pure and applied mathematics. New York: Marcel Dekker; 2001.

[111] Villarreal, R. H., Unmixed bipartite graphs. Rev Colombiana Mat. 2007;41(2):393–395.

[112] Villarreal, R. H., Rees algebras and polyhedral cones of ideals of vertex covers of perfect graphs. J Algebraic Combin. 2008;27:293–305.

[113] Voloshin, V. I., Coloring mixed hypergraphs: theory, algorithms and applications. vol. 17 of Fields institute monographs. Providence, RI: American Mathematical Society; 2002.

[114] Whieldon, G., Jump sequences of edge ideals. Preprint. arXiv:1012.0108v1. 2010.

[115] Woodroofe, R., Chordal and sequentially Cohen–Macaulay clutters. Electron J Combin. 2011;18(1):R208.

[116] Woodroofe, R., Vertex decomposable graphs and obstructions to shellability. Proc Amer Math Soc. 2009;137:3235–3246.

[117] Woodroofe, R., Matchings, coverings, and Castelnuovo–Mumford regularity. Preprint. arXiv:1009.2756. 2010.

[118] Zheng, X., Resolutions of facet ideals. Comm Algebra. 2004;32:2301–2324.

Author Information

Susan Morey, Department of Mathematics, Texas State University, San Marcos, TX, USA.
E-mail: morey@txstate.edu

Rafael H. Villarreal, Departamento de Matemáticas, Centro de Investigación y de Estudios Avanzados del IPN, Mexico City, D.F., Mexico.
E-mail: vila@math.cinvestav.mx

Progress in Commutative Algebra 1, 127–141

Three Simplicial Resolutions

Jeff Mermin

Abstract. We describe the Taylor and Lyubeznik resolutions as simplicial resolutions, and use them to show that the Scarf complex of a monomial ideal is the intersection of all its minimal free resolutions.

Keywords. Simplicial Resolution.

2010 Mathematics Subject Classification. Primary: 13D02; Secondary: 13F20, 05E40, 13-02.

1 Introduction

Let $S = k[x_1, \dots, x_n]$, and let $I \subset S$ be a monomial ideal. An important object in the study of I is its *minimal free resolution*, which encodes essentially all information about I. For example, the Betti numbers of I can be read off as the ranks of the modules in its minimal resolution.

There are computationally intensive algorithms to compute the minimal resolution of an arbitrary ideal (for example, in Macaulay 2 [5], the command `res I` returns the minimal resolution of S/I), but no general description is known, even for monomial ideals. Thus, it is an ongoing problem of considerable interest to find classes of ideals whose minimal resolutions can be described easily. A related problem is to describe non-minimal resolutions which apply to large classes of monomial ideals.

The most general answer to the latter question is Taylor's resolution, a (usually highly non-minimal) resolution which resolves an arbitrary monomial ideal; it is discussed in Section 3.

A very successful approach to both problems in the last decade has been to find combinatorial or topological objects whose structures encode resolutions in some way. This approach began with *simplicial resolutions* [1], and has expanded to involve polytopal complexes [8, 13], cellular complexes [2], CW complexes [3, 15], lattices [7, 12, 11], posets [4], matroids [14], and discrete Morse theory [6].

Resolutions associated to combinatorial objects have distinguished bases, and relationships between the objects lead to relationships between these bases. It thus becomes possible to compare and combine resolutions in all the ways that we can compare or combine combinatorial structures. For example, most of these resolutions turn out to be subcomplexes of the Taylor resolution in a very natural way. The only new result in the paper is Theorem 7.1, which describes the intersection of all the simplicial resolutions of an ideal.

In Section 2, we describe some background material and introduce notation used throughout the paper.

Section 3 introduces the Taylor resolution in a way intended to motivate simplicial resolutions, which are introduced in Section 4.

Section 5 describes the Scarf complex, a simplicial complex which often supports the minimal resolution of a monomial ideal, and otherwise does not support any resolution.

Section 6 defines the family of Lyubeznik resolutions. This section is essentially a special case of an excellent paper of Novik [9], which describes a more general class of resolutions based on so-called "rooting maps".

Section 7 uses the Lyubeznik resolutions to prove Theorem 7.1, that the Scarf complex of an ideal is equal to the intersection of all its simplicial resolutions.

2 Background and Notation

Throughout the paper $S = k[x_1, \dots, x_n]$ is a polynomial ring over an arbitrary field k. In the examples, we use the variables $a, b, c, \dots$ instead of $x_1, x_2, x_3, \dots$.

We depart from the standard notation in two ways, each designed to privilege monomials. First, we write the standard or "fine" multigrading multiplicatively, indexed by monomials, rather than additively, indexed by n-tuples. Second, we index our simplices by monomials rather than natural numbers. Details of both departures, as well as some background on resolutions, are below.

2.1 Algebra

If $I \subset S$ is an ideal, then a *free resolution* of S/I is an exact sequence

$$\mathbb{F}: \quad \cdots \xrightarrow{\phi_n} F_n \xrightarrow{\phi_{n-1}} F_{n-1} \to \cdots \xrightarrow{\phi_0} F_0 \to S/I \to 0$$

where each of the F_i is a free S-module.

We say that $\mathbb{F}$ is *minimal* if each of the modules F_i has minimum possible rank; in this case the ranks are the *Betti numbers* of S/I.

It is not at all obvious a priori that minimal resolutions should exist. For this reason, when I is homogeneous, most standard treatments take the following theorem as the definition instead:

Theorem 2.1. *Let I be a homogeneous ideal, and let $\mathbb{F}$ be a resolution of S/I. Write $\mathbf{m} = (x_1, \dots, x_n)$. Then $\mathbb{F}$ is minimal if and only if $\phi_i(F_{i+1}) \subset \mathbf{m}F_i$ for all i.*

The proof of Theorem 2.1 is technical; see, for example, [10, Section 9].

All the ideals we consider are homogeneous; in fact, they are *monomial* ideals, which is a considerably stronger property.

Definition 2.2. An ideal I is a *monomial ideal* if it has a generating set consisting of monomials. There exists a unique minimal such generating set; we write $\operatorname{gens}(I)$ and call its elements the *generators* of I.

Monomial ideals respect a "multigrading" which refines the usual grading.

Notation 2.3. We write the multigrading multiplicatively. That is, for each monomial m of S, set S_m equal to the k-vector space spanned by m. Then $S = \bigoplus S_m$, and $S_m \cdot S_n = S_{mn}$, so this decomposition is a grading. We say that the monomial m has *multidegree* m. We allow multidegrees to have negative exponents, so, for example, the twisted module $S(m^{-1})$ is a free module with generator in multidegree m, and $S(m^{-1})_n \cong S_{m^{-1}n}$ as a vector space; this is one-dimensional if no exponent of $m^{-1}n$ is negative, and trivial otherwise. Note that $S = S(1)$.

If N and P are multigraded modules, we say that a map $\phi : N \to P$ is *homogeneous of degree* m if $\phi(N_n) \subset P_{mn}$ for all n, and that ϕ is simply *homogeneous* if it is homogeneous of degree 1. We say that a resolution (or, more generally, an algebraic chain complex) is *homogeneous* if all its maps are homogeneous.

The minimal resolution of S/I can be made homogeneous in a unique way by assigning appropriate multidegrees to the generators of its free modules; counting these generators by multidegree yields the *multigraded Betti numbers* of S/I.

2.2 Combinatorics

Let M be a set of monomials (typically, M will be the generators of I). The *simplex on* M is the set of all subsets of M; we denote this by Δ_M. We will sometimes refer to the elements of M as *vertices* of Δ_M.

A *simplicial complex* on M is a subset of Δ_M which is closed under the taking of subsets. If Γ is a simplicial complex on M and $F \in \Gamma$, we say that F is a *face* of Γ. Observe that if F is a face of Γ and $G \subset F$, then G is also a face of Γ. We require that simplicial complexes be nonempty; that is, the empty set must always be a face. (In fact, for our purposes, we may as well assume that every vertex must be a face.)

If F is a face of Γ, we assign F the multidegree $\operatorname{lcm}(m : m \in F)$. Note that the vertex m has multidegree m, and that the empty set has multidegree 1. The *order* of a face F, written $|F|$, is the number of vertices in F; this is one larger than its dimension. If $G \subset F$ and $|G| = |F| - 1$, we say that G is a *facet* of F.

We adopt the convention that the unmodified word "complex" will always mean an algebraic chain complex; simplicial complexes will be referred to with the phrase "simplicial complex". However, recall that every simplicial complex is naturally associated to a chain complex by the following standard construction from algebraic topology:

Construction 2.4. Let Γ be a simplicial complex on M, and impose an order on the monomials of M by writing $M = \{m_1, \dots, m_r\}$. Then we associate to Γ the chain complex $\mathbb{C}_\Gamma$ as follows:

For every face $F \in \Gamma$, we create a formal symbol $[F]$. Write $F = \{m_{i_1}, \dots, m_{i_s}\}$ with increasing indices i_j; then for each facet G of F we may write $G = F \smallsetminus \{m_{i_j}\}$ for some j. We define an orientation by setting ε_G^F equal to 1 if j is odd and to -1 if j is even. For each s, let C_s be the k-vector space spanned by the symbols $[F]$ such that $|F| = s$, and define the map

$$\begin{aligned} \phi_{s-1} : C_s &\to C_{s-1} \\ [F] &\mapsto \sum_{G \text{ is a facet of } F} \varepsilon_G^F [G]. \end{aligned}$$

Then we set $\mathbb{C}_\Gamma$ equal to the complex of vector spaces

$$\mathbb{C}_\Gamma : 0 \to C_r \xrightarrow{\phi_{r-1}} \cdots \xrightarrow{\phi_1} C_1 \xrightarrow{\phi_0} C_0 \to 0.$$

The proof that $\mathbb{C}_\Gamma$ is a chain complex involves a straightforward computation of $\phi^2([m_{i_1}, \dots, m_{i_s}])$. The (reduced) homology of Γ is defined to be the homology of this complex.

In Section 4, we will replace this complex with a homogeneous complex of free S-modules.

3 The Taylor Resolution

Let $I = (m_1, \dots, m_s)$ be a monomial ideal. The Taylor resolution of I is constructed as follows:

Construction 3.1. For a subset F of $\{m_1, \dots, m_r\}$, set $\operatorname{lcm}(F) = \operatorname{lcm}\{m_i : m_i \in F\}$. For each such F, we define a formal symbol $[F]$, called a *Taylor symbol*, with multidegree equal to $\operatorname{lcm}(F)$. For each i, set T_i equal to the free S-module with basis $\{[F] : |F| = i\}$ given by the symbols corresponding to subsets of size i. Note that $T_0 = S[\varnothing]$ is a free module of rank one, and that all other T_i are multigraded modules with generators in multiple multidegrees depending on the symbols $[F]$.

Define $\phi_{-1} : T_0 \to S/I$ by $\phi_{-1}(f[\varnothing]) = f$. Otherwise, we construct $\phi_i : T_{i+1} \to T_i$ as follows.

Given $F = \{m_{j_1}, \dots, m_{j_i}\}$, written with the indices in increasing order, and $G = F \smallsetminus \{m_{j_k}\}$, we set the *sign* ε_G^F equal to 1 if k is odd and to -1 if k is even. Finally, we set

$$\phi_F = \sum_{G = F \smallsetminus \{m_i\},\ \text{some } i} \varepsilon_G^F \frac{\operatorname{lcm}(F)}{\operatorname{lcm}(G)} [G],$$

and define $\phi_i : T_{i+1} \to T_i$ by extending the various ϕ_F. Observe that all of the ϕ_i are homogeneous with multidegree 1.

The *Taylor resolution* of I is the complex

$$\mathbb{T}_I : 0 \to T_r \xrightarrow{\phi_{r-1}} \cdots \xrightarrow{\phi_1} T_1 \xrightarrow{\phi_0} T_0 \to S/I \to 0.$$

It is straightforward to show that the Taylor resolution is a homogeneous chain complex.

The construction of the Taylor resolution is very similar to Construction 2.4; in fact, if Γ is the complete simplex, the only difference is the presence of the lcms in the boundary maps. We will explore this connection in the next section.

Example 3.2. Let $I = (a, b^2, c^3)$. Then the Taylor resolution of I is

$$\mathbb{T}_I\colon\ 0 \to S[a,b^2,c^3] \xrightarrow{\begin{pmatrix} a \\ -b^2 \\ c^3 \end{pmatrix}} \begin{matrix} S[b^2,c^3] \\ \oplus \\ S[a,c^3] \\ \oplus \\ S[a,b^2] \end{matrix} \xrightarrow{\begin{pmatrix} 0 & -c^3 & -b^2 \\ -c^3 & 0 & a \\ b^2 & a & 0 \end{pmatrix}} \begin{matrix} S[a] \\ \oplus \\ S[b^2] \\ \oplus \\ S[c^3] \end{matrix} \xrightarrow{(a\ b^2\ c^3)} S[\varnothing] \to S/I \to 0.$$

Observe that I is a complete intersection and $\mathbb{T}_I$ is its Koszul complex. In fact, these two complexes coincide for all monomial complete intersections.

Example 3.3. Let $I = (a^2, ab, b^3)$. Then the Taylor resolution of I is

$$\mathbb{T}_I\colon\ 0 \to S[a^2,ab,b^3] \xrightarrow{\begin{pmatrix} a \\ -1 \\ b^2 \end{pmatrix}} \begin{matrix} S[ab,b^3] \\ \oplus \\ S[a^2,b^3] \\ \oplus \\ S[a^2,ab] \end{matrix} \xrightarrow{\begin{pmatrix} 0 & -b^3 & -b \\ -b^2 & 0 & a \\ a^2 & a & 0 \end{pmatrix}} \begin{matrix} S[a^2] \\ \oplus \\ S[ab] \\ \oplus \\ S[b^3] \end{matrix} \xrightarrow{(a^2\ ab\ b^3)} S[\varnothing] \to S/I \to 0.$$

This is not a minimal resolution; the Taylor resolution is very rarely minimal.

Theorem 3.4. *The Taylor resolution of I is a resolution of I.*

It is not too difficult to show that $\phi^2 = 0$ in the Taylor complex, but it is not at all clear from the construction that the complex is exact. This seems to be most easily established indirectly by showing that the Taylor resolution is a special case of some more general phenomenon. We will prove Theorem 3.4 in the next section, using the language of simplicial resolutions. Traditionally, one builds the Taylor resolution as an iterated mapping cone; we sketch that argument below.

Sketch of Theorem 3.4. Write $I = (m_1, \dots, m_r)$, and let $J = (m_1, \dots, m_{r-1})$. Consider the short exact sequence

$$0 \to \frac{S}{(J : m_r)} \xrightarrow{m_r} \frac{S}{J} \to \frac{S}{I} \to 0.$$

If $(\mathbb{A}, \alpha)$ and $(\mathbb{B}, \beta)$ are free resolutions of $S/(J : m_r)$ and S/J, respectively, then multiplication by m_r induces a map of complexes $(m_r)_* : \mathbb{A} \to \mathbb{B}$. The *mapping cone complex* $(\mathbb{T}, \gamma)$ is defined by setting $T_i = B_i \oplus A_{i-1}$ and $\gamma|_{\mathbb{B}} = \beta$, $\gamma|_{\mathbb{A}} = (m_r)_* - \alpha$; it is a free resolution of S/I (see, for example, [10, Section 27]).

Inducting on r, $S/(J : m_r)$ is resolved by the Taylor resolution on its (possibly redundant) generating set $\{\frac{\mathrm{lcm}(m_1, m_r)}{m_r}, \dots, \frac{\mathrm{lcm}(m_{r-1}, m_r)}{m_r}\}$, and S/J is resolved by the Taylor resolution on its generators $\{m_1, \dots, m_{r-1}\}$. The resulting mapping cone is the Taylor resolution of I. □

4 Simplicial Resolutions

If $\Gamma = \Delta$, the construction of the Taylor resolution differs from the classical topological construction of the chain complex associated to Γ only by the presence of the monomials $\frac{\mathrm{lcm}(F)}{\mathrm{lcm}(G)}$ in its differential maps. This observation leads naturally to the question of what other simplicial complexes give rise to resolutions in the same way. The resulting resolutions are called *simplicial.* Simplicial resolutions and, more generally, resolutions arising from other topological structures (it seems that the main results can be tweaked to work for anything defined in terms of skeletons and boundaries) have proved to be an instrumental tool in the understanding of monomial ideals. We describe only the foundations of the theory here; for a more detailed treatment, the original paper of Bayer, Peeva, and Sturmfels [1] is a very readable introduction.

Construction 4.1. Let M be a set of monomials, and let Γ be a simplicial complex on M (recall that this means that the vertices of Γ are the monomials in M). Fix an ordering on the elements of M; this induces an orientation ε on Γ. Recall that ε_G^F is either 1 or -1 if G is a facet of F (see Construction 2.4 for the details); it is often convenient to formally set ε_G^F equal to zero when G is not a facet of F.

We assign a multidegree to each face $F \in \Gamma$ by the rule $\mathrm{mdeg}(F) = \mathrm{lcm}(m : m \in F)$ (recall that F is a subset of M, so its elements are monomials).

Now for each face F we create a formal symbol $[F]$ with multidegree $\mathrm{mdeg}(F)$. Let H_s be the free module with basis $\{[F] : |F| = s\}$, and define the differential

$$\phi_{s-1} : H_s \to H_{s-1}$$

$$[F] \mapsto \sum_{G \text{ is a facet of } F} \varepsilon_G^F \frac{\mathrm{mdeg}(F)}{\mathrm{mdeg}(G)} [G].$$

The *complex associated to* Γ is then the algebraic chain complex

$$\mathbb{H}_\Gamma : 0 \to H_r \xrightarrow{\phi_{r-1}} \cdots \xrightarrow{\phi_1} H_1 \xrightarrow{\phi_0} H_0 \to S/I \to 0.$$

Construction 4.1 differs from Construction 2.4 in that it is a complex of free S-modules rather than vector spaces. The boundary maps are identical except for the monomial coefficients, which are necessary to make the complex homogeneous.

Example 4.2. Let I be generated by M, and let Δ be the simplex with vertices M. Then the Taylor resolution of I is the complex associated to Δ.

Example 4.3. Let I be generated by $M = \{a^2, ab, b^3\}$, and let Δ be the full simplex on M, Γ the simplicial complex with facets $\{a^2, ab\}$ and $\{ab, b^3\}$, and Θ the zero-skeleton of Δ. These simplicial complexes, with their faces labeled by multidegree, are pictured in Figure 1.

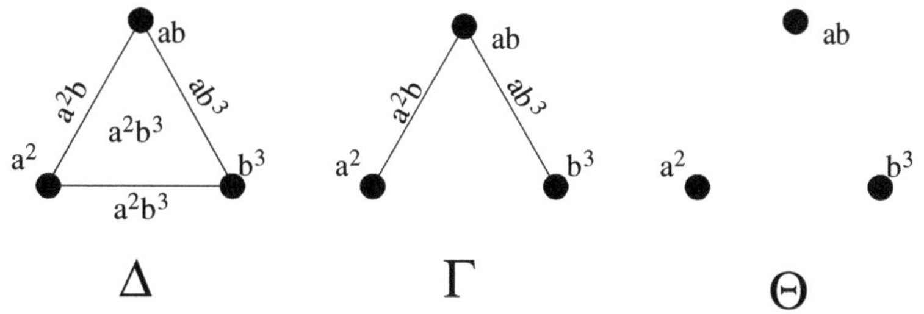

Figure 1. The simplicial complexes Δ, Γ, and Θ of Example 4.3.

The algebraic complex associated to Δ is the Taylor resolution of Example 3.3. The other two associated complexes are

$$\mathbb{H}_\Gamma : 0 \to \begin{matrix} S[a^2,ab] \\ \oplus \\ S[ab,b^3] \end{matrix} \xrightarrow{\begin{pmatrix} -b & 0 \\ a & -b^2 \\ 0 & a \end{pmatrix}} \begin{matrix} S[a^2] \\ \oplus \\ S[ab] \\ \oplus \\ S[b^3] \end{matrix} \xrightarrow{(a^2\ ab\ b^3)} S[\varnothing] \to S/I \to 0$$

and

$$\mathbb{H}_\Theta : 0 \to S[a^2] \oplus S[ab] \oplus S[b^3] \xrightarrow{(a^2\ ab\ b^3)} S[\varnothing] \to S/I \to 0.$$

$\mathbb{H}_\Gamma$ is a resolution (in fact, the minimal resolution) of S/I, and $\mathbb{H}_\Theta$ is not a resolution of I.

The algebraic complex associated to Γ is not always exact; that is, it does not always give rise to a resolution of I. When this complex is exact, we call it a *simplicial resolution*, or *the (simplicial) resolution supported on* Γ. It turns out that there is a topological condition describing whether Γ supports a resolution.

Definition 4.4. Let Γ be a simplicial complex on M, and let μ be a multidegree. We set $\Gamma_{\leq\mu}$ equal to the simplicial subcomplex of Γ consisting of the faces with multidegree divisible by μ,

$$\Gamma_{\leq\mu} = \{F \in \Gamma : \deg(F) \text{ divides } \mu\}.$$

Observe that $\Gamma_{\leq\mu}$ is precisely the faces of Γ whose vertices all divide μ.

Theorem 4.5 (Bayer–Peeva–Sturmfels). *Let Γ be a simplicial complex supported on M, and set $I = (M)$. Then Γ supports a resolution of S/I if and only if, for all μ, the simplicial complex $\Gamma_{\leq\mu}$ has no homology over k.*

Proof. Since $\mathbb{H}_\Gamma$ is homogeneous, it is exact if and only if it is exact (as a complex of vector spaces) in every multidegree. Thus, it suffices to examine the restriction of $\mathbb{H}_\Gamma$ to each multidegree μ.

Observe that $(S[F])_\mu \cong S(\frac{1}{\mathrm{mdeg}(F)})_\mu \cong S_{\frac{\mu}{\mathrm{mdeg}(F)}}$ is a one-dimensional vector space with basis $\frac{\mu}{\mathrm{mdeg}(F)}$ if $\mathrm{mdeg}(F)$ divides μ, and is zero otherwise. Furthermore, since the differential maps ϕ are homogeneous, the monomials appearing in their definition are precisely those which map these basis elements to one another. Thus $(\mathbb{H}_\Gamma)_\mu$ is, with minor abuse of notation, precisely the complex of vector spaces which arises when computing (via Construction 2.4) the homology of the simplicial complex $\{F \in \Gamma : \mathrm{mdeg}(F) \text{ divides } \mu\}$, and this complex is $\Gamma_{\leq\mu}$.

We conclude that Γ supports a resolution of I if and only if $(\mathbb{H}_\Gamma)_\mu$ is exact for every μ, if and only if $(\mathbb{H}_\Gamma)_\mu$ has no homology for every μ, if and only if $\Gamma_{\leq\mu}$ has no homology for every μ. □

Example 4.6. The simplicial complexes $\Gamma_{\leq\mu}$ depend on the underlying monomials M, so it is possible for a simplicial complex to support a resolution of some monomial ideals but not others. For example, the simplicial complex Γ in Example 4.3 supports a resolution of $I = (a^2, ab, b^3)$ because no monomial is divisible by a^2 and b^3 without also being divisible by ab. However, if we were to relabel the vertices with the monomials a, b, and c, the resulting simplicial complex Γ' would not support a resolution of (a, b, c) because $\Gamma'_{\leq ac}$ would consist of two points; this simplicial complex has nontrivial zeroeth homology.

Remark 4.7. Note that the homology of a simplicial complex can depend on the choice of field, so some simplicial complexes support resolutions over some fields but not others. For example, if Γ is a triangulation of a torus, it may support a resolution if the field has characteristic zero, but will not support a resolution in characteristic two. In particular, resolutions of monomial ideals can be characteristic-dependent.

Theorem 4.5 allows us to give a short proof that the Taylor resolution is in fact a resolution.

Proof of Theorem 3.4. Let μ be given. Then $\Delta_{\leq\mu}$ is the simplex with vertices $\{m \in M : m \text{ divides } \mu\}$, which is either empty or contractible. $\square$

5 The Scarf Complex

Unfortunately, the Taylor resolution is usually not minimal. The nonminimality is visible in the nonzero scalars in the differential maps, which occur whenever there exist faces F and G with the same multidegree such that G is in the boundary of F. It is tempting to try to simply remove the nonminimality by removing all such faces; the result is the Scarf complex.

Construction 5.1. Let I be a monomial ideal with generating set M. Let Δ_I be the full simplex on M, and let Σ_I be the simplicial subcomplex of Δ_I consisting of the faces with unique multidegree,

$$\Sigma_I = \{F \in \Delta_I : \mathrm{mdeg}(G) = \mathrm{mdeg}(F) \implies G = F\}.$$

We say that Σ_I is the *Scarf simplicial complex* of I; the associated algebraic chain complex $\mathbb{S}_I$ is called the *Scarf complex* of I. The multidegrees of the faces of Σ_I are called the *Scarf multidegrees* of I.

Remark 5.2. It is not obvious that Σ_I is a simplicial complex. Let $F \in \Sigma_I$; we will show that every subset of F is also in Σ_I. Suppose not; then there exists a minimal $G \subset F$ which shares a multidegree with some other $H \in \Delta_I$. Let E be the symmetric difference of G and H. Then the symmetric difference of E and F has the same multidegree as F.

Example 5.3. Let $I = (a^2, ab, b^3)$. Then the Scarf simplicial complex of I is the complex Γ in Figure 1. The Scarf complex of I is the minimal resolution

$$\mathbb{S}_I : 0 \to \begin{matrix} S[a^2, ab] \\ \oplus \\ S[ab, b^3] \end{matrix} \xrightarrow{\begin{pmatrix} -b & 0 \\ a & -b^2 \\ 0 & a \end{pmatrix}} \begin{matrix} S[a^2] \\ \oplus \\ S[ab] \\ \oplus \\ S[b^3] \end{matrix} \xrightarrow{(a^2\ ab\ b^3)} S[\varnothing] \to S/I \to 0.$$

Example 5.4. Let $I = (ab, ac, bc)$. The Scarf simplicial complex of I consists of three disjoint vertices. The Scarf complex of I is the complex

$$\mathbb{S}_I : 0 \to \begin{matrix} S[ab] \\ \oplus \\ S[ac] \\ \oplus \\ S[bc] \end{matrix} \xrightarrow{(ab\ ac\ bc)} S[\varnothing] \to S/I \to 0.$$

It is not a resolution.

Example 5.4 shows that not every monomial ideal is resolved by its Scarf complex. We say that a monomial ideal is *Scarf* if its Scarf complex is a resolution.

Theorem 5.5. *If the Scarf complex of I is a resolution, then it is minimal.*

Proof. By construction, no nonzero scalars can occur in the differential matrices. □

Bayer, Peeva and Sturmfels [1] call an ideal *generic* if no variable appears with the same nonzero exponent in more than one generator. They show that these "generic" ideals are Scarf.

Unfortunately, most interesting monomial ideals are not Scarf. However, Scarf complexes have proved an important tool in constructing ideals whose resolutions misbehave in various ways [15].

Theorem 5.6. *Let $\mathbb{F}$ be a minimal resolution of I. Then the Scarf complex of I is a subcomplex of $\mathbb{F}$.*

Proof. This is [10, Proposition 59.2]. The proof requires a couple of standard facts about resolutions, but is otherwise sufficiently reliant on the underlying simplicial complexes that we reproduce it anyway.

We know (see, for example, [10, Section 9]) that there is a homogeneous inclusion of complexes from $\mathbb{F}$ to the Taylor complex $\mathbb{T}$. We also know that the multigraded Betti numbers of I, which count the generators of $\mathbb{F}$, can be computed from the homology of the simplicial complexes $\Delta_{\lneq m}$ [10, Section 57]. If $m = \text{mdeg}(G)$ is a Scarf multidegree, then $b_{|G|,m}(S/I) = 1$ and $b_{i,m}(S/I) = 0$ for all other i. If m divides a Scarf multidegree but is not itself a Scarf multidegree, then $b_{i,m}(S/I) = 0$ for all i. In particular, when m is a Scarf multidegree, the Betti numbers of multidegree m also count the number of faces of multidegree m in both Δ_I and Σ_I; these numbers are never greater than one.

By induction on multidegrees, each generator of $\mathbb{F}$ with a Scarf multidegree must (up to a scalar) be mapped under the inclusion to the unique generator of the Taylor resolution with the same multidegree. However, these are exactly the generators of the Scarf complex. Thus, the inclusion from $\mathbb{F}$ to $\mathbb{T}$ induces an inclusion from $\mathbb{S}$ to $\mathbb{F}$. □

6 The Lyubeznik Resolutions

If the Taylor resolution is too large, and the Scarf complex is too small, we might still hope to construct simplicial resolutions somewhere in between. Velasco [15] shows that it is impossible to get the minimal resolution of every ideal in this way, even if we replace simplicial complexes with much more general topological objects. However, there are still classes of simplicial resolutions which are in general much smaller than the Taylor resolution, yet still manage to always be resolutions. One such class is the class of Lyubeznik resolutions, introduced below.

Our construction follows the treatment in an excellent paper of Novik [9], which presents the Lyubeznik resolutions as special cases of resolutions arising from "rooting maps". The only difference between the following construction and Novik's paper is that the extra generality has been removed, and the notation is correspondingly simplified.

Construction 6.1. Let I be a monomial ideal with generating set M, and fix an ordering $\prec$ on the monomials appearing in M. (We do not require that $\prec$ have any special property, such as a term order; any total ordering will do.) Write $M = \{m_1, \dots, m_s\}$ with $m_i \prec m_j$ whenever $i < j$.

Let Δ_I be the full simplex on M; for a monomial $\mu \in I$, set $\min(\mu) = \min_{\prec}\{m_i : m_i \text{ divides } \mu\}$. For a face $F \in \Delta_I$, set $\min(F) = \min(\mathrm{mdeg}(F))$. Thus $\min(F)$ is a monomial. We expect that in fact $\min(F)$ is a vertex of F, but this need not be the case: for example, if $F = \{a^2, b^2\}$, we could have $\min(F) = ab$.

We say that a face F is *rooted* if every nonempty subface $G \subset F$ satisfies $\min(G) \in G$. (Note that in particular $\min(F) \in F$.) By construction, the set $\Lambda_{I,\prec} = \{F \in \Delta_I : F \text{ is rooted}\}$ is a simplicial complex; we call it the *Lyubeznik simplicial complex* associated to I and $\prec$. The associated algebraic chain complex $\mathbb{L}_{I,\prec}$ is called a *Lyubeznik resolution* of I.

Example 6.2. Let $I = (ab, ac, bc)$. Then there are three distinct Lyubeznik resolutions of I, corresponding to the simplicial complexes pictured in Figure 2: Λ_{ab} arises

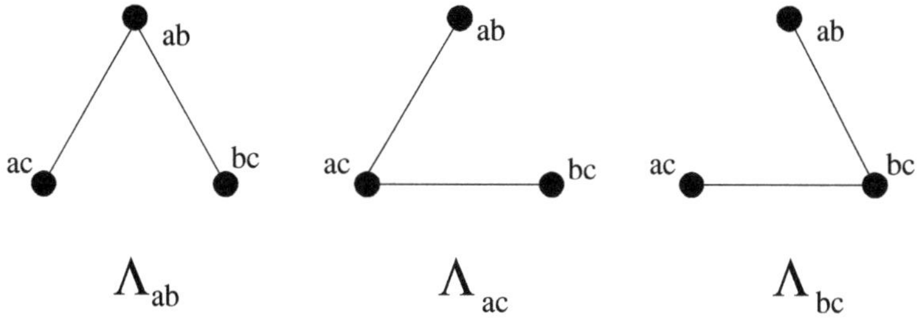

Figure 2. The Lyubeznik resolutions of $I = (ab, ac, bc)$.

from the orders $ab \prec ac \prec bc$ and $ab \prec bc \prec ac$, Λ_{ac} arises from the orders with ac first, and Λ_{bc} arises from the orders with bc first. Each of these resolutions is minimal.

Example 6.3. Let $I = (a^2, ab, b^3)$. There are two Lyubeznik resolutions of I: the Scarf complex, arising from the two orders with ab first, and the Taylor resolution, arising from the other four orders. The corresponding simplicial complexes are pictured in Figure 3.

Remark 6.4. It is unclear how to choose a total ordering on the generators of I which produces a smaller Lyubeznik resolution. Example 6.3 suggests that the obvious

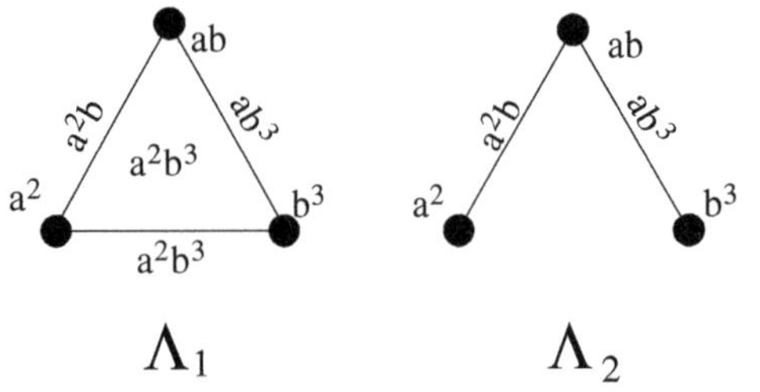

Figure 3. The Lyubeznik resolutions of $I = (a^2, ab, b^3)$.

choice of a term order is a bad one: the lex and graded (reverse) lex orderings all yield the Taylor resolution, while the minimal resolution arises from orderings which cannot be term orders.

We still need to show that, unlike the Scarf complex, the Lyubeznik resolution is actually a resolution.

Theorem 6.5. *The Lyubeznik resolutions of I are resolutions.*

Proof. Let $M = \{m_1, \ldots, m_s\}$ be the generators of I and fix an order $\prec$ on M. For each multidegree μ, we need to show that the simplicial subcomplex $(\Lambda_{I,\prec})_{\leq\mu}$, consisting of the rooted faces with multidegree dividing μ, has no homology.

If $\mu \notin I$, this is the empty complex. If $\mu \in I$, we claim that $(\Lambda_{I,\prec})_{\leq\mu}$ is a cone.

Suppose without loss of generality that $m_1 = \min(\mu)$. We claim that, if F is a face of $(\Lambda_{I,\prec})_{\leq\mu}$, then $F \cup \{m_1\}$ is a face as well. First, note that $\operatorname{mdeg}(F \cup \{m_1\})$ divides μ because both m_1 and $\operatorname{mdeg}(F)$ do. Thus it suffices to show that $F \cup \{m_1\}$ is rooted. Observe that $\min(F \cup \{m_1\}) = m_1$ because $\operatorname{mdeg}(F \cup \{m_1\})$ divides μ and m_1 divides $\operatorname{mdeg}(F \cup \{m_1\})$. If $G \subset F$, then $\min(G) \in G$ because F is rooted, and $\min(G \cup \{m_1\}) = m_1$. Thus $F \cup \{m_1\}$ is rooted.

Hence $(\Lambda_{I,\prec})_{\leq\mu}$ is a simplicial cone on m_1 and is contractible. □

7 Intersections

The only new result of this paper is that the Scarf complex of an ideal I is the intersection of all its minimal resolutions. To make this statement precise, we need to refer to some ambient space that contains all the minimal resolutions; the natural choice is the Taylor resolution.

Theorem 7.1. *Let I be a monomial ideal. Let $\mathbb{D}_I$ be the intersection of all isomorphic embeddings of the minimal resolution of I in its Taylor resolution. Then $\mathbb{D}_I = \mathbb{S}_I$ is the Scarf complex of I.*

Proof. We showed in Theorem 5.6 that the Scarf complex is contained in this intersection. It suffices to show that the intersection of all minimal resolutions lies inside the Scarf complex. We will show that in fact the intersection of all the Lyubeznik resolutions is the Scarf complex. The result will follow because each Lyubeznik resolution contains a copy of its minimal resolution; if the Scarf complex contains the intersection of these embeddings of the minimal resolution, it must contain the (smaller) intersection of all embeddings.

Suppose that F is a face of every Lyubeznik simplicial complex. This means that, regardless of the ordering of the monomial generators of I, the first generator dividing $\mathrm{mdeg}(F)$ appears in F. Equivalently, every generator which divides $\mathrm{mdeg}(F)$ appears as a vertex of F. Thus, F is the complete simplex on the vertices with multidegree dividing $\mathrm{mdeg}(F)$.

Now suppose that there exists another face G with the same multidegree as F. Every vertex of G divides $\mathrm{mdeg}(G) = \mathrm{mdeg}(F)$, so in particular $G \subset F$. But this means that G is also a face of every Lyubeznik simplicial complex, so every generator dividing $\mathrm{mdeg}(G)$ is a vertex of G by the above argument. In particular, $F = G$. This proves that F is the unique face with multidegree $\mathrm{mdeg}(F)$, i.e., F is in the Scarf complex. □

8 Questions

The viewpoint that allows us to discuss Theorem 7.1 as we have, without reference to the gigantic index set in its statement, requires that we consider a resolution together with its basis, so resolutions which are isomorphic as algebraic chain complexes can still be viewed as different objects. The common use of the phrase "*the* minimal resolution" (instead of "*a* minimal resolution") suggests that this this point of view is relatively new, or at any rate has not been deemed significant. In any event, there are some natural questions which would not make sense from a more traditional point of view.

Question 8.1. Let I be a monomial ideal. Are there (interesting) resolutions of I which are not subcomplexes of the Taylor resolution?

It is simple enough to construct uninteresting resolutions which are not subcomplexes of $\mathbb{T}$; for example, one may take the direct sum of $\mathbb{T}$ with a trivial complex $0 \to S \to S \to 0$. (This is only uninteresting when the basis is distinguished, as all non-minimal resolutions are isomorphic to a direct sum of a minimal resolution with trivial complexes. Actually finding the bases for these summands seems to be an intractable problem.) However, all the interesting resolutions I understand are subcomplexes of the Taylor complex in a natural way: their basis elements can be expressed with relative ease as linear combinations of Taylor symbols. Consider for example the edge ideal of a four-cycle, $I = (ab, bc, cd, ad)$. Its minimal resolution occurs inside the Taylor resolution as the subcomplex

$$0 \to \left\langle \begin{matrix} [ab, bc, cd] \\ +[ab, cd, ad] \end{matrix} \right\rangle \to \left\langle \begin{matrix} [ab, bc],\ [bc, cd], \\ [cd, ad],\ [ab, ad] \end{matrix} \right\rangle \to \left\langle \begin{matrix} [ab],\ [bc], \\ [cd],\ [ad] \end{matrix} \right\rangle \to S[\varnothing] \to 0.$$

The generators and first syzygies have bases of pure Taylor symbols, and the second syzygies involve a sparse mixed term. In general, if a resolution is constructed in terms of a topological or combinatorial object, one can find a basis inside the Taylor resolution by triangulating that object.

If we restrict our attention to simplicial resolutions, we can restate Question 8.1 slightly. Supposing that a resolution is a subcomplex of the Taylor resolution, it is simplicial if and only if all its basis elements are Taylor symbols. For a simplicial resolution to fail to be a subcomplex of the Taylor complex, the set of vertices of its underlying simplicial complex must not be a subset of the generators – in other words, the underlying presentation must not be minimal. Generalizing back to arbitrary resolutions, we may ask the following question.

Question 8.2. Let I be a monomial ideal. Are there (interesting) resolutions of I with repeated or non-minimal generators?

My suspicion is that such resolutions may exist, at least for special classes of ideals, and may be useful in the study of homological invariants such as regularity which are interested in the degree, rather than the number, of generators.

Acknowledgments. I thank Ananth Hariharan, Manoj Kummini, Steve Sinnott, and the algebra groups at Cornell and Kansas for inspiration and helpful discussions.

Bibliography

[1] Bayer, D., Peeva, I., Sturmfels, B., Monomial resolutions. Math Res Lett. 1998;5(1-2): 31–46. MR 1618363 (99c:13029)

[2] Bayer, D., Sturmfels, B., Cellular resolutions of monomial modules. J Reine Angew Math. 1998;502:123–140. MR 1647559 (99g:13018)

[3] Batzies, E., Welker, V., Discrete Morse theory for cellular resolutions. J Reine Angew Math. 2002;543:147–168. MR 1887881 (2003b:13017)

[4] Clark, T. B. P., Poset resolutions and lattice-linear monomial ideals. J Algebra. 2010; 323(4):899–919. MR 2578585 (2011b:13037)

[5] Grayson, D. R., Stillman, M. E., Macaulay 2, a software system for research in algebraic geometry. Available from: http://www.math.uiuc.edu/Macaulay2/.

[6] Jöllenbeck, M., Welker, V., Minimal resolutions via algebraic discrete Morse theory. Mem Amer Math Soc. 2009;197(923):vi+74. MR 2488864 (2009m:13017)

[7] Mapes, S., Finite atomic lattices and resolutions of monomial ideals. Preprint. arXiv: 1009.1430. 2010.

[8] Nagel, U., Reiner, V., Betti numbers of monomial ideals and shifted skew shapes. Electron J Combin. 2009;16(2, Special volume in honor of Anders Björner):Research Paper 3, 59. MR 2515766 (2010h:13022)

[9] Novik, I., Lyubeznik's resolution and rooted complexes. J Algebraic Combin. 2002; 16(1):97–101. MR 1941987 (2003j:13021)

[10] Peeva, I., Graded syzygies. vol. 14 of Algebra and applications. London: Springer; 2010.

[11] Peeva, I., Velasco, M., Frames and degenerations of monomial resolutions. Trans Amer Math Soc. 2011;363(4):2029–2046. MR 2746674

[12] Phan, J., Minimal monomial ideals and linear resolutions. Preprint. arXiv:math/0511032v2. 2005.

[13] Sinefakopoulos, A., On Borel fixed ideals generated in one degree. J Algebra. 2008; 319(7):2739–2760. MR 2397405 (2008m:13020)

[14] Tchernev, A. B., Representations of matroids and free resolutions for multigraded modules. Adv Math. 2007;208(1):75–134. MR 2304312 (2008i:13020)

[15] Velasco, M., Minimal free resolutions that are not supported by a CW-complex. J Algebra. 2008;319(1):102–114. MR 2378063 (2008j:13028)

Author Information

Jeff Mermin, Department of Mathematics, Oklahoma State University, Stillwater, OK, USA.
E-mail: mermin@math.okstate.edu

Progress in Commutative Algebra 1, 143–166

A Minimal Poset Resolution of Stable Ideals

Timothy B. P. Clark

Abstract. We give a brief survey of the various topological and combinatorial techniques which have been used to construct the minimal free resolution of a stable monomial ideal in a polynomial ring over a field. The new results appearing in this paper describe a connection between certain topological and combinatorial methods for the description of said minimal resolutions. In particular, we construct a minimal poset resolution of an arbitrary stable monomial ideal by using a poset of Eliahou–Kervaire admissible symbols associated to a stable ideal. The structure of the poset under consideration is quite rich and in related analysis, we exhibit a regular CW complex which supports this resolution.

Keywords. Monomial Ideal, Minimal Poset Resolution, Regular CW Complex, CW Poset, Shellability.

2010 Mathematics Subject Classification. Primary: 13D02; Secondary: 05E40.

1 Introduction

Let $R = \Bbbk[x_1, \dots, x_d]$, where $\Bbbk$ is a field and R is considered with its standard $\mathbb{Z}^d$ grading (multigrading). For a monomial ideal N of R, the *minimal free resolution* of the module R/N is a well-studied invariant whose non-recursive construction using only the field $\Bbbk$ and the unique monomial generators of the ideal is an open problem.

Precisely, a minimal free resolution is an exact sequence of multigraded R-modules connected by multigraded morphisms which encodes the minimal relations between generators of the syzygy modules of R/N. We denote a minimal free resolution of R/N as

$$\mathcal{F} : \ \cdots \longrightarrow \bigoplus_{\alpha\in\mathbb{Z}^n} F_{i,\alpha} \xrightarrow{\partial_i} \bigoplus_{\alpha\in\mathbb{Z}^n} F_{i-1,\alpha} \longrightarrow \cdots \longrightarrow \bigoplus_{\alpha\in\mathbb{Z}^n} F_{1,\alpha} \xrightarrow{\partial_1} \bigoplus_{\alpha\in\mathbb{Z}^n} F_{0,\alpha}$$

where the free module $F_{i,\alpha} = R(-\alpha)^{\beta_{i,\alpha}}$ is of rank $\beta_{i,\alpha}$, the map ∂_i is degree preserving for all i and $\mathrm{Coker}(\partial_1) \cong R/N$.

Structure theorems for the minimal free resolution of several classes of monomial ideals have been developed in the last 20 years, although no technique has proven to be general enough to describe the minimal resolution of an arbitrary monomial ideal. Many of the approaches appearing in the literature associate to an ideal a topological or combinatorial object whose structure is shown to mirror the algebraic structure of a (minimal) resolution. Computationally speaking, a minimal free resolution of R/N

may be constructed recursively by hand or using a computer algebra system such as Macaulay2 [16].

The earliest study of resolutions of monomial ideals was described in the thesis [23] of Diana Taylor, a student of Kaplansky. For a monomial ideal with minimal generating set $\{m_1, \ldots, m_r\}$, Taylor's resolution consists of a free module of rank $\binom{r}{k}$ appearing in homological degree k whose basis elements are in one-to-one correspondence with the cardinality k subsets $I = \{i_1, \cdots, i_k\} \subseteq \{1, \ldots, r\}$ and have multidegree matching the monomial $m_I = \operatorname{lcm}(m_i : i \in I)$. The differential in Taylor's resolution takes the unique basis element e_I to

$$\sum_{j=1}^{r} (-1)^{j+1} \frac{m_I}{m_{I \setminus \{i_j\}}} \cdot e_{I \setminus \{i_j\}}.$$

If one changes perspective of the Taylor resolution only slightly, the underlying vector space structure is easily recognized as the simplicial chain complex of an r-dimensional simplex. This re-interpretation of the Taylor resolution as an object from algebraic topology serves as an organizing example for the use of topological techniques which link the incidence structure of a regular CW complex with the syzygy structure of a monomial ideal.

Bayer and Sturmfels in [3] develop a program for this approach by first establishing a $\mathbb{Z}^n$ grading on a regular CW complex X whose r vertices are each associated with a generator of N. Indeed, for e, a nonempty cell of X, identify e with the set of vertices it contains and label e with the monomial $m_e := \operatorname{lcm}\{m_j : j \in e\}$. Algebraically, the vertices contained in the cell are viewed as a finite subset of the minimal generating set of N.

A complex of multigraded R-modules, $\mathcal{F}_N$, is said to be a *cellular resolution of* R/N if the following three properties are satisfied: for all $i \geq 0$ the free module $(\mathcal{F}_N)_i$ has as its basis the $i - 1$ dimensional cells of X, a basis element $e \in (\mathcal{F}_N)_i$ has multidegree equal to that of the monomial m_e and the differential ∂ of $\mathcal{F}_N$ acts on $e \in (\mathcal{F}_N)_i$ as

$$\partial(e) = \sum_{\substack{e' \subset e \subset X \\ \dim(e) = \dim(e') + 1}} c_{e,e'} \cdot \frac{m_e}{m_{e'}} \cdot e'$$

where $c_{e,e'}$ is the coefficient of the cell e' in the differential of e in the cellular chain complex of X. Bayer and Sturmfels further show that the complex $\mathcal{F}_N$ is a free resolution of N if and only if the subcomplex of X on the vertices whose monomial labels divide m is empty or acyclic over $\mathbb{k}$ for all $m \in R$.

Although this general approach is elegant, the task of determining an appropriate cell complex that supports a minimal free resolution for a monomial ideal is difficult. Moreover, Velasco [24] has constructed a class of monomial ideals whose minimal free resolution cannot be supported on any CW complex. In what follows, we therefore restrict attention to the so-called stable ideals, whose minimal resolution was first constructed explicitly by Eliahou and Kervaire [15] using combinatorial methods. Recently, these ideals were shown to have a minimal cellular resolution separately by Batzies and Welker [1] and Mermin [19]. The connections between the combinatorial technique of Eliahou and Kervaire and these two recent topological approaches comprise the original results appearing in this paper.

Eliahou and Kervaire [15] call a monomial ideal N *stable* if for every monomial $m \in N$, the monomial $m \cdot x_i/x_r \in N$ for each $1 \leq i < r$, where $r = \max\{k : x_k \text{ divides } m\}$. They provide a construction of the minimal free resolution of a stable monomial ideal by identifying basis elements of the free modules (called admissible symbols) and describing how the maps within the resolution act on these symbols. The class of stable ideals has been extensively studied, and several of its subclasses have been shown to have relevant applications, submit to novel analytical techniques, or both.

We recall two refinements of the definition of stability. An ideal is said to be *strongly stable* if whenever $i < j$ and m is a monomial such that $mx_j \in N$, it follows that $mx_i \in N$. Clearly, the stable ideals contain the strongly stable ideals as a subclass. When the characteristic of $\mathbb{k}$ is zero, strongly stable ideals are referred to as *Borel* ideals. The class of Borel ideals have been given much attention due to their importance in Gröbner Basis Theory [13].

Turning to the topological methods which have been used to describe the minimal free resolution of a stable ideal, we focus on the most general constructions appearing in the literature.

Batzies and Welker in [1] develop an application of discrete Morse theory to the reduced cellular chain complex of a $\mathbb{Z}^d$ graded regular CW complex. This technique reduces the length of and number of free modules appearing in a nonminimal cellular resolution by collapsing certain cells of the given CW complex. The (not necessarily regular) CW complex which results from this collapsing procedure is homotopy equivalent to the original regular CW complex but has fewer cells. For the class of stable monomial ideals, this approach results in the reduction of the Taylor resolution of a stable monomial ideal to a minimal cellular resolution which matches the construction of Lyubeznik [18].

Mermin in [19] studies the minimal free resolution of an arbitrary stable ideal by defining a regular CW complex whose cells are built using the variable exchange property which characterizes a stable ideal. This regular CW complex is shown to support the original minimal resolution of Eliahou and Kervaire. Furthermore, the regular CW complex construction of Mermin seems to match that of Batzies and Welker, although the precise connections between these methods have not been studied.

In addition to these general techniques, topological approaches for constructing a minimal resolution of ideals in certain subclasses of the stable ideals are also present in the literature.

Sinefakopoulos in [22] defines a Borel principal ideal as the smallest Borel-fixed monomial ideal having a fixed monomial m in its generating set. The inductive construction which he describes results in a shellable, polyhedral cell complex that supports the minimal free resolution of an arbitrary ideal from this subclass. The cellular incidence structure of this polyhedral complex is significantly different from the CW complexes appearing in [1] and [19].

In [17] Horwitz constructs the minimal free resolution of a Borel ideal which is generated by squarefree quadratic monomials. By reinterpreting said Borel ideal as an edge ideal, the algebraic analogues of certain graph-theoretic techniques prove to be useful in the construction of a minimal resolution. This resolution in fact has a regular cellular structure and has differential maps which coincide with those of the Eliahou–Kervaire resolution.

Using techniques which they first developed in [11], Corso and Nagel [12] recover the result of Horwitz and provide a more explicit construction for the minimal cellular resolution of an arbitrary strongly stable ideal generated in degree two. Their construction associates a strongly stable ideal to a Ferrers tableau which in turn gives rise to an associated polyhedral cell complex. This cell complex is shown to support the minimal free resolution of the strongly stable ideal in question. This technique is generalized to the class of squarefree strongly stable ideals generated in a fixed degree further by Nagel and Reiner in [21].

In this paper, we begin with a combinatorial perspective of stable ideals, whereby we construct a minimal poset resolution of an arbitrary stable ideal N. Precisely, we define a poset $(P_N, <)$ on the admissible symbols of Eliahou and Kervaire by taking advantage of a decomposition property unique to the monomials contained in stable ideals.

In Section 2, we review the fundamentals of poset resolutions and define the poset of admissible symbols P_N. In our first main result, Theorem 2.4, we recover the Eliahou–Kervaire resolution of a stable ideal as a poset resolution. The value of this technique lies in the structural fact that the maps in the resolution act on the basis elements of the free modules in a way that mirrors the covering relations in P_N. Considering the lattice-linear ideals of [9], poset resolutions provide a common perspective from which to view the minimal resolutions of three large and well-studied classes of monomial ideals; stable ideals, Scarf ideals [2] and ideals having a linear resolution [14].

An advantage of the method described herein is that for a fixed stable ideal, the combinatorial information contained in the poset of admissible symbols can be transformed into the topological incidence structure of a regular CW complex. Specifically, the poset of admissible symbols P_N is a *CW poset* in the sense of Björner [5], so that it is the face poset of a regular CW complex X_N.

In our second main result, Theorem 6.4, we show that X_N supports a minimal cellular resolution of the stable ideal N. By using this combinatorial connection, we recover a minimal cellular resolution of N in a manner distinct from two of the previously described methods. Indeed, the cell complex that comes as a consequence of Björner's correspondence coincides with the one produced in the work of Mermin [19] and appears to match the cell complex of Batzies and Welker [1]. The details of these connections are the subject of future research.

2 Poset Resolutions and Stable Ideals

Let $(P, <)$ be a finite poset with set of atoms A and write $\beta \lessdot \alpha$ if $\beta < \alpha$ and there is no $\gamma \in P$ such that $\beta < \gamma < \alpha$. We say that β is *covered by* α in this situation. For $\alpha \in P$, write the order complex of the associated open interval as $\Delta_\alpha = \Delta(\hat{0}, \alpha)$. In [9], the collection of simplicial complexes

$$\{\Delta_\alpha : \alpha \in P\}$$

is used to construct a sequence of vector spaces and vector space maps

$$\mathcal{D}(P) : \quad \cdots \longrightarrow \mathcal{D}_i \xrightarrow{\varphi_i} \mathcal{D}_{i-1} \longrightarrow \cdots \longrightarrow \mathcal{D}_1 \xrightarrow{\varphi_1} \mathcal{D}_0.$$

For $i \geq 1$, the vector space $\mathcal{D}_i$ is defined as

$$\mathcal{D}_i = \bigoplus_{\alpha \in P \setminus \{\hat{0}\}} \mathcal{D}_{i,\alpha},$$

where $\mathcal{D}_{i,\alpha} = \widetilde{H}_{i-2}(\Delta_\alpha, \Bbbk)$. In particular, the vector space $\mathcal{D}_1$ has its basis indexed by the set of atoms A in P. For notational simplicity, when $\lambda \lessdot \alpha$ let $\mathbf{D}_\lambda = \Delta(\hat{0}, \lambda]$ and

$$\Delta_{\alpha,\lambda} = \mathbf{D}_\lambda \cap \Big(\bigcup_{\substack{\beta \lessdot \alpha \\ \lambda \neq \beta}} \mathbf{D}_\beta \Big).$$

When $i \geq 2$, the maps φ_i are defined using the maps in the Mayer–Vietoris long exact sequence in reduced homology associated to the short exact sequence of reduced simplicial chain complexes

$$0 \to \widetilde{\mathcal{C}}_i(\Delta_{\alpha,\lambda}) \to \widetilde{\mathcal{C}}_i(\mathbf{D}_\lambda) \oplus \widetilde{\mathcal{C}}_n\Big(\bigcup_{\substack{\beta \lessdot \alpha \\ \lambda \neq \beta}} \mathbf{D}_\beta \Big) \to \widetilde{\mathcal{C}}_i(\Delta_\alpha) \to 0$$

where the triple under consideration is

$$\Big(\mathbf{D}_\lambda, \bigcup_{\substack{\beta \lessdot \alpha \\ \lambda \neq \beta}} \mathbf{D}_\beta, \Delta_\alpha \Big). \tag{2.1}$$

For $i \geq 2$ we write $\iota : \widetilde{H}_{i-3}(\Delta_{\alpha,\lambda}, \Bbbk) \to \widetilde{H}_{i-3}(\Delta_\lambda, \Bbbk)$ for the map induced in homology by the inclusion map and

$$\delta_{i-2}^{\alpha,\lambda} : \widetilde{H}_{i-2}(\Delta_\alpha, \Bbbk) \to \widetilde{H}_{i-3}(\Delta_{\alpha,\lambda}, \Bbbk)$$

for the connecting homomorphism from the Mayer–Vietoris sequence in homology of (2.1). Set

$$\varphi_i^{\alpha,\lambda} : \mathcal{D}_{i,\alpha} \to \mathcal{D}_{i-1,\lambda}$$

as the composition $\varphi_i^{\alpha,\lambda} = \iota \circ \delta_{i-2}^{\alpha,\lambda}$. The map $\varphi_i : \mathcal{D}_i \to \mathcal{D}_{i-1}$ is then defined componentwise by

$$\varphi_i|_{\mathcal{D}_{i,\alpha}} = \sum_{\lambda \lessdot \alpha} \varphi_i^{\alpha,\lambda}$$

For $i = 0$, we define a one-dimensional vector space as $\mathcal{D}_0 = \widetilde{H}_{-1}(\{\varnothing\}, \Bbbk)$ and define $\varphi_1 : \mathcal{D}_1 \to \mathcal{D}_0$ componentwise as $\varphi_1|_{\mathcal{D}_{1,\alpha}} = \mathrm{id}_{\widetilde{H}_{-1}(\{\varnothing\},\Bbbk)}$.

We now describe the process by which the sequence of vector spaces $\mathcal{D}(P)$ is transformed into a sequence of multigraded modules. For a monomial $m = x_1^{\mathbf{a}_1} \cdots x_d^{\mathbf{a}_d} \in R$ we write $\mathrm{mdeg}(m) = (\mathbf{a}_1, \ldots, \mathbf{a}_d)$ and $\deg_{x_\ell}(m) = \mathbf{a}_\ell$ for $1 \leq \ell \leq d$. Assuming the existence of a map of partially ordered sets $\eta : P \longrightarrow \mathbb{N}^n$, the sequence of vector spaces $\mathcal{D}(P)$ is *homogenized* to produce

$$\mathcal{F}(\eta): \quad \cdots \longrightarrow \mathcal{F}_t \xrightarrow{\partial_t} \mathcal{F}_{t-1} \longrightarrow \cdots \longrightarrow \mathcal{F}_1 \xrightarrow{\partial_1} \mathcal{F}_0,$$

a sequence of free multigraded R-modules and multigraded R-module homomorphisms.

For $i \geq 1$, we set

$$\mathcal{F}_i = \bigoplus_{\hat{0} \neq \lambda \in P} \mathcal{F}_{i,\lambda} = \bigoplus_{\hat{0} \neq \lambda \in P} R \otimes_{\Bbbk} \mathcal{D}_{i,\lambda}$$

where the grading is defined as $\mathrm{mdeg}(x^{\mathbf{a}} \otimes v) = \mathbf{a} + \eta(\lambda)$ for each $v \in \mathcal{D}_{i,\lambda}$.

The differential $\partial_i : \mathcal{F}_i \longrightarrow \mathcal{F}_{i-1}$ in this sequence of multigraded modules is defined as

$$\partial_i|_{\mathcal{F}_{i,\alpha}} = \sum_{\lambda \lessdot \alpha} \partial_i^{\alpha,\lambda}$$

where $\partial_i^{\alpha,\lambda} : \mathcal{F}_{i,\alpha} \longrightarrow \mathcal{F}_{i-1,\lambda}$ takes the form $\partial_i^{\alpha,\lambda} = x^{\eta(\alpha)-\eta(\lambda)} \otimes \varphi_i^{\alpha,\lambda}$ for $\lambda \lessdot \alpha$.

We set $F_0 = R \otimes_{\Bbbk} \mathcal{D}_0$ and multigrade the result with $\mathrm{mdeg}(x^{\mathbf{a}} \otimes v) = \mathbf{a}$ for each $v \in \mathcal{D}_0$. The differential $\partial_1^{\alpha,\lambda} : \mathcal{F}_{1,\alpha} \longrightarrow \mathcal{F}_{0,\lambda}$ is defined componentwise as

$$\partial_1|_{F_{1,\lambda}} = x^{\eta(\lambda)} \otimes \varphi_1|_{\mathcal{D}_{1,\lambda}}.$$

The sequence $\mathcal{F}(\eta)$ approximates a free resolution of the multigraded module R/M where M is the ideal in R generated by the monomials

$$\{x^{\eta(a)} : a \in A\}$$

whose multidegrees are given by the images of the atoms of P.

Definition 2.1 ([9]). If $\mathcal{F}(\eta)$ is an acyclic complex of multigraded modules, then we say that it is a *poset resolution* of the ideal M.

Throughout the remainder of the paper, N will denote a stable monomial ideal in R and we write $G(N)$ as the unique minimal generating set of N. For a monomial $m \in R$ set

$$\max(m) = \max\{k \mid x_k \text{ divides } m\}$$

and

$$\min(m) = \min\{k \mid x_k \text{ divides } m\}.$$

To describe further the class of stable ideals, let $[d-1] = \{1, \dots, d-1\}$, for $I \subseteq [d-1]$ let $\max(I) = \max\{i \mid i \in I\}$ and write $x_I = \prod_{i \in I} x_i$.

In Lemma 1.2 of [15], Eliahou and Kervaire prove that a monomial ideal N is stable if and only if for each monomial $m \in N$ there exists a unique $n \in G(N)$ with the property that $m = n \cdot y$ and $\max(n) \leq \min(y)$. We adopt the language and notation introduced in the paper of Eliahou and Kervaire, and refer to n as the *unique decomposition* of the monomial m. Following their convention, we encode this property in a *decomposition map* $\mathbf{g} : M(N) \longrightarrow G(N)$ where $M(N)$ is the collection of monomials of N and $\mathbf{g}(m) = n$.

Definition 2.2 ([15]). An *admissible symbol* is an ordered pair (I, m) which satisfies $\max(I) < \max(m)$, where $m \in G(N)$ and $I \subseteq [d-1]$.

Definition 2.3. The *poset of admissible symbols* is the set P_N of all admissible symbols associated to N, along with the symbol $\hat{0} = (\varnothing, 1)$ which is defined to be the minimum element of P_N. The partial ordering on P_N is

$$(J, n) \leq (I, m) \iff \begin{array}{l} J \subseteq I \text{ and there exists} \\ C \subseteq I \setminus J \text{ so that } n = \mathbf{g}(x_C m) \end{array}$$

when both symbols are admissible.

In the case when $(J, n) < (I, m)$ and $I = J \cup \{\ell\}$ for some ℓ, then we write $(J, n) \lessdot (I, m)$ to describe the covering that occurs in P_N. As constructed, we have $\hat{0} \lessdot (\varnothing, m)$ for every $m \in G(N)$. We are now in a position to state our first main result.

Theorem 2.4. *Suppose that N is a stable monomial ideal with poset of admissible symbols P_N and define the map $\eta : P_N \longrightarrow \mathbb{N}^n$ so that $(I, m) \mapsto \operatorname{mdeg}(x_I m)$. Then the complex $\mathcal{F}(\eta)$ is a minimal poset resolution of R/N.*

In order to prove Theorem 2.4, we first describe the combinatorial structure of P_N and then exhibit the connection between the complex $\mathcal{F}(\eta)$ and the minimal free resolution of the stable ideal N constructed by Eliahou and Kervaire in [15].

3 The Shellability of P_N

We begin this section by recalling some general facts regarding the shellability of partially ordered sets. Recall that a poset P is called *shellable* if the facets of its order complex $\Delta(P)$ can be arranged in a linear order $F_1, F_2, \ldots, F_t$ in such a way that the subcomplex

$$\Big(\bigcup_{i=1}^{k-1} F_i\Big) \cap F_k$$

is a nonempty union of maximal proper faces of F_k for $k = 2, \ldots, t$. Such an ordering of facets is called a *shelling*.

Definition 3.1. Let $\mathcal{E}(P)$ denote the collection of edges in the Hasse diagram of a poset P. An *edge labeling* of P is a map $\lambda : \mathcal{E}(P) \longrightarrow \Lambda$ where Λ is some poset.

For $\sigma = a_1 \lessdot \cdots \lessdot a_k$, a maximal chain of P, the *edge label* of σ is the sequence of labels $\lambda(\sigma) = (\lambda(a_1 \lessdot a_2), \ldots, \lambda(a_{k-1} \lessdot a_k))$.

Definition 3.2. An edge labeling λ is called an *EL-labeling (edge lexicographical labeling)* if for every interval $[x, y]$ in P,

(i) there is a unique maximal chain σ in $[x, y]$, such that the labels of σ form an increasing sequence in Λ. We call σ the unique increasing maximal chain in $[x, y]$.

(ii) $\lambda(\sigma) < \lambda(\sigma')$ under the lexicographic partial ordering in Λ for all other maximal chains σ' in $[x, y]$.

A graded poset that admits an EL-labeling is said to be *EL-shellable (edge lexicographically shellable)*.

We further recall the following fundamental result of Björner and Wachs.

Theorem 3.3 ([7]). *EL-shellable posets are shellable.*

We now define an edge labeling of the poset of admissible symbols P_N.

Definition 3.4. Let $\lambda : P_N \to \mathbb{Z}$ take the form

$$\lambda\left((J,n) \lessdot (I,m)\right) = \begin{cases} 0 & \text{if } n = 1 \\ -\ell & \text{if } n = m \\ \ell & \text{if } n \neq m, \end{cases}$$

where $\{\ell\} = I \setminus J$.

Example 3.5. The labeled Hasse diagram for the poset of admissible symbols, P_N, of the stable ideal $N = \langle a, b, c\rangle^2 = \langle a^2, ab, ac, b^2, bc, c^2\rangle$ is

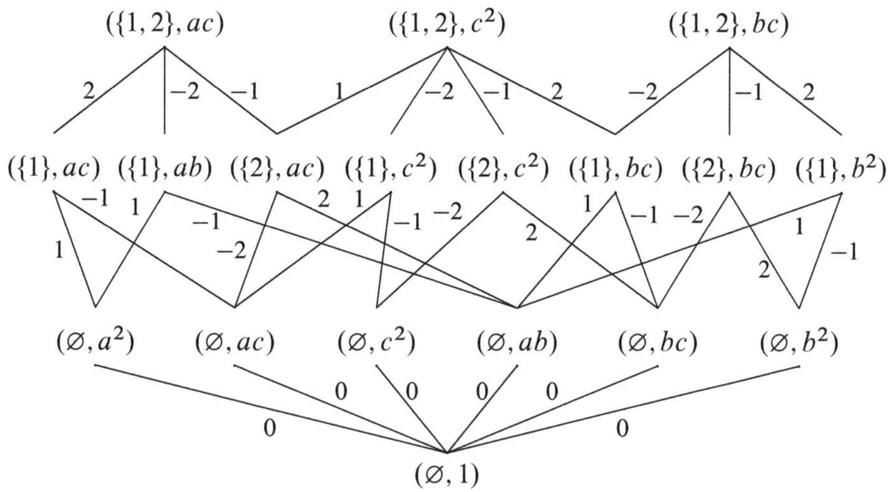

Recall that given a poset P, the *dual poset* P^* has an underlying set identical to that of P, with $x < y$ in P^* if and only if $y < x$ in P. Further, an edge labeling of a poset P may also be viewed as an edge labeling of its dual poset and P is said to be *dual shellable* if P^* is a shellable poset.

Theorem 3.6. *The poset P_N is dual EL-shellable with λ defined as above.*

Before turning to the proof of Theorem 3.6, we discuss some properties of the decomposition map $\mathbf{g}$ and the edge labeling λ.

Remarks 3.1.

(i) [15, Lemma 1.3] For any monomial w and any monomial $m \in N$, we have $\mathbf{g}(w\mathbf{g}(m)) = \mathbf{g}(wm)$ and $\max(\mathbf{g}(wm)) \leq \max(\mathbf{g}(m))$. We refer to the first property as the *associativity* of $\mathbf{g}$.

(ii) Suppose that $[(I,m),(J,n)]$ is a closed interval in the dual poset P_N^*. Given a sequence of labels

$$(l_1,\ldots,l_k)$$

there is at most one maximal chain σ in the closed interval such that

$$\lambda(\sigma) = (l_1,\ldots,l_k).$$

When it exists, this chain must be equal to

$$(I,m) \lessdot (I \setminus \{\ell_1\}, n_1) \lessdot \cdots \lessdot (I \setminus \{\ell_1,\ldots,\ell_{k-1}\}, n_{k-1}) \lessdot (J,n)$$

where $\ell_i = |l_i|$, the set $I \setminus J = \{\ell_1,\ldots,\ell_k\}$ and

$$n_i = \begin{cases} \mathbf{g}(x_{\ell_i} n_{i-1}) & \text{if } l_i > 0 \\ n_{i-1} & \text{if } l_i < 0 \end{cases}$$

for $1 \leq i \leq k$ with $n_0 = m$ and $n_k = n$.

Suppose that $(J, n) < (I, m)$ is a pair of comparable admissible symbols. Then $n = \mathbf{g}(x_{C'}m)$ for some $C' \subseteq I \setminus J$. Let $C = \{c \in C' \mid c \leq \max(n)\}$. Then by associativity and [15, Lemma 1.2] we have $n = \mathbf{g}(x_{C'}m) = \mathbf{g}(x_{C' \setminus C}\mathbf{g}(x_C m)) = \mathbf{g}(x_C m)$. In this way, any representation of $n = \mathbf{g}(x_{C'}m)$ may be reduced to $n = \mathbf{g}(x_C m)$ under the conditions above.

Notation 3.2. Implicit in all subsequent arguments is the convention that a representation $n = \mathbf{g}(x_C m)$ is written in reduced form.

Lemma 3.7. *For a reduced representation of $n = \mathbf{g}(x_C m)$ the set C is the unique subset of minimum cardinality among all $C' \subseteq I \setminus J$ for which $n = \mathbf{g}(x_{C'}m)$.*

Proof. Suppose that C is not the subset of $I \setminus J$ with smallest cardinality, namely that there exists $D \subseteq I \setminus J$ with $|D| < |C|$ and $n = \mathbf{g}(x_D m)$. By definition, $x_C \cdot m = n \cdot y$ and $x_D \cdot m = n \cdot u$ where $\max(n) \leq \min(y)$ and $\max(n) \leq \min(u)$. The assumption of $|D| < |C|$ implies that there exists $c \in C$ such that $c \notin D$. Rearranging and combining the two equations above, we arrive at the equality $x_C \cdot u = x_D \cdot y$. This equality allows us to conclude that x_c divides y since it cannot divide x_D. By definition, $\max(n) \leq \min(y)$ and therefore $\max(n) \leq c$. Further, since we assumed that $n = \mathbf{g}(x_C m)$ possessed the property that $c \leq \max(n)$ we have $c \leq \max(n) \leq c$ so that $\max(n) = c$. This equality also has implications for x_D and u, namely that $c \leq \min(u)$ and $\max(D) \leq \max(n) = c$ so that $\max(D) < c$ since $c \notin D$. However, $c < \max(D) \leq \max(n) = c$ is a contradiction, and our original supposition that such a D exists is false. If C and D are distinct subsets of $I \setminus J$ with $|C| = |D|$ and $n = \mathbf{g}(x_C m) = \mathbf{g}(x_D m)$ then there is a $c \in C$ and $d \in D$ for which $c \notin D$ and $d \notin C$. As before, we use the equality $x_C \cdot u = x_D \cdot y$ and now conclude that x_c divides y and x_d divides u. Therefore, $c \leq \max(C) \leq \max(n) \leq \min(y) \leq c$ and similarly $d \leq \max(D) \leq \max(n) \leq \min(u) \leq d$ so that $\max(n) = c = d$, □

Proof of Theorem 3.6. To prove the dual EL-Shellability of P_N, recall that for the poset of admissible symbols P_N, we have comparability in the dual poset given by $(I, m) < (J, n) \in P_N^*$ if and only if $(J, n) < (I, m) \in P_N$. We proceed with the proof by considering the various types of closed intervals that appear in the dual poset P_N^*.

Case 1: Consider the closed interval $[(I, m), \hat{0}]$. Write $I = \{d_1, \ldots, d_t\}$ so that $d_j < d_{j+1}$, for every $j = 1, \ldots, t$. The maximal chain

$$\sigma = (I, m) \lessdot (I \setminus \{d_t\}, m) \lessdot (I \setminus \{d_t, d_{t-1}\}, m) \lessdot \cdots \lessdot (\varnothing, m) \lessdot \hat{0}$$

has the increasing label

$$(-d_t, -d_{t-1}, \ldots, -d_1, 0).$$

Consider a maximal chain $\tau \in [(I, m), \hat{0}]$ where $\tau \neq \sigma$. If each label in the sequence $\lambda(\tau)$ (except the label of coverings of the form $(\varnothing, n) \lessdot \hat{0}$) is negative, then the sequence $\lambda(\tau)$ cannot be increasing, for it must be a permutation of the sequence $\lambda(\sigma)$ where the rightmost label 0 is fixed. If any label within the sequence $\lambda(\tau)$ is positive, then again $\lambda(\tau)$ cannot be increasing since every maximal chain contains the labeled subchain

$$(\varnothing, n) \overset{0}{\lessdot} \hat{0}$$

for every $(I, m) < (\varnothing, n)$. Therefore, σ is the unique rising chain in the interval $[(I, m), \hat{0}]$. Further, $\lambda(\sigma)$ is lexicographically first among all chains in $[(I, m), \hat{0}]$ since $-d_t < \cdots < -d_1 < 0$.

Case 2: Consider the closed interval $[(I, m), (J, n)]$ of P_N^* where $(J, n) \neq \hat{0}$ and $n = m$. Again write $I \setminus J = \{d_1, \ldots, d_t\}$ such that $d_1 < \cdots < d_t$. Every maximal chain σ in $[(I, m), (J, m)]$ has a label of the form

$$\left(-d_{\rho(t)}, \ldots, -d_{\rho(1)}\right)$$

where $\rho \in \Sigma_t$ is a permutation of the set $\{1, \ldots, t\}$. Therefore, the label

$$(-d_t, \ldots, -d_1)$$

corresponding to the identity permutation is the unique increasing label in the interval $[(I, m), (J, m)]$ and is lexicographically first among all such labels.

Case 3: Consider the closed interval $[(I, m), (J, n)]$ of P_N^* where $(J, n) \neq \hat{0}$ and $m \neq n$. By Lemma 3.7, $n = \mathbf{g}(x_C m)$ for a unique $C \subseteq I \setminus J$ where $\max(C) \leq \max(n)$ and the set C is of minimum cardinality. Writing the set $C = \{c_1, \ldots, c_q\}$ and $(I \setminus J) \setminus C = \{\ell_1, \ldots, \ell_t\}$ where $\ell_1 < \cdots < \ell_t$ and $c_1 < \cdots < c_q$, it follows that the sequence of edge labels $\left(-\ell_t, \ldots, -\ell_1, c_1, \ldots, c_q\right)$ is the increasing label of a maximal chain σ in $[(I, m), (J, n)]$.

Turning to uniqueness, suppose that $\tau \neq \sigma$ is also a chain which has a rising edge label. Then

$$\lambda(\tau) = \left(-d_p, \ldots, -d_1, s_1, \ldots, s_j\right) \tag{3.3}$$

where

$$\{s_1, \ldots, s_j\} \cup \{d_1, \ldots, d_p\} = \{c_1, \ldots, c_q\} \cup \{\ell_1, \ldots, \ell_t\} = I \setminus J,$$

and $-d_p < \ldots < -d_1 < 0 < s_1 < \ldots < s_j$. Since $\tau \neq \sigma$, then $\lambda(\tau) \neq \lambda(\sigma)$ and in particular, $\{d_1, \ldots, d_p\} \neq \{\ell_1, \ldots, \ell_t\}$.

If there exists $\ell \in \{\ell_1, \ldots, \ell_t\}$ with the property that $\ell \notin \{d_1, \ldots, d_p\}$, we must have $\ell \in \{s_1, \ldots, s_j\}$ so that $\ell = s_i$ for some $i < j$ and the label $\lambda(\sigma)$ has the form

$$\left(-d_p, \ldots, -d_1, s_1, \ldots, \ell, \ldots, s_j\right). \tag{3.4}$$

By the definition of $\mathbf{g}$, we have the equalities $x_C \cdot m = n \cdot y$ and $x_S \cdot m = n \cdot u$, which may be combined and simplified to arrive at the equation $x_C \cdot u = x_S \cdot y$. The assumption that $\ell \in S$ and $\ell \notin C$ implies that x_ℓ divides u so that $\max(n) \leq \ell$.

It therefore follows that $\max(n) \leq \ell < s_{i+1} < \cdots < s_j$ when $\ell \neq s_j$ so that $\max(I \setminus \{s_1, \ldots, \ell\}) = s_j > \max(n)$, which contradicts the admissibility of the symbol $(I \setminus \{d_1, \ldots, d_p, s_1, \ldots, \ell\}, n)$. If $\ell = s_j$ then $\max(n) \leq s_j$ and therefore $n = \mathbf{g}(x_{s_1} \cdots x_j m) = \mathbf{g}(x_{s_1} \cdots x_{j-1} m)$ which implies that the symbol $(I \setminus \{s_1, \ldots, s_{j-1}\}, \mathbf{g}(x_{s_1} \cdots x_{j-1} m))$, preceding (J, n) in the chain is not admissible. If there exists $d_g \in \{d_1, \ldots, d_p\}$ with $d_g \notin \{\ell_1, \ldots, \ell_t\}$ then a similar argument again provides a contradiction to admissibility.

We now prove that $\lambda(\sigma)$ is lexicographically smallest among all chains. Aiming for a contradiction, suppose that the label $\lambda(\sigma)$ is not lexicographically smallest so that there exists a maximal dual chain τ with $\lambda(\tau) < \lambda(\sigma)$. Without loss of generality, we may assume that $\lambda(\sigma)$ and $\lambda(\tau)$ differ at their leftmost label $-c$, where $-c < -\ell_t$. Such a c must be an element of the set C since $-\ell_t < \cdots < -\ell_1$ is inherent in the structure of $\lambda(\sigma)$. By construction, $c \in C$ implies that $c \leq \max(n)$ and utilizing the equations $x_C \cdot m = n \cdot y$ and $x_S \cdot m = n \cdot u$, to produce $x_C \cdot u = x_S \cdot y$, it follows that x_c divides y and therefore $c \leq \max(n) \leq \min(y) \leq c$ so that $\max(n) = c$. This forces the element $c = c_q$ for otherwise, the chain σ would contain the subchain $(I \setminus \{\ell_1, \ldots, \ell_t, c_1, \ldots, c\}, n) < (I \setminus J, n)$ where $(I \setminus \{\ell_1, \ldots, \ell_t, c_1, \ldots, c\}, n)$ is not an admissible symbol.

The desired contradiction will be obtained within an investigation of each of the three possibilities for the relationship between $\deg_{x_c}(n)$ and $\deg_{x_c}(m)$.

Suppose $\deg_{x_c}(n) > \deg_{x_c}(m)$ so that $\deg_{x_c}(n) = \deg_{x_c}(m) + 1$, based upon the structure of the set I and the definition of the decomposition map $\mathbf{g}$. In this case, the chain τ cannot end in (J, n) since $-c$, the leftmost label of τ, labels the subchain $(I, m) \lessdot (I \setminus \{c\}, m)$ and the x_c degree of every monomial appearing in the chain τ may not increase.

If $\deg_{x_c}(n) < \deg_{x_c}(m)$ then the unique decomposition $x_C \cdot m = n \cdot u$ implies that x_c divides u , for otherwise $c \in C$ implies that $\deg_{x_c}(n) = \deg_{x_c}(m) + 1$, a contradiction. The conclusion that x_c divides u allows $x_C \cdot m = n \cdot u$ to be simplified to $x_{C'} \cdot m = n \cdot u'$ where $C' = C \setminus \{c\}$ and $u' = u / x_c$. This contradicts the condition that C is the set of smallest cardinality for which $\mathbf{g}(x_C m) = n$.

Lastly, if $\deg_{x_c}(n) = \deg_{x_c}(m)$ we turn to the chain σ, whose rightmost label is c. The subchain with this label is $(I \setminus \{\ell_1, \ldots, \ell_t, c_1, \ldots, c_{j-1}\}, n') < (I \setminus J, n)$ where $x_c \cdot n' = n \cdot y$ where n does not contain this new factor of x_c. The monomial x_c therefore divides y and we can reduce $x_c \cdot n' = n \cdot y$ to $n' = n \cdot u'$ where $u' = u / x_c$, a contradiction to $n' \in G(N)$. This completes the proof. □

With Theorem 3.6 established, we immediately have the following corollary.

Corollary 3.8. *Every interval of P_N which is of the form $[\hat{0}, (I, m)]$ is finite, dual EL-shellable and therefore shellable.*

4 The topology of P_N and properties of $\mathfrak{D}(P_N)$

To establish the connection between the poset P_N and the sequence $\mathfrak{D}(P_N)$, we recall the definition of *CW poset*, due to Björner [5].

Definition 4.1 ([5]). A poset P is called a *CW poset* if

(i) P has a least element $\hat{0}$,

(ii) P is nontrivial (has more than one element),

(iii) For all $x \in P \setminus \{\hat{0}\}$, the open interval $(\hat{0}, x)$ is homeomorphic to a sphere.

After establishing this definition, Björner describes sufficient conditions for a poset to be a CW poset.

Proposition 4.2 ([5, Proposition 2.2]). *Suppose that P is a nontrivial poset such that*

(i) *P has a least element $\hat{0}$,*

(ii) *every interval $[x, y]$ of length two has cardinality four,*

(iii) *For every $x \in P$ the interval $[\hat{0}, x]$ is finite and shellable.*

Then P is a CW poset.

With this proposition in hand, we now may conclude the following about the structure of P_N, the poset of admissible symbols.

Theorem 4.3. *The poset of admissible symbols P_N is a CW poset.*

Proof. The poset P_N has a least element by construction and each of its intervals $[\hat{0}, (I, m)]$ is finite and shellable by Corollary 3.8. Thus, it remains to show that every closed interval in P_N of length two has cardinality four.

Case 1: Let $(J, n) = \hat{0}$ so that the set I is a singleton. It follows that the only poset elements in the open interior of the interval are $(\varnothing, m)$ and $(\varnothing, \mathbf{g}(x_I m))$.

Case 2: Let $(J, n) \neq \hat{0}$ and suppose that $[(J, n), (I, m)]$ is a closed interval of length two in the poset of admissible symbols, P_N. Since the interval is of length two, the set J has the form $I \setminus \{i_0, i_1\}$ for some $i_0 < i_1 \in I$. Further, any poset element in the interval must have either $I \setminus \{i_0\}$ or $I \setminus \{i_1\}$ as its first coordinate, for these sets are the only subsets of I which contain $I \setminus \{i_0, i_1\}$.

Write $m = m' x_{i_2} x_{i_3}$ where $\max(m') \leq i_2 \leq i_3$. We must now consider each of the possible orderings for the elements of the (multi) set $\{i_0, i_1, i_2, i_3\}$ to ascertain the choices available for the monomial n. Our assumptions of the inequalities $i_0 < i_1$ and $i_2 \leq i_3$ together with the admissibility of the symbol (I, m) imply that $i_1 \leq \max(I) < \max(m) \leq i_3$. Hence, determining the number of orderings amounts to producing a count of the number of orderings for elements of the set $\{i_0, i_1, i_2\}$, of which there are three, since $i_0 < i_1$.

Subcase 2.1: Suppose that $i_0 < i_1 < i_2 \leq i_3$.

If $n = m$, then the poset elements which are contained in the open interior of the interval are forced to be $(I \setminus \{i_0\}, m)$ and $(I \setminus \{i_1\}, m)$.

If $n = \mathbf{g}(x_{i_0}m)$ and $\max(I \setminus \{i_0\}) < \max(\mathbf{g}(x_{i_0}m))$ then the symbol $(I \setminus \{i_0\}, \mathbf{g}(x_{i_0}m))$ is admissible, so that it is in the open interior of the interval along with the admissible symbol $(I \setminus \{i_1\}, m)$. The symbol $(I \setminus \{i_0\}, m)$ is not comparable to $(I \setminus \{i_0\}, \mathbf{g}(x_{i_0}m))$ due to the absence of the value i_0. The symbol $(I \setminus \{i_1\}, \mathbf{g}(x_{i_1}m))$ is also not comparable to $(I \setminus \{i_0\}, \mathbf{g}(x_{i_0}m))$, for if it were then either $\mathbf{g}(x_{i_0}m) = \mathbf{g}(x_\varnothing \mathbf{g}(x_{i_1}m)) = \mathbf{g}(x_{i_1}m)$ or $\mathbf{g}(x_{i_0}x_{i_1}m) = n = \mathbf{g}(x_{i_0}m)$. The first equality is impossible since Lemma 3.7 guarantees that $\{i_0\}$ is the unique set containing one element for which $n = \mathbf{g}(x_{i_0}m)$. The second equality also can not occur since Lemma 1.2 of [15] guarantees monomial equality $\mathbf{g}(x_{i_0}x_{i_1}m) = \mathbf{g}(x_{i_0}m)$ if and only if $\max(n) \leq \min(x_{i_1}) = i_1$, which would contradict the assumption that $(I \setminus \{i_0\}, n)$ is an admissible symbol.

If $n = \mathbf{g}(x_{i_0}m)$ and $\max(I \setminus \{i_0\}) \geq \max(\mathbf{g}(x_{i_0}m))$ then the symbol $(I \setminus \{i_0\}, \mathbf{g}(x_{i_0}m))$ is not admissible and is not an element of P_N. However, we are assuming that the symbol $(I \setminus \{i_0, i_1\}, \mathbf{g}(x_{i_0}m))$ is admissible, so that i_1 is the element preventing $(I \setminus \{i_0\}, \mathbf{g}(x_{i_0}m))$ from being admissible and $\max(\mathbf{g}(x_{i_0}m)) \leq i_1$. Lemma 1.2 of [15] therefore guarantees the monomial equality $\mathbf{g}(x_{i_0}\mathbf{g}(x_{i_1}m)) = \mathbf{g}(x_{i_1}m)$ so that $(I \setminus \{i_0\}, \mathbf{g}(x_{i_0}m)) = (I \setminus \{i_0\}, \mathbf{g}(x_{i_0}x_{i_1}m))$ and the symbols $(I \setminus \{i_1\}, m)$ and $(I \setminus \{i_1\}, \mathbf{g}(x_{i_1}m))$ are each contained in the interval. Since $n = \mathbf{g}(x_{i_0}m)$, the symbol $(I \setminus \{i_0\}, m)$ is not comparable to $(I \setminus \{i_0\}, \mathbf{g}(x_{i_0}m))$.

If $n = \mathbf{g}(x_{i_1}m)$ and $\max(I \setminus \{i_1\}) \geq \max(\mathbf{g}(x_{i_1}m))$ then the symbol $(I \setminus \{i_1\}, \mathbf{g}(x_{i_1}m))$ is not admissible and is not an element of P_N. However, we are assuming the admissibility of the symbol $(I \setminus \{i_0, i_1\}, \mathbf{g}(x_{i_1}m))$ and it follows that the element i_0 is preventing the admissibility of $(I \setminus \{i_1\}, \mathbf{g}(x_{i_1}m))$. We therefore have $\max(\mathbf{g}(x_{i_1}m)) \leq i_0 < i_1$ and via Lemma 1.2 of [15], the monomial equality $\mathbf{g}(x_{i_1}m)) = \mathbf{g}(m) = m$. However, $\max(m) \leq i_0 < i_1$ is a contradiction to the admissibility of the symbol (I, m). Hence, $(I \setminus \{i_1\}, \mathbf{g}(x_{i_1}m))$ must be admissible and contained in the open interior of the interval along with the admissible symbol $(I \setminus \{i_0\}, m)$. The symbol $(I \setminus \{i_1\}, m)$ is not comparable to $(I \setminus \{i_1\}, \mathbf{g}(x_{i_1}m))$ due to the absence of the value i_1. The symbol $(I \setminus \{i_0\}, \mathbf{g}(x_{i_0}m))$ is also not comparable to $(I \setminus \{i_1\}, \mathbf{g}(x_{i_1}m))$, for if it were then either $\mathbf{g}(x_{i_1}m) = \mathbf{g}(x_\varnothing \mathbf{g}(x_{i_0}m)) = \mathbf{g}(x_{i_0}m)$ or $\mathbf{g}(x_{i_0}x_{i_1}m) = n = \mathbf{g}(x_{i_1}m)$. The first equality is impossible since Lemma 3.7 guarantees that $\{i_1\}$ is the unique set containing one element for which $n = \mathbf{g}(x_{i_1}m)$. The second equality also can not occur since Lemma 1.2 of [15] guarantees the monomial equality $\mathbf{g}(x_{i_0}x_{i_1}m) = \mathbf{g}(x_{i_1}m)$ if and only if $\max(n) \leq \min(x_{i_0}) = i_0$, which would contradict the assumption that $(I \setminus \{i_1\}, n)$ is an admissible symbol.

If $n = \mathbf{g}(x_{i_0}x_{i_1}m) \neq \mathbf{g}(x_{i_0}m)$ then the symbols $(I \setminus \{i_0\}, \mathbf{g}(x_{i_0}m))$ and $(I \setminus \{i_1\}, \mathbf{g}(x_{i_1}m))$ are admissible and are contained in the open interior of the interval. Clearly, the symbols $(I \setminus \{i_0\}, m)$ and $(I \setminus \{i_1\}, m)$ are not comparable to $(I \setminus \{i_0, i_1\}, \mathbf{g}(x_{i_0}x_{i_1}m))$ in this instance.

For each of these four choices of n, the interval $[(J,n),(I,m)]$ has four elements.

Subcase 2.2: We now consider the two remaining orderings $i_0 < i_2 \leq i_1 < i_3$ and $i_2 \leq i_0 < i_1 < i_3$. Under each of these orderings, we have $\deg_{x_{i_3}}(m) = 1$ and in light of Lemma 1.3 of [15] if $n \neq m$ we have $\max(n) < \max(m) = i_3$ and in turn that $\max(n) \leq i_1$.

If $n = m$, then the poset elements which are contained in the open interior of the interval are forced to be $(I \setminus \{i_0\}, m)$ and $(I \setminus \{i_1\}, m)$.

If $n = \mathbf{g}(x_{i_0}m)$ then $\max(n) \leq i_1$ implies that the symbol $(I \setminus \{i_0\}, \mathbf{g}(x_{i_0}m))$ is not admissible and is not an element of P_N. However, we are assuming that the symbol $(I \setminus \{i_0, i_1\}, \mathbf{g}(x_{i_0}m))$ is admissible, so that $\max(\mathbf{g}(x_{i_0}m)) \leq i_1$ and again using Lemma 1.2 of [15], we have $\mathbf{g}(x_{i_0}\mathbf{g}(x_{i_1}m)) = \mathbf{g}(x_{i_0}m)$. Therefore, $(I \setminus \{i_0\}, \mathbf{g}(x_{i_0}m)) = (I \setminus \{i_0\}, \mathbf{g}(x_{i_0}x_{i_1}m))$ and the symbols $(I \setminus \{i_1\}, m)$ and $(I \setminus \{i_1\}, \mathbf{g}(x_{i_1}m))$ are each contained in the interval. Since $n = \mathbf{g}(x_{i_0}m)$, the symbol $(I \setminus \{i_0\}, m)$ is not comparable to $(I \setminus \{i_0\}, \mathbf{g}(x_{i_0}m))$.

If $n = \mathbf{g}(x_{i_1}m)$ then the symbol $(I \setminus \{i_0\}, m)$ is certainly contained in the closed interval. Further, $(I \setminus \{i_1\}, \mathbf{g}(x_{i_1}m))$ must be admissible for were it not, then the assumption of admissibility for $(I \setminus \{i_0, i_1\}, \mathbf{g}(x_{i_1}m))$ implies that

$$i_0 \geq \max(\mathbf{g}(x_{i_1}m)) \geq \min((\mathbf{g}(x_{i_1}m))) \geq i_1,$$

a contradiction to the initial stipulation that $i_0 < i_1$. The symbol $(I \setminus \{i_1\}, m)$ is incomparable to $(I \setminus \{i_0, i_1\}, \mathbf{g}(x_{i_1}m))$ and were $(I \setminus \{i_0\}, \mathbf{g}(x_{i_0}m))$ comparable to $(I \setminus \{i_0, i_1\}, \mathbf{g}(x_{i_1}m))$, then either $\mathbf{g}(x_{i_1 m}) = \mathbf{g}(x_{\varnothing}\mathbf{g}(x_{i_0}m)) = \mathbf{g}(x_{i_0 m})$ or $\mathbf{g}(x_{i_0}x_{i_1}m) = n = \mathbf{g}(x_{i_0}m)$. The first equality contradicts Lemma 3.7 and the second may be used to arrive at a contradiction to the admissibility of $(I \setminus \{i_1\}, \mathbf{g}(x_{i_1}m))$. These arguments are similar to those used in the case when $n = \mathbf{g}(x_{i_0}m)$ and $\max(I \setminus \{i_0\}) < \max(\mathbf{g}(x_{i_0}m))$.

Again, for each of these three choices of n, the interval has four elements. □

We now analyze the vector spaces which are present in the sequence $\mathcal{D}(P_N)$ at the level of individual poset elements. In order to do so, we recall the following combinatorial results. As is standard, we write $\bar{P} = P \setminus \{\hat{0}, \hat{1}\}$.

Theorem 4.4 ([4, 8]). *If a bounded poset P is EL-shellable, then the lexicographic order of the maximal chains of P is a shelling of $\Delta(P)$. Moreover, the corresponding order of the maximal chains of $\bar{P}$ is a shelling of $\Delta(\bar{P})$.*

Theorem 4.5 ([8]). *Suppose that P is a poset for which $\hat{P} = P \cup \{\hat{0}, \hat{1}\}$ admits an EL-labeling. Then P has the homotopy type of a wedge of spheres. Furthermore, for any fixed EL-labeling:*

(i) $\widetilde{H}_i(\Delta(P), \mathbb{Z}) \cong \mathbb{Z}^{\#\text{falling chains of length } i+2}$,

(ii) *bases for i-dimensional homology (and cohomology) are induced by the falling chains of length $i + 2$.*

In the analysis that follows, we again examine the dual poset P_N^* and focus our attention on the collection of closed intervals of the form $[(I,m),\hat{0}]$, to each of which we apply Theorem 4.5. Indeed, for each admissible symbol $(I,m) \in P_N^*$ where $|I| = q$, the open interval $((I,m),\hat{0})$ is homeomorphic to a sphere of dimension $q-1$ since P_N is a CW poset. Further, the EL-labeling of $[(I,m),\hat{0}]$ guarantees that the unique generator of $\widetilde{H}_{q-1}(\Delta_{I,m},\Bbbk)$ is induced by a unique falling chain of length $q+1$. In the discussion that follows, we use the EL-shelling of Definition 3.4 to produce a canonical generator of $\widetilde{H}_{q-1}(\Delta_{I,m},\Bbbk)$ as a linear combination in which each facet of $\Delta_{I,m}$ occurs with coefficient $+1$ or -1.

To begin, consider a maximal chain $(I,m) \lessdot \sigma \lessdot \hat{0}$ which is of length $q+1$ and appears in the dual closed interval $[(I,m),\hat{0}]$ and write the label of this chain as

$$\left(l_1^\sigma,\ldots,l_q^\sigma,0\right). \tag{4.1}$$

We note that $I = \{|l_1^\sigma|,\ldots,|l_q^\sigma|\}$ and write

$$\varepsilon_\sigma = \operatorname{sgn}(\rho_\sigma)\cdot\operatorname{sgn}\Big(\prod_{t=1}^{q} l_q\Big) \tag{4.2}$$

where $\rho_\sigma \in \Sigma_q$ is the permutation arranging the sequence

$$|l_1^\sigma|,\ldots,|l_q^\sigma|$$

in *increasing* order. We endow the corresponding chain σ in $((I,m),\hat{0})$ with this sign ε_σ and refer to it as the *sign* of σ.

The unique maximal chain τ in $[(I,m),\hat{0}]$ which has a decreasing label is the chain consisting of admissible symbols having at each stage a different monomial as their second coordinate and the sequence of sets

$$I, I\setminus\{i_q\}, I\setminus\{i_{q-1},i_q\},\ldots,\{i_1,i_2\},\{i_1\},\varnothing$$

as their first coordinate. The unique falling chain $\tau \in [(I,m),\hat{0}]$ is therefore

$$(I,m) \lessdot (I_q,m_q) \lessdot (I_{q-1,q},m_{q-1,q}) \lessdot \cdots \lessdot (I_{2,\ldots,q},m_{2,\ldots,q}) \lessdot (\varnothing,m_{1,\ldots,q}) \lessdot \hat{0}$$

where $I = \{i_1,\ldots,i_q\}$ with $i_1 < \ldots < i_q$ and for $j = 1,\ldots,q$, the set $I_{j,\ldots,q} = I\setminus\{i_j,\ldots,i_q\}$ and the monomial $m_{j,\ldots,q} = \mathbf{g}(x_{i_j}\cdots x_{i_q}m)$. The label of the chain τ is therefore

$$\left(i_q,\ldots,i_1,0\right)$$

and is decreasing. If there were another such chain with decreasing label, then such a chain would be counted by Theorem 4.5 and $(\hat{0},(I,m))$ would not have the homotopy type of a sphere, a contradiction to the fact that P_N is a CW poset. In the context of the shelling order produced by the EL-shelling above, the chain τ appears lexicographically last among all maximal chains in the dual interval and is therefore the unique homology facet of $\Delta_{I,m}$.

Definition 4.6. For an admissible symbol (I,m), set

$$f(I,m) = \sum_{\sigma \in ((I,m),\hat{0})} \varepsilon_\sigma \cdot \sigma,$$

the linear combination of all maximal chains of the open interval $((I,m),\hat{0})$ with coefficients given by (4.2).

Viewing the maximal chains of $((I,m),\hat{0})$ as facets in the order complex $\Delta_{I,m}$ we now establish the following.

Lemma 4.7. *The sum $f(I,m)$ is a $(q-1)$-dimensional cycle in $\widetilde{H}_{q-1}(\Delta_{I,m},\Bbbk)$ which is not the boundary of any q-dimensional face.*

Proof. The maximal chains in the open interval $((I,m),\hat{0})$ are each of length $q-1$, so that no q-dimensional faces are present in $\Delta_{I,m}$. Thus, $f(I,m)$ cannot be the boundary of a q-dimensional face of $\Delta_{I,m}$.

We now show that $f(I,m)$ is a $(q-1)$-dimensional cycle. Suppose that σ is a maximal chain in $((I,m),\hat{0})$ and let (J,n) be an element of said chain. We exhibit a unique chain σ' which also appears in $f(I,m)$ and differs from σ only at the element (J,n).

Indeed, consider the chain $(I,m) \lessdot \sigma \lessdot \hat{0}$ along with its subchain $(J_1,n_1) \lessdot (J,n) \lessdot (J_2,n_2)$. In the proof of Theorem 4.3, each closed interval of length two was shown to be of cardinality four, and therefore there exists a unique $(J',n') \in [(J_1,n_1),(J_2,n_2)]$ which is not equal to (J,n). Defining σ' by removing (J,n) and replacing it with (J',n'), we have constructed the desired chain.

We claim that for the chains σ and σ', the associated signs ε_σ and $\varepsilon_{\sigma'}$ are opposite to one another.

If $(J_2,n_2) = \hat{0}$ then $(J_1,n_1) = (\{j\},n_1)$ for some j. Thus, $(J,n) = (\varnothing,n)$ and $(J',n') = (\varnothing,n')$ so that the chains σ and σ' have the same corresponding permutation ρ. Since either $n_1 = n$ or $n_1 = n'$, without loss of generality we assume that $n_1 = n$ so that $n' = \mathbf{g}(x_j n)$. Therefore, the subchain $(\{j\},n) \lessdot (\varnothing,n) \lessdot \hat{0}$ has $-j$ as its label, while $(\{j\},n) \lessdot (\varnothing,n') \lessdot \hat{0}$ has j as its label. This is the only difference in the labels $\lambda(\sigma)$ and $\lambda(\sigma')$ and $\varepsilon_\sigma \neq \varepsilon_{\sigma'}$ is forced.

If $(J_2,n_2) \neq \hat{0}$ then for each case that appears in the classification of intervals of length two described in the proof of Theorem 4.3, we can compute $\varepsilon_\sigma \neq \varepsilon_{\sigma'}$.

When the differential d in the reduced chain complex $\widetilde{\mathcal{C}}_\bullet(\Delta_{I,m})$ is applied to the sum $f(I,m)$, each term appears twice with opposite signs, so that $d(f(I,m)) = 0$ making $f(I,m)$ a $(q-1)$-dimensional cycle in $\widetilde{H}_{q-1}(\Delta_{I,m},\Bbbk)$ as claimed. □

5 Proof of Theorem 2.4

With the choice for the bases of the vector spaces in $\mathcal{D}(P_N)$ established, we now turn to the proof that the poset P_N supports the minimal free resolution of R/N. We first analyze the action of the differential of $\mathcal{D}(P_N)$ when it is applied to an arbitrary basis element $f(I,m)$.

Lemma 5.1. *The map* $\varphi_{q+1}^{(I,m),(J,n)}$ *sends a basic cycle* $f(I,m)$ *to* $(-1)^{p+\delta_{m,n}} \cdot f(J,n)$, *where* $I = \{i_1,\ldots,i_q\}$, *the relationship* $I \setminus J = \{i_p\}$ *holds and*

$$\delta_{m,n} = \begin{cases} 1 & \text{if } m \neq n \\ 0 & \text{otherwise.} \end{cases}$$

Proof. Write d for the simplicial differential in the reduced chain complex $\widetilde{\mathcal{C}}_\bullet(\Delta_{I,m})$. The open interval $((I,m),\hat{0})$ may be realized as the union of half-closed intervals $[(J,n),\hat{0})$, so that the order complex of each half-closed interval is a cone with apex (J,n). Applying the differential to the sum of all facets contained in the interval produces the boundary of the cone, which in this case is the order complex of $((J,n),\hat{0})$. Indeed, when d is applied to the sum

$$\upsilon = \sum_{\sigma \in [(J,n),\hat{0})} \varepsilon_\sigma \cdot \sigma,$$

the faces in which the element (J,n) remains appear twice and have opposite signs as described in the proof of Lemma 4.7. Thus, the only faces that remain in the expansion of $d(\upsilon)$ are of the form $\bar{\sigma} = \sigma \setminus \{(J,n)\}$.

Precisely,

$$\begin{aligned} \varphi_{q+1}^{(I,m),(J,n)}(f(I,m)) &= \Big[d\Big(\sum_{\sigma \in [(J,n),\hat{0})} \varepsilon_\sigma \cdot \sigma\Big)\Big] \\ &= \Big[\sum_{\sigma \in \left[(J,n),\hat{0}\right)} \varepsilon_\sigma \cdot \bar{\sigma}\Big] \\ &= \Big[\sum_{\bar{\sigma} \in ((J,n),\hat{0})} \varepsilon_\sigma \cdot \bar{\sigma}\Big]. \end{aligned} \tag{5.1}$$

The facet $\bar{\sigma}$ has an associated permutation $\rho_{\bar{\sigma}} \in \Sigma_{q-1}$, and using elementary properties of permutation signs, we have $\operatorname{sgn}(\rho_\sigma) = (-1)^{p+1} \cdot \operatorname{sgn}(\rho_{\bar{\sigma}})$, where $I \setminus J = \{i_p\}$. Considering the definition of ε_σ, for each (J,n) for which $(I,m) \lessdot (J,n) \in P_N^*$ we now have

$$\begin{aligned}
\varepsilon_\sigma &= \operatorname{sgn}(\rho_\sigma)\cdot\operatorname{sgn}\Big(\prod_{t=1}^{q} l_q\Big)\\
&= (-1)^{p+1}\cdot\operatorname{sgn}(\rho_{\bar\sigma})\cdot\operatorname{sgn}\Big(\prod_{t=2}^{q} l_q\Big)\cdot\operatorname{sgn}(l_1)\\
&= (-1)^{p+1}\cdot\operatorname{sgn}(l_1)\cdot\varepsilon_{\bar\sigma}\\
&= (-1)^{p+\delta_{m,n}}\cdot\varepsilon_{\bar\sigma}
\end{aligned}$$

since $\operatorname{sgn}(l_1) = 1$ if $n \neq m$ and $\operatorname{sgn}(l_1) = -1$ if $n = m$.

Therefore, (5.1) becomes

$$\begin{aligned}
\varphi_{q+1}^{(I,m),(J,n)}(f(I,m)) &= \Big[\sum_{\bar\sigma\in((J,n),\hat0)} \varepsilon_\sigma\cdot\bar\sigma\Big]\\
&= \Big[\sum_{\bar\sigma\in((J,n),\hat0)} (-1)^{p+\delta_{m,n}}\cdot\varepsilon_{\bar\sigma}\cdot\bar\sigma\Big]\\
&= (-1)^{p+\delta_{m,n}}\cdot\Big[\sum_{\bar\sigma\in((J,n),\hat0)} \varepsilon_{\bar\sigma}\cdot\bar\sigma\Big]\\
&= (-1)^{p+\delta_{m,n}}\cdot f(J,n)
\end{aligned}$$

which proves the lemma. □

As described in Section 2, the map φ_{q+1} is defined componentwise on the one-dimensional $\mathbb{k}$-vectorspace $\mathcal{D}_{q+1,(I,m)}$ for each poset element (I,m). Using the conclusion of Lemma 5.1, we immediately have

$$\varphi_{q+1}|_{\mathcal{D}_{q+1,(I,m)}} = \varphi_{q+1,(I,m)}(f(I,m)) = \sum_{(J,n)\lessdot(I,m)} (-1)^{p+\delta_{m,n}} f(J,n) \qquad (5.2)$$

where $I = \{i_1,\dots,i_q\}$ and $i_1 < \dots < i_q$ and $J = I\setminus\{i_p\}$.

Recall that the poset map $\eta : P_N \longrightarrow \mathbb{N}^n$ is defined as $(I,m)\mapsto \operatorname{mdeg}(x_I m)$, so that we can homogenize the sequence of vector spaces $\mathcal{D}(P_N)$ to produce

$$\mathcal{F}(\eta) : 0 \longrightarrow F_d \xrightarrow{\partial_d^{\mathcal{F}(\eta)}} F_{d-1} \longrightarrow \cdots \longrightarrow F_1 \xrightarrow{\partial_1^{\mathcal{F}(\eta)}} F_0,$$

a sequence of multigraded modules. More precisely, for $q \geq 0$ and a poset element $(I, m) \neq \hat{0}$ where $I = \{i_1, \dots, i_q\}$ and $i_1 < \cdots < i_q$, the differential $\partial^{\mathcal{F}(\eta)}$ acts on a basis element $f(I, m)$ of the free module F_{q+1} via the formula

$$\partial_{q+1}^{\mathcal{F}(\eta)}(f(I,m)) = \sum_{(J,n') \lessdot (I,m)} (-1)^{p+\delta_{m,n'}} x^{\eta(I,m)-\eta(J,n')} \cdot f(J,n')$$
$$= \sum_{(J,m) \lessdot (I,m)} (-1)^p x_{i_p} \cdot f(J,m) - \sum_{(J,n) \lessdot (I,m)} (-1)^p \frac{x_{i_p} m}{\mathbf{g}(x_{i_p} m)} \cdot f(J,n) \tag{5.3}$$

where p takes the same value as in (5.2), so that $I \setminus \{i_p\} = J$.

It remains to show that $\mathcal{F}(\eta)$ is a minimal exact complex, and to do so we identify it as the Eliahou–Kervaire resolution.

Definition 5.2. The *Eliahou–Kervaire minimal free resolution* [15] of a stable ideal N is

$$\mathcal{E} : 0 \longrightarrow E_d \xrightarrow{\partial_d^{\mathcal{E}}} E_{d-1} \longrightarrow \cdots \longrightarrow E_1 \xrightarrow{\partial_1^{\mathcal{E}}} E_0$$

where $E_0 = R$ is the free module of rank one with basis 1 and for $q \geq 0$, E_{q+1} has as basis the admissible symbols

$$\{e(I, m) : I = \{i_1, \dots, i_q\}, \max(I) < \max(m)\}.$$

When applied to a basis element, the differential of $\mathcal{E}$ takes the form

$$\partial_{q+1}^{\mathcal{E}}(e(I,m)) = \sum_{p=1}^{q} (-1)^p x_{i_p} \cdot e(I \setminus \{i_p\}, m)$$
$$- \sum_{p=1}^{q} (-1)^p \frac{x_{i_p} m}{\mathbf{g}(x_{i_p} m)} \cdot e(I \setminus \{i_p\}, \mathbf{g}(x_{i_p} m))$$

where we define $e(I \setminus \{i_p\}, \mathbf{g}(x_p m)) = 0$ when $\max(I \setminus \{i_p\}) \geq \max(\mathbf{g}(x_p m))$ (i.e. the symbol is inadmissible).

We now are in a position to prove the main result of this paper.

Proof of Theorem 2.4. The Eliahou–Kervaire symbols that are admissible index the multigraded free modules in the complexes $\mathcal{E}$ and $\mathcal{F}(\eta)$ and therefore the generators of these modules are in one to one correspondence with one another. Further, comparing Definition 5.2 and (5.3), $\partial^{\mathcal{E}}$ and $\partial^{\mathcal{F}(\eta)}$ have identical behavior on basis elements. The minimality and exactness of $\mathcal{E}$ implies the minimality and exactness of $\mathcal{F}(\eta)$ so that $\mathcal{F}(\eta)$ is a minimal poset resoution of R/N. □

6 A Minimal Cellular Resolution of R/N

The technique which follows is an example of a general approach which interprets cellular resolutions of monomial ideals through the theory of poset resolutions. This approach is described in [10], and is distinct from both the method of [1] concerning stable modules and the method of [19] which is specific to stable ideals. We begin by recalling a fundamental result due to Björner.

Proposition 6.1 ([5, Proposition 3.1]). *A poset P is a CW poset if and only if it is isomorphic to the face poset of a regular CW complex.*

In the case of the poset of admissible symbols P_N, we interpret Björner's proof explicitly to produce the corresponding regular CW complex X_N. On the level of cells, $\hat{0} \in P_N$ corresponds to the empty cell and each admissible symbol (I, m) of P_N corresponds to a closed cell $X_{I,m}$ of dimension $|I|$ for which $P(X_{I,m}) = [\hat{0}, (I, m)]$. Taking $X_N = \bigcup X_{I,m}$ we have an isomorphism of posets $P(X_N) \cong P_N$. The regular CW complex X_N also comes equipped with a $\mathbb{Z}^n$ grading by realizing the map $\eta : P_N \longrightarrow \mathbb{N}^n$ of Theorem 2.4 as a map $\eta : X_N \longrightarrow \mathbb{N}^n$ where a cell $X_{I,m} \mapsto \eta(I, m) = \text{mdeg}(x_I m)$.

Example 6.2. The stable ideal $N = \langle a, b, c\rangle^2 = \langle a^2, ab, ac, b^2, bc, c^2\rangle$ has minimal resolution supported by X_N, the regular CW complex depicted below, which has six 0-cells, eight 1-cells and three 2-cells. The face poset of this cell complex $P(X_N)$ is isomorphic to the poset of admissible symbols P_N given in Example 3.5.

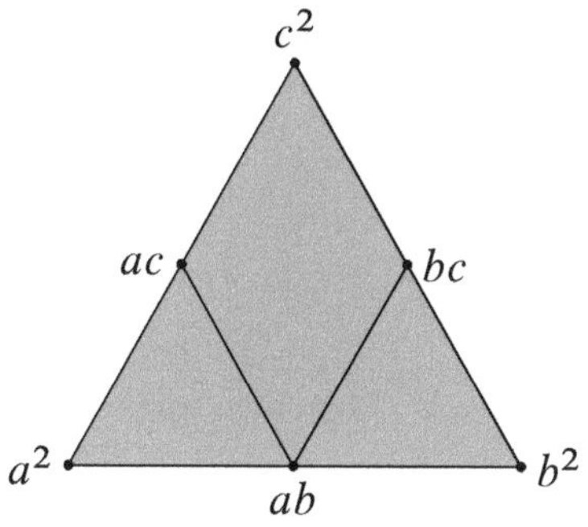

We recall the following well-known definition to which we incorporate the information given by the poset map η. For a more comprehensive view of cellular and CW resolutions, see [1, 3, 24].

Definition 6.3. A complex of multigraded R-modules, $\mathcal{F}_N$, is said to be a *cellular resolution of R/N* if there exists an $\mathbb{N}^n$-graded regular CW complex X such that:

(i) For all $i \geq 0$, the free module $(\mathcal{F}_N)_i$ has as its basis the $i - 1$ dimensional cells of X.

(ii) For a basis element $e \in (\mathcal{F}_N)_i$, one has $\operatorname{mdeg}(e) = \eta(e)$,

(iii) The differential ∂ of $\mathcal{F}_N$ acts on a basis element $e \in (\mathcal{F}_N)_i$ as

$$\partial(e) = \sum_{\substack{e' \subset e \subset X \\ \dim(e) = \dim(e')+1}} c_{e,e'} \cdot x^{\eta(e)-\eta(e')} \cdot e'$$

where $c_{e,e'}$ is the coefficient of the cell e' in the differential of e in the cellular chain complex of X.

With this definition in hand, we are now able to reinterpret Theorem 2.4 in our final result.

Theorem 6.4. *Suppose that N is a stable monomial ideal. Then the minimal free resolution $\mathcal{F}(\eta)$ is a minimal cellular resolution of R/N.*

Proof. Conditions 1 and 2 of Definition 6.3 are clear from the structure of X_N, its correspondence to the poset P_N and the construction of the resolution $\mathcal{F}(\eta)$. It therefore remains to verify that condition 3 is satisfied. The main result in [10] provides a canonical isomorphism between the complex $\mathcal{D}(P_N)$ and $\mathcal{C}(X_N)$, the cellular chain complex of X_N. Therefore, the differential of $\mathcal{F}(\eta)$ satisfies condition 3. □

Acknowledgments. I am grateful to Amanda Beecher and Alexandre Tchernev for helpful questions and discussions during this project. Thanks to Jeff Mermin for careful reading of an early version of this document.

Bibliography

[1] Batzies, E., Welker, V., Discrete Morse theory for cellular resolutions. J Reine Angew Math. 2002;543:147–168.

[2] Bayer, D., Peeva, I., Sturmfels, B., Monomial resolutions. Math Res Lett. 1998;5(1–2): 31–46.

[3] Bayer, D., Sturmfels, B., Cellular resolutions of monomial modules. J Reine Angew Math. 1998;502:123–140.

[4] Björner, A., Shellable and Cohen–Macaulay partially ordered sets. Trans Amer Math Soc. 1980;260(1):159–183.

[5] Björner, A., Posets, regular CW complexes and Bruhat order. European J Combin. 1984; 5(1):7–16.

[6] Björner, A., The homology and shellability of matroids and geometric lattices. Matroid applications. vol. 40 of Encyclopedia of mathematics and its applications. Cambridge: Cambridge University Press; 1992. p. 226–283.

[7] Björner, A., Wachs, M., On lexicographically shellable posets. Trans Amer Math Soc. 1983;277(1):323–341.

[8] Björner, A., Wachs, M., Shellable nonpure complexes and posets. I. Trans Amer Math Soc. 1996;348(4):1299–1327.

[9] Clark, T., Poset resolutions and lattice-linear monomial ideals. J Algebra. 2010;323(4): 899–919.

[10] Clark, T., Tchernev, A., Regular CW complexes and poset resolutions. In preparation.

[11] Corso, A., Nagel, U., Specializations of Ferrers ideals. J Algebraic Combin. 2008;28(3): 425–437.

[12] Corso, A., Nagel, U., Monomial and toric ideals associated to Ferrers graphs. Trans Amer Math Soc. 2009;361(3):1371–1395.

[13] Eisenbud, D., Commutative algebra with a view toward algebraic geometry. vol. 150 of Graduate texts in mathematics. New York: Springer-Verlag; 1995.

[14] Eisenbud, D., The geometry of syzygies. A second course in commutative algebra and algebraic geometry. vol. 229 of Graduate texts in mathematics. New York: Springer-Verlag; 2005.

[15] Eliahou, S., Kervaire, M., Minimal resolutions of some monomial ideals. J Algebra. 1990;129(1):1–25.

[16] Grayson, D., Stillman, M., Macaulay2, a software system for research in algebraic geometry, Available from: http://www.math.uiuc.edu/Macaulay2/.

[17] Horwitz, N., Linear resolutions of quadratic monomial ideals. J Algebra. 2007;318(2): 981–1001.

[18] Lyubeznik, G., A new explicit finite free resolution of ideals generated by monomials in an R-sequence. J Pure Appl Algebra. 1988;51(1–2):193–195.

[19] Mermin, J., The Eliahou–Kervaire resolution is cellular. J Commut Algebra. 2010;2(1): 55–78.

[20] Munkres, J., Elements of algebraic topology. Menlo Park, CA: Addison-Wesley Publishing Company; 1984.

[21] Nagel, U., Reiner, V., Betti numbers of monomial ideals and shifted skew shapes. Electron J Combin. 2009;16(2, Special volume in honor of Anders Björner):Research Paper 3.

[22] Sinefakopoulos, A., On Borel fixed ideals generated in one degree. J Algebra. 2008; 319(7):2739–2760.

[23] Taylor, D., Ideals generated by monomials in an R-sequence [PhD thesis]. University of Chicago; 1966.

[24] Velasco, M., Minimal free resolutions that are not supported by a CW-complex. J Algebra. 2008;319(1):102–114.

Author Information

Timothy B. P. Clark, Department of Mathematics and Statistics, Loyola University Maryland, Baltimore, MD, USA.
E-mail: tbclark@loyola.edu

Progress in Commutative Algebra 1, 167–198

Subsets of Complete Intersections and the EGH Conjecture

Susan M. Cooper

Abstract. We give a "Macaulay-type" characterization of the Hilbert functions of finite sets of distinct points which are subsets of complete intersections in projective space $\mathbb{P}^2$ and a family of complete intersections in $\mathbb{P}^3$. Doing so, we prove that the Hilbert functions of subsets of complete intersections of type $\{d_1, d_2\}, \{2, d, d\}$ and $\{3, d, d\}$ are the same as those for *rectangular complete intersections* of the same type. This verifies special cases of the Eisenbud–Green–Harris Conjecture, which is a statement that *lex-plus-powers ideals* exhibit extremal conditions among all homogeneous ideals containing a homogeneous regular sequence in fixed degrees. The characterization is then applied to give a family of points which has the *Cayley–Bacharach Property*.

Keywords. Hilbert Functions, O-sequences, Regular Sequences, Cayley–Bacharach Property.

2010 Mathematics Subject Classification. Primary: 13D40, 14M10; Secondary: 13C40.

1 Introduction

A finite set $\mathbb{X}$ of distinct, reduced points in projective space $\mathbb{P}^n$ is contained in some (in fact, many) complete intersections, if we have the freedom to pick the degrees of the defining polynomials. This paper focuses on the following natural question which arises by fixing the degrees which define the complete intersection:

Question 1.1. Fix integers $1 \leq d_1 \leq d_2 \leq \cdots \leq d_n$ and let $\mathcal{H}$ be the Hilbert function of some finite set of distinct points in $\mathbb{P}^n$. Do there exist finite sets of distinct, reduced points $\mathbb{X}$ and $\mathbb{Y}$ such that: (i) $\mathbb{X} \subseteq \mathbb{Y}$; (ii) the Hilbert function of $\mathbb{X}$ is $\mathcal{H}$; and (iii) $\mathbb{Y}$ is a complete intersection of type $\{d_1, \ldots, d_n\}$?

In a purely algebraic sense, Question 1.1 is connected to the *Eisenbud–Green–Harris Conjecture* (denoted EGH Conjecture) in the following way. Let $S = k[x_1, \ldots, x_n]$, where k is an algebraically closed field of characteristic zero and each variable x_i has degree 1. F. S. Macaulay characterized the sequences (called *O-sequences*) which occur as the Hilbert function of any quotient S/I, where $I \subseteq S$ is a homogeneous ideal (for details see [2, 19, 23]). Clements–Lindström [4], Greene–Kleitman [16],

The author acknowledges funding from the Natural Sciences and Engineering Research Council of Canada during parts of this project.

Kruskal [18], and Katona [17] give combinatorial results which can be used to generalize Macaulay's Theorem to Hilbert functions of quotients T/J where J is a monomial ideal in the ring $T = S/(x_1^{d_n}, x_2^{d_{n-1}}, \dots, x_n^{d_1})$. Cooper and Roberts [5] extend the work of Clements–Lindström to include non-monomial ideals in T. It is natural to try to extend this to obtain a "Macaulay-type" characterization for the Hilbert functions of standard graded k-algebras S/I where I is *any* homogeneous ideal containing a regular sequence in fixed degrees. If true, the EGH Conjecture implies that the growth bounds of Clements–Lindström also characterize such Hilbert functions, and hence Hilbert functions of subsets of complete intersections in projective space.

Unfortunately, the EGH Conjecture is only known to be true in some exceptional cases (see Section 2.1). When $n = 3$ the outstanding cases for the EGH Conjecture are the "tight degrees" $\mathbb{D} = \{d_1, d_2, d_3\}$ where $d_1 \leq d_2 \leq d_3 \leq d_1 + d_2 - 2$. These remaining cases have proven to be very difficult to verify. In this paper, we study the cases $\mathbb{D} = \{2, d_2, d_3\}$ and $\mathbb{D} = \{3, d_2, d_3\}$ where $d_3 = d_2$ in a geometric setting by exploiting the idea of O-sequences.

This paper is organized as follows. Section 2 extracts work from [5] which applies the results of Clements–Lindström and Greene–Kleitman to achieve bounds for Hilbert functions of quotients of $S/(x_1^{d_n}, x_2^{d_{n-1}}, \dots, x_n^{d_1})$. In Section 3 we define *rectangular complete intersections* and characterize the Hilbert functions of subsets of these point sets. This naturally leads to the conjecture that the Hilbert functions of subsets of complete intersections are completely characterized by the Hilbert functions of subsets of rectangular complete intersections. In Section 4 we give a brief overview of the tools which will be applied in later sections. Sections 5, 6 and 7 are dedicated to answering Question 1.1 for a family of complete intersections. Finally, we conclude in Section 8 with an application to the open problem of characterizing the Hilbert functions of points having the Cayley-Bacharach Property.

The proofs in this paper are done in a geometric setting. However, many of the details also work in a purely algebraic setting by replacing Theorem 4.1 with linkage of Artinian complete intersections (see [6, Theorem 3]).

2 Preliminary Definitions and Results

From now on we fix k to be an algebraically closed field of characteristic zero. We also let $S = k[x_1, \dots, x_n]$ and $R = k[x_0, x_1, \dots, x_n]$ with the standard grading. Let $>_{\text{lex}}$ denote the degree-lexicographic order with $x_0 >_{\text{lex}} x_1 >_{\text{lex}} \cdots >_{\text{lex}} x_n$. In addition, we fix integers $1 \leq d_1 \leq d_2 \leq \cdots \leq d_n$ and let $\mathbb{D} = \{d_1, d_2, \dots, d_n\}$.

2.1 The Eisenbud–Green–Harris Conjecture and Complete Intersections

Let S_t denote the k-vector space spanned by the monomials of degree t in S. If $I \subseteq S$ is a homogeneous ideal then $A = S/I$ has an obvious gradation $A = \bigoplus_{t \geq 0} A_t$, where $A_t = S_t/I_t$ is a finite-dimensional vector space. The *Hilbert function* of A is the function $H(A) : \mathbb{Z}_{\geq 0} \to \mathbb{Z}_{\geq 0}$ defined by

$$H(A,t) = \dim_k(A_t).$$

It is often convenient to record the integers $H(A,t)$ in a sequence and write $H(A) = \{H(A,t)\}_{t\geq 0}$.

Hilbert functions have been extensively studied. The most celebrated result is *Macaulay's Theorem* which implies that $H(A)$ can be described using *lex ideals*. Recall that a monomial ideal $L \subseteq S$ is said to be a *lex ideal* if for all monomials $u \in L$ and $v >_{\text{lex}} u$ with $\deg(u) = \deg(v)$ one has $v \in L$.

Theorem 2.1 (Macaulay's Theorem [19, 23]). *Let $I \subseteq S$ be a homogeneous ideal. Then there exists a lex ideal L such that $H(S/I) = H(S/L)$.*

The Eisenbud–Green–Harris Conjecture is an attempt to generalize Macaulay's Theorem to ideals containing regular sequences. Indeed, this conjecture has gained much recent attention. In this generalization, lex-plus-powers ideals play the analogous role of the lex-ideals.

Definition 2.2. A monomial ideal $L \subseteq S$ is said to be a *lex-plus-powers ideal with respect to $\mathbb{D}$* if there is a lex ideal $L' \subseteq S$ such that $L = L' + (x_1^{d_1}, \dots, x_n^{d_n})$.

For example, let $\mathbb{D} = \{2,3,3\}$. Then $L = (x_1^2, x_2^3, x_3^3, x_1x_2^2, x_1x_2x_3)$ is a lex-plus-powers ideal with respect to $\mathbb{D}$, but $\tilde{L} = (x_1^2, x_2^3, x_3^3, x_1x_2^2, x_1x_2x_3, x_2^2x_3)$ is not since $x_1x_3^2 >_{\text{lex}} x_2^2x_3$ and $x_1x_3^2$ is not in $\tilde{L}$. We have the following natural conjecture:

Conjecture 2.3 ([3]). Let $I \subseteq S$ be a homogeneous ideal containing a regular sequence $F_1, \dots, F_n$ of degrees $\deg(F_i) = d_i$. Then there exists a homogeneous ideal J containing $\{x_1^{d_1}, x_2^{d_2}, \dots, x_n^{d_n}\}$ such that $H(S/I) = H(S/J)$.

Clements–Lindström [4] show (in a combinatorial fashion) that for any *monomial* ideal $M \subseteq S$ containing $\{x_1^{d_1}, x_2^{d_2}, \dots, x_n^{d_n}\}$ there is a lex-plus-powers ideal L with respect to $\mathbb{D}$ with $H(S/M) = H(S/L)$. This generalizes to non-monomial ideals.

Lemma 2.4 ([5]). *If $I \subseteq S$ is any homogeneous ideal containing $\{x_1^{d_1}, x_2^{d_2}, \dots, x_n^{d_n}\}$, then there is a lex-plus-powers ideal L with respect to $\mathbb{D}$ such that $H(S/I) = H(S/L)$.*

Thus, Conjecture 2.3 can be restated as follows.

Conjecture 2.5 (Eisenbud–Green–Harris (EGH) Conjecture [3, 8, 9, 11]). If $I \subseteq S$ is a homogeneous ideal containing a regular sequence $F_1, F_2, \dots, F_n$ of degrees $\deg(F_i) = d_i$, then there is a lex-plus-powers ideal L with respect to $\mathbb{D}$ such that $H(S/I) = H(S/L)$.

Despite much effort, the EGH Conjecture is known to be true only in some exceptional cases. It was originally stated in the case when each $d_i = 2$. The conjecture has been proven in the cases where L is an *almost complete intersection* [10], and when $n = 2$ [22]. Mermin–Murai [20] and Mermin–Peeva–Stillman [21] have interesting results related to the conjecture for ideals containing a monomial regular sequence. Recently, Caviglia–Maclagan [3] have proven that the EGH Conjecture is true if $d_j > \sum_{i=1}^{j-1}(d_i - 1)$ for $j = 1, \ldots, n$. Thus, when $n = 3$ the outstanding cases for the EGH Conjecture are the "tight degrees" $\mathbb{D} = \{d_1, d_2, d_3\}$ where $d_1 \leq d_2 \leq d_3 \leq d_1 + d_2 - 2$.

In this paper we focus on the geometric side of the EGH Conjecture. If $\mathbb{X}$ is a finite set of distinct points in $\mathbb{P}^n$, then the *Hilbert function of* $\mathbb{X}$ is simply the sequence $H(\mathbb{X}) = H(R/\mathbf{I}(\mathbb{X}))$, where $\mathbf{I}(\mathbb{X}) \subseteq R = k[x_0, x_1, \ldots, x_n]$ is the homogeneous ideal of $\mathbb{X}$. Hilbert functions of ideals of finite sets of distinct points are well-studied. We will see a "Macaulay-type" characterization in the next section. This characterization uses the first difference operator Δ. Let $\mathcal{P} = \{c_i\}_{i \geq 0}$ be a sequence of non-negative integers. We define the *first difference sequence* of $\mathcal{P}$ to be the sequence $\Delta\mathcal{P} = \{e_i\}_{i \geq 0}$ where $\Delta\mathcal{P}(0) = e_0 = c_0 = \mathcal{P}(0)$ and $\Delta\mathcal{P}(i) = e_i = c_i - c_{i-1}$ for $i \geq 1$.

The main object of our study are complete intersections.

Definition 2.6. Let $\mathbb{Y}$ be a finite set of distinct points in $\mathbb{P}^n$. Then $\mathbb{Y}$ is a *complete intersection of type* $\{d_1, \ldots, d_n\}$ if $\mathbf{I}(\mathbb{Y})$ can be generated by exactly n homogeneous polynomials $F_1, \ldots, F_n \in R = k[x_0, \ldots, x_n]$, where $F_1, \ldots, F_n$ is a regular sequence and $\deg(F_i) = d_i$.

Notation 2.7. We denote by C.I.$(d_1, \ldots, d_n)$ all the finite sets in $\mathbb{P}^n$ of $d_1 d_2 \cdots d_n$ distinct points which are complete intersections of type $\{d_1, \ldots, d_n\}$.

Hilbert functions of complete intersections have especially nice properties and are well-known.

Facts 2.8. Let $\mathbb{Y} \in$ C.I.$(d_1, \ldots, d_n) \subseteq \mathbb{P}^n$. Then the following hold:

(i) $\Delta H(\mathbb{Y})$ is the Hilbert function $H(k[x_1, \ldots, x_n]/(x_1^{d_1}, \ldots, x_n^{d_n}))$;

(ii) The least integer t for which $\Delta H(\mathbb{Y}, t) = 0$ is $d_1 + d_2 + \cdots + d_n - (n-1)$;

(iii) $\Delta H(\mathbb{Y}, t)$ is symmetric, i.e. $\Delta H(\mathbb{Y}, t) = \Delta H(\mathbb{Y}, d_1 + d_2 + \cdots + d_n - n - t)$ (where $\Delta H(\mathbb{Y}, s) = 0$ for all negative s).

Notation 2.9. (i) By Facts 2.8, sets in C.I.$(d_1, \ldots, d_n)$ have Hilbert functions which depend on the degrees $d_1, \ldots, d_n$, rather than on the location of the points. That is, all sets in C.I.$(d_1, \ldots, d_n)$ have the same Hilbert function. From this point on we will denote this special Hilbert function by $H_{\text{C.I.}(d_1,\ldots,d_n)}$.

(ii) We will often compare Hilbert functions of finite sets $\mathbb{X} \subseteq \mathbb{Y} \subseteq \mathbb{P}^n$. Given two sequences $\mathcal{P} = \{c_i\}_{i \geq 0}$ and $\mathcal{P}' = \{e_i\}_{i \geq 0}$, we will write $\mathcal{P} \leq \mathcal{P}'$ if $c_i \leq e_i$ for each i.

2.2 Some Enumeration

It will be useful to consider Macaulay's Theorem in a combinatorial setting. Let I be a homogeneous ideal in $S = k[x_1, \dots, x_n]$. The growth from one degree to the next of $H(S/I)$ can be explicitly described using the well-known *Macaulay's function* ${}^{<i>} : \mathbb{Z}_{\geq 0} \to \mathbb{Z}_{\geq 0}$.

Definitions-Proposition 2.10 ([2, Lemma 4.2.6]). Let h and i be positive integers. Then h can be written uniquely in the form

$$h = \binom{m_i}{i} + \binom{m_{i-1}}{i-1} + \cdots + \binom{m_j}{j},$$

where $m_i > m_{i-1} > \cdots > m_j \geq j \geq 1$. We call this expression for h the *i-binomial expansion* of h and define

$$h^{<i>} = \binom{m_i + 1}{i+1} + \binom{m_{i-1} + 1}{i} + \cdots + \binom{m_j + 1}{j+1}.$$

By convention, we define $0^{<i>} = 0$.

For example, the 3-binomial expansion of 16 is $16 = \binom{5}{3} + \binom{4}{2}$ and so $16^{<3>} = \binom{6}{4} + \binom{5}{3} = 25$. We will often use the fact that if h and i are positive integers such that $i \geq h$, then $h^{<i>} = h$.

Definitions 2.11. Let $\mathcal{P} := \{c_i\}_{i \geq 0}$ be a sequence of non-negative integers.

(i) $\mathcal{P}$ is called an *O-sequence* if $c_0 = 1$ and $c_{i+1} \leq c_i^{<i>}$ for all $i \geq 1$.

(ii) $\mathcal{P}$ is called a *differentiable O-sequence* if $\Delta\mathcal{P}$ is an O-sequence (in which case, $\mathcal{P}$ is also an O-sequence).

Macaulay's description of the Hilbert functions of quotients S/I can be stated alternatively as follows.

Theorem 2.12 (Macaulay's Theorem [19, 23]). *Let $\mathcal{H} = \{c_i\}_{i \geq 0}$ be a sequence of non-negative integers. The following are equivalent:*

(i) *$\mathcal{H}$ is an O-sequence;*

(ii) *$\mathcal{H}$ is the Hilbert function of some quotient S/I where I is a homogeneous ideal.*

This has been generalized to Hilbert functions of points as follows.

Theorem 2.13 ([14]). *Let $\mathcal{H} = \{c_i\}_{i \geq 0}$ be a sequence of non-negative integers. The following are equivalent:*

(i) *$\mathcal{H}$ is a differentiable O-sequence, $c_1 \leq n + 1$, and $c_i = s$ for $i \gg 0$;*

(ii) *$\mathcal{H}$ is the Hilbert function of some s distinct points in $\mathbb{P}^n$.*

The idea of O-sequences has been extended to the setting of complete intersections. We state the main results here; for full details see [4, 5, 16]. We begin with the analog to Macaulay's function.

Definitions-Proposition 2.14. Fix $\mathbb{D} = \{d_1, \ldots, d_n\}$ where the d_i are integers such that $1 \leq d_1 \leq d_2 \leq \cdots \leq d_n$. Let $t, h \geq 1$ be fixed integers with the property that $h \leq \Delta H_{\text{C.I.}(d_1,\ldots,d_n)}(t)$. Then h can be written uniquely in the form

$$h = \Delta H_{\text{C.I.}(d_{b_t},\ldots,d_n)}(t) + \Delta H_{\text{C.I.}(d_{b_{t-1}},\ldots,d_n)}(t-1) + \cdots + \Delta H_{\text{C.I.}(d_{b_i},\ldots,d_n)}(i)$$

where $1 \leq b_t \leq b_{t-1} \leq \cdots \leq b_i \leq n$, terms of the form $\Delta H_{\text{C.I.}(d_1,\ldots,d_n)}(j)$ appear at most once, and terms of the form $\Delta H_{\text{C.I.}(d_i,\ldots,d_n)}(j)$ appear at most $d_{i-1} - 1$ times for $2 \leq i \leq n$. In addition, if $t \neq 0$ then $i > 0$. We call such a decomposition the *t-$\mathbb{D}$-binomial expansion of h*. With this expansion, we define $h^{(t)}$ to be the number

$$h^{(t)} = \Delta H_{\text{C.I.}(d_{b_t},\ldots,d_n)}(t+1) + \Delta H_{\text{C.I.}(d_{b_{t-1}},\ldots,d_n)}(t) + \cdots + \Delta H_{\text{C.I.}(d_{b_i},\ldots,d_n)}(i+1).$$

By convention, we define $0^{(t)} = 0$.

The function $h^{(t)}$ depends on h, t, and $\{d_1, \ldots, d_n\}$. Our notation does not imply the dependence on $\{d_1, \ldots, d_n\}$. However, it will be implicit in our discussions what $d_1, \ldots, d_n$ are (see Example 2.16).

Below is the "Macaulay-type" characterization for the Hilbert functions of graded quotients of $k[x_1, \ldots, x_n]/(x_1^{d_n}, \ldots, x_n^{d_1}) \cong k[x_1, \ldots, x_n]/(x_1^{d_1}, \ldots, x_n^{d_n})$. Note that the proof requires the highest degree to be on the variable x_1 in order to work with order ideals; see [5, Remark 3.18] for a concrete example.

Theorem 2.15 ([5]). *Let I be a homogeneous ideal of $S_{\mathbb{D}} = S/(x_1^{d_n}, \ldots, x_n^{d_1})$ where $H(S_{\mathbb{D}}/I, t) = h_t$. Then $h_{t+1} \leq h_t^{(t)}$ for all $t \geq 1$.*

The following computational scheme can be used to find the bound described in Theorem 2.15. Fix a degree t and let h be a positive integer such that $h \leq H(S_{\mathbb{D}}, t)$. In order to find the t-$\mathbb{D}$-binomial expansion of h we construct "Pascal's Table" whose i^{th} row is

$$\begin{aligned}\Delta H_{\text{C.I.}(d_{n-i+1},\ldots,d_n)} &= H(k[x_1, \ldots, x_i]/(x_1^{d_n}, \ldots, x_i^{d_{n-i+1}})) \\ &= H(k[x_1, \ldots, x_i]/(x_1^{d_{n-i+1}}, \ldots, x_i^{d_n})).\end{aligned}$$

Number the columns starting from 0. Decompose h as follows: select the largest number in column t which does not exceed h, call this number α_1. If $\alpha_1 = h$ then we are done. If $\alpha_1 < h$ then select the largest number in column $(t-1)$ which does not exceed $h - \alpha_1$, call this number α_2. If $\alpha_1 + \alpha_2 = h$ then we are done. If $\alpha_1 + \alpha_2 < h$ then we repeat by decomposing $h - \alpha_1 - \alpha_2$, starting in column $(t-2)$. Continue in this fashion. It turns out that $h^{(t)}$ is the number obtained by shifting each entry obtained in decomposing h one unit to the right in the table.

Example 2.16. Let $d_1 = d_2 = 4, d_3 = 7$ and $S_{\mathbb{D}} = k[x_1, x_2, x_3]/(x_1^7, x_2^4, x_3^4)$. We use the "Pascal's Table":

degree:	0	1	2	3	4	5	6	7	8	9	10	11	12	13	···
$\Delta H_{\text{C.I.}(7)}$:	1	1	1	1	1	1	1	0	0	0	0	0	0	0	→
$\Delta H_{\text{C.I.}(4,7)}$:	1	2	3	4	4	4	4	3	2	1	0	0	0	0	→
$\Delta H_{\text{C.I.}(4,4,7)}$:	1	3	6	10	13	15	16	15	13	10	6	3	1	0	→

Suppose that $I \subseteq S_{\mathbb{D}}$ is a homogeneous ideal with $H(S_{\mathbb{D}}/I, 3) = 8$. The 3-$\{4,4,7\}$-binomial expansion of 8 is

$$8 = \Delta H_{\text{C.I.}(4,7)}(3) + \Delta H_{\text{C.I.}(4,7)}(2) + \Delta H_{\text{C.I.}(7)}(1) = 4 + 3 + 1.$$

Hence,

$$8^{(3)} = \Delta H_{\text{C.I.}(4,7)}(4) + \Delta H_{\text{C.I.}(4,7)}(3) + \Delta H_{\text{C.I.}(7)}(2) = 4 + 4 + 1 = 9.$$

We conclude that $H(S_{\mathbb{D}}/I, 4) \leq 8^{(3)} = 9$.

If we extend the definition of the d_i to be "∞" then Definitions-Proposition 2.10 coincides with Definitions-Proposition 2.14. In this case, the "Pascal's Table" constructed is simply Pascal's Triangle and the decompositions are the usual binomial expansions.

Theorem 2.15 allows us to restate the EGH Conjecture in a combinatorial fashion.

Conjecture 2.17 (Eisenbud–Green–Harris (EGH) Conjecture II [3, 8, 9, 11]). Let $1 \leq d_1 \leq d_2 \leq \cdots \leq d_n$ be integers. Suppose that $I \subseteq S = k[x_1, \ldots, x_n]$ is a homogeneous ideal which contains a regular sequence $F_1, F_2, \ldots, F_n$ of degrees $\deg(F_i) = d_i$. If $H(S/I, t) = h_t$, then $h_{t+1} \leq h_t^{(t)}$ for all $t \geq 1$.

3 Rectangular Complete Intersections

We now study *rectangular complete intersections*. These special sets play the analogous role to lex-plus-powers ideals in the algebraic setting of the EGH Conjecture, and hence play a crucial role in classifying the Hilbert functions of subsets of arbitrary complete intersections.

Assumption 3.1. Fix integers $d_1 \leq d_2 \leq \cdots \leq d_n$. From now on we assume that all elements of C.I.$(d_1, \ldots, d_n)$ are in $\mathbb{P}^n$ and no smaller projective space, i.e. we assume that $d_1 \geq 2$. We will label the coordinates of points in $\mathbb{P}^n$ as $[x_0 : x_1 : \ldots : x_n]$.

Definitions-Proposition 3.2. Fix positive integers $2 \leq d_1 \leq d_2 \leq \cdots \leq d_n$. Let $\mathbb{Y} \subseteq \mathbb{P}^n$ be the following set of $d_1 d_2 \cdots d_n$ distinct points with integer coordinates:

$$\{[1 : b_1 : \ldots : b_n] \mid b_i \in \mathbb{Z}, 0 \leq b_1 \leq d_n - 1, 0 \leq b_2 \leq d_{n-1} - 1, \ldots, 0 \leq b_n \leq d_1 - 1\}.$$

Then $\mathbb{Y} \in$ C.I.$(d_1, \ldots, d_n)$. The set $\mathbb{Y}$ is called the *rectangular complete intersection of type* $\{d_1, \ldots, d_n\}$, and is denoted $\mathbb{Y} =$ Rect.C.I.$(d_1, \ldots, d_n)$ (see Example 3.3).

Onc can vcrify that Rect.C.I.$(d_1, \ldots, d_n)$ is the intersection of the hypersurfaces $\mathbb{H}_1, \ldots, \mathbb{H}_n$, where $\mathbb{H}_i$ is the zero set of h_i with:

$$\begin{aligned} h_1 &= x_1(x_1 - 1x_0) \cdots (x_1 - (d_n - 1)x_0) \\ h_2 &= x_2(x_2 - 1x_0) \cdots (x_2 - (d_{n-1} - 1)x_0) \\ &\vdots \\ h_n &= x_n(x_n - 1x_0) \cdots (x_n - (d_1 - 1)x_0). \end{aligned}$$

Although the inequalities $b_1 \leq d_n - 1, b_2 \leq d_{n-1} - 1, \ldots, b_n \leq d_1 - 1$ seem backwards and awkward, the definition of a rectangular complete intersection is made in this fashion for the following important reason: it has become the standard in the literature to use the assumed ordering $d_1 \leq d_2 \leq \cdots \leq d_n$ when dealing with the EGH Conjecture yet in order to apply the proof of Theorem 2.15 we need to work with order ideals in which the lowest indexed variables have the largest exponents. As the next example demonstrates, using $b_1 \leq d_n - 1, b_2 \leq d_{n-1} - 1, \ldots, b_n \leq d_1 - 1$ gives a convenient bijection between the points in Rect.C.I.$(d_1, \ldots, d_n)$ and the desired order ideals.

Example 3.3. Rect.C.I.$(2, 3, 4)$ can be visualized as the 24 dots ($\bullet$) in the following rectangle. Starting from the origin, we use 4 units in the x_1-direction, 3 units in the x_2-direction, and 2 units in the x_3-direction, i.e. Rect.C.I.$(2, 3, 4)$ is visualized as a 3×4 block on two levels.

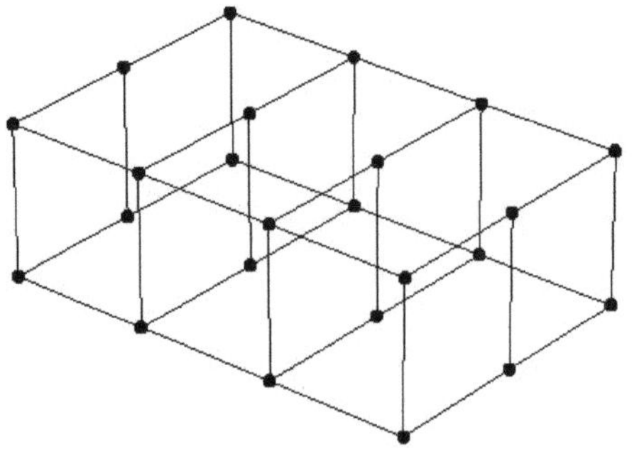

The points in the bottom layer of Rect.C.I.(2, 3, 4) are:

$$\{[1 : b_1 : b_2 : 0] \mid b_1, b_2 \in \mathbb{Z}, 0 \leq b_1 \leq 3, 0 \leq b_2 \leq 2\}$$

and the points in the top layer of Rect.C.I.(2, 3, 4) are:

$$\{[1 : b_1 : b_2 : 1] \mid b_1, b_2 \in \mathbb{Z}, 0 \leq b_1 \leq 3, 0 \leq b_2 \leq 2\}.$$

Note that we have the bijections

$$\begin{aligned}\text{Rect.C.I.}(2,3,4) &\leftrightarrow \{(b_1, b_2, b_3) \in \mathbb{Z}^3 \mid 0 \leq b_i \leq 4 - i\} \\ &\leftrightarrow \{x_1^{b_1} x_2^{b_2} x_3^{b_3} \in k[x_1, x_2, x_3] \mid 0 \leq b_i \leq 4 - i\},\end{aligned}$$

by

$$[1 : b_1 : b_2 : b_3] \leftrightarrow (b_1, b_2, b_3) \leftrightarrow x_1^{b_1} x_2^{b_2} x_3^{b_3}.$$

Theorem 3.4. *Let $\Delta\mathcal{H} = \{h_t\}_{t\geq 0}$ be the first difference Hilbert function for some finite set of distinct points in $\mathbb{P}^n$ such that $\Delta\mathcal{H} \leq \Delta H_{\text{C.I.}(d_1,\ldots,d_n)}$, where $2 \leq d_1 \leq \cdots \leq d_n$. Then $\Delta\mathcal{H}$ is the first difference Hilbert function for some subset of* Rect.C.I.$(d_1, \ldots, d_n)$ *if and only if $h_{t+1} \leq h_t^{(t)}$ for all $t \geq 1$.*

Proof. By [14, Lemma 2.3], if there is a subset of Rect.C.I.$(d_1, \ldots, d_n)$ with first difference Hilbert function $\Delta\mathcal{H}$ then we must have $\Delta\mathcal{H} \leq \Delta H_{\text{C.I.}(d_1,\ldots,d_n)}$. First suppose that $h_{t+1} \leq h_t^{(t)}$ for all $t \geq 1$. We construct a subset $\mathbb{X}$ of Rect.C.I.$(d_1, \ldots, d_n)$ such that $\Delta H(\mathbb{X}) = \Delta\mathcal{H}$. Let

$$T = \{x_1^{b_1} x_2^{b_2} \cdots x_n^{b_n} \mid b_i \in \mathbb{Z}, 0 \leq b_i \leq d_{n-i+1} - 1\}$$

and

$$T_i = \{x_1^{b_1} x_2^{b_2} \cdots x_n^{b_n} \in T \mid \deg(x_1^{b_1} \cdots x_n^{b_n}) = b_1 + b_2 + \cdots + b_n = i\}.$$

For each i, we also let K_i be the h_i largest monomials of T_i with respect to the degree-reverse-lexicographic ordering and $K = \cup_{i=1} K_i$.

Since $h_{t+1} \leq h_t^{(t)}$ for all $t \geq 1$, the set K is an order ideal of monomials (see [5] for complete details). Now let $\mathcal{M}$ denote the set of all monomials in $S = k[x_1, \ldots, x_n]$. Then $\mathcal{M} \setminus K$ generates a monomial ideal J of S. Suppose J is minimally generated by the monomials $f_l = x_1^{a_{1l}} x_2^{a_{2l}} \cdots x_n^{a_{nl}}$. Let $I \subseteq R = k[x_0, \ldots, x_n]$ be the ideal generated by the homogeneous polynomials

$$\overline{f_l} = \prod_{j=1}^{n} \Big(\prod_{i=0}^{a_{jl}-1} (x_j - i x_0) \Big).$$

By [12, Theorem 2.2], J *lifts* to I. By definition, this means that I is a radical ideal of R; x_0 is not a zero-divisor on R/I; and $(I, x_0)/(x_0)$ is isomorphic to J. One of the main steps in the verification that J lifts to I (as described in the proof of [12, Theorem 2.2]) is to show that $I = \mathbf{I}(\mathbb{X})$ where

$$\mathbb{X} = \{[1 : c_1 : \ldots : c_n] \mid x_1^{c_1} x_2^{c_2} \cdots x_n^{c_n} \in K\} \subset \mathbb{P}^n$$

(the reader is encouraged to see [12] for more details). We see immediately that $\mathbb{X} \subseteq$ Rect.C.I.$(d_1, \ldots, d_n)$ and $\Delta H(\mathbb{X}) = \Delta\mathcal{H}$.

Conversely, now suppose there is a subset $\mathbb{X}$ of Rect.C.I.$(d_1, \ldots, d_n)$ with first difference Hilbert function $\Delta\mathcal{H}$. Let $I \subseteq S = k[x_1, \ldots, x_n]$ be the ideal obtained from $\mathbf{I}(\mathbb{X})$ after moding out by x_0. Then $H(S/I) = \Delta\mathcal{H}$. Using the 1-1 correspondence between the points of Rect.C.I.$(d_1, \ldots, d_n)$ and the monomials of S, Theorem 2.15 is applied to obtain $h_{t+1} \leq h_t^{(t)}$ for all $t \geq 1$. □

Example 3.5. Consider Rect.C.I.(3,4,4). Our "Pascal's Table" is:

degree:	0	1	2	3	4	5	6	7	8	9	10	11	12	13	$\cdots$
$\Delta H_{\text{C.I.}(4)}$:	1	1	1	1	0	0	0	0	0	0	0	0	0	0	$\rightarrow$
$\Delta H_{\text{C.I.}(4,4)}$:	1	2	3	4	3	2	1	0	0	0	0	0	0	0	$\rightarrow$
$\Delta H_{\text{C.I.}(3,4,4)}$:	1	3	6	9	10	9	6	3	1	0	0	0	0	0	$\rightarrow$

Let $\Delta\mathcal{H} = (1, 3, 6, 8, 7, 1, 0, 0, \ldots)$. Clearly $\Delta\mathcal{H} \leq \Delta H_{\text{C.I.}(3,4,4)}$. It is easy to verify that $\Delta\mathcal{H}(t+1) \leq \Delta\mathcal{H}(t)^{(t)}$ for all $t \geq 1$. For example, the 5-$\{3, 4, 4\}$-binomial expansion for $\Delta\mathcal{H}(5)$ is

$$\Delta\mathcal{H}(5) = 1 = \Delta H_{\text{C.I.}(4)}(5) + \Delta H_{\text{C.I.}(4)}(4) + \Delta H_{\text{C.I.}(4)}(3) = 0 + 0 + 1 \text{ and}$$

$$\Delta\mathcal{H}(6) = 0 \leq 1^{(5)} = \Delta H_{\text{C.I.}(4)}(6) + \Delta H_{\text{C.I.}(4)}(5) + \Delta H_{\text{C.I.}(4)}(4) = 0.$$

We see that the order ideal K from the proof of Theorem 3.4 has pieces:

$$\begin{aligned}
K_0 &= \{x_1^0 x_2^0 x_3^0\} \\
K_1 &= \{x_1, x_2, x_3\} \\
K_2 &= \{x_1^2, x_1x_2, x_1x_3, x_2^2, x_2x_3, x_3^2\} \\
K_3 &= \{x_1^3, x_1^2x_2, x_1^2x_3, x_1x_2^2, x_1x_2x_3, x_1x_3^2, x_2^3, x_2^2x_3\} \\
K_4 &= \{x_1^3x_2, x_1^3x_3, x_1^2x_2^2, x_1^2x_2x_3, x_1x_2^3, x_1x_2^2x_3, x_2^3x_3\} \\
K_5 &= \{x_1^3x_2^2\}
\end{aligned}$$

and the set $\mathbb{X}$ constructed consists of the points with integer coordinates:

$$\begin{aligned}
&\{[1 : i : j : k] \mid 0 \leq i, j, k \leq 1, i + j + k = 1\}, \\
&\{[1 : i : j : k] \mid 0 \leq i, j, k \leq 2, i + j + k = 2\}, \\
&\{[1 : i : j : k] \mid 0 \leq i, j \leq 3, 0 \leq k \leq 1, i + j + k = 3\}, \\
&\{[1 : i : j : k] \mid 0 \leq i, j \leq 3, 0 \leq k \leq 1, i + j + k = 4\}, \\
&[1 : 0 : 0 : 0],\ [1 : 1 : 0 : 2],\ [1 : 3 : 2 : 0].
\end{aligned}$$

Theorem 3.4 naturally leads to the following conjecture:

Conjecture 3.6. The sets

$$\{\mathcal{H} \mid \text{there exist sets } \mathbb{X} \subseteq \mathbb{Y} \in \text{C.I.}(d_1, \ldots, d_n) \text{ such that } H(\mathbb{X}) = \mathcal{H}\}$$

and

$$\{\mathcal{H} \mid \text{there exist sets } \mathbb{X} \subseteq \mathbb{Y} \in \text{Rect.C.I.}(d_1, \ldots, d_n) \text{ such that } H(\mathbb{X}) = \mathcal{H}\}$$

are equal. Equivalently, if $\Delta\mathcal{H} = \{h_t\}_{t \geq 0}$ is the first difference Hilbert function of some finite set of distinct points, then there exist sets $\mathbb{X} \subseteq \mathbb{Y} \in \text{C.I.}(d_1, \ldots, d_n)$ with $\Delta H(\mathbb{X}) = \Delta\mathcal{H}$ if and only if $h_{t+1} \leq h_t^{(t)}$ for all $t \geq 1$.

4 Some Key Tools

In this section we briefly collect some of the tools which will be used in later sections when proving special cases of Conjecture 3.6. Unless otherwise stated, we will continue to let $R = k[x_0, \ldots, x_n]$ and $S = k[x_1, \ldots, x_n]$. We also fix integers $2 \leq d_1 \leq d_2 \leq \cdots \leq d_n$.

4.1 Pairs of Hilbert Functions and Maximal Growth

When characterizing Hilbert functions of subsets of complete intersections, we can apply results coming from linkage theory. In particular, the *Generalized Cayley–Bacharach Theorem* below gives a formula relating the Hilbert functions of subsets of complete intersections. That is, Hilbert functions of subsets of complete intersections come in pairs.

Theorem 4.1 ([6, Theorem 3]). *Let* $\mathbb{Y} \in \text{C.I.}(d_1, \ldots, d_n) \subseteq \mathbb{P}^n$ *and let* $\mathbb{X} \subseteq \mathbb{Y}$. *Then*

$$\Delta H(\mathbb{Y}, t) = \Delta H(\mathbb{X}, t) + \Delta H(\mathbb{Y} \setminus \mathbb{X}, d_1 + d_2 + \cdots + d_n - n - t).$$

In addition to the Cayley–Bacharach Theorem, we have a collection of powerful results involving extremal behavior of Hilbert functions. Gotzmann [15] considers the situation when a Hilbert function attains the maximal growth permitted by Macaulay's Theorem. Also, Bigatti–Geramita–Migliore [1] study algebraic and geometric consequences of Hilbert functions attaining maximal growth as described by Macaulay's function.

Definition 4.2. Let I be a homogeneous ideal in the polynomial ring R. We say that $H(R/I)$ has *maximal growth in degree* d if $H(R/I, d+1) = H(R/I, d)^{<d>}$.

Remark 4.3. If $H(R/I)$ has maximal growth in degree d then I has a minimal generating set which has no element in degree $d + 1$.

Using maximal growth, Bigatti–Geramita–Migliore gave situations in which homogeneous ideals are guaranteed to have a greatest common divisor (GCD) in certain degrees. We will apply these results and refer the reader to [1] for the full details.

4.2 Ideals Containing Regular Sequences

In this section we present two useful facts of ideals containing regular sequences. Our first fact concerns regular sequences and minimal generating sets.

Proposition 4.4. *Let $I \subseteq S$ be a proper, non-zero homogeneous ideal generated in degrees less than or equal to d. Suppose that I contains a homogeneous regular sequence of length r. Then there exists a minimal generating set for I containing r homogeneous polynomials in degrees less than or equal to d which are a regular sequence.*

Proof. We begin by listing a minimal set of generators of I in order of increasing degree, say $F_1, F_2, F_3, \ldots, F_x$, where F_i is a homogeneous polynomial of degree d_i. Fix $J_i = (F_1, F_2, \ldots, F_i)$ for all $i \geq 1$. Then clearly $\mathrm{ht}(J_1) \leq \mathrm{ht}(J_2) \leq \cdots \leq \mathrm{ht}(J_x)$ and $\mathrm{ht}(J_x) \geq r$.

We start our regular sequence with F_1. If $r = 1$ then we are done. So we may as well suppose $r \geq 2$. Let $\wp_1, \ldots, \wp_l$ be the associated prime ideals of (F_1). Then each $\wp_i$ is a minimal associated prime ideal of (F_1) and has height 1. Let s be the smallest integer i such that $\mathrm{ht}(J_i) \geq 2$. Now if $(J_s)_{d_s} \subseteq (\wp_1)_{d_s} \cup (\wp_2)_{d_s} \cup \cdots \cup (\wp_l)_{d_s} \cup (J_{s-1})_{d_s}$ then either $(J_s)_{d_s} \subseteq (\wp_i)_{d_s}$ for some i, or $(J_s)_{d_s} \subseteq (J_{s-1})_{d_s}$. (In general, if V and $W_1, \ldots, W_g$ are finite-dimensional vector spaces over an infinite field such that $V \subseteq W_1 \cup \cdots \cup W_g$, then it follows from elementary properties that $V \subseteq W_j$ for some $1 \leq j \leq g$.) But since $J_{s-1} = (F_1, \ldots, F_{s-1})$ and $J_s = (F_1, \ldots, F_{s-1}, F_s)$, we see that $(J_{s-1})_{d_s}$ is a proper vector subspace of $(J_s)_{d_s}$. In addition, $\mathrm{ht}(J_s) \geq 2$ but for each i we have that $\mathrm{ht}(\wp_i) = 1$, and so $(J_s)_{d_s} \nsubseteq (\wp_i)_{d_s}$ for any i. Therefore, $(J_s)_{d_s} \nsubseteq (\wp_1)_{d_s} \cup (\wp_2)_{d_s} \cup \cdots \cup (\wp_l)_{d_s} \cup (J_{s-1})_{d_s}$, and so there exists $H_1 \in (J_s)_{d_s}$ such that $H_1 \notin \wp_1 \cup \wp_2 \cup \cdots \cup \wp_l \cup J_{s-1}$. Since H_1 is not in the union of the associated prime ideals of (F_1), we see that F_1, H_1 is a regular sequence. Further, since $H_1 \notin J_{s-1}$, we must have that F_1, H_1 is part of a minimal generating set of I. We are done if $r = 2$.

The proof is completed by repeatedly applying the above argument. For example, if $r = 3$ we work with a minimal generating set which includes F_1, H_1, say $F_1, H_1, T_1, \ldots, T_l$, where F_1 is a form of degree d_1, H_1 is a form of degree d_s and T_i is a form of degree $d_i \geq d_s$. □

Our second fact involves building regular sequences in certain degrees.

Lemma 4.5. *Let $I \subseteq S$ be a homogeneous ideal containing a homogeneous regular sequence $F_1, \ldots, F_n$. There exist general linear forms $L_1, L_2, \ldots, L_n \in S$ such that $F_1 L_1^{e_1}, \ldots, F_n L_n^{e_n}$ is a regular sequence for any integers $e_1, \ldots, e_n \geq 0$.*

We will only use Lemma 4.5 in the case of three variables. For concreteness we prove the lemma in this special case; the proof for the general case follows the same outline.

Proof. Let $e_1, e_2, e_3 \geq 0$ be integers. Since F_1, F_2, F_3 is a regular sequence, we know that $S/(F_1, F_2)$ has a non-zero-divisor. Hence, there exists a general linear form $L_3 \in S$ such that L_3 is a non-zero-divisor on $S/(F_1, F_2)$. By assumption, F_3 is also a non-zero-divisor on $S/(F_1, F_2)$. So $F_3 L_3^{e_3}$ is a non-zero-divisor on $S/(F_1, F_2)$. That is, $F_1, F_2, F_3 L_3^{e_3}$ is a regular sequence. We can repeat this argument on the regular sequence $F_3 L_3^{e_3}, F_1, F_2$ to obtain a regular sequence $F_3 L_3^{e_3}, F_1, F_2 L_2^{e_2}$ where L_2 is a non-zero-divisor on $S/(F_3 L_3^{e_3}, F_1)$. Repeating the argument a third time gives the desired regular sequence $F_1 L_1^{e_1}, F_2 L_2^{e_2}, F_3 L_3^{e_3}$. □

5 Subsets of Complete Intersections in $\mathbb{P}^2$

We begin our study of Conjecture 3.6 in $\mathbb{P}^2$. We provide a self-contained proof here, but remark that the result also follows from the known case of the EGH Conjecture when $n = 2$. The following proposition describes the desired bounds.

Proposition 5.1. *Let c and t be positive integers such that $c < \Delta H_{\text{C.I.}(d_1,d_2)}(t)$, where $d_1 \geq 2$.*

(i) *If $t \leq d_2 - 2$, then $c^{(t)} = c$.*

(ii) *If $t \geq d_2 - 1$, then $c^{(t)} = c - 1$.*

Proof. Let $\mathbb{D} = \{d_1, d_2\}$. In order to calculate the t-$\mathbb{D}$-binomial expansion of c we use the "Pascal's Table":

degree:	0	1	$\cdots$	$d_1 - 1$	$\cdots$	$d_2 - 1$	d_2	$\cdots$	$d_1 + d_2 - 1$	$\cdots$
$\Delta H_{\text{C.I.}(d_2)}$:	1	1	$\cdots$	1	$\cdots$	1	0	$\cdots$	0	$\rightarrow$
$\Delta H_{\text{C.I.}(d_1,d_2)}$:	1	2	$\cdots$	d_1	$\cdots$	d_1	$d_1 - 1$	$\cdots$	0	$\rightarrow$

Since $c < \Delta H_{\text{C.I.}(d_1,d_2)}(t)$, note that when finding the t-$\mathbb{D}$-binomial expansion of c we will only use entries from the row labeled $\Delta H_{\text{C.I.}(d_2)}$. If the expansion uses terms from degrees $\geq d_2 - 1$ then, after shifting each entry one unit to the right in the table, the resulting number $c^{(t)}$ will equal $c - 1$; otherwise, $c^{(t)} = c$. □

Notation 5.2. Let $\mathcal{H}$ be the Hilbert function of some finite set of distinct points in $\mathbb{P}^n$. We let $\sigma(\mathcal{H}) = \min\{t \geq 1 \mid \Delta\mathcal{H}(t) = 0\}$.

Our next proposition is a key fact in verifying Conjecture 3.6 for $\mathbb{P}^2$.

Proposition 5.3. *If there exist finite sets of distinct points* $\mathbb{X} \subseteq \mathbb{Y} \in \text{C.I.}(d_1, d_2)$ *such that* $\sigma(H(\mathbb{X})) > d_2 \geq d_1 \geq 2$, *then* $\Delta H(\mathbb{X}, l) \neq \Delta H(\mathbb{X}, l+1)$ *for any degree* $l \in \{d_2 - 1, \ldots, \sigma(H(\mathbb{X})) - 2\}$.

Proof. Suppose $\Delta H(\mathbb{X}, l) = \Delta H(\mathbb{X}, l+1)$ for some $l \in \{d_2 - 1, \ldots, \sigma(H(\mathbb{X})) - 2\}$. Let j, where $d_2 - 1 \leq j \leq \sigma(H(\mathbb{X})) - 2$, be the least degree s such that $\Delta H(\mathbb{X}, s) = \Delta H(\mathbb{X}, s+1)$. Fix $\mathbb{W} = \mathbb{Y} \setminus \mathbb{X}$. Then, by Facts 2.8 and Theorem 4.1, we have

$$\begin{aligned}\Delta H(\mathbb{W}, t) &= \Delta H(\mathbb{Y}, t) - \Delta H(\mathbb{X}, d_1 + d_2 - 2 - t) \\ &= \Delta H(\mathbb{Y}, d_1 + d_2 - 2 - t) - \Delta H(\mathbb{X}, d_1 + d_2 - 2 - t).\end{aligned}$$

Fix $\Gamma = \sigma(H(\mathbb{Y})) - \sigma(H(\mathbb{X})) = d_1 + d_2 - 1 - \sigma(H(\mathbb{X}))$. If $\Gamma = 0$ then $\Delta H(\mathbb{W}, 0) = 0$, and so $\Delta H(\mathbb{W})$ is not an O-sequence, a contradiction. So we may as well assume that $\Gamma \neq 0$. Then

$$\begin{aligned}&\Delta H(\mathbb{W}, t) = \Delta H(\mathbb{Y}, d_1 + d_2 - 2 - t), \quad \text{for } 0 \leq t \leq \Gamma - 1 \\ &\Delta H(\mathbb{W}, t) \leq \Delta H(\mathbb{Y}, d_1 + d_2 - 2 - t) - 1 \leq t, \quad \text{for } \Gamma \leq t \leq d_1 + d_2 - 2.\end{aligned}$$

Since $\Delta H(\mathbb{Y}, t) = \Delta H(\mathbb{Y}, t+1) + 1$ for $d_2 - 1 \leq t \leq \sigma(H(\mathbb{Y})) - 1$ and $\Delta H(\mathbb{X}, j) = \Delta H(\mathbb{X}, j+1)$, we have

$$\Delta H(\mathbb{W}, d_1 + d_2 - 2 - j) = \Delta H(\mathbb{W}, d_1 + d_2 - 2 - (j+1)) + 1.$$

In addition, we have

$$\begin{aligned}d_1 + d_2 - 2 - (j+1) &= d_1 + d_2 - 3 - j \\ &\geq d_1 + d_2 - 3 - (\sigma(H(\mathbb{X})) - 2) \\ &= d_1 + d_2 - 1 - \sigma(H(\mathbb{X})) \\ &= \Gamma.\end{aligned}$$

Hence $c = \Delta H(\mathbb{W}, d_1 + d_2 - 2 - (j+1)) \leq d_1 + d_2 - 2 - (j+1)$, implying $c^{<d_1+d_2-2-(j+1)>} = c$. But

$$\Delta H(\mathbb{W}, d_1 + d_2 - 2 - j) = c + 1 > c = c^{<d_1+d_2-2-(j+1)>}.$$

We conclude that $\Delta H(\mathbb{W})$ is not an O-sequence, a contradiction. □

We are now in a situation to prove Conjecture 3.6 for $\mathbb{P}^2$.

Theorem 5.4. *Let* $\Delta\mathcal{H} = \{h_t\}_{t \geq 0}$ *be the first difference Hilbert function of some finite set of distinct points in* $\mathbb{P}^2$ *such that* $\Delta\mathcal{H} \leq \Delta H_{\text{C.I.}(d_1, d_2)}$, *where* $d_1 \geq 2$. *Then there exist finite sets of distinct points* $\mathbb{X} \subseteq \mathbb{Y} \in \text{C.I.}(d_1, d_2)$ *such that* $\Delta H(\mathbb{X}) = \Delta\mathcal{H}$ *if and only if* $h_{t+1} \leq h_t^{(t)}$ *for all* $t \geq 1$.

Proof. If $h_{t+1} \le h_t^{(t)}$ for all $t \ge 1$, then Theorem 3.4 gives that there exists a subset $\mathbb{X}$ of Rect.C.I.(d_1, d_2) with $\Delta H(\mathbb{X}) = \Delta \mathcal{H}$.

Now suppose there exist finite sets of distinct points $\mathbb{X} \subseteq \mathbb{Y} \in$ C.I.(d_1, d_2) such that $\Delta H(\mathbb{X}) = \Delta \mathcal{H}$. We may as well assume $1 \le h_t < \Delta H_{\text{C.I.}(d_1,d_2)}(t)$ and $h_{t+1} \ne 0$, otherwise the inequality $h_{t+1} \le h_t^{(t)}$ is trivially satisfied.

Assume $t \le d_2 - 2$. Then $h_t \le t$, and hence $h_t^{<t>} = h_t$. Macaulay's Theorem implies $h_{t+1} \le h_t^{<t>} = h_t$. By Proposition 5.1, $h_t^{(t)} = h_t$, and so $h_{t+1} \le h_t^{(t)}$.

Now let $t \ge d_2 - 1$. Then $h_t \le d_1 - 1 \le d_2 - 1 \le t$, and so $h_t^{<t>} = h_t$. As above, $h_{t+1} \le h_t^{<t>} = h_t$. If $h_{t+1} = h_t$ then we have a contradiction to Proposition 5.3. So $h_{t+1} \le h_t - 1$ and we are done by Proposition 5.1. □

6 Subsets of C.I.$(2, d_2, d_3)$ with $d_2 = d_3$

Recall that when $n = 3$ the outstanding cases for the EGH Conjecture are the "tight degrees" $\mathbb{D} = \{d_1, d_2, d_3\}$ where $d_1 \le d_2 \le d_3 \le d_1 + d_2 - 2$. In the geometric setting of Conjecture 3.6, we now focus our attention on the cases when $d_1 = 2$ and $d_2 = d_3$. For ease of notation, in this section we set $d_2 = d_3 = d \ge 2$.

We first describe the necessary bounds of Clements–Lindström. In the following arguments, we use the summation notation $\sum$ where, by convention, if $j > c$ then we set $\sum_j^c = 0$.

Proposition 6.1. *Let b and t be positive integers such that $b < \Delta H_{\text{C.I.}(2,d,d)}(t)$.*

(i) *Assume $t \le d - 2$.*

  (a) *If $1 \le b \le t$, then $b^{(t)} = b$.*

  (b) *If $b \ge (t + 1)$, then $b^{(t)} = b + 1$.*

(ii) *If $t = d - 1$, then $b^{(d-1)} = b - 1$.*

(iii) *Assume $d \le t \le 2d - 3$.*

  (a) *If $b \le 2d - t - 1$, then $b^{(t)} = b - 1$.*

  (b) *If $b > 2d - t - 1$, then $b^{(t)} = b - 2$.*

Proof. Let $\mathbb{D} = \{2, d, d\}$. To calculate the t-$\mathbb{D}$-binomial expansion of b we use the "Pascal's Table":

degree:	0	1	$\cdots$	$d-1$	d	$d+1$	$d+2$	$\cdots$	$2d-2$	$2d-1$	$\cdots$
$\Delta H_{\text{C.I.}(d)}$:	1	1	$\cdots$	1	0	0	0	$\cdots$	0	0	$\rightarrow$
$\Delta H_{\text{C.I.}(d,d)}$:	1	2	$\cdots$	d	$d-1$	$d-2$	$d-3$	$\cdots$	1	0	$\rightarrow$
$\Delta H_{\text{C.I.}(2,d,d)}$:	1	3	$\cdots$	$2d-1$	$2d-1$	$2d-3$	$2d-5$	$\cdots$	$\cdots$	$\cdots$	$\cdots$

Since $b < \Delta H_{\text{C.I.}(2,d,d)}(t)$, note that when finding the t-$\mathbb{D}$-binomial expansion of b we will only use entries from the rows labeled $\Delta H_{\text{C.I.}(d)}$ and $\Delta H_{\text{C.I.}(d,d)}$. Moreover, the expansion can involve at most one entry from the row labeled $\Delta H_{\text{C.I.}(d,d)}$. Observe that when we shift each summand x in the binomial expansion of b one place to the right in the table, the resulting corresponding summand for $b^{(t)}$ is either $x - 1$ or x or $x + 1$ depending on the degree x is associated to. We discuss each assertion separately.

(i) Note that $\Delta H_{\text{C.I.}(d,d)}(i) = i + 1$ for $1 \leq i \leq d - 1$.

(a) Suppose that $1 \leq b \leq t \leq d - 2$. Then the t-$\mathbb{D}$-binomial expansion of b involves only entries from the row labeled $\Delta H_{\text{C.I.}(d)}$. Since $t \leq d - 2$, when a number $x = 1$ in the expansion is shifted one place to the right in the table the resulting number is $x = 1$. Thus, $b^{(t)}$ equals b.

(b) Assume that $b \geq t + 1$. Then the t-$\mathbb{D}$-binomial expansion of b involves exactly $\Delta H_{\text{C.I.}(d,d)}(t) = t + 1$ and entries from the row labeled $\Delta H_{\text{C.I.}(d)}$. When the entry $\Delta H_{\text{C.I.}(d,d)}(t) = t + 1$ is shifted one place to the right in the table one obtains $t + 2$. Since $t \leq d - 2$, the shifting of the remaining entries in the expansion results in the same number of 1's as there were to begin with. Thus, $b^{(t)} = b + 1$.

(ii) Suppose that $t = d - 1$. If $b < d$ then the $(d - 1)$-$\mathbb{D}$-binomial expansion of b is

$$b = \Delta H_{\text{C.I.}(d)}(d - 1) + \Delta H_{\text{C.I.}(d)}(d - 2) + \cdots + \Delta H_{\text{C.I.}(d)}(d - b) = \sum_{i=1}^{b} 1.$$

If $b \geq d$, then we can write $b = d + a$ where $0 \leq a < \Delta H_{\text{C.I.}(d,d)}(d - 2) = d - 1$ (since the expansion can involve at most one entry from the row labeled $\Delta H_{\text{C.I.}(d,d)}$). Thus the $(d - 1)$- $\mathbb{D}$-binomial expansion of b is

$$b = \Delta H_{\text{C.I.}(d,d)}(d - 1) + \sum_{i=2}^{a+1} (\Delta H_{\text{C.I.}(d)}(d - i)) = d + \sum_{i=1}^{a} 1.$$

When shifting the entries in the expansion of b one place to the right in the table, we see that shifting the number x from the degree $d - 1$ column results in $x - 1$ and shifting the other entries results in the same numbers as before shifting. Thus, $b^{(d-1)} = b - 1$.

(iii) Suppose $d \leq t \leq 2d - 3$.

(a) If $b = 2d - t - 1$, then the t-$\mathbb{D}$-binomial expansion of b is

$$b = \Delta H_{\text{C.I.}(d,d)}(t) = 2d - t - 1$$

giving

$$b^{(t)} = \Delta H_{\text{C.I.}(d,d)}(t + 1) = 2d - t - 2 = b - 1.$$

If $b < 2d - t - 1$, then the t-$\mathbb{D}$-binomial expansion of b involves only entries from the row labeled $\Delta H_{\text{C.I.}(d)}$. Since $d \leq t \leq 2d-3$, the shifting of the entry in the degree $d-1$ column to the right guarantees $b^{(t)} = b-1$.

(b) If $b > 2d - t - 1$, then the t-$\mathbb{D}$-binomial expansion of b involves exactly $\Delta H_{\text{C.I.}(d,d)}(t)$ and a sum of entries from the row labeled $\Delta H_{\text{C.I.}(d)}$. Since $d \leq t \leq 2d - 3$, the shifting of the entries $\Delta H_{\text{C.I.}(d,d)}(t) = 2d - t - 1$ and $\Delta H_{\text{C.I.}(d)}(d-1) = 1$ to the right results in $b^{(t)} = b - 2$. □

We now verify the desired Hilbert function bounds of Conjecture 3.6 degree-by-degree.

Lemma 6.2. *Suppose that $\mathbb{X} \subseteq \mathbb{Y} \in$ C.I.$(2,d,d)$ are finite sets of distinct points where $\Delta H(\mathbb{X}) = \{h_t\}_{t\geq 0}$. Then $h_{t+1} \leq h_t^{(t)}$ for $1 \leq t \leq d-2$.*

Proof. As argued in Theorem 5.4, we can assume $0 < h_t < \Delta H_{\text{C.I.}(2,d,d)}(t)$.

Case 1: Assume $1 \leq h_t \leq t$. By Proposition 6.1, $h_t^{(t)} = h_t$. But we have $h_t^{<t>} = h_t$ and so, since $\Delta H(\mathbb{X})$ is an O-sequence, $h_{t+1} \leq h_t^{<t>} = h_t = h_t^{(t)}$.

Case 2: Now assume $h_t = (t+1)+a$ for some non-negative integer a. By Proposition 6.1 and its proof we can assume $0 \leq a < \Delta H_{\text{C.I.}(d,d)}(t-1) = t$ and have $h_t^{(t)} = h_t + 1$. Note that the t-binomial expansion of h_t is

$$h_t = \binom{t+1}{t} + \binom{t-1}{t-1} + \cdots + \binom{t-a}{t-a}.$$

So $h_t^{<t>} = h_t + 1$. By Macaulay's Theorem, $h_{t+1} \leq h_t + 1 = h_t^{(t)}$. □

The next lemma will be helpful in verifying Conjecture 3.6 for degree $t = d - 1$ both for complete intersections of type $\{2,d,d\}$ and $\{3,d,d\}$.

Lemma 6.3. *If $\mathbb{T}$ is a finite set of distinct points in $\mathbb{P}^3$ such that $\Delta H(\mathbb{T}, d-1) = \Delta H(\mathbb{T}, d)$ and $\Delta H(\mathbb{T}, d-1) \leq d-1$, then $\mathbb{T}$ cannot be contained in any complete intersection of type $\{2,d,d\}$ or $\{3,d,d\}$.*

Proof. Assume $\mathbb{T} \subseteq \mathbb{K} \in$ C.I.$(2,d,d)$. Without loss of generality, we can assume that no point of $\mathbb{T}$ lies on the hyperplane $x_0 = 0$. Let $I \subseteq k[x_1,x_2,x_3]$ be the image of $\mathbf{I}(\mathbb{T})$ obtained by moding out by x_0. Then I contains a regular sequence F, G, H where $\deg(F) = 2$ and $\deg(G) = \deg(H) = d$, and $H(k[x_1,x_2,x_3]/I) = \Delta H(\mathbb{T})$. Since $\Delta H(\mathbb{T})$ has maximal growth in degree $(d-1)$, $\mathbf{I}(\mathbb{T})$ has no generator of degree d and $\mathbf{I}(\mathbb{K}) \subseteq (\mathbf{I}(\mathbb{T})_{\leq d-1})$. This implies that I has a regular sequence of length 3 in degrees $\leq d-1$. Further, if we let $\Delta H(\mathbb{T}, d-1) = h_{d-1}$ then observe that the $(d-1)$-binomial expansion of h_{d-1} is

$$h_{d-1} = \binom{d-1}{d-1} + \binom{d-2}{d-2} + \cdots + \binom{d-h_{d-1}}{d-h_{d-1}}.$$

Thus, by Gotzmann's Persistence Theorem [15], the Hilbert polynomial of the quotient $k[x_1, x_2, x_3]/(I_{\leq d-1})$ is

$$P(t) = \binom{t}{0} + \binom{t-1}{0} + \cdots + \binom{t+1-h_{d-1}}{0}.$$

We see that $(I_{\leq d-1})$ defines a scheme of dimension 0 in $\mathbb{P}^2$. Thus, the length of the maximal regular sequence contained in I using elements of degree $\leq d-1$ is exactly 2, a contradiction. The same argument holds if $\mathbb{T}$ is contained in a complete intersection of type $\{3, d, d\}$. □

We can now prove Conjecture 3.6 in the case of complete intersections of type $\{2, d, d\}$ in degree $t = d-1$.

Lemma 6.4. *Suppose that* $\mathbb{X} \subseteq \mathbb{Y} \in$ C.I.$(2, d, d)$ *are finite sets of distinct points where* $\Delta H(\mathbb{X}) = \{h_t\}_{t \geq 0}$. *Then* $h_d \leq h_{d-1}^{(d-1)}$.

Proof. As in Lemma 6.2, we can assume $0 < h_{d-1} < \Delta H_{\text{C.I.}(2,d,d)}(d-1) = 2d-1$. By Proposition 6.1, we need to show $h_d \leq h_{d-1}^{(d-1)} = h_{d-1} - 1$.

Case 1: Assume first that $h_{d-1} \leq d-1$. Since $\Delta H(\mathbb{X})$ is an O-sequence, $h_d \leq h_{d-1}^{<d-1>} = h_{d-1}$. The inequality $h_d \leq h_{d-1} - 1$ now follows immediately from Lemma 6.3.

Case 2: Suppose now that $d \leq h_{d-1} \leq 2d-2$. We assume that $h_d = h_{d-1} + i$, where $i \geq 0$. By Theorem 4.1,

$$\Delta H(\mathbb{Y} \setminus \mathbb{X}, d-1) = 2d - 1 - h_{d-1} - i$$
$$\Delta H(\mathbb{Y} \setminus \mathbb{X}, d) = 2d - 1 - h_{d-1}.$$

Since $2d - 1 - i - h_{d-1} \leq d-1$, we have $(2d - 1 - h_{d-1} - i)^{<d-1>} = 2d - 1 - h_{d-1} - i$. But if $i \geq 1$, then $2d - 1 - h_{d-1} > 2d - 1 - h_{d-1} - i$ contradicting the fact that $\Delta H(\mathbb{Y} \setminus \mathbb{X})$ is an O-sequence. Thus $i = 0$, and the argument used in Case 1 but applied to $\Delta H(\mathbb{Y} \setminus \mathbb{X})$ completes the proof. □

The next lemma verifies Conjecture 3.6 for the tail-end degrees. It takes advantage of the fact that Hilbert functions of subsets of complete intersections come in pairs by reducing the situation to the beginning degrees in the complementary Hilbert function.

Lemma 6.5. *Suppose that* $\mathbb{X} \subseteq \mathbb{Y} \in \text{C.I.}(2, d, d)$ *are finite sets of distinct points where* $\Delta H(\mathbb{X}) = \{h_t\}_{t \geq 0}$. *Then* $h_{t+1} \leq h_t^{(t)}$ *for* $d \leq t < 2d - 2$.

Proof. As in Lemma 6.2, we can assume that $0 < h_t < \Delta H_{\text{C.I.}(2,d,d)}(t)$. By Proposition 6.1, if $h_t \leq 2d - t - 1$ then $h_t^{(t)} = h_t - 1$, otherwise $h_t^{(t)} = h_t - 2$.

We work with $\Delta H(\mathbb{Y} \setminus \mathbb{X})$. Note that (by Theorem 4.1)

$$\Delta H(\mathbb{Y} \setminus \mathbb{X}, 2d - t - 2) = 4d - 2t - 3 - h_{t+1}$$
$$\Delta H(\mathbb{Y} \setminus \mathbb{X}, 2d - t - 1) = 4d - 2t - 1 - h_t.$$

Since $d \leq t < 2d - 2$, we must have $0 < 2d - t - 2 \leq d - 2$ and thus $\Delta H(\mathbb{Y} \setminus \mathbb{X})$ satisfies the hypotheses of Lemma 6.2. Further, we can assume $1 \leq \Delta H(\mathbb{Y} \setminus \mathbb{X}, 2d - t - 2) < \Delta H_{\text{C.I.}(2,d,d)}(2d - t - 2)$.

Case 1: Assume $h_{t+1} > 2d - t - 2$. Then $\Delta H(\mathbb{Y} \setminus \mathbb{X}, 2d - t - 2) < 2d - t - 1$. By Proposition 6.1 and Lemma 6.2, $\Delta H(\mathbb{Y} \setminus \mathbb{X}, 2d - t - 1) \leq \Delta H(\mathbb{Y} \setminus \mathbb{X}, 2d - t - 2)$ and hence $h_{t+1} \leq h_t - 2 < h_t - 1$. We conclude that $h_{t+1} \leq h_t^{(t)}$.

Case 2: Assume $h_{t+1} \leq 2d - t - 2$. Then $\Delta H(\mathbb{Y} \setminus \mathbb{X}, 2d - t - 2) \geq 2d - t - 1$. By Proposition 6.1 and Lemma 6.2, $\Delta H(\mathbb{Y} \setminus \mathbb{X}, 2d - t - 1) \leq \Delta H(\mathbb{Y} \setminus \mathbb{X}, 2d - t - 2) + 1$, and so $h_{t+1} \leq h_t - 1$. If $h_t \leq 2d - t - 1$, then $h_{t+1} \leq h_t - 1 = h_t^{(t)}$ and we are done.

In the case that $h_t \geq 2d - t$, assume $h_{t+1} = h_t - 1$. Then

$$\Delta H(\mathbb{Y} \setminus \mathbb{X}, 2d - t - 1) > \Delta H(\mathbb{Y} \setminus \mathbb{X}, 2d - t - 2) = (\Delta H(\mathbb{Y} \setminus \mathbb{X}, 2d - t - 2))^{<2d-t-2>},$$

a contradiction. Thus, if $h_t \geq 2d - t$, we must have $h_{t+1} \leq h_t - 2 = h_t^{(t)}$. □

Remark 6.6. As noted by an anonymous referee, one can in spirit simplify Lemma 6.5 with the more general statement: Suppose $\mathbb{X} \subseteq \mathbb{Y} \in \text{C.I.}(d_1, \ldots, d_n)$ where $\Delta H(\mathbb{X}) = \{h_l\}_{l \geq 0}$. Let $s = d_1 + \cdots + d_n - n$ and t be a non-negative integer such that $t \leq s$. Then if the inequality $h_{t+1} \leq h_t^{(t)}$ holds we have that $h_{s-t} \leq h_{s-t-1}^{(s-t-1)}$ also holds. In order to rigorously prove this statement for complete intersections of type $\{2, d, d\}$, for example, one would have to work out the details from the proof of part (iii) Proposition 6.1 and Lemma 6.5.

We are now ready to prove the main theorem of this section. That is, we now prove Conjecture 3.6 for complete intersections of type $\{2, d_2, d_3\}$.

Theorem 6.7. *Let* $\Delta\mathcal{H} = \{h_i\}_{i \geq 0}$ *be the first difference Hilbert function for some finite set of distinct points in* $\mathbb{P}^3$ *such that* $\Delta\mathcal{H} \leq \Delta H_{\text{C.I.}(2,d_2,d_3)}$. *Then there exist finite sets of distinct points* $\mathbb{X} \subseteq \mathbb{Y} \in \text{C.I.}(2, d_2, d_3)$ *such that* $\Delta H(\mathbb{X}) = \Delta\mathcal{H}$ *if and only if* $h_{t+1} \leq h_t^{(t)}$ *for all* $t \geq 1$.

Proof. Using the results of Caviglia–Maclagan [3], the only case remaining to consider is: C.I.$(2, d_2, d_3)$ where $d_2 = d_3$. Let $d = d_2 = d_3$. Suppose that $h_{t+1} \le h_t^{(t)}$ for all $t \ge 1$. By Theorem 3.4 there exists a subset $\mathbb{X}$ of Rect.C.I.$(2, d, d)$ such that $\Delta H(\mathbb{X}) = \Delta \mathcal{H}$.

Conversely, suppose $\mathbb{X} \subseteq \mathbb{Y} \in$ C.I.$(2, d, d)$ such that $\Delta H(\mathbb{X}) = \Delta \mathcal{H}$. The assertion is obviously true if the subset $\mathbb{X}$ is of cardinality 1, 2, or $2d^2$. Using the symmetry of Theorem 4.1, the assertion is also obviously true if $\mathbb{X}$ is of cardinality $2d^2 - 1$ or $2d^2 - 2$. Thus we can assume that the cardinality of $\mathbb{X}$ is strictly between 2 and $2d^2 - 2$.

The inequality $h_{t+1} \le h_t^{(t)}$ is clearly satisfied if $h_{t+1} = 0$. Since we have at least 3 points in our subset $\mathbb{X}$, in order for $\Delta H(\mathbb{Y} \setminus \mathbb{X})$ to be an O-sequence, Theorem 4.1 guarantees that $\Delta \mathcal{H}(s) = 0$ for $s \ge 2d - 1$. Thus we need only verify $h_{t+1} \le h_t^{(t)}$ for $t \le 2d - 3$. All of these cases are verified in Lemmas 6.2, 6.4, and 6.5. □

7 Subsets of C.I.$(3, d_2, d_3)$ with $d_3 = d_2$

In this section we prove Conjecture 3.6 for complete intersections of type $\{3, d_2, d_3\}$ where $d_2 = d_3$. As before, for ease of notation, we set $d_2 = d_3 = d$.

Proposition 7.1. *Let b and t be positive integers such that $b < \Delta H_{\text{C.I.}(3,d,d)}(t)$.*

(i) *Assume $t \le d - 2$.*

- (a) *If $1 \le b \le t$, then $b^{(t)} = b$.*
- (b) *If $t + 1 \le b \le 2t$, then $b^{(t)} = b + 1$.*
- (c) *If $b \ge 2t + 1$, then $b^{(t)} = b + 2$.*

(ii) *Assume $t = d - 1$.*

- (a) *If $b \le 2d - 2$, then $b^{(d-1)} = b - 1$.*
- (b) *If $b \ge 2d - 1$, then $b^{(d-1)} = b$.*

(iii) *Assume $t = d$.*

- (a) *If $b \le d - 1$, then $b^{(d)} = b - 1$.*
- (b) *If $b \ge d$, then $b^{(d)} = b - 2$.*

(iv) *Assume $d + 1 \le t \le 2d - 2$.*

- (a) *If $b \le 2d - t - 1$, then $b^{(t)} = b - 1$.*
- (b) *If $2d - t \le b \le 4d - 2t - 1$, then $b^{(t)} = b - 2$.*
- (c) *If $b \ge 4d - 2t$, then $b^{(t)} = b - 3$.*

Proof. Let $\mathbb{D} = \{3, d, d\}$. The "Pascal's Table" used to find the t-$\mathbb{D}$-binomial expansion of b is:

degree:	$0\ 1\ \cdots$	$d-1$	d	$d+1$	$d+2$	$\cdots$	$2d-2$	$2d-1$	$\cdots$
$\Delta H_{\text{C.I.}(d)}$:	$1\ 1\ \cdots$	1	0	0	0	$\cdots$	0	0	$\to$
$\Delta H_{\text{C.I.}(d,d)}$:	$1\ 2\ \cdots$	d	$d-1$	$d-2$	$d-3$	$\cdots$	1	0	$\to$
$\Delta H_{\text{C.I.}(3,d,d)}$:	$1\ 3\ \cdots$	$3d-3$	$3d-2$	$3d-3$	$3d-6$	$\cdots$	$\cdots$	$\cdots$	$\cdots$

Since $b < \Delta H_{\text{C.I.}(3,d,d)}(t)$, note that when finding the t-$\mathbb{D}$-binomial expansion of b we will only use entries from the rows labeled $\Delta H_{\text{C.I.}(d)}$ and $\Delta H_{\text{C.I.}(d,d)}$. Moreover, the expansion can involve at most two entries from the row labeled $\Delta H_{\text{C.I.}(d,d)}$. As with Proposition 6.1, observe that when we shift each summand x in the binomial expansion of b one place to the right in the table, the resulting corresponding summand for $b^{(t)}$ is either $x-1$ or x or $x+1$ depending on the degree x is associated to.

(i) Note that $\Delta H_{\text{C.I.}(d,d)}(i) = i + 1$ for $1 \le i \le d-1$.

(a) The proof is identical to that of Case (i) (a) of Proposition 6.1.

(b) This proof is identical to that of Case (ii) (b) of Proposition 6.1.

(c) If $b \ge 2t + 1$, then the t-$\mathbb{D}$-binomial expansion of b involves

$$\Delta H_{\text{C.I.}(d,d)}(t) = t + 1, \Delta H_{\text{C.I.}(d,d)}(t-1) = t$$

and entries from the row labeled $\Delta H_{\text{C.I.}(d)}$. When the entry

$$\Delta H_{\text{C.I.}(d,d)}(t) = t + 1$$

is shifted one place to the right in the table one obtains $t+2$, and when $\Delta H_{\text{C.I.}(d,d)}(t-1) = t$ is shifted one obtains $t+1$. Since $t \le d-2$, the shifting of the remaining entries in the expansion results in the same number of 1's as there were to begin with. Thus, $b^{(t)} = b + 2$.

(ii) Suppose that $t = d - 1$.

(a) Suppose $b \le 2d-2$. The proof of this case is identical to that of Case 2 of Proposition 6.1.

(b) Suppose $b \ge 2d - 1$. Write $b = d + (d-1) + a$ for some non-negative integer a. Since the expansion can involve at most two entries from the row labeled $\Delta H_{\text{C.I.}(d,d)}$, we can assume $0 \le a < \Delta H_{\text{C.I.}(d,d)}(d-3) = d-2$. In this case the $(d-1)$-$\mathbb{D}$-binomial expansion of b involves exactly $\Delta H_{\text{C.I.}(d,d)}(d-1) = d$ and $\Delta H_{\text{C.I.}(d,d)}(d-2) = d-1$ and entries from the row labeled $\Delta H_{\text{C.I.}(d)}$. When the entry $\Delta H_{\text{C.I.}(d,d)}(d-1) = d$ is shifted one place to the right in the table one obtains $d-1$, and when the entry $\Delta H_{\text{C.I.}(d,d)}(d-2) = d-1$ is shifted one obtains d. Since $t = d-1$, the shifting of the remaining entries in the expansion results in the same number of 1's as there were to begin with. Thus, $b^{(t)} = b$.

(iii) Assume $t = d$.

(a) Assume $b \leq d - 1$. If $b = d - 1$, then the d-$\mathbb{D}$-binomial expansion of b is $b = \Delta H_{\mathrm{C.I.}(d,d)}(d)$ giving $b^{(d)} = \Delta H_{\mathrm{C.I.}(d,d)}(d+1) = d - 2 = b - 1$. Otherwise, if $b < d - 1$, then the d-$\mathbb{D}$-binomial expansion of b involves only entries from the row labeled $\Delta H_{\mathrm{C.I.}(d)}$. Since $t = d$, the shifting of the entry $\Delta H_{\mathrm{C.I.}(d)}(d-1)$ one place to the right results in $b^{(t)} = b - 1$.

(b) Assume $b \geq d$. We have two cases to consider.

Case A: Suppose $b < 2d - 1$. Then the d-$\mathbb{D}$-binomial expansion of b involves $\Delta H_{\mathrm{C.I.}(d,d)}(d) = d - 1$ and entries from the row labeled $\Delta H_{\mathrm{C.I.}(d)}$. When we shift the entries one place to the right, the entry $\Delta H_{\mathrm{C.I.}(d,d)}(d) = d - 1$ results in $d - 2$, the entry $\Delta H_{\mathrm{C.I.}(d)}(d-1) = 1$ results in 0 and the remaining entries produce no change. So, $b^{(d)} = b - 2$.

Case B: Suppose $b \geq 2d - 1$. In this case the d-$\mathbb{D}$-binomial expansion of b involves exactly $\Delta H_{\mathrm{C.I.}(d,d)}(d) = d - 1$ and $\Delta H_{\mathrm{C.I.}(d,d)}(d-1) = d$ and entries from the row labeled $\Delta H_{\mathrm{C.I.}(d)}$. When we shift the entries one place to the right, the entry $\Delta H_{\mathrm{C.I.}(d,d)}(d) = d - 1$ results in $d - 2$, the entry $\Delta H_{\mathrm{C.I.}(d,d)}(d-1) = d$ results in $d - 1$ and the remaining entries produce no change. So, $b^{(d)} = b - 2$.

(iv) Assume $d + 1 \leq t \leq 2d - 2$.

(a) Suppose $b \leq 2d - t - 1$. If $b = 2d - t - 1$, then the t-$\mathbb{D}$-binomial expansion of b is $b = \Delta H_{\mathrm{C.I.}(d,d)}(t) = 2d - t - 1$ and so $b^{(t)} = \Delta H_{\mathrm{C.I.}(d,d)}(t+1) = 2d - t - 2 = b - 1$. Otherwise, if $b < 2d - t - 1$ then the t-$\mathbb{D}$-binomial expansion of b involves only entries from the row labeled $\Delta H_{\mathrm{C.I.}(d)}$. Since $d + 1 \leq t \leq 2d - 2$, the shifting of the entry from the column $d - 1$ guarantees $b^{(t)} = b - 1$.

(b) Suppose $2d - t \leq b \leq 4d - 2t - 1$. First note that if $b = 4d - 2t - 1$, then the t-$\mathbb{D}$-binomial expansion of b is

$$b = \Delta H_{\mathrm{C.I.}(d,d)}(t) + \Delta H_{\mathrm{C.I.}(d,d)}(t-1) = (2d - t - 1) + (2d - t)$$

and so

$$\begin{aligned} b^{(t)} &= \Delta H_{\mathrm{C.I.}(d,d)}(t+1) + \Delta H_{\mathrm{C.I.}(d,d)}(t) \\ &= (2d - t - 2) + (2d - t - 1) = b - 2. \end{aligned}$$

Otherwise, if $b < 4d - 2t - 1$, then the t-$\mathbb{D}$-binomial expansion of b involves exactly $\Delta H_{\mathrm{C.I.}(d,d)}(t)$ and entries from the row labeled $\Delta H_{\mathrm{C.I.}(d)}$. In this case, the shifting of the entries $\Delta H_{\mathrm{C.I.}(d,d)}(t)$ and $\Delta H_{\mathrm{C.I.}(d)}(d-1)$ gives $b^{(t)} = b - 2$.

(c) Assume $b \geq 4d-2t$. In this case the t-$\mathbb{D}$-binomial expansion of b involves $\Delta H_{\mathrm{C.I.}(d,d)}(t) = 2d-t-1$, $\Delta H_{\mathrm{C.I.}(d,d)}(t-1) = 2d-t-2$ and a sum of entries from the row labeled $\Delta H_{\mathrm{C.I.}(d)}$. Since $d+1 \leq t \leq 2d-2$, the shifting of the entries $\Delta H_{\mathrm{C.I.}(d,d)}(t)$, $\Delta H_{\mathrm{C.I.}(d,d)}(t-1)$, and $\Delta H_{\mathrm{C.I.}(d)}(d-1) = 1$ one place to the right results in $b^{(t)} = b - 3$. □

We can now verify Conjecture 3.6 in the case of $d_1 = 3 \leq d_2 = d_3$. As in Section 6, we proceed with a series of lemmas that argue degree-by-degree.

Lemma 7.2. *Suppose that $\mathbb{X} \subseteq \mathbb{Y} \in \mathrm{C.I.}(3,d,d)$ are finite sets of distinct points where $\Delta H(\mathbb{X}) = \{h_t\}_{t\geq 0}$. Then $h_{t+1} \leq h_t^{(t)}$ for $1 \leq t \leq d-2$.*

Proof. As argued in Theorem 5.4, we can assume $0 < h_t < \Delta H_{\mathrm{C.I.}(3,d,d)}(t)$. If $h_t \leq 2t$ then the argument is identical to that seen for Lemma 6.2. So we may as well assume $h_t \geq 2t+1$ and write $h_t = (t+1)+t+a$ for some non-negative integer a. By Proposition 7.1 and its proof we can assume $0 \leq a < \Delta H_{\mathrm{C.I.}(d,d)}(t-2) = t-1$ and have $h_t^{(t)} = h_t + 2$. Note that the t-binomial expansion of h_t is

$$h_t = \binom{t+1}{t} + \binom{t}{t-1} + \binom{t-2}{t-2} + \cdots + \binom{t-(a+1)}{t-(a+1)}.$$

So $h_t^{<t>} = h_t + 2$. By Macaulay's Theorem, $h_{t+1} \leq h_t + 2 = h_t^{(t)}$ as desired. □

Lemma 7.3. *Suppose that $\mathbb{X} \subseteq \mathbb{Y} \in \mathrm{C.I.}(3,d,d)$ are finite sets of distinct points where $\Delta H(\mathbb{X}) = \{h_t\}_{t\geq 0}$. Then $h_d \leq h_{d-1}^{(d-1)}$.*

Proof. As in Lemma 6.2, we can assume $0 < h_{d-1} < \Delta H_{\mathrm{C.I.}(3,d,d)}(d-1) = 3d-3$.

Case 1: Suppose $h_{d-1} \leq 2d-2 = 2(d-1)$. We are claiming that $h_d \leq h_{d-1}^{(d-1)} = h_{d-1} - 1$. Suppose, on the contrary, $h_d = h_{d-1} - 1 + i$ where $i \geq 1$.

Case 1a: Assume $h_{d-1} \leq d-1$. Then $h_{d-1} - 1 + i = h_d \leq h_{d-1}^{<d-1>} = h_{d-1}$. Thus, $i = 1$ and so $h_d = h_{d-1}$, a contradiction to Lemma 6.3 which implies that $h_d \neq h_{d-1}$. We conclude that $h_d \leq h_{d-1} - 1$ as desired.

Case 1b: We now assume $d \leq h_{d-1} \leq 2(d-1)$. Then the $(d-1)$-binomial expansion of h_{d-1} is

$$h_{d-1} = \binom{d}{d-1} + \binom{d-2}{d-2} + \binom{d-3}{d-3} + \cdots + \binom{d-(h_{d-1}-d)-1}{d-(h_{d-1}-d)-1}.$$

Thus $h_{d-1} - 1 + i = h_d \leq h_{d-1}^{<d-1>} = h_{d-1} + 1$. Since $i \geq 1$, we have that either $h_d = h_{d-1}$ or $h_d = h_{d-1} + 1$. Now, by [14, Theorem 2.5], there exists a subset

$\mathbb{W} \subseteq \mathbb{X} \subseteq \mathbb{Y}$ such that

$$\Delta H(\mathbb{W}, j) = \begin{cases} h_j, & \text{for } j \leq d-2; \\ h_{d-1}, & \text{for } j = d-1; \\ h_{d-1}, & \text{for } j = d; \\ 0, & \text{for } j \geq d+1. \end{cases}$$

Note that $\mathbf{I}(\mathbb{W})_{\leq d-1} = \mathbf{I}(\mathbb{X})_{\leq d-1}$, but that $\Delta H(\mathbb{W}, d)$ and $\Delta H(\mathbb{X}, d)$, and hence $\mathbf{I}(\mathbb{W})_d$ and $\mathbf{I}(\mathbb{X})_d$, may not be equal. However, working with $\mathbb{W}$ will lead to the desired contradiction.

We can assume that no point of $\mathbb{Y}$ lies on the hyperplane $x_0 = 0$. Let $I \subseteq k[x_1, x_2, x_3]$ be the image of $\mathbf{I}(\mathbb{W})$ obtained by moding out by x_0. Then I contains a regular sequence $F, G, H \in k[x_1, x_2, x_3]$, where $\deg(F) = 3 \leq \deg(G) = d = \deg(H)$, and $H(k[x_1, x_2, x_3]/I) = \Delta H(\mathbb{W})$. Further, by assumption, $\Delta H(\mathbb{W}, d-1) = h_{d-1} = \Delta H(\mathbb{W}, d)$. Since $\Delta H(\mathbb{W})$ has maximal growth off by one in degree $(d-1)$, $\mathbf{I}(\mathbb{W})$ (and hence I) has at most one generator of degree d.

Claim 1: We can assume that $\dim_k(\mathbf{I}(\mathbb{W})_1) = \dim_k(\mathbf{I}(\mathbb{W})_2) = 0$.

Proof of Claim 1: If $\dim_k(\mathbf{I}(\mathbb{W})_1) \neq 0$ then $\mathbb{W}$ is contained in a complete intersection of type $\{1, m, n\}$ where $1 \leq m \leq n \leq d$. But $\Delta H_{\text{C.I.}(1,m,n)}$ reaches the value d at most once while $\Delta H(\mathbb{W}, d-1) = \Delta H(\mathbb{W}, d) = h_{d-1} \geq d$, a contradiction. Thus we assume $\dim_k(\mathbf{I}(\mathbb{W})_1) = 0$.

If $\dim_k(\mathbf{I}(\mathbb{W})_2) \neq 0$ then $\mathbb{W} \subseteq \mathbb{Y}'' \in \text{C.I.}(2, m, n)$ where $2 \leq m \leq n \leq d$. By adding points to $\mathbb{Y}''$, we see that $\mathbb{W}$ is contained in a complete intersection of type $\{2, m, m\}$. Proposition 6.1 and Theorem 6.7 imply that this is impossible. Hence we also can assume that $\dim_k(\mathbf{I}(\mathbb{W})_2) = 0$.

This finishes the proof of Claim 1.

Claim 2: $\mathbf{I}(\mathbb{W})$ has exactly one generator of degree d.

Proof of Claim 2: Since we know that $\mathbf{I}(\mathbb{W})$ has at most one generator of degree d, it suffices to show that $\mathbf{I}(\mathbb{W})$ has at least one generator of degree d. Suppose, on the contrary, that $\mathbf{I}(\mathbb{W})$ has no generator of degree d. Then $\mathbf{I}(\mathbb{Y}) \subseteq (\mathbf{I}(\mathbb{W})_{\leq d-1})$, and so $\mathbb{W} \subseteq \mathbb{Y}' \in \text{C.I.}(l, m, n)$ where $1 \leq l \leq m \leq n \leq d-1$ and $l \leq 3$. If $l = 1$ or 2 we obtain a contradiction from Claim 1. Thus we assume that $l = 3$. First suppose that $m = n = d-1$ and let $\mathbf{I}(\mathbb{Y}') = (F_1, F_2, F_3)$ where $\deg(F_1) = 3 \leq \deg(F_2) = \deg(F_3) = d-1$. By Theorem 4.1, $\Delta H(\mathbb{Y}' \setminus \mathbb{W})$ is an O-sequence and

$$\Delta H(\mathbb{Y}' \setminus \mathbb{W}, d-2) = 3d - h_{d-1} - 6$$
$$\Delta H(\mathbb{Y}' \setminus \mathbb{W}, d-1) = 3d - h_{d-1} - 5.$$

Now if $3d - 6 - h_{d-1} \leq d-2$, then $3d - h_{d-1} - 5 > (3d - 6 - h_{d-1})^{<d-2>} = 3d - 6 - h_{d-1}$, a contradiction. So, we assume that $3d - 6 - h_{d-1} > d-2$. In this case,

$3d-6-h_{d-1} \leq 2d-6$ and so the $(d-2)$-binomial expansion of $\Delta H(\mathbb{Y}'\setminus\mathbb{W}, d-2) = 3d-6-h_{d-1}$ is

$$3d-6-h_{d-1} = \binom{d-1}{d-2} + \binom{d-3}{d-3} + \binom{d-4}{d-4} + \cdots + \binom{r}{r},$$

where there are $2d-h_{d-1}-5$ terms of the form $\binom{j}{j}$. Thus,

$$(3d-6-h_{d-1})^{<d-2>} = \binom{d}{d-1} + (2d-h_{d-1}-5) = 3d-h_{d-1}-5.$$

We see that $\Delta H(\mathbb{Y}' \setminus \mathbb{W})$ has maximal growth in degree $d-2$. Thus $\mathbf{I}(\mathbb{Y}' \setminus \mathbb{W})$ has no generators of degree $(d-1)$, and so $(F_1, F_2, F_3) \subseteq (\mathbf{I}(\mathbb{Y}' \setminus \mathbb{W})_{\leq d-2})$. By Proposition 4.4, $(\mathbf{I}(\mathbb{Y}' \setminus \mathbb{W})_{\leq d-2})$ contains a regular sequence G_1, G_2, G_3, where $\deg(G_1) \leq \deg(G_2) \leq \deg(G_3) \leq d-2$. We can assume, without loss of generality, that no point of $\mathbb{Y}'$ lies on the hypersurface $\mathbf{V}(x_0)$. Let $\overline{J} \subseteq R/(x_0)$ be the canonical image of $\mathbf{I}(\mathbb{Y}'\setminus\mathbb{W})$. Then $\Delta H(\mathbb{Y}'\setminus\mathbb{W}) = H(k[x_1,x_2,x_3]/\overline{J})$ and $\overline{J}$ contains 3 forms in degrees $\leq d-2$ which are a regular sequence. By Lemma 4.5, $\overline{J}$ contains 3 forms in degree $d-2$ which are a regular sequence. But $H(k[x_1,x_2,x_3]/\overline{J}, d-2) \geq d-1$ and $H(k[x_1,x_2,x_3]/\overline{J})$ has maximal growth in degree $d-2$. Hence, by [1, Proposition 2.7], $\overline{J}_{d-2}$ and $\overline{J}_{d-1}$ have a greatest common divisor (GCD) of positive degree, a contradiction.

If m or n is strictly less than $d-1$ we can add points to $\mathbb{Y}'$ so that $\mathbb{W}$ is contained in a complete intersection of type $\{3, d-1, d-1\}$. By above, we see that this leads to a contradiction. Therefore, $\mathbf{I}(\mathbb{W})$ must have a generator of degree d, completing the proof of Claim 2.

By Proposition 4.4, we know that I contains a regular sequence F, G where $3 = \deg(F) \leq e = \deg(G) \leq d-1$. By Lemma 4.5, there exist linear forms $L_1, L_2 \in k[x_1,x_2,x_3]$ such that $FL_1^{d-3}, GL_2^{d-e} \in k[x_1,x_2,x_3]$ are again a regular sequence. Consider $N = (I_{\leq d-1}) \subseteq A = k[x_1,x_2,x_3]$. Then, since I has exactly one generator of degree d, $H(A/N, d-1) = h_{d-1}$ and $H(A/N, d) = h_{d-1}+1$, which is maximal growth in degree $(d-1)$. Further, by Gotzmann's Persistence Theorem [15], the Hilbert polynomial of A/N is

$$P(t) = \binom{t+1}{t} + \text{(lower terms)}$$

which has degree 1. Thus, N has height 1, a contradiction since $FL_1^{d-3}, GL_2^{d-e} \in N_d$ is a regular sequence.

Case 2: Now suppose that $2d-1 \leq h_{d-1} < 3d-3$. We are claiming that $h_d \leq h_{d-1}^{(d-1)} = h_{d-1}$. Suppose that $h_d = h_{d-1}+i$ where $i \geq 1$. We work with $\Delta H(\mathbb{Y}\setminus\mathbb{X})$

to arrive at a contradiction. By Theorem 4.1,

$$\Delta H(\mathbb{Y} \setminus \mathbb{X}, d) = 3d - 2 - h_{d-1} - i$$
$$\Delta H(\mathbb{Y} \setminus \mathbb{X}, d+1) = 3d - 3 - h_{d-1}.$$

We have $3d - 2 - h_{d-1} - i \leq d - 1 - i < d$. Thus $(3d - 2 - h_{d-1} - i)^{<d>} = 3d - 2 - h_{d-1} - i$. We see that if $i > 1$, then $3d - 3 - h_{d-1} > 3d - 2 - h_{d-1} - i$ which implies that $\Delta H(\mathbb{Y} \setminus \mathbb{X})$ is not an O-sequence. So from this point on we may as well assume $i = 1$. The argument is now completed as in Case 1a but with replacing $\Delta H(\mathbb{X})$ with $\Delta H(\mathbb{Y} \setminus \mathbb{X})$ and using degrees $\leq d$ when defining the ideal $\overline{J}$ (i.e., $\Delta H(\mathbb{Y} \setminus \mathbb{X})$ has maximal growth in degree d). □

Lemma 7.4. *Suppose that* $\mathbb{X} \subseteq \mathbb{Y} \in$ C.I.$(3, d, d)$ *are finite sets of distinct points where* $\Delta H(\mathbb{X}) = \{h_t\}_{t \geq 0}$. *Then* $h_{d+1} \leq h_d^{(d)}$.

Proof. As in past cases, we can assume $0 < h_d < \Delta H_{\text{C.I.}(3,d,d)}(d)$. By Proposition 7.1, if $h_d \leq d - 1$, then $h_d^{(d)} = h_d - 1$; and if $d \leq h_d \leq 3d - 3$, then $h_d^{(d)} = h_d - 2$.

Now, by Theorem 4.1,

$$\Delta H(\mathbb{Y} \setminus \mathbb{X}, d-1) = 3d - 3 - h_{d+1}$$
$$\Delta H(\mathbb{Y} \setminus \mathbb{X}, d) = 3d - 2 - h_d.$$

In what follows we apply Lemma 7.3 to $\Delta H(\mathbb{Y} \setminus \mathbb{X})$. Without loss of generality we may as well assume $1 \leq \Delta H(\mathbb{Y} \setminus \mathbb{X}, d-1) < 3d - 3 = \Delta H_{\text{C.I.}(3,d,d)}(d-1)$.

Case 1: Assume $3d - 3 - h_{d+1} \geq 2d - 1$. By Proposition 7.1 and Lemma 7.3, we must have

$$3d - 2 - h_d \leq (3d - 3 - h_{d+1})^{(d-1)} = 3d - 3 - h_{d+1},$$

and so $h_{d+1} \leq h_d - 1$. Thus we are done if $h_d \leq d - 1$.

If $d \leq h_d \leq 3d - 3$, then we need only rule out the case where $h_{d+1} = h_d - 1$. Assuming we are in this case, we see that $h_{d+1} = h_d - 1 \geq d - 1$. But, by assumption, $3d - 3 - h_{d+1} \geq 2d - 1$, and so $h_{d+1} \leq d - 2$, a contradiction.

Case 2: Now suppose that $1 \leq 3d - 3 - h_{d+1} \leq 2d - 2$. Then, by Proposition 7.1 and Lemma 7.3,

$$3d - 2 - h_d \leq (3d - 3 - h_{d+1})^{(d-1)} = 3d - 4 - h_{d+1}$$

and so $h_{d+1} \leq h_d - 2 < h_d - 1$, as desired. □

As in Lemma 6.5, the following lemma takes advantage of the fact that Hilbert functions of subsets of complete intersections come in pairs by reducing the situation of the tail-end-degrees to the beginning degrees in the complementary Hilbert function.

Lemma 7.5. *Suppose* $\mathbb{X} \subseteq \mathbb{Y} \in$ C.I.$(3, d, d)$ *are finite sets of distinct points where* $\Delta H(\mathbb{X}) = \{h_t\}_{t\geq 0}$. *Then* $h_{t+1} \leq h_t^{(t)}$ *for* $d + 1 \leq t \leq 2d - 2$.

Proof. As argued in Lemma 6.2, we can assume $1 \leq h_t < \Delta H_{\text{C.I.}(3,d,d)}(t)$. We show that we can apply Lemma 7.2 to $\Delta H(\mathbb{Y} \setminus \mathbb{X})$ to obtain the desired inequality. We know that $\Delta H(\mathbb{Y} \setminus \mathbb{X})$ is an O-sequence and

$$\Delta H(\mathbb{Y} \setminus \mathbb{X}, 2d - t - 1) = 6d - 3t - 3 - h_{t+1}$$
$$\Delta H(\mathbb{Y} \setminus \mathbb{X}, 2d - t) = 6d - 3t - h_t.$$

Note that since $d + 1 \leq t \leq 2d - 2$, we have $1 \leq 2d - t - 1 \leq d - 2$. We may as well assume that $1 \leq \Delta H(\mathbb{Y} \setminus \mathbb{X}, 2d - t - 1) < \Delta H_{\text{C.I.}(3,d,d)}(2d - t - 1) = 6d - 3t - 3$.

Case 1: Suppose first that $1 \leq h_{t+1} \leq m + n - t - 2$. Then

$$2(m + n - t - 1) + 1 \leq 3m + 3n - 3t - 3 - h_{t+1} < 3(m + n - t - 1).$$

So, by Proposition 7.1 and Lemma 7.2, $\Delta H(\mathbb{Y}\setminus\mathbb{X}, 2d-t) \leq \Delta H(\mathbb{Y}\setminus\mathbb{X}, 2d-t-1)+2$. Thus $h_{t+1} \leq h_t - 1$. By Proposition 7.1, if $1 \leq h_t \leq 2d - t - 1$, then $h_t^{(t)} = h_t - 1$ and we are done. We consider two possible cases.

Case 1a: Assume that $2d - t \leq h_t \leq 4d - 2t - 1$. By Proposition 7.1, $h_t^{(t)} = h_t - 2$. Since $h_{t+1} \leq h_t - 1$, we need only consider the situation where $h_{t+1} = h_t - 1$. Assume $h_{t+1} = h_t - 1$. Then

$$2d - t - 1 \leq \Delta H(\mathbb{Y} \setminus \mathbb{X}, 2d - t - 1) = 6d - 3t - 2 - h_t \leq 2(2d - t - 1).$$

Thus the $(2d - t - 1)$-binomial expansion of $6d - 3t - h_t - 2$ can take one of two forms: if $6d - 3t - h_t - 2 \neq 2d - t - 1$, then

$$6d - 3t - h_t - 2 = \binom{2d - t}{2d - t - 1} + \binom{2d - t - 2}{2d - t - 2} + \cdots + \binom{r}{r},$$

with $4d - 2t - h_t - 2$ terms $\binom{l}{l}$ and so $(6d - 3t - h_t - 2)^{<2d-t-1>} = 6d - 3t - h_t - 1$; if $6d - 3t - h_t - 2 = 2d - t - 1$, then $(6d - 3t - h_t - 2)^{<2d-t-1>} = 6d - 3t - h_t - 2 = 2d - t - 1$.

For either of the expansions, $\Delta H(\mathbb{Y} \setminus \mathbb{X}, 2d - t) > (6d - 3t - h_t - 2)^{<2d-t-1>}$ and so $\Delta H(\mathbb{Y} \setminus \mathbb{X})$ is not an O-sequence, a contradiction. Therefore, $h_{t+1} \leq h_t - 2$.

Case 1b: Now assume that $4d - 2t \leq h_t < 6d - 3t = \Delta H_{\text{C.I.}(3,d,d)}(t)$. By Proposition 7.1, $h_t^{(t)} = h_t - 3$. Since $h_{t+1} \leq h_t - 1$, we need only consider the situations where $h_{t+1} = h_t - 1$ or $h_{t+1} = h_t - 2$. If $h_{t+1} = h_t - 1$ or $h_t - 2$ then $\Delta H(\mathbb{Y} \setminus \mathbb{X}, 2d - t - 1) \leq 2d - t - 1$, and so $(\Delta H(\mathbb{Y} \setminus \mathbb{X}, 2d - t - 1))^{<2d-t-1>} = \Delta H(\mathbb{Y} \setminus \mathbb{X}, 2d - t - 1)$. But $\Delta H(\mathbb{Y} \setminus \mathbb{X}, 2d - t) > \Delta H(\mathbb{Y} \setminus \mathbb{X}, 2d - t - 1)$, a contradiction to the fact that $\Delta H(\mathbb{Y} \setminus \mathbb{X})$ is an O-sequence.

Case 2: Now suppose $2d - t - 1 \leq h_{t+1} \leq 4d - 2t - 3$. Then

$$2d - t \leq \Delta H(\mathbb{Y} \setminus \mathbb{X}, 2d - t - 1) = 6d - 3t - 3 - h_{t+1} \leq 2(2d - t - 1).$$

By Proposition 7.1 and Lemma 7.2, $\Delta H(\mathbb{Y} \setminus \mathbb{X}, 2d - t) \leq \Delta H(\mathbb{Y} \setminus \mathbb{X}, 2d - t - 1) + 1$. As a consequence we see that $h_{t+1} \leq h_t - 2$. By Proposition 7.1, if $h_t \leq 4d - 2t - 1$, then $h_t^{(t)} = h_t - 2$ and we are done. So assume that $4d - 2t \leq h_t < 6d - 3t = \Delta H_{\text{C.I.}(3,d,d)}(t)$. By Proposition 7.1, $h_t^{(t)} = h_t - 3$. Since $h_{t+1} \leq h_t - 2$, we need only rule out the case where $h_{t+1} = h_t - 2$. In this case the same argument used as in Case 1b gives a contradiction.

Case 3: If $h_{t+1} \geq 4d - 2t - 2$, then Proposition 7.1 says that $h_t^{(t)} = h_t - 3$. In this case we have $\Delta H(\mathbb{Y} \setminus \mathbb{X}, 2d - t - 1) \leq 2d - t - 1$. By Proposition 7.1 and Lemma 7.2, $\Delta H(\mathbb{Y} \setminus \mathbb{X}, 2d - t) \leq \Delta H(\mathbb{Y} \setminus \mathbb{X}, 2d - t - 1)$ and hence $h_{t+1} \leq h_t - 3$ as claimed. □

We can now prove Conjecture 3.6 for complete intersections of type $\{3, d, d\}$.

Theorem 7.6. *Let $\Delta\mathcal{H} = \{h_i\}_{i \geq 0}$ be the first difference Hilbert function for some finite set of distinct points in $\mathbb{P}^3$ such that $\Delta\mathcal{H} \leq \Delta H_{\text{C.I.}(3,d,d)}$. Then there exist finite sets of distinct points $\mathbb{X} \subseteq \mathbb{Y} \in$ C.I.$(3, d, d)$ such that $\Delta H(\mathbb{X}) = \Delta\mathcal{H}$ if and only if $h_{t+1} \leq h_t^{(t)}$ for all $t \geq 1$.*

Proof. Suppose that $h_{t+1} \leq h_t^{(t)}$ for all $t \geq 1$. By Theorem 3.4 there exists a subset $\mathbb{X}$ of Rect.C.I.$(3, d, d)$ such that $\Delta H(\mathbb{X}) = \Delta\mathcal{H}$.

Conversely, suppose $\mathbb{X} \subseteq \mathbb{Y} \in$ C.I.$(3, d, d)$ such that $\Delta H(\mathbb{X}) = \Delta\mathcal{H}$. The assertion is obviously true if the subset $\mathbb{X}$ is of cardinality 1, 2, or $3d^2$. Using the symmetry of Theorem 4.1, the assertion is also obviously true if $\mathbb{X}$ is of cardinality $3d^2 - 1$ or $3d^2 - 2$. Thus we can assume that the cardinality of $\mathbb{X}$ is strictly between 2 and $3d^2 - 2$.

The inequality $h_{t+1} \leq h_t^{(t)}$ is clearly satisfied if $h_{t+1} = 0$. Since we have at least 3 points in our subset $\mathbb{X}$, in order for $\Delta H(\mathbb{Y} \setminus \mathbb{X})$ to be an O-sequence, Theorem 4.1 guarantees that $\Delta\mathcal{H}(s) = 0$ for $s \geq 2d$. Thus we need only verify $h_{t+1} \leq h_t^{(t)}$ for $t \leq 2d - 2$. All of these cases are verified in Lemmas 7.2, 7.3, 7.4, and 7.5. □

8 An Application: The Cayley–Bacharach Property

Under the hypothesis that Conjecture 3.6 is true, we conclude by presenting a family of point sets which are guaranteed to have the *Cayley–Bacharach Property.*

Definition 8.1. Let $\mathbb{X}$ be a finite set of r distinct points in $\mathbb{P}^n$. $\mathbb{X}$ is said to have the *Cayley–Bacharach Property*, denoted *CBP*, if every subset of $(r - 1)$ points of $\mathbb{X}$ has the same Hilbert function.

We have the following fact:

Lemma 8.2 ([14, Theorem 2.5]). *Let* $\mathbb{X} = \{P_1, \ldots, P_r\}$ *be a finite set of distinct points in* $\mathbb{P}^n$. *Fix an integer* s, $1 \leq s < r$. *Then there exists a subset* $\mathbb{W}$ *of* $\mathbb{X}$ *of exactly* s *points such that* $H(\mathbb{W}, t) = \min\{H(\mathbb{X}, t), s\}$ *for all* $t \geq 0$.

We refer to $H(\mathbb{W})$ in Lemma 8.2 as the *truncated Hilbert function of* $H(\mathbb{X})$ *at* s. For example, if $\mathbb{X} \subset \mathbb{P}^2$ has Hilbert function $H(\mathbb{X}) = (1, 3, 6, 9, 9, \ldots)$, then the truncated Hilbert function of $H(\mathbb{X})$ at $s = 5$ is $(1, 3, 5, 5, \ldots)$. Using Theorem 2.13, we see that if $\mathbb{X}$ is a set of r distinct points in $\mathbb{P}^n$ with $H(\mathbb{X}) = (1, h_1, h_2, \ldots, h_{l-1}, r, r, r, \ldots)$ where $l-1$ is the largest degree t with $H(\mathbb{X}, t) \neq r$, then the truncated Hilbert function of $H(\mathbb{X})$ at $(r-1)$ is $(1, h_1, h_2, \ldots, h_{l-1}, r-1, r-1, \ldots)$. Lemma 8.2 immediately gives:

Lemma 8.3. *A set* $\mathbb{X} = \{P_1, \ldots, P_r\} \subseteq \mathbb{P}^n$ *of* r *distinct points has the CBP if and only if the Hilbert function of every subset of* $(r-1)$ *points of* $\mathbb{X}$ *is precisely the truncated Hilbert function of* $H(\mathbb{X})$ *at* $(r-1)$.

Much effort has been put into characterizing the Hilbert functions of sets with the Cayley–Bacharach Property. We now present a special family of subsets of complete intersections which have the CBP. We first need to define *minimal* $\{d_1, \ldots, d_n\}$-*functions.*

Definition 8.4. Fix integers $t, h \geq 1$ and $1 \leq d_1 \leq d_2 \leq \cdots \leq d_n$ with the property that $h \leq \Delta H_{\text{C.I.}(d_1,\ldots,d_n)}(t)$. Let $\mathbb{D} = \{d_1, \ldots, d_n\}$ and suppose that the t-$\mathbb{D}$-binomial expansion of h is

$$h = \Delta H_{\text{C.I.}(d_{a_t},\ldots,d_n)}(t) + \Delta H_{\text{C.I.}(d_{a_{t-1}},\ldots,d_n)}(t-1) + \cdots + \Delta H_{\text{C.I.}(d_{a_i},\ldots,d_n)}(i).$$

We define $h_{(t)}$ to be the number

$$\begin{aligned} h_{(t)} &= \Delta H_{\text{C.I.}(d_{a_t},\ldots,d_n)}(t-1) + \Delta H_{\text{C.I.}(d_{a_{t-1}},\ldots,d_n)}(t-2) \\ &\quad + \cdots + \Delta H_{\text{C.I.}(d_{a_i},\ldots,d_n)}(i-1). \end{aligned}$$

Remarks 8.5. Let $h, t, d_1, \ldots, d_n$ be as in Definition 8.4. The following assertions are straightforward to verify.

(i) $h_{(t)}$ depends on h, t, and $\{d_1, \ldots, d_n\}$. It will be clear from our discussions what $d_1, \ldots, d_n$ are.

(ii) The decomposition for $h_{(t)}$ is the $(t-1)$-$\mathbb{D}$-binomial expansion of $h_{(t)}$.

(iii) $h_{(t)}^{(t-1)} = h$.

(iv) Note that $h_{(t)}$ is the least integer g, $1 \leq g \leq \Delta H_{\text{C.I.}(d_1,\ldots,d_n)}(t-1)$, such that $h \leq g^{(t-1)}$.

Definition 8.6. Let $\{h_i\}_{i\geq 0} \leq \Delta H_{\text{C.I.}(d_1,\dots,d_n)}$ be an O-sequence and $\mathbb{D} = \{d_1, \dots, d_n\}$. We say that $\{h_i\}_{i\geq 0}$ is a *minimal* $\mathbb{D}$*-function* if $h_{t-1} = (h_t)_{(t)}$ for all $t \geq 1$ such that $h_t \neq 0$.

Proposition 8.7. *Assume that Conjecture* 3.6 *is true for the integers* $\mathbb{D} = \{d_1, \dots, d_n\}$ *and let* $\mathbb{X}$ *be a subset of a complete intersection of type* $\{d_1, \dots, d_n\}$. *If* $\Delta H(\mathbb{X})$ *is a minimal* $\mathbb{D}$*-function, then* $\mathbb{X}$ *has the CBP.*

Proof. Suppose $\mathbb{X} = \{P_1, \dots, P_r\} \subseteq \mathbb{Y}$ where $\mathbb{Y}$ is a complete intersection as in the hypothesis. Using Lemma 8.3, we need to show that the Hilbert function of every subset of $(r-1)$ points of $\mathbb{X}$ is the truncated Hilbert function of $H(\mathbb{X})$ at $(r-1)$. Since $\Delta H(\mathbb{X}) = \{h_t\}_{t\geq 0}$ is assumed to be a minimal $\mathbb{D}$-function, $h_{t-1} = (h_t)_{(t)}$ for all $t \geq 1$.

Let $\mathbb{W}$ be a subset of $(r-1)$ points of $\mathbb{X}$. Then $\mathbb{W} \subseteq \mathbb{Y}$. Let $\Delta H(\mathbb{W}) = \{g_t\}_{t\geq 0}$. Since $\mathbb{W}$ and $\mathbb{X}$ are subsets of a complete intersection of type $\{d_1, \dots, d_n\}$, the assumption that Conjecture 3.6 holds true gives that $h_{t+1} \leq h_t^{(t)}$ and $g_{t+1} \leq g_t^{(t)}$ for all $t \geq 1$.

By [13, Theorem 3.4] and [14, Lemma 2.3], $g_t \leq h_t$ for all $t \geq 0$. Since $\mathbb{W}$ is a subset of $(r-1)$ points of $\mathbb{X}$, there is a positive integer e such that $g_t = h_t$ for $t \leq e-1$ and $g_e < h_e$. Let $\sigma(H(\mathbb{X}))$ be the least degree t for which $h_t = 0$. Clearly $1 \leq e \leq \sigma(H(\mathbb{X}))-1$. Suppose for a moment that $e < \sigma(H(\mathbb{X}))-1$. Then, taking Remarks 8.5 into account, we see that $g_t < h_t$ for all $e \leq t \leq \sigma(H(\mathbb{X}))-1$. But $\sum_{j=1}^{\infty} g_j = (r-1)$ and $\sum_{j=1}^{\infty} h_j = r$, a contradiction. We conclude $e = \sigma(H(\mathbb{X})) - 1$, and so $H(\mathbb{W})$ must indeed be the truncated Hilbert function of $H(\mathbb{X})$ at $(r-1)$. Therefore $\mathbb{X}$ has the CBP. □

It is well-known [7, 6] that a finite set of points $\mathbb{X} \subseteq \mathbb{P}^2$ is a complete intersection if and only if $\mathbb{X}$ has the CBP and $\Delta H(\mathbb{X})$ is symmetric, i.e. $\Delta H(\mathbb{X}, t) = \Delta H(\mathbb{X}, N-t)$ for $0 \leq t \leq N$, where $N =$ (the least integer where $\Delta H(\mathbb{X}) = 0) - 1$. Thus, when $n = 2$ we can use Proposition 8.7 to guarantee that certain sets of points are complete intersections.

Lemma 8.8. *Let* $\mathbb{X}$ *be a subset of a complete intersection of type* $\{d_1, d_2\}$, *where* $d_1 \geq 2$. *Suppose* $t \geq 1$ *is the least degree in which* $\Delta H(\mathbb{X})$ *is non-zero. If* $\Delta H(\mathbb{X})$ *is a minimal* $\mathbb{D}$*-function, where* $\mathbb{D} = \{d_1, d_2\}$, *such that* $h_t = 1$, *then* $\Delta H(\mathbb{X})$ *is symmetric.*

Proof. Let $\Delta H(\mathbb{X}) = \{h_s\}_{s\geq 0}$. If $h_t = \Delta H_{\text{C.I.}(d_1,d_2)}(t)$, then we must have $\Delta H(\mathbb{X}) = \Delta H_{\text{C.I.}(d_1,d_2)}$.

If $s \leq t$ and $h_s < \Delta H_{\text{C.I.}(d_1,d_2)}(t)$, then the s-$\mathbb{D}$-binomial expansion of h_s has the form

$$h_s = \Delta H_{\text{C.I.}(d_2)}(s) + \Delta H_{\text{C.I.}(d_2)}(s-1) + \cdots + \Delta H_{\text{C.I.}(d_2)}(s - h_s + 1).$$

Thus,

$$(h_s)_{(s)} = \Delta H_{\text{C.I.}(d_2)}(s-1) + \Delta H_{\text{C.I.}(d_2)}(s-2) + \cdots + \Delta H_{\text{C.I.}(d_2)}(s-h_s).$$

Observe that if $t < d_2$, then this means $h_s = 1$ for all $s \leq t$. Further, if $t \geq d_2$, then we must have $h_{s+1} = h_s - 1$ for $s \geq d_2 - 1$, $h_s = h_{d_2-1}$ for $d_1 - 1 \leq s \leq h_{d_2-1}$ and $h_s = h_{s-1} + 1$ for $s \leq h_{d_2-1} - 1$. □

Corollary 8.9. *Let $\mathbb{X}$ be a subset of a complete intersection $\{d_1, d_2\}$, where $d_1 \geq 2$, such that $\Delta H(\mathbb{X})$ satisfies the hypotheses of Lemma 8.8. Then $\mathbb{X}$ is a complete intersection in $\mathbb{P}^2$.*

Acknowledgments. Sincerest thanks goes to the author's Ph.D. advisors A. V. Geramita and L. G. Roberts for their guidance. The author is also very grateful to J. Migliore and M. Roth for their valuable feedback.

Bibliography

[1] Bigatti, A., Geramita, A. V., Migliore, J. C., Geometric consequences of extremal behavior in a theorem of Macaulay. Trans Amer Math Soc. 1994;346(1):203–235.

[2] Bruns, W., Herzog, J., Cohen–Macaulay rings. Cambridge: Cambridge University Press; 1993.

[3] Caviglia, G., Maclagan, D., Some cases of the Eisenbud–Green–Harris conjecture. Math Res Lett. 2008;15(3):427–433.

[4] Clements, G. F., Lindström, B., A generalization of a combinatorial theorem of Macaulay. J Combin Theory. 1969;7:230–238.

[5] Cooper, S. M., Roberts, L. G., Algebraic interpretation of a theorem of Clements and Lindström. J Commut Algebra. 2009;1(3):361–380.

[6] Davis, E. D., Geramita, A. V., Orecchia, F., Gorenstein algebras and the Cayley–Bacharach theorem. Proc Amer Math Soc. 1985;93:593–597.

[7] Davis, E., Maroscia, P., Complete intersections in $\mathbb{P}^2$. Proceedings, Conference on Complete Intersections; Jun 1983; Acireale. Lecture Notes in Mathematics.

[8] Eisenbud, D., Green, M., Harris, J., Higher Castelnuovo theory. Astérisque. 1993;218: 187–202.

[9] Eisenbud, D., Green, M., Harris, J., Cayley–Bacharach theorems and conjectures. Bull Amer Math Soc. 1996;33(3):295–324.

[10] Francisco, C., Almost complete intersections and the Lex-plus-powers conjecture. J Algebra. 2004;276(2):737–760.

[11] Francisco, C., Richert, B., Lex-plus-powers ideals in syzygies and Hilbert functions. Peeva, I., editor. vol. 254 of Lecture Notes in Pure and Applied Mathematics. Chapman & Hall/CRC; 2007. p. 113–144.

[12] Geramita, A. V., Gregory, D., Roberts, L. G., Monomial ideals and points in projective space. J Pure Appl Algebra. 1986;40:33–62.

[13] Geramita, A. V., Maroscia, P., Roberts, L. G., The Hilbert function of a reduced k-algebra. In: The Curves Seminar at Queen's, Vol. II. vol. 61 of Queen's papers in pure and applied mathematics. 1982. p. C1–C63.

[14] Geramita, A. V., Maroscia, P., Roberts, L. G., The Hilbert function of a reduced k-algebra. J Lond Math Soc. 1983;28(2):443–452.

[15] Gotzmann, G., Eine Bedingung für die Flachheit und das Hilbertpolynom eines graduierten Ringes. Math Z. 1978;158:61–70.

[16] Greene, C., Kleitman, D., Proof techniques in the theory of finite sets. In: Rota, G.-C., editor. Studies in combinatorics. vol. 17 of MAA Studies in Mathematics. Mathematical Association of America; 1978. p. 22–79.

[17] Katona, G., A Theorem of Finite Sets. In: Erdős, P., Katona, G., editors. Theory of Graphs. Proceedings, Tihany Conference; 1966. New York, London: Academic Press; 1968.

[18] Kruskal, J., The number of simplicies in a complex. In: Mathematical Optimization Techniques. Berkeley, Los Angeles: University of California Press; 1963.

[19] Macaulay, F. S., Some properties of enumeration in the theory of modular forms. Proc Lond Math Soc. 1927;26:531–555.

[20] Mermin, J., Murai, S., The lex-plus-powers conjecture holds for pure powers. Adv Math. 2011;226(4):3511–3539.

[21] Mermin, J., Peeva, I., Stillman, M., Ideals containing the squares of the variables. Adv Math. 2008;217(5):2206–2230.

[22] Richert, B. P., A study of the lex plus powers conjecture. J Pure Appl Alg. 2004;186(2): 169–183.

[23] Stanley, R. P., Hilbert functions of graded algebras. Adv Math. 1978;28:57–83.

Author Information

Susan M. Cooper, Department of Mathematics, Central Michigan University, Mt. Pleasant, MI, USA.
E-mail: s.cooper@cmich.edu

Progress in Commutative Algebra 1, 199–230

The Homological Conjectures

Paul C. Roberts

Abstract. We describe the conjectures in Commutative Algebra known as the Homological Conjectures, outline their history over the past forty years or so, and give a summary of their current status, including a list of the main conjectures that are open at the present time.

Keywords. Homological Conjectures, Multiplicities, Cohen–Macaulay Modules, Direct Summand Conjecture.

2010 Mathematics Subject Classification. Primary: 13D22, 13D02; Secondary: 13C10, 13C14, 13H05, 14C17.

1 Introduction

The term "Homological Conjectures" is used here to refer to a certain set of related conjectures about homological properties of commutative rings. While there are numerous conjectures in this area, the ones discussed here are those collected in a monograph of Mel Hochster in 1975 entitled "Topics in the homological theory of modules over commutative rings" [24], as well as several ones that have developed out of them. In this monograph Hochster stated a number of earlier conjectures, added a few of his own, and solved several of them. Since then new ones have been added and some of them have been settled. It is the aim of this article to outline this history, starting at the beginning and ending by giving an idea of the present situation. We will attempt to give some idea of the methods and concepts behind the various advances, and give references to more complete accounts.

The article is organized as follows. In each of the first few sections we discuss a set of related conjectures on Hochster's diagram and follow their development up to the present. These sections follow a roughly chronological order as far as the origins of the conjectures are concerned, beginning with Serre's multiplicity conjectures which were one of the major influences behind the whole subject. However, there has been recent progress even on some of the earliest conjectures, and we will discuss, for example, recent developments on Serre's original conjectures before getting to generalizations of these conjectures which came much earlier. In addition to the major advances, we will mention many other developments, but there are a lot of them and we have not attempted to cover them all.

This research was supported by NSF grant 0500588.

We give Hochster's 1975 diagram below. The conjectures appearing in the diagram will then be stated in the following sections. We give a table of contents below, including the numbers from the diagram that are defined in each section. Those denoted M_0, M_1, and M_2 are parts of (8), the Serre Multiplicity Conjectures, and (9), the Strong Multiplicity Conjectures.

Here is the outline.

(i) The Serre Multiplicity Conjectures ((1), (8)).

(ii) The Peskine–Szpiro Intersection Conjecture ((2), (3), (4), (5)).

(iii) Generalizations of the Multiplicity Conjectures ((9), (12), (13)).

(iv) The Monomial, Direct Summand, and Canonical Element Conjectures ((10), (11))

(v) Cohen–Macaulay Modules and Algebras ((6), (7)).

(vi) The Syzygy Conjecture and the Improved New Intersection Conjecture.

(vii) Tight Closure Theory

(viii) The Strong Direct Summand Conjecture.

(ix) Almost Cohen–Macaulay Algebras.

(x) A Summary of Open Questions.

There have been several summaries of progress on these conjectures over the years, including two in the last decade. Jan Strooker [63] has an book on the state of the Homological Conjectures in 1990; it also includes a lot of the necessary background in Commutative Algebra. There is also a set of notes coming from a Minicourse on Classical Questions in Commutative Algebra at the University of Utah which covered many aspects of the subject. These notes can be found at the University of Utah website at http://www.math.utah.edu/vigre/minicourses/2004.html#b. Hochster also has a summary from the conference in honor of Phil Griffith which talks about some of the recent developments [30].

We next give the diagram of implications between various conjectures from Hochster's monograph from 1975. The conjectures will be stated in detail in subsequent sections. Implications are indicated in the diagram by arrows, and the diagram is set up so that, for the most part, the stronger conjectures are toward the top and the left side. Since 1975 some of the conjectures have been proven or counterexamples have been found, and in addition some new implications have been found and new conjectures added to the list. We will give a more recent version, with conjectures which have been settled taken off the list and new ones added, later in the paper.

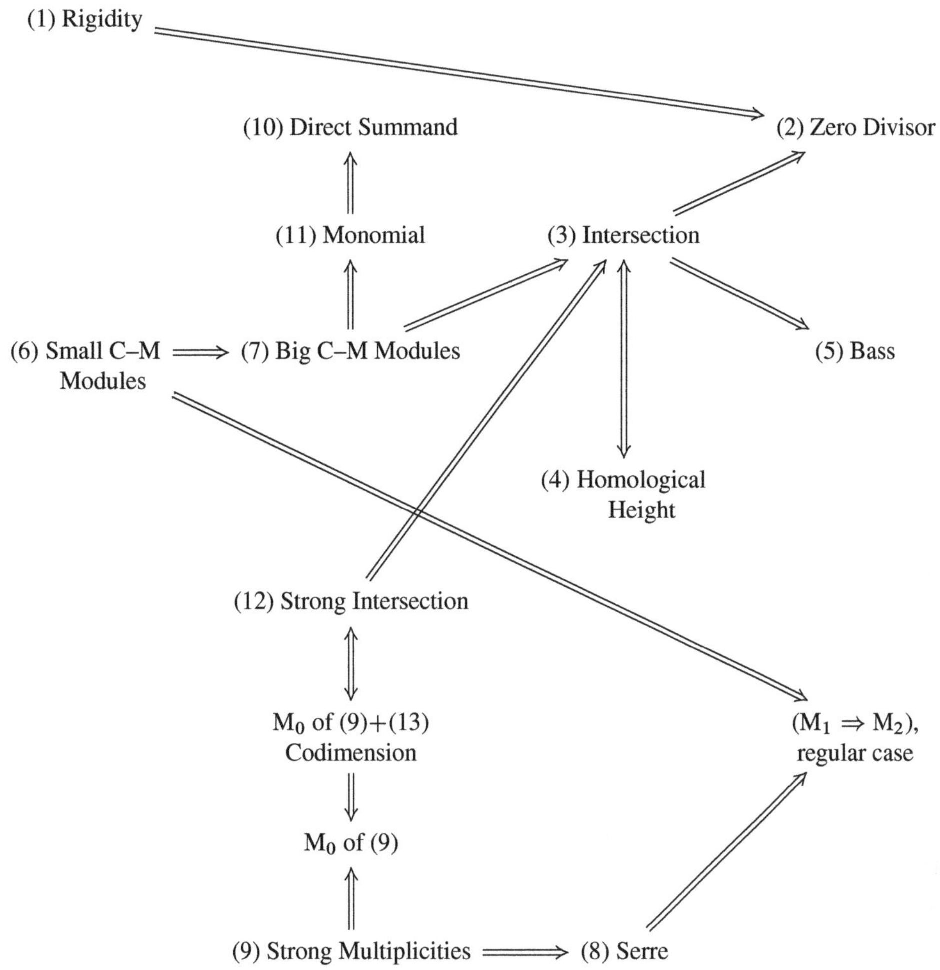

Figure 1. Hochster's diagram of homological conjectures in 1975.

2 The Serre Multiplicity Conjectures

Among the earliest conjectures in this subject were those of Jean-Pierre Serre which arose from his theory of intersection multiplicities using homological methods. The idea was to extend the algebraic theory from classical methods that worked, say, for intersections of curves in the plane, to a more general situation.

We look briefly at the case of the intersection of two curves in the affine plane over an algebraically closed field k. In this case each curve is defined by one polynomial in two variables, say x and y, so we have polynomials f and g defining the two curves. The condition that a point p in the plane corresponding to a maximal ideal $\mathfrak{m}$

of $A = k[x, y]$ is an isolated point of intersection means that the ideal (f, g) generated by f and g is primary to the maximal ideal of the local ring $A_{\mathfrak{m}}$. The intersection multiplicity is then defined simply to be the length of the quotient $A_{\mathfrak{m}}/(f, g)$, or, equivalently, its dimension over the field k.

There is more than one way to generalize this to an arbitrary dimension d. First, one can take the intersection of d hyperplanes; in the case of affine space, for example, this can be done in the same way as curves in the plane. One can also define the intersection of two subvarieties (or subschemes). These subvarieties will be defined locally by ideals I and J at a point of intersection corresponding to a maximal ideal $\mathfrak{m}$. However, in this case, defining the intersection multiplicity to be the length of $A_{\mathfrak{m}}/(I, J)$ does not work; for example, Bézout's Theorem in projective space would not hold with this definition. What Serre did was to correct this definition by taking an Euler characteristic involving higher Tor modules. He defined the intersection multiplicity for any pair of modules M and N over a regular local ring A such that $M \otimes N$ has finite length as follows.

$$\chi(M, N) = \sum_{i=0}^{d} (-1)^i \operatorname{length}(\operatorname{Tor}_i^R(M, N)).$$

The case of subvarieties above is where $M = A_{\mathfrak{m}}/I$ and $N = A_{\mathfrak{m}}/J$. In this case, letting $A_{\mathfrak{m}} = R$, we have $\operatorname{Tor}_0^R(M, N) = \operatorname{Tor}_0^R(R/I, R/J) = R/I \otimes_R R/J = R/(I, J)$, so that the previous definition appears as the first term in this alternating sum. Serre's definition has many nice properties, such as additivity in each variable, but now some conditions which were clear before, such as the fact that it is nonnegative, are not so clear. Serre stated three conjectures which are equivalent to the four we give here. The notation M_i refers to Hochster's diagram.

Conjecture 1. (i) (M_0) $\dim(M) + \dim(N) \leq \dim(R)$.

(ii) (M_1: Vanishing) If $\dim(M) + \dim(N) < \dim(R)$, then $\chi(M, N) = 0$.

(iii) (Nonnegativity) $\chi(M, N) \geq 0$.

(iv) (M_2: Positivity) If $\dim(M) + \dim(N) = \dim(R)$, then $\chi(M, N) > 0$.

Serre's original conjectures, as stated in Serre [62], V.B.3 and V.B.4, were nonnegativity, M_0, and that $\dim(M) + \dim(N) = \dim(R)$ if and only if $\chi(M, N) > 0$. The reason for stating them the way we did comes from later developments.

Before continuing, it will be good to go over some of the issues that arose in studying these questions, since they have been part of this subject ever since. First, there are three basic cases. Since R is a regular local ring it is an integral domain and has a maximal ideal $\mathfrak{m}$ and residue field k. The cases are

(i) Characteristic zero: R contains a field of characteristic zero.

(ii) Positive characteristic: R contains a field of positive characteristic p for some p.

(iii) Mixed characteristic: R has characteristic zero but k has positive characteristic p for some p.

The first two cases are called the equicharacteristic case. The mixed characteristic case can be further divided into the unramified case, in which the prime p is not in $\mathfrak{m}^2$, and the ramified case, in which it is in $\mathfrak{m}^2$. The most difficult case for these conjectures and many others is the ramified case in mixed characteristic.

The method of proof used by Serre in the equicharacteristic case, called "reduction to the diagonal", goes roughly as follows. First, one shows that the statments hold if one of the modules, say M, is of the form R/I, where I is an ideal generated by a regular sequence (see Serre [62, Section IV.A.3]). If R is a complete equicharacteristic regular local ring, the Cohen structure theorem says that it is a power series ring over a field. If we now have arbitrary modules M and N over a power series ring $k[[X_1, \ldots, X_d]]$, we think of N as being a module over another power series ring $k[[Y_1, \ldots, Y_d]]$ and consider the "complete tensor product" $M \hat{\otimes}_k N$ as a module over $k[[X_1, \ldots, X_d, Y_1, \ldots, Y_d]]$; notice that the tensor product is taken over the subfield k. Let I denote the ideal of $k[[X_1, \ldots, X_d, Y_1, \ldots, Y_d]]$ generated by $(X_1 - Y_1, \ldots, X_d - Y_d)$; these elements form a regular sequence and define the diagonal subscheme in $\mathrm{Spec}(k[[X_i, Y_i]])$. Then one shows that

$$M \otimes_R N \cong (M \hat{\otimes}_k N) \otimes_{k[[X_i,Y_i]]} (k[[X_i, Y_i]]/I)$$

and similarly for higher Tors. This reduces the question to one of the form a regular ring modulo a regular sequence, where it all works. Needless to say there are a lot of details omitted here; the complete story can be found in Serre [62].

In addition to the equicharacteristic case, Serre proved these results in the case of an unramified ring of mixed characteristic, and he proved the first statement for general regular local rings.

Serre also stated conjectures about partial Euler characteristics; that is, sums of the form

$$\chi_i(M, N) = \sum_{j=i}^{d} (-1)^{j-i} \mathrm{length}(\mathrm{Tor}_i(M, N)).$$

We note that this gives the alternating sum of lengths of Tor, starting now with the term $\mathrm{Tor}_i(M, N)$ with a positive sign instead of $\mathrm{Tor}_0(M, N) = M \otimes_R N$. Serre proved that in the equicharacteristic case, we have $\chi_i(M, N) \geq 0$ for all $i \geq 0$, and, in fact, if $\mathrm{Tor}_i(M, N) \neq 0$ and $i > 0$, then $\chi_i(M, N) > 0$. This implies in particular the following for equicharacteristic rings in the case where $M \otimes_R N$ has finite length. In fact, this was a result of Auslander [1] for all pairs of modules over unramified regular local rings, and it was conjectured to be true in general.

Conjecture 2 (The Rigidity of Tor (1)). Let M and N be finitely generated modules over a regular local ring. Then if $\mathrm{Tor}_i(M, N) = 0$ for some $i > 0$, then $\mathrm{Tor}_j(M, N) = 0$ for all $j \geq i$.

The general case of Rigidity (R is still assumed regular) was proven by Lichtenbaum in [43]. He also extended Serre's results on partial Euler characteristics to the unramified case for $i \geq 2$ or when M and N are torsion-free; Hochster [26] completed the proof in the unramified case. The conjecture on partial Euler characteristics is still open for ramified regular local rings of mixed characteristic.

In the remainder of this section we discuss later developments on these conjectures. R is always assumed to be a regular local ring.

2.1 The Vanishing Conjecture

The first of the multiplicity conjectures to be proven was the Vanishing Conjecture. This was proven independently in Roberts [51] (see also [53]) and by Gillet and Soulé in [18] (see also [19]). Both of the proofs involved new machinery in either Algebraic Geometry or K-theory. Before discussing these developments we put them into a more recent context.

Let A be a local ring, and let M be a module of finite projective dimension. Then M has a finite free resolution

$$0 \to F_k \to F_{k-1} \to \cdots \to F_0 \to M \to 0.$$

It is often more convenient to replace M with its resolution

$$0 \to F_k \to F_{k-1} \to \cdots \to F_0 \to 0.$$

This is a *perfect complex*, which means a bounded complex of finitely generated free modules. If M and N are both modules of finite projective dimension, and

$$0 \to G_t \to G_{t-1} \to \cdots \to G_0 \to N \to 0$$

is a free resolution of N, then the tensor product of complexes $F_\bullet \otimes G_\bullet$ gives a complex with homology $\mathrm{Tor}_i(M, N)$. Since all modules over a regular local ring have finite projective dimension, this means that Serre's multiplicity conjectures can be formulated in terms of perfect complexes.

We now let $K_0(A)$ denote the K-group of perfect complexes over a local ring A. $K_0(A)$ is defined to be the free abelian group with generators isomorphism classes $[F_\bullet]$ of perfect complexes with relations given by

(i) $[F_\bullet] = [F'_\bullet] + [F''_\bullet]$ if there is a short exact sequence of complexes

$$0 \to F'_\bullet \to F_\bullet \to F''_\bullet \to 0.$$

(ii) $[F_\bullet] = [G_\bullet]$ if there exists a map of complexes $F_\bullet \to G_\bullet$ that induces an isomorphism on homology modules.

A map of complexes that induces an isomorphism on homology modules is called a *quasi-isomorphism*.

In addition to its structure as an abelian group, $K_0(A)$ has a product defined by the tensor product of complexes, which we have already seen is related to intersection multiplicities. We define the *support* of a complex to be the union of the supports of its homology modules, or, equivalently, the set of prime ideals $\mathfrak{p}$ for which the localization of the complex at $\mathfrak{p}$ is not exact. If $F_\bullet$ is a perfect complex with support W and $G_\bullet$ one with support Z, then it is not hard to show that the support of $F_\bullet \otimes G_\bullet$ is $W \cap Z$. Putting this together, we can see that $K_0(A)$ has a filtration by support and this filtration is compatible with the product structure.

As mentioned above, the Vanishing Conjecture for regular rings was proven around 1985; there were two independent proofs using different methods. However, in both cases the main idea was to replace the above filtration by a grading with good properties. Suppose that we could give $K_0(R)$, for a regular local ring R, a grading by codimension, so that we had $K_0(R) = \oplus_{i=0}^{d} G_i$, where G_i gave the component representing elements with support of codimension i, and satisfying the condition that the intersection pairing mapped $G_i \times G_j$ to $G_{i+j} \bigoplus G_{i+j+1} \oplus \cdots$. Then if M and N were modules (or perfect complexes) with $\dim M + \dim N < \dim R$, they would be represented by sums of elements of G_i and G_j respectively with $i + j > d$, so the intersection product would be zero. This, roughly, is what each of the proofs did.

In the proof be Gillet and Soulé the grading was given by eigenspaces of Adams operations on $K_0(R)$; see [19] for details.

In Roberts [51] the grading was given by a map to the rational Chow group, which we define briefly. For a Noetherian ring A, we define the ith graded piece of the rational Chow group, denoted $\mathrm{CH}_i(A)_{\mathbb{Q}}$, to be the $\mathbb{Q}$ vector space on generators $[\mathfrak{p}]$, where $\mathfrak{p}$ is a prime ideal such that the dimension of $A/\mathfrak{p}$ is i modulo an equivalence relation called rational equivalence. Rational equivalence is defined by setting $\mathrm{div}(\mathfrak{q}, x)$ to zero in $\mathrm{CH}_i(A)$, where $\mathfrak{q}$ is a prime ideal such that $A/\mathfrak{q}$ has dimension $i + 1$, x is an element of A not in $\mathfrak{q}$, and, letting $B = A/\mathfrak{q}$,

$$\mathrm{div}(\mathfrak{q}, x) = \sum_{\mathfrak{p}} \mathrm{length}(B_{\mathfrak{p}}/xB_{\mathfrak{p}})[\mathfrak{p}],$$

where the sum is over $\mathfrak{p}$ with $\dim(A/\mathfrak{p}) = i$. There is then a map τ from $K_0(A)$ to operators on $\mathrm{CH}_i(A)_{\mathbb{Q}}$. If A is a regular local ring, we can replace dimension by codimension and obtain a grading with the properties above. For details see Fulton [16] and Roberts [53] and [56].

These techniques allowed one to prove the Vanishing Conjecture also in the case in which M and N are modules of finite projective dimension over a complete intersection. We will discuss other generalizations in a later section.

2.2 Gabber's Proof of the Nonnegativity Conjecture

The third of Serre's conjectures, Nonnegativity, was proven by Gabber around 1996. Gabber never published the proof, but a brief summary appears in Berthelot [3], and more extensive versions can be found in Hochster [27] and Roberts [57]. Again there was a new ingredient; this time it was a theorem of de Jong on the existence of regular alterations [39]. We give here a special case of this theorem which applies to this problem.

Theorem 1 (A. J. de Jong). *Let A be a local integral domain which is essentially of finite type over a discrete valuation ring. Then there exists a scheme X with a projective map $X \to Spec(A)$ such that*

(i) *X is an integral regular scheme (that is, all the local rings of X are regular).*
(ii) *The field of rational functions $k(X)$ is a finite extension of the field of fractions of A.*

There is some work involved in reducing to the case in which A is essentially of finite type over a discrete valuation ring, and even in this case the proof is quite nontrivial. We mention briefly where de Jong's theorem is applied. It suffices to show that $\chi(A/\mathfrak{p}, A/\mathfrak{q}) \geq 0$ for prime ideals $\mathfrak{p}$ and $\mathfrak{q}$ such that $(\mathfrak{p}, \mathfrak{q})$ is primary to the maximal ideal. The theorem is applied to one of the quotients, say $A/\mathfrak{p}$. The machinery of intersection theory must be extended to perfect complexes on schemes and projective morphisms as well as over commutative rings. One curious feature of the proof is that at one point it is necessary to assume that the original local ring is ramified; it is easy to reduce to this case but unexpected that it would be useful. The proof also gives a new proof of the Vanishing Conjecture. We refer to the references above for descriptions of the proof.

2.3 The Positivity Conjecture

The positivity conjecture remains open. There have been several approaches to it, and we mention two.

One approach is based on the following. Let M and N be two modules over a regular local ring R such that $\dim M + \dim N = \dim R$. If M is Cohen–Macaulay, its minimal free resolution has length $\dim R - \dim M$, and if N is also Cohen–Macaulay, then the condition on the length of the resolution of M implies that $\mathrm{Tor}_i(M, N) = 0$ for $i > 0$. Thus $\chi(M, N)$ is the length of $M \otimes_R N$, which is clearly positive. Hence if we can reduce to the case in which M and N are Cohen–Macaulay, we are done.

A method for reducing to this case is, first, to reduce to the case where M and N are of the form $A/\mathfrak{p}$ by taking filtrations with quotients of this form; since the Vanishing Conjecture holds, we can reduce to modules of this form. If we could now find an $A/\mathfrak{p}$-module of the dimension of $A/\mathfrak{p}$ which was Cohen–Macaulay for any $\mathfrak{p}$, we could, again using vanishing, reduce to the case in which M and N are Cohen–Macaulay and complete the proof. The missing fact is the existence of what are called "small Cohen-Macauly modules"; these will be discussed in a later section. (What we have just described is the arrow from "Small C–M Modules" to "($M_1 \Rightarrow M_2$), regular case" in Hochster's diagram.)

The other, more recent, attempts to prove the Positivity Conjecture use Gabber's construction. Kurano and Roberts [42] give a criterion for positivity to hold using this construction. Dutta [10] gives a formula for intersection multiplicities using the blow-up of the maximal ideal of a regular local ring, again using Gabber's ideas. It is not clear whether any of these methods will lead to a proof of positivity, however, and that conjecture remains open.

3 The Peskine–Szpiro Intersection Conjecture

Serre's introduction of homological methods into intersection theory created much more interest in questions on homological algebra, and, in particular, properties of modules of finite projective dimension. The Auslander–Buchsbaum–Serre theorem states that every R-module has finite projective dimension if and only if the ring R is regular, so one point of view is that properties of modules over regular local rings should extend to properties of modules of finite projective dimension over arbitrary local rings. One direction was to generalize the multiplicity properties themselves; this will be considered in the next section. A different direction was started by Peskine and Szpiro with their "Intersection Theorem". This was a main theorem of their paper *Dimension projective finie et cohomologie locale*, which was one of the most important papers in the development of the Homological Conjectures.

The Peskine–Szpiro Intersection Conjecture states:

Conjecture 3. Let A be a local ring, let M be an A-module of finite projective dimension, and let N be a module such that $M \otimes N$ has finite length. Then the Krull dimension of N is less than or equal to the projective dimension of M.

They stated this result as a theorem rather than a conjecture, since it was a theorem for rings of positive characteristic and rings essentially of finite type over a field of characteristic zero. We discuss this in more detail below.

In some ways this conjecture is analogous to Serre's conjectures. By the Auslander-Buchsbaum Theorem, the depth of a module is related to the projective dimension (if finite) by

$$\operatorname{projdim} M + \operatorname{depth} M = \operatorname{depth} A.$$

or

$$\operatorname{projdim} M = \operatorname{depth} A - \operatorname{depth} M.$$

Thus the Peskine–Szpiro Theorem can be stated that

$$\dim N + \operatorname{depth} M \leq \operatorname{depth} A.$$

This is analogous, but certainly not equivalent, to the Serre theorem. Its interest lies in the fact that it implies several other conjectures from that time, of which we state two.

Conjecture 4 (Bass). If a ring A has a finitely generated nonzero module of finite injective dimension, then A is Cohen–Macaulay.

Conjecture 5 (Auslander). Let M be a finitely generated module of finite projective dimension. If $a \in A$ is a nonzerodivisor on M, then a is a nonzerodivisor on A.

We refer to the paper of Peskine and Szpiro [47] for proofs that these conjectures are implied by the Intersection Conjecture.

A newer version of the Intersection Conjecture was introduced shortly thereafter; it is in the spirit of generalizing from modules to complexes referred to in the previous section.

Conjecture 6 (The New Intersection Conjecture). Let A be a local ring of dimension d. If

$$0 \to F_k \to F_{k-1} \to \cdots \to F_0 \to 0$$

is a complex of finitely generated free modules such that $H_i(F_\bullet)$ has finite length for each i and $H_0(F_\bullet) \neq 0$, then $k \geq d$.

That this implies the original conjecture can be seen by applying the New Intersection Conjecture to a projective resolution of M tensored with a suitable module of the form $A/\mathfrak{p}$ for $\mathfrak{p}$ a prime ideal of A in the support of N.

In addition to stating this conjecture and several others, Peskine and Szpiro introduced two methods that are still very much in use in this area. Perhaps the most important is the use of the Frobenius map and reduction to positive characteristic. We briefly recall how this works.

Let A be a ring of positive characteristic p. Then the Frobenius map, which we denote F, is the ring homomorphism defined by $F(a) = a^p$; it is a ring homomorphism since $p = 0$ on A so $(a + b)^p = a^p + b^p$ for all a and b in A. The basic idea of using this map to prove conjectures is to assume that there is a counterexample, and then to take a limit over powers of the Frobenius map to obtain a contradiction. A simpler method, which works sometimes, is to show that a high enough power of the Frobenius map produces an example that can be shown not to exist.

The second step in this process is to reduce the characteristic zero case to the case of positive characteristic. Peskine and Szpiro introduced this method for this kind of problem, and it was completed by Hochster.

The procedure is fairly complicated, but one step, reduction from finitely generated over a field to positive characteristic, goes something like this. Given a counterexample over a ring that is a finitely generated ring over a field of characteristic zero, one first, using the fact that there are only finitely many elements to consider, reduces to the case of a ring that is finitely generated over the rational numbers, and then one reduces further to an example over a ring finitely generated over the integers. Finally, one shows that for all but finitely many primes p, the reduction modulo p and gives a counterexample in characteristic p. Peskine and Szpiro used this method to prove the Intersection Conjecture in the case of a local ring essentially of finite type over a field of characteristic zero, and they also used the Artin approximation Theorem to extend this to the case of a ring whose completion was the completion of a ring essentially of finite type over a field of characteristic zero. Shortly thereafter Hochster was able to extend this method to the general equicharacteristic case.

3.1 Hochster's Metatheorem

One of the main results of Hochster [24] was the following.

Theorem 2. *Let ξ be a system of polynomial equations in $d+q$ variables $X_1, \dots, X_d$ and $Y_1, \dots, Y_q$ over $\mathbb{Z}$, say*

$$F_1(X_1, \dots, X_d, Y_1, \dots, Y_q) = 0$$

$$\vdots$$

$$F_t(X_1, \dots, X_d, Y_1, \dots, Y_q) = 0.$$

Suppose that ξ has a solution in a local ring R which contains a field of characteristic zero such that $\dim(R) = d$ and the values $x_1, \dots, x_d$ for $X_1, \dots, X_d$ form a system of parameters for R.

Then there exists a local ring S containing a field of characteristic $p > 0$ such that $\dim(S) = d$ and there is a solution of ξ such that the values $x'_1, \dots, x'_d$ for $X_1, \dots, X_d$ form a system of parameters for S.

The proof of this theorem used Artin Approximation, and it finished the characteristic zero case of several of the conjectures, including the Intersection Conjecture and various others that we will discuss below.

The case of mixed characteristic was proven in Roberts [52]. Like the Serre vanishing theorem, this used the theory of local Chern characters. Another essential ingredient was a theorem relating Chern characters in positive characteristic to limits over the Frobenius map. Details of this and more can be found in Roberts [53].

The other main technique introduced in the paper of Peskine and Szpiro was local cohomology. As this is a topic that is still extremely important in this area, we will review some of the important points. For more complete introductions to the subject we refer to Brodman and Sharp [5] and *Twenty-Four Hours of Local Cohomology* [38].

Let A be, as usual, a commutative Noetherian ring, and let I be an ideal of A. For any A-module M, we define the submodule $\Gamma_I(M)$ to be the set of $m \in M$ that are annihilated by a power of I. It is easy to see that this is indeed a submodule of M and that Γ_I defines a left exact functor from the category of A-modules to itself. The functor is not right exact, however, and the right derived functors of Γ_I applied to a module M, denoted $H_I^i(M)$, are the local cohomology modules of M with support in I.

The most important case as far as the Homological Conjectures are concerned is the case in which A is local and I is the maximal ideal $\mathfrak{m}$. We note that if I and J are ideals with the same support, so that we have $J^n \subseteq I$ and $I^m \subseteq J$ for some m and n, then it is clear that $\Gamma_I(M) = \Gamma_J(M)$ for all modules M, and thus the local cohomology modules with supports in I and J are the same. If $\mathfrak{m}$ is the maximal ideal of a local ring A of dimension d, we can thus replace $\mathfrak{m}$ by an ideal I generated by a system of parameters $(x_1, \dots, x_d)$. Now given a set of generators for I, there are two standard methods for computing the local cohomology modules.

First, we let $C^\bullet$ denote the complex

$$0 \to A \to \prod_i A_{x_i} \to \prod_{i<j} A_{x_i x_j} \to \cdots \to A_{x_1 x_2 \cdots x_d} \to 0,$$

where the A at the left has degree zero and the $A_{x_1 x_2 \cdots x_d}$ at the right has degree d. The maps are given by the inclusions with appropriate signs. Then it can be shown that

$$H_I^i(M) = H^i(M \otimes_A C^\bullet)$$

for any A-module M.

The second method is as a direct limit. For each n we take the Koszul complex $K^\bullet(x_1^n, \dots, x_d^n)$. For $m > n$ there is a map of complexes from $K^\bullet(x_1^n, \dots, x_d^n)$ to $K^\bullet(x_1^m, \dots x_d^m)$. The limit of these is, in fact the above complex and tensoring with M again gives local cohomology.

For a ring of positive characteristic one can also define local cohomology as a limit over powers of the Frobenius map, and this was one of the methods introduced by Peskine and Szpiro. We will not discuss this further here, but we will return to the topic of local cohomology in later sections.

One of the facts that is used over and over in studying these conjectures is the following. If we assume that A is a complete domain of dimension d, then there is an element $c \neq 0$ that annihilates the local cohomology modules $H_\mathfrak{m}^i(A)$ for $i < d$. This was proven in Roberts [50], where the element c was taken to be in a product of annihilators of the cohomology of a dualizing complex, and in Hochster and Huneke [31], where c was taken to be an element such that the localization $A_c = A[1/c]$ is Cohen–Macaulay. Keeping in mind that the ring A is Cohen–Macaulay if and only if the local cohomology modules $H_\mathfrak{m}^i$ are zero for $i < d$, it is not surprising that this fact is useful for approaching these conjectures in the non-Cohen–Macaulay case; this method works especially well when combined with the use of the Frobenius map.

To conclude this section we note that Avramov, Buchweitz, and Iyengar have formulated a generalization of the New Intersection Conjecture, called the "Class Inequality", to differential modules. A complex is a special case of a differential module; the differential module is the direct sum of the modules in the complex with differential given by the sum of the boundary maps. They prove this inequality in the equicharacteristic case; the case of mixed characteristic is still open. We refer to [2] for details.

4 Generalizations of the Multiplicity Conjectures

As has already been mentioned, one direction of research on these conjectures was to generalize Serre's conjectures to nonregular rings. We recall that we had defined the intersection multiplicity $\chi(M, N)$, where M and N are two finitely generated modules over a regular local ring R with $M \otimes_R N$ of finite length to be

$$\chi(M, N) = \sum_{i=0}^{d} (-1)^i \text{length}(\text{Tor}_i(M, N)).$$

As stated, this would be defined over any local ring; however, the fact that R is regular implies that higher Tors are zero, which, using the long exact sequence of Tors, implies that $\chi(M, N)$ is additive in M and in N. Over a nonregular ring we need an extra condition, and the weakest condition which makes this work is that one of the modules, say M, has finite projective dimension; its projective dimension will still be at most d. We now restate the conjectures with this assumption.

Conjecture 7. Let A be a local ring, and let M and N be finitely generated modules such that M has finite projective dimension and $M \otimes_A N$ has finite length. Then
(i) (M_0) $\dim(M) + \dim(N) \leq \dim(A)$.
(ii) (M_1: Vanishing) If $\dim(M) + \dim(N) < \dim(A)$, then $\chi(M, N) = 0$.
(iii) (M_2: Positivity) If $\dim(M) + \dim(N) = \dim(A)$, then $\chi(M, N) > 0$.

It is a rather remarkable fact that the first of these conjectures, which appears to be the most basic, is still open in this generality. It holds for many examples of modules of finite projective dimension, and there are many easy counterexamples if neither module has finite projective dimension, but it is not known in the case stated here, in spite of the fact that it is a rather simple statement about the nature of the support of a module of finite projective dimension.

4.1 The Graded Case

One of the remarkable results of Peskine and Szpiro was a Comptes Rendus article [48] in which they proved some of the conjectures for the graded case. More precisely, they assumed that A is a standard graded ring over an Artinian local ring (such as a field), M is a graded module of finite projective dimension, and N is another graded module. In this case, M has a finite free resolution by modules that are direct sums of $A(n_{ij})$, the graded module A with grading shifted by n_{ij}, for various n_{ij}. They gave a formula which allows one to compute the intersection multiplicities in terms of the n_{ij} in such a way that they could prove all three parts of this conjecture.

In addition, they proved the following conjecture in the graded case:

Conjecture 8. Let M be an A-module of finite projective dimension. Then

$$\text{grade}(M) = \dim(A) - \dim(M).$$

We recall that the grade of a module is the longest possible length of a regular sequence contained in the annihilator of M. This is the Codimension Conjecture (13) of Hochster's 1975 diagram; the word codimension was once used for what we now call grade. This conjecture is a statement about the prime ideals in the support of a module of finite projective dimension and holds, for example, for an equidimensional ring, but it is still open in general. For a discussion of this conjecture we refer to [55].

We make one final remark about the methods of this paper on the graded case. The authors said at the time that their method of computing intersection multiplicities through numerical invariants was a kind of "Riemann–Roch Theorem". This was in fact one of the main inspirations for later work on finding a Riemann–Roch Theorem in general. On the other hand, the question of whether this is really a version of the Riemann–Roch Theorem of Hirzebruch was not raised until later, and a direct proof that they agree was only given recently (see [58]).

We now return to the main topic of the Strong Multiplicity Conjectures.

As mentioned above, the first of these conjectures is still open. The second two, however, are false. This was an example of Dutta, Hochster, and McLaughlin [11] which was one of the turning points in research in this area. We present an outline of this example, leaving out the details.

Let k be a field, and let A be $k[X, Y, Z, Y]/(XY - ZW)$ localized at the maximal ideal (X, Y, Z, W) (or $k[[X, Y, Z, W]]/(XY - ZW)$ if you prefer). Let $N = A/(X, Z)$. We note that since (X, Z) contains $XY - ZW$, N has dimension 2. The problem is to construct a module of finite length and finite projective dimension such that $\chi(M, N) \neq 0$. This is carried out by a detailed computation of a set of matrices representing the action of X, Y, Z, and W on a finite dimensional vector space; the authors determine the precise conditions these matrices must satisfy and produce a set of large matrices satisfying them.

This counterexample also had influence on the theory of local Chern characters, showing that they did not vanish where predicted. More on this approach to the question can be found in Szpiro [65] and Roberts [53] and [56].

We mention a result of Sather-Wagstaff [61] which is similar to statement M_0 above but where the hypothesis of finite projective dimension is replaced by a condition on multiplicity.

Theorem 3. *Let A be an excellent quasi-unmixed Cohen–Macaulay local ring that contains a field. Let $\mathfrak{p}$ and $\mathfrak{q}$ be prime ideals such that $A/\mathfrak{p} \otimes_A A/\mathfrak{q}$ has finite length and the multiplicity of $A_\mathfrak{p}$ is equal to the multiplicity of A. Then* $\dim(A/\mathfrak{p}) + \dim(A/\mathfrak{q}) \leq \dim(A)$.

The interesting point here is that the condition on multiplicities is automatic for regular local rings, since the localization of a regular local ring is regular so both multiplicities are one, like the condition in M_0 that M have finite projective dimension.

We now return to a discussion of further developments on counterexamples to this generalization of the Vanishing Conjecture. As mentioned above, the example of

Dutta, Hochster, and McLaughlin had implications for local Chern characters. Let A be a local domain of dimension d which is either complete or essentially of finite type over a field (this is sufficient so that Chow groups and local Chern characters are defined). Associated to A are two elements of the Chow group of $\mathrm{CH}_*(A)$. The first is the class $[A]$; since A is a domain, 0 is a prime ideal and this defines an element of $\mathrm{CH}_d(A)$. The second is the local Todd class, denote $\tau(A)$, which is equal to $[A]$ up to elements of lower dimension, and which is what is used in formulas for multiplicities. If A is a complete intersection, it can be shown that $\tau(A) = [A]$; there are no lower terms. The counterexample to vanishing enables one to construct an example of a Cohen–Macaulay domain A of dimension 3 for which the dimension 2 component of $\tau(A)$, denoted $\tau_2(A)$, is not zero and (more important), there is a module of finite length and finite projective dimension whose local Chern character does not vanish on $\tau_2(A)$. This left open the question of whether there was a similar example where A is Gorenstein. If A is Gorenstein of dimension d, then it can be shown that $\tau_{d-1}(A) = 0$, so any nonvanishing component would have to be of higher codimension.

First Kurano [41] provided an example of a Gorenstein ring of dimension 5 for which $\tau_3(A) \neq 0$. C. Miller and Singh [46] then gave an example, also Gorenstein of dimension 5, for which there exists a module of finite length and finite projective dimension whose local Chern character does not vanish on $\tau_3(A)$. In Roberts and Srinivas [60], a general theorem was proven for local rings A which are localizations at the maximal ideal of a standard graded ring such that the associated projective scheme X is smooth (this includes all the above examples). In the nice case in which the Chow group of X is essentially the same as the cohomology of X (which is also true in the above examples), the main theorem states that if η is any cohomology class that is zero when intersected with the hyperplane section, intersection with η can be represented by a module of finite length and finite projective dimension. This implies in particular that there is such a module for Kurano's example. It also means that counterexamples of this sort are quite natural when seen from the point of view of intersection theory in Algebraic Geometry.

In all of the discussion in this section, we have only assumed that one of the modules M and N has finite projective dimension. If we assume that both modules have finite projective dimension, the conjectures are still open. If the ring is a complete intersection, then the Vanishing Conjecture is known in this case. There is an example in Roberts [54] of two perfect complexes which define positive cycles for which the intersection multiplicity is negative (this cannot happen over regular local rings), which may suggest that the Positivity Conjecture does not hold in this generality.

However, there are no indications that the Vanishing Conjecture for two modules of finite projective dimension is not true, and this is one of the main open questions in this area at the present time. In the counterexamples described above, the module M of finite projective dimension has finite length and the module N has dimension less than the dimension of the ring. If N also has finite projective dimension, the fact that its dimension is less than that of the ring implies that the alternating sum of ranks of modules in its resolution is zero, and it follows that $\chi(M, N) = 0$. Thus one must look elsewhere if one hopes to find a counterexample to vanishing with both modules of finite projective dimension.

4.2 The Generalized Rigidity Conjecture

The conjecture on the Rigidity of Tor was also generalized from the regular case to the case in which one module had finite projective dimension. This was disproven by Heitmann in [20].

It could be thought that with the regular case proven and the generalized case false, that would be the end of the story for the question of Rigidity of Tor. However, there have been several further results in this area, particularly for modules over hypersurfaces. We give two examples.

First, we have the following theorem due to Huneke and R. Wiegand [37] (their actual theorem is a little stronger than this).

Theorem 4. *Let $A = R/(f)$ be a hypersurface of dimension d, where R is an unramified regular local ring of dimension $d + 1$. Let M and N be A-modules such that*

(i) *$M \otimes_A N$ has finite length.*

(ii) $\dim(M) + \dim(N) \leq d$.

Then if $\mathrm{Tor}_i(M, N) = 0$ for some $i \geq 0$, then $\mathrm{Tor}_j(M, N) = 0$ for $j \geq i$.

A more recent result on this topic is due to Hailong Dao [6]. This uses a construction of Hochster for hypersurfaces which had been introduced earlier to study these conjectures. Let $A = R/(f)$ be a hypersurface, and suppose also that A is an isolated singuarity. Then a resolution of a finitely generate module is eventually periodic of period 2 by results of Eisenbud [12], and the $\mathrm{Tor}_i(M, N)$ are eventually of finite length since A has an isolated singularity. Hochster defined

$$\theta(M, N) = \mathrm{length}(\mathrm{Tor}_{2i}(M, N)) - \mathrm{length}(\mathrm{Tor}_{2i+1}(M, N)).$$

Dao proved the following theorem. Here A is a hypersurface of the form $R/(f)$, but in addition to R being regular, it must also be a power series ring over a field or a discrete valuation ring, so that in particular all of Serre's multiplicity conjectures hold.

Theorem 5. *Let A be as above, and let M and N be two finitely generated A-modules. Assume that $\theta(M, N) = 0$. If $\mathrm{Tor}_i(M, N) = 0$ for some $i \geq 0$, then $\mathrm{Tor}_j(M, N) = 0$ for $j \geq i$.*

We also want to mention an example of Dutta [9], which shows that the partial Euler characteristic $\chi_2(M, N)$ can be negative for two modules of finite projective dimension over a Gorenstein ring. While the original counterexample to vanishing shows that Serre's conjectures on partial Euler characteristics cannot be extended in general, Dutta's example is interesting in that it shows that they can fail even in a case where vanishing holds.

5 The Monomial, Direct Summand, and Canonical Element Conjectures

The first two of these conjectures, the Monomial and Direct Summand Conjectures, were introduced by Hochster and are listed in his diagram as consequences of the existence of big Cohen–Macaulay modules (which we will discuss below). They can all be proven in the equicharacteristic case by reduction to positive characteristic as outlined in the previous section.

The Direct Summand Conjecture states:

Conjecture 9 (Direct Summand Conjecture). If R is a regular local ring and S is a module-finite extension of R, then R is a direct summand of S as an R-module.

The Monomial Conjecture states:

Conjecture 10 (Monomial Conjecture). If $x_1, \dots, x_d$ is a system of parameters for a local ring R, then

$$x_1^t x_2^t \cdots x_d^t \notin (x_1^{t+1}, \dots, x_d^{t+1}).$$

It is not too difficult to show that these two conjectures are equivalent. Shortly thereafter Hochster formulated the Canonical Element Conjecture. There are several versions of this conjecture, and we state three. The first shows why it is called the "Canonical Element" conjecture, and the second and third are easier to compute. In all three statements we let A be a local ring of dimension d with maximal ideal $\mathfrak{m}$ and residue field k.

Conjecture 11. Canonical Element Conjecture I: Let $\mathrm{syz}^d(k)$ be the dth syzygy module of k, so that there is an exact sequence

$$0 \to \mathrm{syz}^d(k) \to F_{d-1} \to \cdots \to F_0 \to k \to 0,$$

where the F_i are free modules. Using the Yoneda definition of Ext, this d-fold extension defines an element of $\mathrm{Ext}^d(k, \mathrm{syz}^d(k))$, and hence, mapping to the limit, an element η of

$$\varinjlim_n \mathrm{Ext}^d(A/\mathfrak{m}^n, \mathrm{syz}^d(k)) = H^d_{\mathfrak{m}}(\mathrm{syz}^d(k)).$$

Then $\eta \neq 0$. (η is the "canonical element").

Canonical Element Conjecture II: Let $x_1, \ldots, x_d$ be a system of parameters for A. Let K_* be the Koszul complex on $x_1, \ldots, x_d$ and let F_* be a free resolution of k. Suppose we have

$$\begin{array}{ccc} K_* & \longrightarrow & A/(x_1, \ldots, x_d) \\ \downarrow{\scriptstyle \phi_*} & & \downarrow \\ F_* & \longrightarrow & k. \end{array}$$

Then $\phi_d \neq 0$.

Canonical Element Conjecture III: Let $x_1, \ldots, x_d$ be a system of parameters for A. Let K_* be the Koszul complex on $x_1, \ldots, x_d$ and let F_* be a free resolution of $A/(x_1, \ldots, x_d)$. Suppose we have

$$\begin{array}{ccc} K_* & \longrightarrow & A/(x_1, \ldots, x_d) \\ \downarrow{\scriptstyle \phi_*} & & \downarrow \\ F_* & \longrightarrow & A/(x_1, \ldots, x_d). \end{array}$$

Then the image of ϕ_d is not contained in $\mathfrak{m} F_d$.

These three are not obviously equivalent, and proofs of their equivalence and the fact that they are also equivalent to the Direct Summand and Monomial Conjectures can be found in Hochster [25] and Dutta [7]. It should be pointed out that the fact that the Monomial Conjecture or Direct Summand Conjecture implies the Canonical Element Conjecture is quite nontrivial for rings of positive or mixed characteristic; in characteristic zero the Direct Summand Conjecture is trivial and holds for any normal domain since the trace map can be divided by the degree of the extension of quotient fields. It is equivalent to the Canonical Element Conjecture in characteristic zero only in the sense that both are known to be true.

These three conjectures have been among the most seriously studied during the years since their formulation. They follow from the existence of big Cohen–Macaulay modules, and they can be proven directly by the method of reduction to positive characteristic outlined above in the equicharacteristic case. Since Cohen–Macaulay modules exist in dimension at most two, the conjectures have been known since the beginning in any characteristic in dimension less than three. The reference Hochster [25] also contains many more interesting results on this topic, and it includes the fact that to prove the direct summand conjecture it suffices to prove it in the case in which R is an *unramified* regular local ring, a condition that is often assumed in studying the problem.

Although the canonical element of the conjecture looks somewhat mysterious, there are a number of conjectures similar to the ones we are discussing that involve the canonical module or, more generally, the dualizing complex of a local ring. For examples, we refer to Strooker and Stückrad [64] and Dutta [8], where the Monomial Conjecture is related to properties of a dualizing module.

A major breakthrough on these conjectures came in 2003, when Heitmann [22] proved the Direct Summand Conjecture (and therefore several others) in dimension three in mixed characteristic. This proof did not involve new machinery, but rather it showed, by prodigious computations, that if one had a non-Cohen–Macaulay ring of mixed characteristic of dimension three for which p, x, y is a system of parameters, and if we have a relation $ap^N \in (x, y)$, then for any integer n, in some finite extension we have that $ap^{1/n} \in (x, y)$. Thus we do not get $a \in (x, y)$ (which would of course be true in a Cohen–Macaulay ring), but something close, and Heitmann proved that this is enough to prove the Direct Summand Conjecture. A little later Heitmann [23] showed that the system of parameters p, x, y can be replaced by any system of parameters; it is not necessary to assume that one of them is p. We will discuss this further below.

6 Cohen–Macaulay Modules and Algebras

The importance of finding Cohen–Macaulay modules was clear from the beginnings of this subject. Serre had already shown, as we mentioned above, that if M and N are Cohen–Macaulay in the situation of his positivity conjecture, then $\mathrm{Tor}_i(M, N) = 0$ for all $i > 0$, so that $\chi(M, N)$ is simply the length of $M \otimes N$, which is clearly positive. It is also not difficult to show that most of the conjectures we have discussed over a ring A of dimension d will follow if there exists a finitely generated Cohen–Macaulay module of dimension d. Such a module is called a "small Cohen–Macaulay module" (it is also sometimes called a "maximal Cohen–Macaulay module", which admittedly is not terribly consistent terminology).

There are rings which cannot have small Cohen–Macaulay modules, such as noncatenary rings, but these can be considered pathological. In addition, most of the conjectures we have been discussing can be reduced to the complete case, and it would suffice to show that small Cohen–Macaulay modules exist for complete domains.

Conjecture 12. Every complete local domain has a small Cohen–Macaulay module.

This conjecture is easy if the dimension of A is at most two, since in dimension one any domain is Cohen–Macaulay, and in dimension two one can take the normalization, which is Cohen–Macaulay. However, very little is known beyond that case. There is an example for graded rings attributed independently to Peskine and Szpiro, Hartshorne, and Hochster; they showed that small Cohen–Macaulay modules exist for graded domains of positive characteristic in dimension three (for a proof see Hochster [29]). Dan Katz [40] proved that there is such a module for extensions obtained by adjoining a pth root to an unramified regular local ring. On the other hand, there are non-Cohen–Macaulay unique factorization domains, which cannot have small Cohen–Macaulay modules of rank one [4]. But basically this question is completely open.

One of the new developments in Hochster's 1975 paper was to introduce a weaker version of Cohen–Macaulay modules, called "big" Cohen–Macaulay modules. Their existence does not imply the implication in Serre's conjecture, but it does imply the Intersection Conjecture, the Canonical Element Conjecture, and several others.

Let A be a local ring with system of parameters $x_1, \ldots, x_d$. A *big Cohen–Macaulay module* is an A-module M such that

(i) $x_1, \ldots, x_d$ form a regular sequence on M.

(ii) $M/(x_1, \ldots, x_d)M \neq 0$.

The second condition is crucial; there are numerous infinitely generated modules that satisfy the first condition but not the second, and without this condition none of the stated implications hold. If M is a small Cohen–Macaulay module, however, Nakayama's Lemma implies that condition 2 holds.

Conjecture 13. Every local ring has a big Cohen–Macaulay module.

Like the conjectures in the previous section, this conjecture is known in the equicharacteristic case and for rings of dimension at most 3. The basic method used by Hochster in [24] was to kill any bad relations as follows. If M is not Cohen–Macaulay, there exists an $m \in M$ such that $x_i m \in (x_1, \ldots, x_{i-1})$ but $m \notin (x_1, \ldots, x_{i-1})$ for some i. We then extend M to $M' = M \oplus A^{i-1}$ modulo the relation $(m, x_1, \ldots, x_{i-1})$; this puts the image of m into the submodule $(x_1, \ldots, x_{i-1})M'$. We then take a huge and carefully constructed limit, and it is then easy to see that the limit will satisfy the first condition. The problem is to show that the second condition also holds. The original proof in the equicharacteristic case involves the Frobenius map and Hochster's Metatheorem. The proof in dimension 3 uses Heitmann's results.

6.1 Weakly Functorial Big Cohen–Macaulay Algebras

A further development in this area was the introduction of big Cohen–Macaulay algebras. One method for construction such an algebra is similar to that of big Cohen–Macaulay modules mentioned earlier, but instead of taking a free module in the extension and dividing by the relation as above, one takes a free commutative algebra; that is, a polynomial ring and again divides by an appropriate relation and takes a limit. For the applications one would like it to be functorial; this does not seem possible, but when they exist they can be made "weakly functorial", which is enough for many applications. We give the definitions.

Let R be a local ring with system of parameters $x_1, \ldots, x_d$. A *big Cohen–Macaulay algebra* is an algebra A over R such that

(i) $x_1, \ldots, x_d$ form a regular sequence on A.

(ii) $A/(x_1, \ldots, x_d)A \neq 0$.

"Weakly Functorial" means that given $R \to S$, one can find Cohen–Macaulay algebras A and B and a diagram

$$\begin{array}{ccc} R & \longrightarrow & S \\ \downarrow & & \downarrow \\ A & \longrightarrow & B \end{array}$$

Conjecture 14. Every local ring has a big Cohen–Macaulay algebra, and for any map of local rings they can be chosen to be weakly functorial in the sense given above.

It can be seen that this is a considerably stronger conjecture than the existence of big Cohen–Macaulay modules, which in turn is stronger than the conjectures of the previous section. However, it has so far been the case that once methods had been developed to prove one of these conjectures, it can be applied to prove the existence of weakly functorial big Cohen–Macaulay algebras. An example is the case of dimension 3 in mixed characteristic, where the results of Heitmann's proof of the Direct Summand conjecture were used by Hochster to prove this conjecture as well [28].

A remarkable theorem appeared in 1990, with the proof by Hochster and Huneke that R^+ is Cohen–Macaulay for R a domain of positive characteristic [33]. Here R^+ is the *absolute integral closure* of R, which means the integral closure in the algebraic closure of its quotient field. This was later given a much simpler proof by Huneke and Lyubeznik [36]. This is better than just the existence, since it gives a specific construction in the positive characteristic case.

The existence of weakly functorial big Cohen–Macaulay algebras has many applications; for example, they imply the conjectures on the vanishing of maps of Tor and that direct summands of regular local rings are Cohen–Macaulay that we will state below. For more details on the existence and applications of such algebras we refer to Hochster and Huneke [34].

7 The Syzygy Conjecture and the Improved New Intersection Conjecture

Evans and Griffiths proved the following theorem for rings containing a field [13].

Theorem 6. *Let A be a Cohen–Macaulay local ring containing a field, and let M be a finitely generated kth module of syzygies that has finite projective dimension. If M is not free, then M has rank at least k.*

In proving this conjecture it turned out that a stronger version of the Intersection Conjecture was one of the key points in the proof. This was named the "Improved New Intersection Conjecture".

Conjecture 15. Let A be a local ring of dimension d, and let

$$0 \to F_k \to \cdots \to F_1 \to F_0 \to 0$$

be a complex of finitely generated free modules such that $H_i(F_\bullet)$ has finite length for $i \geq 1$ and the cokernel of $F_1 \to F_0$ has a minimal generator annihilated by a power of the maximal ideal. Then $k \geq d$.

The original New Intersection Conjecture is the case where the cokernel of $F_1 \to F_0$ is itself of finite length (and nonzero). While this is a version of the Intersection Conjecture, it is in fact stronger, and is equivalent to the Canonical Element Conjecture and the others in that group. Thus it is now known in the equicharacteristic case and in dimension at most 3.

Recently Evans and Griffith have proven their Syzygy Theorem for certain graded modules of mixed characteristic [15]. They also have a more extensive account of problems concerning syzygies in [14].

8 Tight Closure Theory

In 1985 Hochster and Huneke introduced the concept of tight closure. It is defined for equicharacteristic rings; to keep the discussion simple we will give the definition for integral domains of positive characteristic.

Definition 1. Let I be an ideal of an integral domain A of positive characteristic p. The *tight closure* of I, denoted I^*, is the set of $a \in A$ for which there is an element $c \neq 0$ in A such that $ca^{p^e} \in I^{[p^e]}$ for all $e \geq 0$.

Here $I^{[p^e]}$ is the ideal generated by i^{p^e} for all $i \in I$. Tight closure is also defined for rings of characteristic zero using a method of reduction to positive characteristic. We refer to Huneke's notes from the Fargo conference [35] and its bibliography for much more information about tight closure. We will mention some connections to the problems we have been discussing here.

First of all, tight closure made it possible to give nicer proofs of some of the Homological Conjectures, such as the Monomial Conjecture and the existence of big Cohen–Macaulay modules, in the equicharacteristic case, although the basic idea, reduction to positive characteristic and the use of the Frobenius map, was similar to methods used earlier. One of the first ideas that arose from this was to attempt to find a similar closure operation that would work in mixed characteristic. A list of the desired properties of such a closure operation can be found, for example, in the Introduction of the notes of Huneke cited above; for the purposes of these conjectures, one of the main ones is "colon-capturing", which states that if $x_1, \ldots, x_d$ is a system of parameters, and if $ax_i \in (x_1, \ldots, x_{i-1})$ for some i, then a is in the closure of $(x_1, \ldots, x_{i-1})$. A closure operation with all the right properties has not been found; however, this did inspire some new methods; in particular, Heitmann's proof of the Direct Summand Conjecture in dimension three was motivated in part by an attempt to show that "full extended plus closure" satisfies the colon-capturing condition in mixed characteristic. In this part of the discussion we assume that A is a complete local domain and recall that A^+ is the integral closure of A in the algebraic closure of its quotient field.

Definition 2. If $x \in A$, then x is in the full extended plus closure of I if there exists $c \neq 0 \in A$ such that for every positive integer n, $c^{1/n}x \in (I, p^n)A^+$. We write $x \in I^{\text{epf}}$.

A similar closure operation had been defined by Hochster and Huneke in [32]. They defined the "dagger closure" to as follows (with the same assumptions on A). In this definition we fix a valuation v on A^+ with values in $\mathbb{Q} \cup \{\infty\}$ which is nonnegative on A^+ and positive on the maximal ideal of A^+.

Definition 3. If $x \in A$, then x is in the dagger closure of I if there exist elements $u \in A^+$ of arbitrarily small positive order with $ux \in IA^+$. We write $x \in I^\dagger$.

It is easy to see that $I^{\text{epf}} \subseteq I^\dagger$. The main result of Hochster and Huneke was that dagger closure and tight closure are the same in positive characteristic, so it made sense to try to show that dagger closure satisfies the colon-capturing property. Heitmann's results show that this is true in dimension three.

In addition to leading to these developments on the original homological conjectures, the connections that tight closure demonstrated with other areas inspired some new conjectures.

Conjecture 16 (Vanishing of Maps of Tors). Let R be a regular ring, A a module finite torsion-free extension of R, and T a regular local ring with a map ϕ from A to T. Then for every R-module M and every $i \geq 1$, the map induced by ϕ from $\operatorname{Tor}_i^R(M, A)$ to $\operatorname{Tor}_i^R(M, T)$ is zero.

This conjecture has a similar flavor to some of the previous ones, particularly in the case where T is a finite A-module, and it implies several of them. However, this one is much more general; T could be an infinite extension, or on the other hand it could be the residue field of A if A is a local ring. It is known in the equicharacteristic case. We refer to Hochster [30] for a more complete discussion of this conjecture and its relation to other ones.

Another result of tight closure was to give a simple proof in characteristic zero that invariants of certain group actions on regular rings are Cohen–Macaulay. They proved, in fact, that a direct summand of a regular ring in equal characteristic is Cohen–Macaulay; it is a conjecture in mixed characteristic.

Conjecture 17. A direct summand of a regular ring is Cohen–Macaulay.

If we apply this to the R-module $M = R/(x_1^{t+1}, \dots, x_d^{t+1}, x_1^t x_2^t \cdots x_d^t)$ where R is (regular) local of dimension d and $x_1, \dots, x_d$ is a system of parameters, it is not hard to see that this conjecture implies the Monomial Conjecture and hence also the Direct Summand Conjecture. In fact, it is equivalent to a stronger version of this conjecture.

9 The Strong Direct Summand Conjecture

In this section we discuss several recent variations on conjectures related to Direct Summands.

Conjecture 18 (Strong Direct Summand Conjecture). Let R be a regular local ring and let A be a finite extension of R. Let Q be a height one prime ideal of A containing xR, where x is a minimal generator of the maximal ideal of R. Then xR is a direct summand of Q.

At first sight this appears to be a rather gratuitous generalization of the Direct Summand Conjecture. It is indeed a generalization, since if this holds, then since xA is contained in Q, the splitting map from Q to xR induces one from xA to xR, and dividing by x we obtain one from A to R. Its importance comes from the surprising fact that it is equivalent to the Vanishing Conjecture for maps of Tors. This was proven by N. Ranganathan in [49].

She also had a strong version of the Monomial Conjecture:

Conjecture 19 (Strong Monomial Conjecture). Let A be a local domain with system of parameters $(x_1, \dots, x_d)$. Let Q be a height one prime of A containing x_1. Then

$$x_1(x_1x_2 \cdots x_d)^t \notin (x_1^{t+1}, \dots, x_d^{t+1})Q$$

for all $t > 0$.

A much more complete discussion of the conjectures of the last two sections and relations between them can be found in Hochster [29] and [30].

Recent work on the Strong Monomial Conjecture can be found in McCullough [45].

We include here the updated version of Hochster's diagram from the 2004 Minicourse at the University of Utah. For the most part, the conjectures in the new diagram have been stated since the earlier diagram; the main exceptions are some of those in the lower right. Even here some of the implications are new (since 1975). Also, the Intersection and New Intersection conjectures have actually been proven, but they are included here to show how they fit with some of the more recent conjectures.

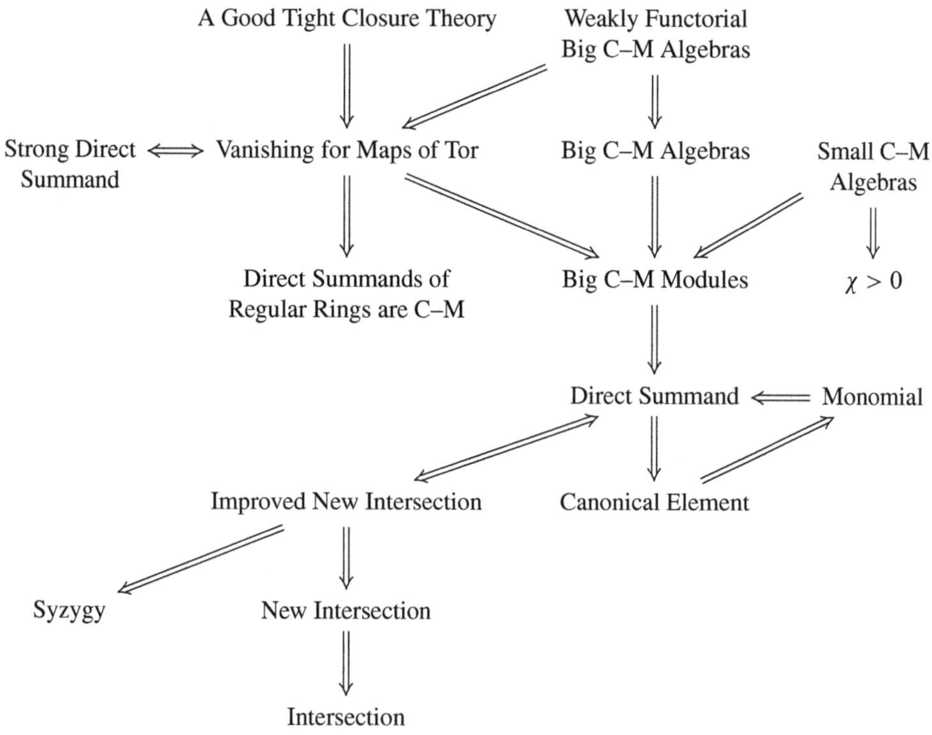

Figure 2. Hochster's diagram of homological conjectures in 2004.

10 Almost Cohen–Macaulay Algebras

As outlined above, Heitmann's proof of the Direct Summand Conjecture in dimension three introduced a new method for attacking many of the Homological Conjectures in mixed characteristic. In this section we will go into more detail about this method and questions that it raised.

What Heitmann showed originally was that if A is a complete normal local domain of dimension 3 and of mixed characteristic, if p, x, y is a system of parameters, and if $ap^N \in (x, y)$ for some $a \in A$, then for any integer $n > 0$, there is a finite extension B of A such that $ap^{1/n} \in (x, y)B$. This implies that the local cohomology $H^2_{\mathfrak{m}}(A^+)$ is annihilated by $p^{1/n}$ for all $n > 0$, where, as usual, A^+ is the integral closure of A in the algebraic closure of its quotient field. In a later paper ([23]) he extended this to show that p can be replaced by any u in the maximal ideal of A, and, using the fact that the condition that A is a normal domain of dimension 3, so that $H^2_{\mathfrak{m}}(A)$ has finite length, it is easy to see that this implies that $H^2_{\mathfrak{m}}(A^+)$ is annihilated by the maximal ideal of A^+. Thus it is a vector space over the field $A^+/\mathfrak{m}_{A^+}$. It is still an open question whether it is actually zero.

As we also described earlier, the result, for example, that if $ap^N \in (x,y)$ in A then $ap^{1/n} \in (x,y)$ in A^+ can be stated by the fact that certain closure operations have the colon-capturing property in this case. The closure operations are full extended plus closure of Heitmann [23] and dagger closure of Hochster and Huneke [32]. While the fact that full extended plus closure has this property is a stronger result, for the remainder of this section we will only consider dagger closure, since the fact that it has the colon-capturing property is enough to prove, for example, the Direct Summand Conjecture. We describe this in more detail.

Let A be a ring as above; we take a valuation v on A with values in the ordered abelian group $\mathbb{R}$ of real numbers, Then v is a function from A to $\mathbb{R} \cup \{\infty\}$ satisfying

(i) $v(ab) = v(a) + v(b)$ for $a, b \in A$.

(ii) $v(a+b) \geq \min\{v(a), v(b)\}$ for $a, b \in A$.

(iii) $v(a) = \infty$ if and only if $a = 0$.

We will assume also that $v(a) \geq 0$ for $a \in A$ and that $v(a) > 0$ for a in the maximal ideal of A. The existence of such a valuation follows from standard facts on extensions of valuations, see for example Zariski–Samuel [66], Chapter VI.

If I is an ideal of a local domain A with a valuation v satisfying the above properties, then a is in the dagger closure $I^\dagger$ of I if there exist elements $u \in A^+$ of arbitrarily small positive order, with $ux \in aA^+$. It follows from Heitmann's result that in mixed characteristic in dimension three dagger closure has the colon-capturing property. It also follows that, still in dimension three, the local cohomology module $H^2_{\mathfrak{m}}(A^+)$ is annihilated by arbitrarily small elements. To generalize this we make the following definitions.

We say that an A-module M is *almost zero* with respect to v if for all $m \in M$ and for all $\epsilon > 0$, there exists an $a \in A$ with $v(a) < \epsilon$ and $am = 0$.

This terminology comes from a paper of Faltings [16], where he proves that certain local cohomology groups are almost zero. The topic of almost zero modules was developed in much more detail by Gabber and Ramero [17].

Definition 4. An A-algebra B is *almost Cohen–Macaulay* if

(i) $H^i_{(x)}(B)$ is almost zero for $i = 0, \ldots, d-1$.

(ii) $B/(x_1, \ldots, x_d)B$ is not almost zero.

An alternative definition of almost Cohen–Macaulay can be obtained by defining a sequence $x_1, \ldots, x_d$ to be *almost regular* if $\{a | ax_i \in (x_1, \ldots, x_{i-1})\}/(x_1, \ldots, x_{i-1})$ is almost zero for $i = 1, \ldots, d$ and defining A to be almost Cohen–Macaulay if a system of parameters is almost regular (together with condition (2)). Standard methods show that this definition implies the former one (see for example Matsumura [44], Theorem 16.5 (i)).

Question 1. Let A be a complete Noetherian local domain. Is A^+ almost Cohen–Macaulay?

Of course, the result of Hochster–Huneke and Huneke–Lyubeznik that we referred to above says that if A has positive characteristic, then A^+ is actually Cohen–Macaulay. However, this is not true in characteristic zero, since if we have a normal non-Cohen–Macaulay domain A, since A is a direct summand of every finite extension using the trace map, a nontrivial element of local cohomology cannot go to zero in A^+. There is little evidence that this would be true in general, but there are some examples in characteristic zero in dimension 3 by Roberts, Singh, and Srinivas [59], and Heitmann, as we have seen, showed that it is true in mixed characteristic in dimension 3. As we have said, it is still open whether A^+ is Cohen–Macaulay in that case.

This question can also be generalized further. Instead of the class of almost zero modules defined above, we can take other classes. To make the theory work we should take a class $\mathcal{C}$ of almost zero modules satisfying the following conditions.

(i) If $0 \to M' \to M \to M'' \to 0$ is a short exact sequence, then $M \in \mathcal{C}$ if and only if M' and M'' are in $\mathcal{C}$.

(ii) $\mathcal{C}$ is closed under direct limits.

Question 2. Let A be a local ring. Does there exist an almost Cohen–Macaulay algebra over A for some class of almost zero modules?

11 A Summary of Open Questions

We summarize some of the main questions which remain open. Since they have varying degrees of likelihood of being true, we simply label them all as "Questions".

11.1 The Serre Positivity Conjecture

Question 3. Let R be a ramified regular local ring of mixed characteristic, and let M and N be R-modules such that $M \otimes_R N$ has finite length. If $\dim M + \dim N = \dim R$, is $\chi(M, N) > 0$?

This conjecture would follow from the existence of small Cohen–Macaulay modules. There has been some recent work to attempt to use Gabber's proof of the Nonnegativity Conjecture to prove this, but so far it has not been successful.

11.2 Partial Euler Characteristics

Question 4. If R is a ramified regular local ring of mixed characteristic of dimension d and M and N are R-modules such that $M \otimes_R N$ has finite length, is

$$\chi_i(M, N) = \sum_{j=i}^{d} (-1)^{i+j} \text{length}(\text{Tor}(M, N)) \geq 0?$$

11.3 Strong Multiplicity Conjectures

Question 5. Let A be a local ring, and let M and N be A-modules such that $M \otimes_A N$ has finite length and M has finite projective dimension. Is $\dim(M) + \dim(N) \leq \dim(A)$?

Question 6. Let A be a local ring, and let M and N be A-modules such that $M \otimes_A N$ has finite length and both M and N have finite projective dimension.

(i) If $\dim(M) + \dim(N) < \dim A$, is $\chi(M, N) = 0$?

(ii) If $\dim(M) + \dim(N) = \dim A$, is $\chi(M, N) > 0$?

This question has been studied over the years, although it is still very much open. There are complexes which define positive cycles for which the positivity part fails, which may be a sign that the positivity is not true in this generality.

11.4 Cohen–Macaulay Modules and Related Conjectures

Question 7 (Small Cohen–Macaulay modules). Let A be a complete local domain of dimension d. Does there exist a finitely generated A-module of depth d?

No one has yet succeeded in coming up with a way to approach this question in dimension 3 or greater.

Question 8 (Big Cohen–Macaulay modules). Let A be a local domain of mixed characteristic of dimension d with system of parameters $x_1, \ldots, x_d$. Does there exist an A-module M for which

(i) $x_1, \ldots, x_d$ is a regular sequence on M.

(ii) $M/(x_1, \ldots, x_d)M \neq 0$.

As discussed at length, there are numerous conjectures which follow from this one, many of which are equivalent. Out of these we will state two, one because it is quite concrete, and the other because it is the strongest of these conjectures. Both of these are open in the case where A has mixed characteristic and dimension greater than three.

Question 9 (Monomial Conjecture). Let A be a local ring with system of parameters $x_1, \ldots, x_d$. Is $x_1^t x_2^t \cdots x_d^t$ in the ideal $(x_1^{t+1}, x_2^{t+1}, \ldots, x_d^{t+1})$?

Question 10. Can one construct weakly functorial big Cohen–Macaulay algebras?

We refer to Section 6.1 for a precise statement of what this means.

11.5 Almost Cohen–Macaulay Algebras

Question 11. Let A be a local ring. Does A have an almost Cohen–Macaulay algebra?

We refer to the previous section for a precise statement of this question.

Bibliography

[1] Auslander, M., Modules over unramified regular local rings. Proceedings, International Congress Stockholm; 1962. p. 230–233.

[2] Avramov, L., Buchweitz, R., Iyengar, S., Class and rank of differential modules. Invent Math. 2007;169:1–35.

[3] Berthelot, P., Altérations de variétés algébriques [d'après A. J.de Jong]. Sémin Bourbaki. 1996:exposé 815.

[4] Bertin, M.-J., Anneaux d'invariants d'anneaux de polynomes, en caractéristique p. C R Acad Sci Paris Sér A–B 1967;264:A653–A656.

[5] Brodman, M. P., Sharp, R. Y., Local cohomology: an algebraic introduction with geometric applications. vol. 60 of Cambridge studies in advanced mathematics. Cambridge: Cambridge University Press; 1998.

[6] Dao, H., Decency and Tor-rigidity for modules over hypersurfaces. arXiv: math.AC/0611568.

[7] Dutta, S. P., On the canonical element conjecture. Trans Amer Math Soc. 1987;299:803–811.

[8] Dutta, S. P., A note on the monomial conjecture. Trans Amer Math Soc. 1998;350:2871–2878.

[9] Dutta, S. P., On negativity of higher Euler characteristics. Amer J Math. 2004;126:1341–1354.

[10] Dutta, S. P., Intersection multiplicity of Serre on regular schemes. J Algebra. 2008;319: 1530–1554.

[11] Dutta, S. P., Hochster, M., McLaughlin, J. E., Modules of finite projective dimension with negative intersection multiplicities. Invent Math. 1985;79:253–291.

[12] Eisenbud, D., Homological algebra on a complete intersection, with an application to group representations. Trans Amer Math Soc. 1980;260:35–64.

[13] Evans, E. G., Griffith, P., The syzygy problem. Ann of Math (2). 1981;114:323–333.

[14] Evans, E. G., Griffith, P., Syzygies. vol. 106 of London Mathematical Society lecture note series. Cambridge: Cambridge University Press; 1985.

[15] Evans, E. G., Griffith, P., A graded syzygy theorem in mixed characteristic. Math Res Lett. 2001;8:605-611.

[16] Faltings, G., Almost étale extensions. Astérisque 2002;297:185–270.

[17] Gabber, O., Ramero, L., Almost ring theory, vol. 1800 of Lecture notes. Springer.

[18] Gillet, H., Soulé, C. K-théorie et nullité des multiplicités d'intersection. C R Acad Sci Paris Sér I. 1985;300(3):71–74.

[19] Gillet, H., Soulé, C., Intersection theory using Adams operations. Invent Math. 1987;90: 243–277.

[20] Heitmann, R., A counterexample to the rigidity conjecture for rings. Bull. Amer Math Soc (N S). 1993;29:94–97.

[21] Heitmann, R., The plus closure in mixed characteristic. J Algebra. 1997;193:688–708.

[22] Heitmann, R. C., The direct summand conjecture in dimension three. Ann of Math (2). 2002;156:695–712.

[23] Heitmann, R. C., Extended plus closure and colon-capturing. J Algebra. 2005;293:407–426.

[24] Hochster, M., Topics in the homological theory of modules over commutative rings. vol. 24 of CBMS regional conference series in mathematics. 1975.

[25] Hochster, M., Canonical elements in local cohomology modules and the direct summand conjecture. J Algebra. 1983;84:503–553.

[26] Hochster, M., Euler characteristics over unramified regular local rings. Illinois J Math. 1984;28:281–285.

[27] Hochster, M., Nonnegativity of intersection multiplicities in ramified regular local rings following Gabber/De Jong/Berthelot [unpublished notes]. Available from: http://www.math.lsa.umich.edu/~hochster/mult.ps.

[28] Hochster, M., Big Cohen–Macaulay algebras in dimension three via Heitmann's theorem. J Algebra. 2002;254:395–408.

[29] Hochster, M., Notes from the 2004 Microprogram at the University of Utah. Available from: http://www.math.utah.edu/vigre/minicourses/algebra/hochster.pdf.

[30] Hochster, M., Homological conjectures, old and new. Illinois J Math. 2007;51(1):151–169.

[31] Hochster, M., Huneke, C., Tight closure, invariant theory, and the Briançon–Skoda theorem. J Amer Math Soc. 1990;3:31–116.

[32] Hochster, M., Huneke, C., Tight closure and elements of small order in integral extensions. J Pure Appl Algebra. 1991;71:233–247.

[33] Hochster, M., Huneke, C., Infinite integral extensions and big Cohen–Macaulay algebras. Ann of Math (2). 1992;135:53–89.

[34] Hochster, M., Huneke, C., Applications of the existence of big Cohen–Macaulay algebras. Adv Math. 1995;113:45–117.

[35] Huneke, C., Tight closure and its applications. vol. 88 of CBMS regional conference series in mathematics. 1996.

[36] Huneke, C., Lyubeznik, G., Absolute integral closure in positive characteristic. Adv Math. 2007;210:498–504.

[37] Huneke, C., Wiegand, R., Tensor products of modules and the rigidity of Tor. Math Ann. 1994;299:449–476.

[38] Iyengar, S., Leuschke, G., Leykin, A., Miller, C., Miller, E., Singh, A., Walther, U., Twenty-four hours of local cohomology. vol. 87 of graduate studies in mathematics. Providence, RI: American Mathematical Society; 2007.

[39] de Jong, A. J., Smoothness, semi-stability, and alterations. Publ Math Inst Hautes Études Sci. 1996;83:51–93.

[40] Katz, D., On the existence of maximal Cohen–Macaulay modules over pth root extensions. Proc Amer Math Soc. 1999;127:2601–2609.

[41] Kurano, K., A remark on the Riemann–Roch formula on affine schemes associated with Noetherian local rings. J Math Soc Japan 1993;45:369–390.

[42] Kurano, K., Roberts, P., The positivity of intersection multiplicities and symbolic powers of prime ideals. Compos Math. 2000;122:165–182.

[43] Lichtenbaum, S., On the vanishing of Tor in regular local rings. Illinois J Math. 1966;10: 220–226.

[44] Matsumura, H., Commutative rings. Cambridge University Press; 1986.

[45] McCullough, J., A note on the strong direct summand conjecture. Proc Amer Math Soc 2009;127:2857–2864.

[46] Miller, C., Singh, A., Intersection multiplicities over Gorenstein rings. Math Ann. 2000; 317:155–171.

[47] Peskine, C., Szpiro, L., Dimension projective finie et cohomologie locale. Publ Math Inst Hautes Études Sci. 1973;42:49–119.

[48] Peskine, C., Szpiro, L., Syzygies et multiplicités. C R Acad Sci Paris Sér A 1974;278: 1421–1424.

[49] Ranganathan, N., Splitting in module-finite extensions and the vanishing conjecture for maps of Tor [thesis]. University of Michigan; 2000.

[50] Roberts, P., Two applications of dualizing complexes over local rings. Ann Sci Éc Norm Supér (4). 1976;9:103–106.

[51] Roberts, P., The vanishing of intersection multiplicities of perfect complexes. Bull Amer Math Soc. 1985;13:127–130.

[52] Roberts, P., Le théorème d'intersection. C R Acad Sci Paris Sér I. 1987;304(7):177–180.

[53] Roberts, P., Intersection theorems. In: Commutative Algebra. Proceedings of an MSRI Microprogram. New York: Springer-Verlag; 1989. p. 417–436.

[54] Roberts, P., Negative intersection multiplicities on singular varieties. Contemp Math. 1991;123(Proceedings of the Zeuthen Conference 1989):213–222.

[55] Roberts, P., The homological conjectures. In: Free resolutions in commutative algebra and algebraic geometry. Proceedings, Sundance 90. vol. 2 of Research notes in mathematics. 1992.

[56] Roberts, P., Multiplicities and Chern classes in local algebra. vol. 133 of Tracts in mathematics. Cambridge University Press; 1998.

[57] Roberts, P., Recent developments on Serre's multiplicity conjectures: Gabber's proof of the nonnegativity conjecture. Enseign Math (2). 1998;44:305–324.

[58] Roberts, P., Singh, A., Reconciling Riemann–Roch results. To appear.

[59] Roberts, P., Singh, A., Srinivas, V., Annihilators of local cohomology in characteristic zero. Illinois J Math. 2007;51:237–254.

[60] Roberts, P., Srinivas, V., Modules of finite length and finite projective dimension. Invent Math. 2003;151:1–27.

[61] Sather-Wagstaff, S., A dimension inequality for Cohen–Macaulay rings. Trans Amer Math Soc. 2002;354:993–1005.

[62] Serre, J.-P., Local algebra. Springer monographs in mathematics, Berlin: Springer-Verlag; 2000.

[63] Strooker, J. R., Homological questions in commutative algebra. vol. 106 of London Mathematical Society lecture note series. Cambridge: Cambridge University Press; 1990.

[64] Strooker, J. R., Stückrad, J., Monomial conjecture and complete intersections. Manuscripta Math. 1993;79:153–159.

[65] Szpiro, L., Sur la théorie des complexes parfaits. In: Commutative Algebra. Proceedings, Durham 1981. vol. 72 of London Mathematical Society lecture notes series. 1982. 83–90.

[66] Zariski, O., Samuel, P., Commutative algebra II. vol. 29 of Graduate texts in mathematics. New York, Heidelberg: Springer-Verlag; 1975.

Author Information

Paul C. Roberts, Department of Mathematics, University of Utah, Salt Lake City, UT, USA.
E-mail: `roberts@math.utah.edu`

Progress in Commutative Algebra 1, 231–269

The Compatibility, Independence, and Linear Growth Properties

Yongwei Yao

Abstract. The first part is about primary decomposition. After reviewing the basic definitions, we survey the compatibility, independence, and linear growth properties that have been known. Then, we prove the linear growth property of primary decomposition for a new family of modules.

In the remaining sections, we study secondary representation, which can be viewed as a dual of primary decomposition. Correspondingly, we study the compatibility, independence, and linear growth properties of secondary representations.

Keywords. Primary Decomposition, Weak Primary Decomposition, Secondary Representation, Associated Prime, Attached Prime, Compatibility, Independence, Linear Growth.

2010 Mathematics Subject Classification. Primary: 13E05; Secondary: 13C99, 13H99.

1 Introduction

Throughout this paper, all rings are assumed to be commutative with one; and they are not necessarily Noetherian unless we state so explicitly.

Sections 2–6 are dedicated to the theory of primary decomposition. In its classic form, it states that every ideal in a Noetherian ring can be expressed as an intersection of finitely many primary ideals. Later, the theory of primary decomposition was developed for modules. In particular, if a module is Noetherian, then every submodule is decomposable.

Although the primary decompositions are not unique in general, there are certain uniqueness properties governing the primary decompositions.

In Section 2, basic definitions and properties in the theory of primary decomposition are reviewed. In Section 3, we go over the compatibility property, which says that primary components from different primary decompositions of a fixed submodule can be put together and the resulting intersection is still a primary decomposition of the submodule. Maximal primary components are studied in Section 4. In Section 5, the linear growth property of primary decomposition is reviewed. We establish the linear growth property for a new family of modules in Section 6.

In Sections 7–12, we study the secondary representation theory. This can be viewed as a dual of the primary decomposition theory. In this theory, a module is representable

The author was partially supported by the National Science Foundation (DMS-0700554).

if it can be expressed as a finite sum of secondary submodules. It turns out that every Artinian module has a secondary representation.

Many of the results in the theory of secondary representation have their dual forms in the theory of primary decomposition. Because of this, one often draws inspiration from one theory and then applies it to the other. In this note, the theory of secondary representation is presented in a way that would make the duality between the two theories evident.

In Section 7, we go over the fundamentals of the theory of secondary representation. In the subsequent sections, we study and prove the compatibility, minimal components, independence, and linear growth properties of secondary representation. In Section 9, we discuss a result of Sharp [9] that makes the classic Matlis duality applicable to Artinian modules even if the ring is not Noetherian. This allows us to establish results on secondary representation by reducing them to the dual results in the theory of primary decomposition.

Many of the results in Sections 8–12 were obtained in [18].

2 Primary decomposition

In this section, we give a brief introduction to the notions of associated prime and primary decomposition. Systematic treatments of primary decomposition can be found in many textbooks, for example, [1], [2], [3] or [7].

Let R be a ring (not necessarily Noetherian) and M an R-module.

We say that a prime ideal $P \in \operatorname{Spec}(R)$ is *associated* to M if there exists $x \in M$ such that $(0 :_R x) = P$. The set of all primes associated to M is denoted $\operatorname{Ass}_R(M)$, or simply $\operatorname{Ass}(M)$ when R is understood from the context.

Following [1], we say that a prime ideal $P \in \operatorname{Spec}(R)$ *belongs* to M if there exists $x \in M$ such that $\sqrt{(0 :_R x)} = P$. (In fact, the terminology "*P belongs to* 0 *in M*" was used in [1].) The set of all primes belonging to M is denoted $\operatorname{Ass}'_R(M)$, or simply $\operatorname{Ass}'(M)$ when R is understood from the context.

We say that M is *coprimary* (over R) if, for every $r \in R$, either r is M-regular (i.e., $(0 :_M r) = 0$) or $r \in \sqrt{\operatorname{Ann}(M)}$. (Under this definition, 0 is a coprimary module.) It turns out that, if $M \neq 0$ is coprimary and if we let $P = \sqrt{\operatorname{Ann}(M)}$, then $P \in \operatorname{Spec}(R)$; in this case, we say M is P-coprimary. (This definition recovers the definition of primary ideals in that an ideal Q is P-primary (in R) if and only if R/Q is P-coprimary as an R-module.)

We also define $\operatorname{Ass}''_R(M) := \{P \in \operatorname{Spec}(R) \mid \exists K \subseteq M,\ K \text{ is } P\text{-coprimary}\}$; or, equivalently, $\operatorname{Ass}''_R(M) := \{P \in \operatorname{Spec}(R) \mid \exists x \in M,\ R/(0 :_R x) \text{ is } P\text{-coprimary}\}$. This notion Ass'' and the notion Att (to be defined in Section 7) are dual to each other.

Quite generally, if M is P-coprimary, then $\operatorname{Ass}'_R(M) = \{P\} = \operatorname{Ass}''(M)$.

For R-modules $Q \subseteq M$, we say that Q is $(P$-$)$primary if M/Q is $(P$-$)$coprimary. For R-modules $N \subsetneq M$, we say that N is *decomposable* in M (over R) if there exist

R-submodules Q_i that are P_i-primary in M, for $i = 1, \dots, s$, such that

$$N = Q_1 \cap \cdots \cap Q_s.$$

This intersection is called a *primary decomposition* of N in M (over R). One can always convert a primary decomposition to a *minimal* one in the sense that $P_i \neq P_j$ for all $i \neq j$ and $N \neq \bigcap_{i \neq k} Q_i$ for every $k = 1, \dots, s$. So from now on, as a general rule, all primary decompositions are assumed to be minimal unless stated otherwise explicitly.

For every R-module M, we agree that M is decomposable in M with $M = M$ being the unique primary decomposition of M in M.

Given R-modules $N \subseteq M$, N is decomposable in M if and only if 0 is decomposable in M/N; and the primary decompositions of N in M are in one-to-one correspondence with the primary decompositions of 0 in M/N.

Similarly, let $N \subseteq M$ be R-modules and let I be an ideal of R such that $I \subseteq \operatorname{Ann}(M)$, so that $N \subseteq M$ can be naturally viewed as modules over R/I. Then N is decomposable in M as R-modules if and only if N is decomposable in M as (R/I)-modules.

Next, we list some properties of primary decomposition. We need to introduce some notation that will be used in the sequel: Given an R-module M, we use $\operatorname{Min}(M)$ to denote the set of all the minimal primes over $\operatorname{Ann}(M)$. For a multiplicative subset $U \subseteq R$, we use $M[U^{-1}]$ to denote the module of fractions after inverting all the elements in U, so that $M[U^{-1}] \cong M \otimes_R R[U^{-1}]$.

Theorem 2.1. *Let $N \subseteq M$ be R-modules and suppose $N = Q_1 \cap \cdots \cap Q_s$ is a (minimal) primary decomposition of N in M in which Q_i is P_i-primary.*

(i) *We have $\{P_1, \dots, P_s\} = \operatorname{Ass}'_R(M/N) = \operatorname{Ass}''_R(M/N)$, which is independent of the particular (minimal) primary decompositions in M.*

(i′) *We have $\operatorname{Min}(M/N) \subseteq \{P_1, \dots, P_s\}$. In fact, $\operatorname{Min}(M/N)$ equals the set of the minimal members of $\{P_1, \dots, P_s\}$ (under inclusion).*

(ii) *If P_i is minimal in $\operatorname{Ass}'_R(M/N)$, then Q_i is uniquely determined as $Q_i = \operatorname{Ker}\big(M \to (M/N)_{P_i}\big)$. See* (iv) *below.*

(iii) *Let $h\colon A \to R$ be a ring homomorphism, so that $N \subseteq M$ may be viewed as A-modules. Let K be an A-submodule of M such that $N \subseteq K$ (e.g., $K = M$). Then N is decomposable in K as A-modules. If $N \subsetneq K$, then*

$$N = \bigcap_{Q_i \not\supseteq K} (Q_i \cap K)$$

is a (not necessarily minimal) primary decomposition of N in K over A, in which $Q_i \cap K$ is $h^{-1}(P_i)$-primary in K provided that $Q_i \not\supseteq K$.

(iii′) *In particular, $\operatorname{Ass}'_A(M) = h^*(\operatorname{Ass}'_R(M))$, in which $h^*\colon \operatorname{Spec}(R) \to \operatorname{Spec}(A)$ is the continuous map naturally induced by h.*

(iv) *Let* $U \subseteq R$ *be a multiplicative set. Then* $N[U^{-1}] = \bigcap_{U \cap P_i = \varnothing} Q_i[U^{-1}]$ *is a primary decomposition in* $M[U^{-1}]$, *in which* $Q_i[U^{-1}]$ *is* $P_i[U^{-1}]$*-primary in* $M[U^{-1}]$; *and* $\operatorname{Ker}(M \to (M/N)[U^{-1}]) = \bigcap_{U \cap P_i = \varnothing} Q_i$.

(v) *For any finitely generated ideal* I *of* R, $\bigcap_{I \not\subseteq P_i} Q_i = (N :_M I^n)$ *for* $n \gg 0$.

(v′) *For any non-empty subset* I *of* R, $\bigcap_{I \not\subseteq P_i} Q_i = \bigcap_{r \in I} (\bigcup_{n \in \mathbb{N}} (N :_M r^n))$.

Remark 2.2. In [2, Chapter IV], the notion of primary decomposition is generalized to *weak primary decomposition*. (This was simply called *primary decomposition* in [2]. We add the word "weak" into the terminology in order to distinguish it from the notion of (ordinary) primary decomposition.) For an R-module M and $P \in \operatorname{Spec}(R)$, we say that P is *weakly associated* to M if P is minimal over the ideal $\operatorname{Ann}_R(x)$ (i.e., $P \in \operatorname{Min}(Rx)$) for some $x \in M$. Denote by $\operatorname{Ass_f}(_R M)$, or simply $\operatorname{Ass_f}(M)$, the set of all the prime ideals weakly associated to M (cf. [2, page 289, Chapter IV, §1, Exercise 17]). It is clear $M = 0 \iff \operatorname{Ass_f}(M) = \varnothing$.

We say that M is *weakly coprimary* if, for all $r \in R$, either $(0 :_M r) = 0$ or $\bigcup_{n \geq 0} (0 :_M r^n) = M$. If $M \neq 0$ is coprimary, it follows that $\{r \in R \mid (0 :_M r) \neq 0\} =: P$ is prime, and we say that M is weakly P-coprimary. It turns out that M is weakly P-coprimary if and only if $\operatorname{Ass_f}(M) = \{P\}$. See [2, page 292, Chapter IV, §2, Exercises 11, 12].

Given R-modules $Q \subseteq M$, we say that Q is *weakly P-primary* in M if M/Q is weakly P-coprimary, i.e., $\operatorname{Ass_f}(M/Q) = \{P\}$. Now, for $N \subseteq M$, we say that N is *weakly decomposable* in M if there exist weakly P_i-primary submodules Q_i in M, $i = 1, \dots, s$, such that $N = Q_1 \cap \dots \cap Q_s$. If such decompositions exist, we can make them minimal. Weak primary decompositions enjoy many of the properties of primary decompositions; see [2, page 294, Chapter IV, §2, Exercise 20] and Theorem 2.3 below. Conversely, if Q is P-primary in M then Q is weakly P-primary in M; thus every primary decomposition is a weak primary decomposition.

In [13], some of the basic properties of $\operatorname{Ass_f}$ and weak primary decomposition were worked out in detail via elementary techniques.

We state the following weak-primary-decomposition analogue of Theorem 2.1.

Theorem 2.3. *Suppose* $N = Q_1 \cap \dots \cap Q_s$ *is a minimal weak primary decomposition of* N *in* M, *in which* Q_i *is weakly* P_i*-primary.*

(i) *We have* $\{P_1, \dots, P_s\} = \operatorname{Ass}'(M/N) = \operatorname{Ass}''(M/N) = \operatorname{Ass_f}(M/N)$, *which is independent of the particular (minimal) primary decompositions in* M.

(i′) *We have* $\operatorname{Min}(K/N) \subseteq \{P_1, \dots, P_s\}$ *for all* R*-submodule* K *satisfying* $N \subseteq K \subseteq M$ *and* K/N *is finitely generated.*

(ii) *If* P_i *is minimal in* $\{P_1, \dots, P_s\} = \operatorname{Ass_f}(M/N)$ *(under inclusion), then* Q_i *is uniquely determined as* $Q_i = \operatorname{Ker}\big(M \to (M/N)_{P_i}\big)$. *See* (iv) *below.*

(iii) *Let $h\colon A \to R$ be a ring homomorphism, so that $N \subseteq M$ may be viewed as A-modules. Let K be an A-submodule of M such that $N \subseteq K$ (e.g., $K = M$). Then N is weakly decomposable in K as A-modules. If $N \subsetneq K$, then*

$$N = \bigcap_{Q_i \not\supseteq K} (Q_i \cap K)$$

is a (not necessarily minimal) weak primary decomposition of N in K over A, in which $Q_i \cap K$ is weakly $h^{-1}(P_i)$-primary in K provided that $Q_i \not\supseteq K$.

(iii′) *In particular, $\mathrm{Ass_f}({}_AM) = h^*(\mathrm{Ass_f}({}_RM))$, in which $h^*\colon \mathrm{Spec}(R) \to \mathrm{Spec}(A)$ is the continuous map naturally induced by h.*

(iv) *Let $U \subseteq R$ be a multiplicative set. Then $N[U^{-1}] = \bigcap_{U \cap P_i = \varnothing} Q_i[U^{-1}]$ is a weak primary decomposition in $M[U^{-1}]$, in which $Q_i[U^{-1}]$ is $P_i[U^{-1}]$-primary in $M[U^{-1}]$; and $\mathrm{Ker}(M \to (M/N)[U^{-1}]) = \bigcap_{U \cap P_i = \varnothing} Q_i$.*

(v) *For any finitely generated ideal I of R, $\bigcap_{I \not\subseteq P_i} Q_i = \bigcup_{n \in \mathbb{N}} (N :_M I^n)$.*

(v′) *For any non-empty subset I of R, $\bigcap_{I \not\subseteq P_i} Q_i = \bigcap_{r \in I} (\bigcup_{n \in \mathbb{N}} (N :_M r^n))$.*

It is well known that if $0 \to M_1 \to M_2 \to M_3$ is an exact sequence of R-modules then $\mathrm{Ass}(M_1) \subseteq \mathrm{Ass}(M_2) \subseteq \mathrm{Ass}(M_1) \cup \mathrm{Ass}(M_3)$; and $\mathrm{Ass}(\bigoplus_{i \in \Delta} K_i) = \bigcup_{i \in \Delta} \mathrm{Ass}(K_i)$ for any family $\{K_i\}_{i \in \Delta}$ of R-modules. The analogue also holds if we replace Ass with Ass′, Ass″ or $\mathrm{Ass_f}$. (See [2, page 289, Chapter IV, Section 1, Example 17 (c)] for the $\mathrm{Ass_f}$-analogue.) Here we present the Ass′-analogue, as it will be referred to in the proof of Lemma 3.2.

Lemma 2.4. *Let $0 \to M_1 \to M_2 \to M_3$ be an exact sequence of modules over a ring R. Then $\mathrm{Ass}'(M_1) \subseteq \mathrm{Ass}'(M_2) \subseteq \mathrm{Ass}'(M_1) \cup \mathrm{Ass}'(M_3)$.*

Moreover, $\mathrm{Ass}'(\bigoplus_{i \in \Delta} K_i) = \bigcup_{i \in \Delta} \mathrm{Ass}'(K_i)$ for any family $\{K_i\}_{i \in \Delta}$ of R-modules.

Proof. We sketch a proof of the first claim. Without loss of generality, assume $M_1 \subseteq M_2$ and $M_2/M_1 \subseteq M_3$. As $\mathrm{Ass}'(M_1) \subseteq \mathrm{Ass}'(M_2)$ is clear, it remains to show $\mathrm{Ass}'(M_2) \subseteq \mathrm{Ass}'(M_1) \cup \mathrm{Ass}'(M_2/M_1)$. Let $P \in \mathrm{Ass}'(M_2)$, so that $P = \sqrt{(0 :_R x)}$ for some $x \in M_2$. If there exists $r \in R \setminus P$ such that $rx \in M_1$, then it is straightforward to see that $P = \sqrt{(0 :_R rx)}$ and hence $P \in \mathrm{Ass}'(M_1)$. If $rx \notin M_1$ for all $r \in R \setminus P$, then it follows that $P = \sqrt{(0 :_R \overline{x})}$, where $\overline{x} = x + M_1 \in M_2/M_1$, and hence $P \in \mathrm{Ass}'(M_2/M_1$.

The second claim follows from the first when Δ is finite. In the general case, it is easy to see $\mathrm{Ass}'(\bigoplus_{i \in \Delta} K_i) \supseteq \bigcup_{i \in \Delta} \mathrm{Ass}'(K_i)$. Conversely, if $P \in \mathrm{Ass}'(\bigoplus_{i \in \Delta} K_i)$, then there exists a finite subset $\Delta' \subseteq \Delta$ such that $P \in \mathrm{Ass}'(\bigoplus_{i \in \Delta'} K_i)$. It then follows that $P \in \bigcup_{i \in \Delta'} \mathrm{Ass}'(K_i) \subseteq \bigcup_{i \in \Delta} \mathrm{Ass}'(K_i)$. □

We end this section with some basic facts concerning various kinds of associated prime ideals as well as decomposability. Let M be an R-module. It is clear that $\mathrm{Ass}(M) \subseteq \mathrm{Ass}''(M) \subseteq \mathrm{Ass}'(M) \subseteq \mathrm{Ass_f}(M) \subseteq \mathrm{Spec}(R)$. Consequently, as there is the Zariski topology on $\mathrm{Spec}(R)$, all the others are topological (sub)spaces. Quite generally, for any subset X of $\mathrm{Spec}(R)$, the Zariski topology on $\mathrm{Spec}(R)$ induces a topological structure on X in such a way that the closed sets of X are of the form $V_X(I) := \{P \in X \mid P \supseteq I\}$ with $I \subseteq R$.

If R is Noetherian or M is Noetherian over R, then $\mathrm{Ass}_R(M) = \mathrm{Ass}'_R(M) = \mathrm{Ass}''(M) = \mathrm{Ass_f}(M)$, and $\mathrm{Ass}_R(M) = \varnothing \iff M = 0$.

If M is Noetherian, then M is P-coprimary $\iff \mathrm{Ass}_R(M) = \{P\}$.

If $N \subseteq M$ are R-modules such that the quotient M/N is Noetherian over R, then N is decomposable in M. This is a classic result due to E. Noether.

There are more definitions of associated primes in the literature. See a list of these definitions in [15, Remark 3.11].

3 Compatibility of Primary Components

Throughout this section, let R be a (not necessarily Noetherian) ring and let $N \subseteq M$ be R-modules such that N is decomposable in M.

Notation 3.1. Let $X \subseteq \mathrm{Ass}'(M/N)$. Say

$$X = \{P_1, \dots, P_r\} \subseteq \{P_1, \dots, P_r, \dots, P_s\} = \mathrm{Ass}'(M/N).$$

(i) If $N = Q_1 \cap \cdots \cap Q_r \cap \cdots \cap Q_s$ is a primary decomposition of N in M with Q_i being P_i-primary, then we say $Q = Q_1 \cap Q_2 \cap \cdots \cap Q_r$ is an *X-primary component* (or a *primary component* over X) of $N \subseteq M$. If $X = \varnothing$, then we agree that M is the only X-primary component of $N \subseteq M$.

(ii) We call an X-primary component of $N \subseteq M$ maximal if it is not properly contained in any X-primary component of $N \subseteq M$.

(iii) We use $\Lambda_X(N \subseteq M)$ to denote the set of all possible X-primary components of N in M.

(iv) We use $\mathring{\Lambda}_X(N \subseteq M)$ to denote the set of all maximal X-primary components of N in M. (Note that $\mathring{\Lambda}_X(N \subseteq M) \neq \varnothing$ if M/N is Noetherian.)

(v) In case $X = \{P\} \subseteq \mathrm{Ass}(M/N)$, we may simply write Λ_P and $\mathring{\Lambda}_P$ instead of $\Lambda_{\{P\}}$ and $\mathring{\Lambda}_{\{P\}}$ respectively.

Note that, for $P \in \mathrm{Ass}'(M/N)$, the P-primary components are not necessarily unique in general (cf. Corollary 4.4). The compatibility property (see Theorem 3.3) says that if one takes a P-primary component of $N \subseteq M$ for each $P \in \mathrm{Ass}'(M/N)$ (from possibly different decompositions), then they are "compatible" in the sense that their intersection is exactly N, thus producing a primary decomposition of N in M. This was proved in [16] and [17] under the Noetherian assumption (but see [16, Remark 1.2]). Here we state the results more generally.

Lemma 3.2 (Cf. [17, Lemma 1.1]). *Let $N \subseteq M$ be R-modules such that N is decomposable in M, and $X \subseteq \mathrm{Ass}'(M/N)$. For an R-module Q such that $N \subseteq Q \subseteq M$, the following are equivalent:*

(i) *Q is an X-primary component of $N \subseteq M$, i.e., $Q \in \Lambda_X(N \subseteq M)$.*

(ii) *Q is decomposable in M, $\mathrm{Ass}'\left(\frac{M}{Q}\right) \subseteq X$ and $\mathrm{Ass}'\left(\frac{Q}{N}\right) \subseteq \mathrm{Ass}'\left(\frac{M}{N}\right) \setminus X$.*

(iii) *Q is decomposable in M, $\mathrm{Ass}'\left(\frac{M}{Q}\right) = X$ and $\mathrm{Ass}'\left(\frac{Q}{N}\right) = \mathrm{Ass}'\left(\frac{M}{N}\right) \setminus X$.*

Proof. The proof of [17, Lemma 1.1], with Ass′ instead of Ass, should work here, in light of Lemma 2.4 and the fact that N is automatically decomposable in Q. □

Theorem 3.3 (Compatibility). *Let $N \subseteq M$ be R-modules such that N is decomposable in M. Let $X_i \subseteq \mathrm{Ass}'(M/N)$ and $Q_{X_i} \in \Lambda_{X_i}(N \subseteq M)$ for $1 \le i \le n$.*

(i) *Then $\bigcap_{i=1}^n Q_{X_i} \in \Lambda_X(N \subseteq M)$, where $X = \bigcup_{i=1}^n X_i$.*

(ii) *In particular, suppose $\mathrm{Ass}'(M/N) = \{P_1, \dots, P_s\}$ and $Q_i \in \Lambda_{P_i}(N \subseteq M)$ for each $i = 1, 2, \dots, s$. Then $N = Q_1 \cap Q_2 \cap \dots \cap Q_s$, which is necessarily a minimal primary decomposition of $N \subseteq M$.*

Proof. The proof of [17, Corollary 1.2], with Ass′ instead of Ass, should work here. Note that, by construction, $\bigcap_{i=1}^n Q_{X_i}$ is decomposable in M. □

Remark 3.4. As noted in [16, Remark 1.2], the compatibility property is also shared by weak primary decompositions (cf. Remark 2.2). In fact, the analogues of Lemmas 2.4, 3.2 and Theorem 3.3 hold after every Ass′ is replaced with $\mathrm{Ass_f}$, "decomposable" with "weakly decomposable", and after $\Lambda_X(N \subseteq M)$ is interpreted accordingly. In [13], Stalvey presented a detailed proof of the compatibility for weak primary decomposition, following the proof given in [16, Theorem 1.1].

4 Maximal Primary Components, Independence

In this section, let $N \subseteq M$ be R-modules such that N is decomposable in M and $X \subseteq \mathrm{Ass}'(M/N)$. Note that $\mathrm{Ass}'(M/N)$ is a topological space in Zariski topology.

Notation 4.1. Let $N \subseteq M$ be as above. Since $\mathrm{Ass}'(M/N)$ is finite, every subset $X \subseteq \mathrm{Ass}'(M/N)$ has a unique minimal open superset in $\mathrm{Ass}'(M/N)$, which we denote by $\mathrm{o}(X)$. (Although this notation does not reflect the ambient space $\mathrm{Ass}'(M/N)$, there should be no danger of ambiguity.) For any $P \in \mathrm{Ass}'(M/N)$, we may simply write $\mathrm{o}(\{P\})$ as $\mathrm{o}(P)$. In fact, $\mathrm{o}(X) = \{P \in \mathrm{Ass}'(M/N) \mid P \subseteq \bigcup_{P' \in X} P'\}$, by prime avoidance.

Note that, if X is open in $\mathrm{Ass}'(M/N)$ (i.e., $X = \mathrm{o}(X)$), then there is a unique X-primary component of $N \subseteq M$, which is determined as $\mathrm{Ker}(M \to (M/N)[U^{-1}])$ with $U = R\backslash(\bigcup_{P\in X} P)$ (cf. Theorem 2.1 (iv)). This inspires the following definition.

Definition 4.2. Let R be a ring, $N \subseteq M$ be R-modules such that N is decomposable in M, and $X \subseteq \mathrm{Ass}'_R(M/N)$. We say that the primary decompositions of N in M are *independent* over X, or X-independent, if $\Lambda_X(N \subseteq M)$ consists of exactly one component, i.e., $|\Lambda^*_X(M)| = 1$.

Now assume that M/N is Noetherian over R. Thus, for any $X \subseteq \mathrm{Ass}(M/N) = \mathrm{Ass}'(M/N)$, maximal X-primary components exist. (When studying primary decompositions of N in M, we may simply study the primary decompositions of 0 in M/N as modules over $R/\mathrm{Ann}(M/N)$. Note that $R/\mathrm{Ann}(M/N)$ is Noetherian under the current assumption.)

In case $(R, \mathfrak{m})$ is local, maximal $\mathfrak{m}$-primary components were studied in [4]. In [17, Theorem 1.3], maximal X-primary components of $N \subseteq M$ were studied for general $X \subseteq \mathrm{Ass}(M/N)$. This is stated below.

Theorem 4.3. *Let $N \subseteq M$ be R-modules such that M/N is Noetherian over R, and $X \subseteq \mathrm{Ass}(M/N)$. Say $X = \{P_1, P_2, \dots, P_r\}$ and set $U = R \setminus (\bigcup_{i=1}^r P_i)$. Then*

(i) $\mathring{\Lambda}_X(N \subseteq M) = \{\bigcap_{i=1}^r Q_i \mid Q_i \in \mathring{\Lambda}_{P_i}(N \subseteq M),\ 1 \le i \le r\}$.

Consequently, we also have the following:

(i) *For every $Q \in \Lambda_X(N \subseteq M)$, $Q = \bigcap\{Q' \mid Q' \in \mathring{\Lambda}_X(N \subseteq M),\ Q \subseteq Q'\}$. In fact, every $Q \in \Lambda_X(N \subseteq M)$ is an intersection of finitely many $Q' \in \mathring{\Lambda}_X(N \subseteq M)$.*

(ii) *The intersection $\bigcap\{Q \mid Q \in \mathring{\Lambda}_X(N \subseteq M)\} = \bigcap\{Q \mid Q \in \Lambda_X(N \subseteq M)\}$ is equal to $\mathrm{Ker}\left(M \to (M/N)[U^{-1}]\right)$, which is the unique $\mathrm{o}(X)$-primary component in $\Lambda_{\mathrm{o}(X)}(N \subseteq M)$.*

Proof. We may assume $N = 0$. Then M/N can be viewed as a finitely generated module over the Noetherian ring $R/\mathrm{Ann}(M/N)$; and the same proof of [17, Theorem 1.3] works here. □

Now we study the property of X-independence. Quite generally, X-independence holds when X is open in $\mathrm{Ass}'(M/N)$ by Theorem 4.3 (iii) (also see [1, Theorem 4.10]). In fact, Theorem 4.3 implies that the primary decompositions of $N \subseteq M$ are independent over X if and only if X is open in $\mathrm{Ass}'(M/N) = \mathrm{Ass}(M/N)$ under the assumption that M/N) is Noetherian.

Theorem 4.4 ([17, Corollary 1.5]). *Let $N \subseteq M$ be R-modules such that M/N is Noetherian over R, and $X \subseteq \mathrm{Ass}(M/N)$. The following are equivalent:*

(i) *X is open in $\mathrm{Ass}(M/N)$.*

(ii) *$\Lambda_X(N \subseteq M)$ consists of only one X-primary component.*

(iii) $\Lambda_X(N \subseteq M)$ *is finite.*

(iv) $\mathring{\Lambda}_X(N \subseteq M)$ *is finite.*

Proof. This follows from Theorem 4.3; or see the proof of [17, Corollary 1.5]. □

Remark 4.5. As above, assume that M/N is Noetherian over R. By Theorem 4.4, there are infinitely many P-primary components of N in M if $P \in \text{Ass}(M/N)$ is an embedded prime.

5 Linear growth of primary components

Swanson showed the following *linear growth* property concerning the primary decompositions of I^n in R:

Theorem 5.1 ([14]). *Let R be a Noetherian ring and I an ideal of R. Then there exists $k \in \mathbb{N} := \{0, 1, 2, \ldots\}$ such that, for every $n \in \mathbb{N}$, there exists a primary decomposition (of I^n in R)*

$$I^n = Q_{n,1} \cap Q_{n,2} \cap \cdots \cap Q_{n,s(n)} \quad \textit{(with } Q_{n,i} \textit{ being } P_{n,i}\textit{-primary in } R\textit{)}$$

such that $(P_{n,i})^{kn} \subseteq Q_{n,i}$ for all $i = 1, 2, \ldots, s(n)$.

This was later generalized to any Noetherian R-module M together with several ideals in [12] via a study of injective modules. The same result was also later obtained in [16, 17] via different methods. In [17], this kind of property was also proved for families of Tor and Ext modules. (See Theorem 5.4 for the precise statements.)

Inspired by the above, we formulate the following definition of the linear growth property of primary decomposition.

Definition 5.2. Given a family $\mathcal{F} = \{M_a \mid a = (a_1, a_2, \ldots, a_r) \in \mathbb{N}^r\}$ consisting of R-modules, we say $\mathcal{F}$ satisfies the *linear growth* property of primary decomposition (over R) if there exists $k \in \mathbb{N}$ such that, for every $a = (a_1, a_2, \ldots, a_r) \in \mathbb{N}^r$ such that $M_a \neq 0$, there exists a primary decomposition of 0 in M_a,

$$0 = Q_{a,1} \cap Q_{a,2} \cap \cdots \cap Q_{a,s(a)} \quad \text{(with } Q_{a,i} \text{ being } P_{a,i}\text{-primary in } M_a\text{)}$$

such that $(P_{a,i})^{k|a|} M_a \subseteq Q_{a,i}$ for all $i = 1, 2, \ldots, s(a)$, where $|a| = a_1 + \cdots + a_r$.

When the above occurs, we refer to k as a *slope* of $\mathcal{F}$. (Clearly, if k is a slope of $\mathcal{F}$, then all the integers greater than k are also slopes of $\mathcal{F}$.)

The linear growth property is a measure of the 'sizes' of the primary components as $a \in \mathbb{N}^r$ varies. Roughly speaking, it says that there are primary decompositions in which the primary components are "not too small".

Next, we set up some notation, which will also be used in Section 6 and Section 12.

Notation 5.3. Let R be a ring, I_i, J_j ideals of R and X_i, Y_j indeterminates, for $i \in \{1, \ldots, s\}$ and $j \in \{1, \ldots, t\}$ with s and t positive integers.

(i) By $m \in \mathbb{Z}^s$, we mean $m := (m_1, \ldots, m_s) \in \mathbb{Z}^s$; similarly for $n \in \mathbb{Z}^t$.

(ii) For $m \in \mathbb{Z}^s$ and $n \in \mathbb{Z}^t$, denote $(m, n) := (m_1, \ldots, m_s, n_1, \ldots, n_t) \in \mathbb{Z}^{s+t}$.

(iii) For any ideal I of R, $I^e = R$ if $e \leq 0$.

(iv) For $m \in \mathbb{Z}^s$ and $n \in \mathbb{Z}^t$, denote $I^m := I_1^{m_1} \cdots I_s^{m_s}$ and $J^n := J_1^{n_1} \cdots J_t^{n_t}$.

(v) For $m \in \mathbb{Z}^s$ and $n \in \mathbb{Z}^t$, denote $X^m := X_1^{m_1} \cdots X_s^{m_s}$ and $Y^n := Y_1^{n_1} \cdots Y_t^{n_t}$.

(vi) Denote $\mathbb{N} = \{i \mid i \in \mathbb{Z}, i \geq 0\} = \{0, 1, 2, \ldots\}$.

(vii) For all $m \in \mathbb{N}^s$ and $n \in \mathbb{N}^t$ (so that $(m, n) \in \mathbb{N}^{s+t}$), denote $|m| = \sum_{i=1}^s m_i$, $|n| = \sum_{j=1}^t n_j$ and $|(m, n)| = |m| + |n|$.

(viii) By $0 \in \mathbb{Z}^s$, we mean $0 := (0, \ldots, 0) \in \mathbb{Z}^s$; similarly for $0 \in \mathbb{Z}^t$.

(ix) Denote $e_i := (0, \ldots, 0, 1, 0, \ldots, 0) \in \mathbb{Z}^s$, with the i-th component 1.

(x) Denote $f_j := (0, \ldots, 0, 1, 0, \ldots, 0) \in \mathbb{Z}^t$, with the j-th component 1.

We list some results on the linear growth property, including [14], as follows:

Theorem 5.4. *Let A be a Noetherian ring, M a finitely generated A-module, R an A-algebra, N a Noetherian R-module, and $J_1, \ldots, J_t$ ideals of R. Then each of the following families of R-modules has the linear growth property for primary decomposition (over R):*

(i) *The family* $\{N/J^n N \mid n \in \mathbb{N}^t\}$; *see* [14, 12, 16, 17].

(ii) *The family* $\{R/\overline{J^n} \mid n \in \mathbb{N}^t\}$ *if R is Noetherian; see* [11].

(iii) *The family* $\{\operatorname{Tor}_c^A(M, N/J^n N) \mid n \in \mathbb{N}^t\}$; *see* [17].

(iv) *The family* $\{\operatorname{Ext}_A^c(M, N/J^n N) \mid n \in \mathbb{N}^t\}$; *see* [17].

Note that, in Theorem 5.4, N is a finitely generated module over $R/\operatorname{Ann}_R(N)$, which is a Noetherian A-algebra. Also note that each of (iii) and (iv) recovers (i) as a special case. In fact, both (iii) and (iv) are direct consequences of the following:

Theorem 5.5 ([17, Theorem 3.2]). *Let A be a ring and R an A-algebra. Let N be any Noetherian R-module, $J_1, \ldots, J_t$ fixed ideals of R, and $c \in \mathbb{Z}$. Fix any complex*

$$F_\bullet : \quad \cdots \longrightarrow F_{c+1} \longrightarrow F_c \longrightarrow F_{c-1} \longrightarrow \cdots$$

of finitely generated flat A-modules. For any $n \in \mathbb{N}^t$, denote

$$E_n = \mathrm{H}^c\left(\operatorname{Hom}_A(F_\bullet, \tfrac{N}{J^n N})\right) \quad \text{and} \quad T_n = \mathrm{H}_c\left(F_\bullet \otimes_A \tfrac{N}{J^n N}\right),$$

the c-th cohomology and homology of the respective complexes. Then the family $\{E_n \mid n \in \mathbb{N}^t\}$ and the family $\{T_n \mid n \in \mathbb{N}^t\}$, both consisting of finitely generated R-modules, satisfy the linear growth property of primary decomposition over R.

Proof. This was essentially proved in [17, Theorem 3.2]: By replacing R with $R/\operatorname{Ann}_R(N)$, we may assume R is Noetherian. Then, for each i, $F_i \otimes_A R$ is flat and finitely presented over R. Hence $F_\bullet \otimes_A R$ is a complex of finitely generated projective modules over R. By Hom-$\otimes$ adjointness and associativity of tensor,

$$E_n \cong \mathrm{H}^c\left(\operatorname{Hom}_R(F_\bullet \otimes_A R, \tfrac{N}{J^n N})\right) \quad \text{and} \quad T_n \cong \mathrm{H}_c\left((F_\bullet \otimes_A R) \otimes_R \tfrac{N}{J^n N}\right).$$

Now [17, Theorem 3.2] applies, which completes the proof. □

Theorem 5.5 will be used to prove the linear growth property of primary decompositions for $\{\operatorname{Tor}_c^R(\frac{M}{I^m M}, \frac{N}{J^n N}) \mid (m,n) \in \mathbb{N}^s \times \mathbb{N}^t\}$ in the next Section 6.

We end this section with an easy fact concerning the linear growth property of primary decomposition.

Lemma 5.6. *Let $h\colon A \to R$ be a ring homomorphism, $\{M_n \mid n \in \mathbb{N}^t\}$ a family of R-modules, $\{K_n \mid n \in \mathbb{N}^t\}$ a family of A-modules such that $K_n \subseteq M_n$ as A-modules for all $n \in \mathbb{N}^t$, and U a multiplicative subset of R.*

If $\{M_n \mid n \in \mathbb{N}^t\}$ satisfies the linear growth property of primary decomposition over R with a slope k, then $\{K_n \mid n \in \mathbb{N}^t\}$ and $\{M_n[U^{-1}] \mid n \in \mathbb{N}^t\}$ satisfy the linear growth property of primary decomposition over A and $R[U^{-1}]$ respectively with the same slope k.

Proof. This follows (almost immediately) from Theorem 2.1 (iii) and (iv). □

6 Linear Growth of $\{\operatorname{Tor}_c^R(\frac{M}{I^m M}, \frac{N}{J^n N})\}$

Assume that R is a Noetherian ring, $I_1, \ldots, I_s, J_1, \ldots, J_t$ are ideals of R, M and N are finitely generated R-modules, and $c \in \mathbb{Z}$. For all $m \in \mathbb{N}^s$ and all $n \in \mathbb{N}^t$, denote (cf. Notation 5.3)

$$T_{(m,n)} := \operatorname{Tor}_c^R\left(\frac{M}{I^m M}, \frac{N}{J^n N}\right) \quad \text{and} \quad E_{(m,n)} := \operatorname{Ext}_R^c\left(\frac{M}{I^m M}, \frac{N}{J^n N}\right).$$

The families $\{T_{(m,n)} \mid (m,n) \in \mathbb{N}^s \times \mathbb{N}^t\}$ and $\{E_{(m,n)} \mid (m,n) \in \mathbb{N}^s \times \mathbb{N}^t\}$ consist of finitely generated R-modules indexed by $\mathbb{N}^s \times \mathbb{N}^t = \mathbb{N}^{s+t}$.

In [17], the author asked whether the family $\{T_{(m,n)}\}$ or $\{E_{(m,n)}\}$ could satisfy the linear growth property of primary decomposition. Although this is still open for $\{E_{(m,n)} \mid (m,n) \in \mathbb{N}^s \times \mathbb{N}^t\}$ (see Question 6.6), we are going to establish this for $\{T_{(m,n)} \mid (m,n) \in \mathbb{N}^s \times \mathbb{N}^t\}$ in this section. In fact, it is a corollary of the following theorem.

Theorem 6.1. *Let R be a ring, A and B flat R-algebras such that A, B and $A \otimes_R B$ are all Noetherian rings. Let A' and B' be homomorphic images (i.e., quotient rings)* *of A and B respectively, M a finitely generated A'-module, $I_1, \ldots, I_s$ ideals of A', N a finitely generated B'-module, and $J_1, \ldots, J_t$ ideals of B'. Fix any $c \in \mathbb{Z}$.*

Then $\{\operatorname{Tor}_c^R(\frac{M}{I^m M}, \frac{N}{J^n N}) \mid (m,n) \in \mathbb{N}^s \times \mathbb{N}^t = \mathbb{N}^{s+t}\}$ satisfies the linear growth property of primary decomposition over (the Noetherian ring) $A' \otimes_R B'$.

Proof. It suffices to prove the linear growth property over $A \otimes_R B$, which maps onto $A' \otimes_R B'$. Thus, without loss of generality, we may assume $A = A'$ and $B = B'$.

There exists $g \in \mathbb{N}$, large enough, such that

$$I_i = (x_{i1}, \ldots, x_{ig})A \qquad \text{and} \qquad J_j = (y_{j1}, \ldots, y_{jg})B$$

in which $x_{ik} \in A$ and $y_{jk} \in B$ for all $i \in \{1, \ldots, s\}$, all $j \in \{1, \ldots, t\}$ and all $k \in \{1, \ldots, g\}$. (We pick a uniform g only to make the notation simpler.)

Define the following ($\mathbb{Z}^s$-graded) rings and module (cf. Notation 5.3):

$$\mathscr{A} := A[X_{ik}, X_i \mid 1 \le k \le g,\ 1 \le i \le s],$$
$$\mathcal{A} := \bigoplus_{m \in \mathbb{Z}^s} I^m X^{-m} = A[I_i X_i^{-1}, X_i \mid 1 \le i \le s] \subseteq A[X_i^{-1}, X_i \mid 1 \le i \le s],$$
$$\mathcal{M} := \bigoplus_{m \in \mathbb{Z}^s} I^m M X^{-m},$$

in which X_{ik} and X_i are (independent) variables. Both $\mathscr{A}$ and $\mathcal{A}$ are naturally rings via the polynomial operations, and $\mathcal{M}$ is naturally an $\mathcal{A}$-module, which is finitely generated. Moreover, we make all of them $\mathbb{Z}^s$-graded by assigning degrees as follows (cf. Notation 5.3):

$$\deg(A) = \deg(M) = 0 := (0, \ldots, 0) \in \mathbb{Z}^s,$$
$$\deg(X_{ik}) = \deg(X_i^{-1}) = e_i := (0, \ldots, 0, 1, 0, \ldots, 0) \in \mathbb{Z}^s,$$
$$\deg(X_i) = -e_i = (0, \ldots, 0, -1, 0, \ldots, 0) \in \mathbb{Z}^s.$$

Under the gradings, $\mathcal{M}$ is a graded $\mathcal{A}$-module. There is a surjective homogeneous A-algebra homomorphism $\phi\colon \mathscr{A} \twoheadrightarrow \mathcal{A}$ determined by

$$X_{ik} \longmapsto x_{ik} X_i^{-1}, \quad X_i \longmapsto X_i.$$

This makes $\mathcal{M}$ a finitely generated graded module over $\mathscr{A}$. (Clearly, both $\mathscr{A}$ and $\mathcal{A}$ are finitely generated A-algebras and hence Noetherian.)

Similarly, we define the following $\mathbb{Z}^t$-graded rings and module (cf. Notation 5.3):

$$\mathscr{B} := B[Y_{jk}, Y_j \mid 1 \le k \le g,\ 1 \le j \le t],$$
$$\mathcal{B} := \bigoplus_{n \in \mathbb{Z}^t} J^n Y^{-n} = B[J_j Y_j^{-1}, Y_j \mid 1 \le j \le t] \subseteq B[Y_j^{-1}, Y_j \mid 1 \le j \le t],$$
$$\mathcal{N} := \bigoplus_{n \in \mathbb{Z}^t} J^n N Y^{-n},$$

with Y_{jk} and Y_j variables and with the gradings given by (cf. Notation 5.3)

$$\deg(B) = \deg(N) = 0 := (0,\dots,0) \in \mathbb{Z}^t,$$
$$\deg(Y_{jk}) = \deg(Y_j^{-1}) = f_j := (0,\dots,0,1,0,\dots,0) \in \mathbb{Z}^t,$$
$$\deg(Y_j) = -f_j = (0,\dots,0,-1,0,\dots,0) \in \mathbb{Z}^t.$$

There is a surjective homogeneous B-algebra homomorphism $\psi\colon \mathscr{B} \twoheadrightarrow \mathcal{B}$ given by

$$Y_{jk} \longmapsto y_{jk}Y_j^{-1}, \quad Y_j \longmapsto Y_j.$$

This makes $\mathcal{N}$ a finitely generated graded module over $\mathscr{B}$, since $\mathcal{N}$ is (naturally) a finitely generated graded module over $\mathcal{B}$. (Clearly, both $\mathscr{B}$ and $\mathcal{B}$ are finitely generated B-algebras and hence Noetherian.)

We now consider $\mathscr{C} := \mathscr{A} \otimes_R \mathscr{B}$ and $\mathfrak{C} := \mathcal{A} \otimes_R \mathcal{B}$, which are clearly Noetherian (since they are finitely generated algebras over $A \otimes_R B$). In the sequel, we use $[-]_h$ to denote the h-th homogeneous component of a graded module. (For example, $[\mathscr{A}]_\alpha$ stands for the homogeneous component of $\mathscr{A}$ of degree α with the understanding that $\alpha = (\alpha_1,\dots,\alpha_s) \in \mathbb{Z}^s$, since $\mathscr{A}$ is $\mathbb{Z}^s$-graded.) Keeping this in mind, we observe that both $\mathscr{C}$ and $\mathfrak{C}$ are naturally $\mathbb{Z}^{s+t}$-graded rings with

$$[\mathscr{C}]_{(\alpha,\beta)} = [\mathscr{A} \otimes_R \mathscr{B}]_{(\alpha,\beta)} = [\mathscr{A}]_\alpha \otimes_R [\mathscr{B}]_\beta \quad \text{and}$$
$$[\mathfrak{C}]_{(\alpha,\beta)} = [\mathcal{A} \otimes_R \mathcal{B}]_{(\alpha,\beta)} = [\mathcal{A}]_\alpha \otimes_R [\mathcal{B}]_\beta$$

for all $(\alpha,\beta) \in \mathbb{Z}^s \times \mathbb{Z}^t = \mathbb{Z}^{s+t}$. In particular, for $(0,0) \in \mathbb{Z}^s \times \mathbb{N}^t$, we have

$$[\mathscr{C}]_{(0,0)} = [\mathscr{A} \otimes_R \mathscr{B}]_{(0,0)} = [\mathscr{A}]_0 \otimes_R [\mathscr{B}]_0 = A \otimes_R B \quad \text{and}$$
$$[\mathfrak{C}]_{(0,0)} = [\mathcal{A} \otimes_R \mathcal{B}]_{(0,0)} = [\mathcal{A}]_0 \otimes_R [\mathcal{B}]_0 = A \otimes_R B.$$

Moreover, the surjective homogeneous R-algebra homomorphisms $\phi\colon \mathscr{A} \twoheadrightarrow \mathcal{A}$ and $\psi\colon \mathscr{B} \twoheadrightarrow \mathcal{B}$ induce an surjective homogeneous R-algebra homomorphism

$$\phi \otimes \psi\colon \mathscr{C} \twoheadrightarrow \mathfrak{C}.$$

Write down graded free resolutions of $\mathcal{M}$ over $\mathscr{A}$ and of $\mathcal{N}$ over $\mathscr{B}$ respectively by (free) modules of finite ranks (over $\mathscr{A}$ and over $\mathscr{B}$ respectively)

$$\mathscr{F}_\bullet: \ \cdots \longrightarrow \mathscr{F}_i \longrightarrow \mathscr{F}_{i-1} \longrightarrow \cdots \longrightarrow \mathscr{F}_1 \longrightarrow \mathscr{F}_0 \ (\longrightarrow \mathcal{M}) \longrightarrow 0,$$
$$\mathscr{G}_\bullet: \ \cdots \longrightarrow \mathscr{G}_j \longrightarrow \mathscr{G}_{j-1} \longrightarrow \cdots \longrightarrow \mathscr{G}_1 \longrightarrow \mathscr{G}_0 \ (\longrightarrow \mathcal{N}) \longrightarrow 0.$$

Then $\mathscr{F}_\bullet \otimes_R \mathscr{G}_\bullet$ is (naturally) a $\mathbb{Z}^{s+t}$-graded complex composed of finitely generated free $\mathscr{C}$-modules over the $\mathbb{Z}^{s+t}$-graded ring $\mathscr{A} \otimes_R \mathscr{B} =: \mathscr{C}$.

By abuse of notation, we use X^mY^n to denote $(X^m \otimes 1)(1 \otimes Y^n) = X^m \otimes Y^n \in \mathscr{C}$. By Theorem 5.5, the following family (of $\mathscr{C}$-modules)

$$\left\{ \mathrm{H}_c\left((\mathscr{F}_\bullet \otimes_R \mathscr{G}_\bullet) \otimes_{\mathscr{C}} \frac{\mathscr{C}}{(X^mY^n)} \right) \;\middle|\; (m,n) \in \mathbb{N}^s \times \mathbb{N}^t = \mathbb{N}^{s+t} \right\}$$

has the linear growth property of primary decomposition over $\mathscr{C} = \mathscr{A} \otimes_R \mathscr{B}$.

We are going to show that the above linear growth property implies the linear growth property of $\{\mathrm{Tor}_c^R(\frac{M}{I^m M}, \frac{N}{J^n N}) \mid (m,n) \in \mathbb{N}^s \times \mathbb{N}^t = \mathbb{N}^{s+t}\}$ over $A \otimes_R B$.

Firstly, for all $(m,n) \in \mathbb{N}^s \times \mathbb{N}^t$, the modules $\mathrm{H}_c((\mathscr{F}_\bullet \otimes_R \mathscr{G}_\bullet) \otimes_{\mathscr{C}} \frac{\mathscr{C}}{(X^m Y^n)})$ are annihilated by $\mathrm{Ker}(\phi \otimes \psi)$; so, naturally, they are all graded modules over $\mathcal{C} = \mathcal{A} \otimes_R \mathcal{B}$. (This follows directly from how $\mathcal{M}$, $\mathcal{N}$, $\mathscr{F}_\bullet$ and $\mathscr{G}_\bullet$ are constructed: Multiplication by every element in $\mathrm{Ker}(\phi)$ (resp. $\mathrm{Ker}(\psi)$) is homotopic to 0 on $\mathscr{F}_\bullet$ (resp. $\mathscr{G}_\bullet$); and $\mathrm{Ker}(\phi \otimes \psi)$ is generated by $\mathrm{Ker}(\phi)$ and $\mathrm{Ker}(\psi)$ since both ϕ and ψ are surjective.) Hence $\{\mathrm{H}_c((\mathscr{F}_\bullet \otimes_R \mathscr{G}_\bullet) \otimes_{\mathscr{C}} \frac{\mathscr{C}}{(X^m Y^n)}) \mid (m,n) \in \mathbb{N}^s \times \mathbb{N}^t = \mathbb{N}^{s+t}\}$ has the linear growth property of primary decomposition over $\mathcal{C}$.

Secondly, for every $m \in \mathbb{N}^s$ and $n \in \mathbb{N}^t$, there is a canonical homogeneous isomorphism of $(\mathscr{A} \otimes_R \mathscr{B})$-complexes

$$(\mathscr{F}_\bullet \otimes_R \mathscr{G}_\bullet) \otimes_{\mathscr{C}} \frac{\mathscr{C}}{(X^m Y^n)} \cong \left(\mathscr{F}_\bullet \otimes_{\mathscr{A}} \frac{\mathscr{A}}{(X^m)}\right) \otimes_R \left(\mathscr{G}_\bullet \otimes_{\mathscr{B}} \frac{\mathscr{B}}{(Y^n)}\right).$$

Therefore, for each $(\alpha,\beta) \in \mathbb{Z}^s \times \mathbb{Z}^t$, there is an isomorphism between the following $(\mathscr{A}_0 \otimes_R \mathscr{B}_0)$-complexes

$$\left[(\mathscr{F}_\bullet \otimes_R \mathscr{G}_\bullet) \otimes_{\mathscr{C}} \frac{\mathscr{C}}{(X^m Y^n)}\right]_{(\alpha,\beta)} \cong \left[\left(\mathscr{F}_\bullet \otimes_{\mathscr{A}} \frac{\mathscr{A}}{(X^m)}\right)\right]_\alpha \otimes_R \left[\left(\mathscr{G}_\bullet \otimes_{\mathscr{B}} \frac{\mathscr{B}}{(Y^n)}\right)\right]_\beta.$$

Thirdly, observe that X_i is regular on both $\mathcal{M}$ and $\mathscr{A}$ for every $i \in \{1,\dots,s\}$ while Y_j is regular on both $\mathcal{N}$ and $\mathscr{B}$ for every $j \in \{1,\dots,t\}$. Thus X^m is regular on both $\mathcal{M}$ and $\mathscr{A}$ while Y^n is regular on both $\mathcal{N}$ and $\mathscr{B}$ for every $m \in \mathbb{N}^s$ and $n \in \mathbb{N}^t$. Consequently,

(a) $\mathscr{F}_\bullet \otimes_{\mathscr{A}} \frac{\mathscr{A}}{(X^m)}$ is a graded free resolution of $\frac{\mathcal{M}}{X^m \mathcal{M}}$ over graded ring $\frac{\mathscr{A}}{(X^m)}$;

(b) $\mathscr{G}_\bullet \otimes_{\mathscr{B}} \frac{\mathscr{B}}{(Y^n)}$ is a graded free resolution of $\frac{\mathcal{N}}{Y^n \mathcal{N}}$ over graded ring $\frac{\mathscr{B}}{(Y^n)}$.

Moreover, by the construction of $\mathscr{A}$ and $\mathscr{B}$, all of their homogeneous components are free A-modules and free B-modules respectively; so they are all flat R-modules. It follows that all of the homogeneous components of $\frac{\mathscr{A}}{(X^m)}$ and $\frac{\mathscr{B}}{(Y^n)}$ are free A-modules and free B-modules respectively and hence flat over R, for all $m \in \mathbb{N}^s$ and $n \in \mathbb{N}^t$. In light of this, statements (a) and (b) above imply the following (for all $m \in \mathbb{N}^s$, $n \in \mathbb{N}^t$, $\alpha \in \mathbb{Z}^s$ and $\beta \in \mathbb{Z}^t$):

($[\mathrm{a}]_\alpha$) $[\mathscr{F}_\bullet \otimes_{\mathscr{A}} \frac{\mathscr{A}}{(X^m)}]_\alpha$ is a flat resolution of $[\frac{\mathcal{M}}{X^m \mathcal{M}}]_\alpha = \frac{I^\alpha M}{I^{\alpha+m} M}$ over R;

($[\mathrm{b}]_\beta$) $[\mathscr{G}_\bullet \otimes_{\mathscr{B}} \frac{\mathscr{B}}{(Y^n)}]_\beta$ is a flat resolution of $[\frac{\mathcal{N}}{Y^n \mathcal{N}}]_\beta = \frac{J^\beta N}{J^{\beta+n} N}$ over R.

In particular, for $\alpha = 0 \in \mathbb{Z}^s$ and $\beta = 0 \in \mathbb{Z}^t$, we have (for all $m \in \mathbb{N}^s$ and $n \in \mathbb{N}^t$)

($[\mathrm{a}]_0$) $[\mathscr{F}_\bullet \otimes_{\mathscr{A}} \frac{\mathscr{A}}{(X^m)}]_0$ is a flat resolution of $[\frac{\mathcal{M}}{X^m \mathcal{M}}]_0 = \frac{M}{I^m M}$ over R;

($[\mathrm{b}]_0$) $[\mathscr{G}_\bullet \otimes_{\mathscr{B}} \frac{\mathscr{B}}{(Y^n)}]_0$ is a flat resolution of $[\frac{\mathcal{N}}{Y^n \mathcal{N}}]_0 = \frac{N}{J^n N}$ over R.

Now we study $\mathrm{H}_c((\mathscr{F}_\bullet \otimes_R \mathscr{G}_\bullet) \otimes_{\mathscr{C}} \frac{\mathscr{C}}{(X^m Y^n)})$, which is a $\mathbb{Z}^{s+t}$-graded module, in terms of its homogeneous components. Recall that $[\mathscr{C}]_{(0,0)} = A \otimes_R B = [\mathfrak{C}]_{(0,0)}$. Combining the three paragraphs above, we obtain the following isomorphisms over $A \otimes_R B$:

$$\begin{aligned}
&\mathrm{H}_c\left((\mathscr{F}_\bullet \otimes_R \mathscr{G}_\bullet) \otimes_{\mathscr{C}} \frac{\mathscr{C}}{(X^m Y^n)}\right) \\
&\quad = \bigoplus_{(\alpha,\beta)\in\mathbb{Z}^s\times\mathbb{Z}^t} \left[\mathrm{H}_c\left((\mathscr{F}_\bullet \otimes_R \mathscr{G}_\bullet) \otimes_{\mathscr{C}} \frac{\mathscr{C}}{(X^m Y^n)}\right)\right]_{(\alpha,\beta)} \\
&\quad = \bigoplus_{(\alpha,\beta)\in\mathbb{Z}^s\times\mathbb{Z}^t} \mathrm{H}_c\left(\left[(\mathscr{F}_\bullet \otimes_R \mathscr{G}_\bullet) \otimes_{\mathscr{C}} \frac{\mathscr{C}}{(X^m Y^n)}\right]_{(\alpha,\beta)}\right) \\
&\quad \cong \bigoplus_{(\alpha,\beta)\in\mathbb{Z}^s\times\mathbb{Z}^t} \mathrm{H}_c\left(\left[\mathscr{F}_\bullet \otimes_{\mathscr{A}} \frac{\mathscr{A}}{(X^m)}\right]_\alpha \otimes_R \left[\mathscr{G}_\bullet \otimes_{\mathscr{B}} \frac{\mathscr{B}}{(Y^n)}\right]_\beta\right) \\
&\quad = \bigoplus_{(\alpha,\beta)\in\mathbb{Z}^s\times\mathbb{Z}^t} \mathrm{Tor}_c^R\left(\frac{I^\alpha M}{I^{\alpha+m} M}, \frac{J^\beta N}{J^{\beta+n} N}\right)
\end{aligned}$$

for all $(m,n) \in \mathbb{N}^s \times \mathbb{N}^t$. In particular, for all $(m,n) \in \mathbb{N}^s \times \mathbb{N}^t$,

$$\begin{aligned}
\mathrm{Tor}_c^R\left(\frac{M}{I^m M}, \frac{N}{J^n N}\right) &\cong \left[\mathrm{H}_c\left((\mathscr{F}_\bullet \otimes_R \mathscr{G}_\bullet) \otimes_{\mathscr{C}} \frac{\mathscr{C}}{(X^m Y^n)}\right)\right]_{(0,0)} \\
&\subseteq \mathrm{H}_c\left((\mathscr{F}_\bullet \otimes_R \mathscr{G}_\bullet) \otimes_{\mathscr{C}} \frac{\mathscr{C}}{(X^m Y^n)}\right)
\end{aligned}$$

as $(A \otimes_R B)$-modules.

In summary, the family $\{\mathrm{H}_c((\mathscr{F}_\bullet \otimes_R \mathscr{G}_\bullet) \otimes_{\mathscr{C}} \frac{\mathscr{C}}{(X^m Y^n)}) \mid (m,n) \in \mathbb{N}^s \times \mathbb{N}^t\}$ satisfies the linear growth property of primary decomposition over the graded ring $\mathfrak{C}$ with $[\mathfrak{C}]_{(0,0)]} = A \otimes_R B$; and for each $(m,n) \in \mathbb{N}^s \times \mathbb{N}^t$, $\mathrm{Tor}_c^R(\frac{M}{I^m M}, \frac{N}{J^n N})$ is an $(A \otimes_R B)$-submodule of $\mathrm{H}_c((\mathscr{F}_\bullet \otimes_R \mathscr{G}_\bullet) \otimes_{\mathscr{C}} \frac{\mathscr{C}}{(X^m Y^n)})$ up to isomorphism.

Finally, by Lemma 5.6, the family

$$\left\{\mathrm{Tor}_c^R\left(\frac{M}{I^m M}, \frac{N}{J^n N}\right) \;\middle|\; (m,n) \in \mathbb{N}^s \times \mathbb{N}^t = \mathbb{N}^{s+t}\right\}$$

satisfies the linear growth of primary decomposition over $A \otimes_R B$. □

In fact, the proof of Theorem 6.1 implies the following (apparently) stronger result concerning infinitely many families and a *uniform* slope (cf. Definition 5.2).

Theorem 6.2. *Keep the notation and the assumptions in Theorem* 6.1.
Then there exists k such that for all $(\alpha,\beta)\in\mathbb{Z}^s\times\mathbb{Z}^t$, *the family*

$$\mathcal{T}^{(\alpha,\beta)} := \left\{\operatorname{Tor}_c^R\left(\frac{I^\alpha M}{I^{\alpha+m}M},\frac{J^\beta N}{J^{\beta+n}N}\right) \,\middle|\, (m,n)\in\mathbb{N}^s\times\mathbb{N}^t=\mathbb{N}^{s+t}\right\}$$

satisfies the linear growth property of primary decomposition over $A'\otimes_R B'$ *with the uniform slope k. More explicitly, for every* $(\alpha,\beta)\in\mathbb{Z}^{s+t}$ *and* $(m,n)\in\mathbb{N}^{s+t}$ *such that* $\operatorname{Tor}_c^R\left(\frac{I^\alpha M}{I^{\alpha+m}M},\frac{J^\beta N}{J^{\beta+n}N}\right)\neq 0$, *there exists a primary decomposition of* 0 *in* $\operatorname{Tor}_c^R\left(\frac{I^\alpha M}{I^{\alpha+m}M},\frac{J^\beta N}{J^{\beta+n}N}\right)$ *over* $A'\otimes_R B'$,

$$0=Q_{\alpha,\beta,m,n,1}\cap Q_{\alpha,\beta,m,n,2}\cap\cdots\cap Q_{\alpha,\beta,m,n,s(\alpha,\beta,m,n)},$$

with $Q_{\alpha,\beta,m,n,i}$ *being* $P_{\alpha,\beta,m,n,i}$*-primary in* $\operatorname{Tor}_c^R\left(\frac{I^\alpha M}{I^{\alpha+m}M},\frac{J^\beta N}{J^{\beta+n}N}\right)$, *such that*

$$(P_{\alpha,\beta,m,n,i})^{k|(m,n)|}\operatorname{Tor}_c^R\left(\frac{I^\alpha M}{I^{\alpha+m}M},\frac{J^\beta N}{J^{\beta+n}N}\right)\subseteq Q_{\alpha,\beta,m,n,i}$$

for all $i=1,2,\ldots,s(\alpha,\beta,m,n)$.

Proof. As seen in the proof of Theorem 6.1, for all $(m,n)\in\mathbb{N}^s\times\mathbb{N}^t$ and for all $(\alpha,\beta)\in\mathbb{Z}^s\times\mathbb{Z}^t$, we have

$$\begin{aligned}\operatorname{Tor}_c^R\left(\frac{I^\alpha M}{I^{\alpha+m}M},\frac{J^\beta N}{J^{\beta+n}N}\right) &\cong \left[\mathrm{H}_c\left((\mathscr{F}_\bullet\otimes_R\mathscr{G}_\bullet)\otimes_{\mathscr{C}}\frac{\mathscr{C}}{(X^mY^n)}\right)\right]_{(\alpha,\beta)}\\ &\subseteq \mathrm{H}_c\left((\mathscr{F}_\bullet\otimes_R\mathscr{G}_\bullet)\otimes_{\mathscr{C}}\frac{\mathscr{C}}{(X^mY^n)}\right)\end{aligned}$$

as $(A\otimes_R B)$-modules.

Say k is a slope for $\{\mathrm{H}_c((\mathscr{F}_\bullet\otimes_R\mathscr{G}_\bullet)\otimes_{\mathscr{C}}\frac{\mathscr{C}}{(X^mY^n)})\mid (m,n)\in\mathbb{N}^s\times\mathbb{N}^t=\mathbb{N}^{s+t}\}$ over $\mathfrak{C}$. By Lemma 5.6, all the families $\mathcal{T}^{(\alpha,\beta)}$, $(\alpha,\beta)\in\mathbb{Z}^s\times\mathbb{Z}^t$, satisfy the linear growth property of primary decomposition over $A\otimes_R B$ with the same slope k. □

Remark 6.3. Recall that an R-algebra S is said to be *essentially of finite type over R* if $S\cong T[U^{-1}]$ with T a finitely generated R-algebra and U a multiplicative subset of T. We remark that Theorems 6.1 and 6.2 apply when A' and B' are essentially of finite type over R. This is because one can then let A and B be of the form $T[U^{-1}]$ with T being a polynomial ring over R (hence flat over R) with finitely many variables.

Remark 6.4. Note that Theorems 6.1 and 6.2 include the case of $s=0$ or $t=0$. For example, when $s=0$, Theorem 6.1 states that the family $\{\operatorname{Tor}_c^R(M,\frac{N}{J^nN})\mid n\in\mathbb{N}^t\}$ satisfies the linear growth property of primary decomposition over $A'\otimes_R B'$, which is slightly different from Theorem 5.4 (iii); and Theorem 6.2 says that, for all $\beta\in\mathbb{Z}^t$, the families $\{\mathrm{Tor}_c^R(M, \frac{J^\beta N}{J^{\beta+n}N}) \mid n \in \mathbb{N}^t\}$ satisfy the linear growth property of primary decomposition over $A' \otimes_R B'$ with a uniform slope.

In fact, if $s = 0$, we can relax the condition on A and M by assuming that A is any R-algebra such that $A \otimes_R B$ is Noetherian and M is any finitely generated A-module, while the other assumptions remain the same. The proof is similar, but we construct $\mathscr{G}_\bullet$ only. By Theorem 5.5, the family $\{\mathrm{H}_c((A \otimes_R \mathscr{G}_\bullet) \otimes_{A\otimes_R \mathscr{B}} \frac{M\otimes_R \mathscr{B}}{Y^n(M\otimes_R \mathscr{B})}) \mid n \in \mathbb{N}^t\}$ has linear growth property of primary decomposition over $A \otimes_R \mathscr{B}$. The rest follows in a similar way, by considering the homogeneous components graded by $\mathbb{Z}^t$. It might be helpful to note the natural homogeneous $(A \otimes_R \mathscr{B})$-isomorphisms

$$(A \otimes_R \mathscr{G}_\bullet) \otimes_{A\otimes_R \mathscr{B}} \frac{M \otimes_R \mathscr{B}}{Y^n(M \otimes_R \mathscr{B})} \cong M \otimes_R \left(\mathscr{G}_\bullet \otimes_{\mathscr{B}} \frac{\mathscr{B}}{(Y^n)}\right) \quad \text{and}$$

$$\left[\mathrm{H}_c\left(M \otimes_R \left(\mathscr{G}_\bullet \otimes_{\mathscr{B}} \frac{\mathscr{B}}{(Y^n)}\right)\right)\right]_\beta = \mathrm{H}_c\left(M \otimes_R \left[\mathscr{G}_\bullet \otimes_{\mathscr{B}} \frac{\mathscr{B}}{(Y^n)}\right]_\beta\right)$$
$$\cong \mathrm{Tor}_c^R\left(M, \frac{J^\beta N}{J^{\beta+n}N}\right)$$

over $A \otimes_R B = [A \otimes_R \mathscr{B}]_0$.

As promised, we state the following corollary (when $A = R = B$).

Corollary 6.5. *Let R be a Noetherian ring, M and N finitely generated R-modules, $I_1, \ldots, I_s, J_1, \ldots, J_t$ ideals of R, and $c \in \mathbb{Z}$.*

Then $\{\mathrm{Tor}_c^R(\frac{M}{I^m M}, \frac{N}{J^n N}) \mid (m,n) \in \mathbb{N}^s \times \mathbb{N}^t = \mathbb{N}^{s+t}\}$ satisfies the linear growth property of primary decomposition over R.

More generally, the families $\{\mathrm{Tor}_c^R(\frac{I^\alpha M}{I^{\alpha+m}M}, \frac{J^\beta N}{J^{\beta+n}N}) \mid (m,n) \in \mathbb{N}^s \times \mathbb{N}^t\}$, for all $(\alpha, \beta) \in \mathbb{Z}^s \times \mathbb{Z}^t$, satisfy the linear growth property of primary decomposition over R with a uniform slope.

Question 6.6. Keep the notation and the assumptions in Corollary 6.5. Does the family $\{\mathrm{Ext}_R^c(\frac{M}{I^m M}, \frac{N}{J^n N}) \mid (m,n) \in \mathbb{N}^s \times \mathbb{N}^t\}$ satisfy the linear growth property?

When $c \leq 0$, the linear growth property of primary decomposition can be easily established for $\{\mathrm{Ext}_R^c(\frac{M}{I^m M}, \frac{N}{J^n N}) \mid (m,n) \in \mathbb{N}^s \times \mathbb{N}^t\}$. For general c, the question is open even for the family $\{\mathrm{Ext}_R^c(\frac{M}{I^m M}, N) \mid m \in \mathbb{N}^s\}$.

7 Secondary Representation

Secondary representations were first studied by I. G. Macdonald [6] and D. Kirby [5]. The theory can be viewed as a dual of the theory of primary decomposition. (See Theorem 9.2 and Observation 9.4 for example, where this duality is demonstrated explicitly.) For this reason, it was called *coprimary decomposition* in [5]. Systematic treatment of secondary representation can be found in many sources, for example, see [5], [6], [7], and [8] as well as many other papers authored or co-authored by R. Y. Sharp.

Assume that R is a ring (not necessarily Noetherian) and M is an R-module. In this section, we briefly review some of the basic definitions and properties.

We say that M is *secondary* if, for all $r \in R$, either $rM = M$ or $r \in \sqrt{\operatorname{Ann}(M)}$. (Note that, under this definition, 0 is a secondary module.) If $M \neq 0$ is secondary, then $P := \sqrt{\operatorname{Ann}(M)}$ is a prime ideal; and we say M is P-secondary in this case.

It is easy to see that if M is P-secondary then, for any multiplicatively closed subset U of R and any finitely generated ideal I of R, we have

$$[U]M = \begin{cases} M & \text{if } P \cap U = \varnothing \\ 0 & \text{if } P \cap U \neq \varnothing \end{cases} \quad \text{and} \quad \bigcap_{i \in \mathbb{N}} (I^i M) = \begin{cases} M & \text{if } I \not\subseteq P \\ 0 & \text{if } I \subseteq P \end{cases}$$

in which $[U]M := \bigcap_{u \in U} (uM)$.

For a general R-module M and a prime ideal P, we say that P is *attached* to M if there is an R-submodule N of M such that M/N is P-secondary, that is, a homomorphic image of M is P-secondary. The set of all the primes attached to M is denoted $\operatorname{Att}_R(M)$, or simply $\operatorname{Att}(M)$ if R is understood. (Note that Att and Ass$''$ are dual to each other.)

As $\operatorname{Att}(M) \subseteq \operatorname{Spec}(R)$, there is a topology on $\operatorname{Att}(M)$ that is induced by the Zariski topology on $\operatorname{Spec}(R)$.

If M is P-secondary, then $\operatorname{Att}(M) = \{P\}$. If M is an Artinian R-module, then M is P-secondary $\iff \operatorname{Att}(M) = \{P\}$, and $M = 0 \iff \operatorname{Att}(M) = \varnothing$.

Example 7.1. Let $(R, \mathfrak{m})$ be any Noetherian local domain, not necessarily complete. Then $\operatorname{E}_R(R/\mathfrak{m})$, the injective hull of the residue field $R/\mathfrak{m}$, is 0-secondary; so that $\operatorname{Att}_R(\operatorname{E}_R(R/\mathfrak{m})) = \{0\}$. Note that $\operatorname{E}_R(R/\mathfrak{m})$ is Artinian, and the zero ideal 0 is not the maximal ideal $\mathfrak{m}$ if $\dim(R) > 0$. (However, $\operatorname{Ass}_R(\operatorname{E}_R(R/\mathfrak{m})) = \{\mathfrak{m}\}$.)

We also note that $R/\mathfrak{m}$ is both $\mathfrak{m}$-secondary and $\mathfrak{m}$-coprimary as an R-module with $\operatorname{Att}_R(R/\mathfrak{m}) = \{\mathfrak{m}\} = \operatorname{Ass}_R(R/\mathfrak{m})$.

For an R-module $M \neq 0$, we say M is *representable* (over R) if there exist submodules Q_i that are P_i-secondary, for $i = 1, \dots, s$, such that

$$M = Q_1 + \cdots + Q_s.$$

This summation is called a *secondary representation* of M. One can always convert a secondary representation to a *minimal* one in the sense that $P_i \neq P_j$ for all $i \neq j$ and $M \neq \sum_{i \neq k} Q_i$ for every $k = 1, \dots, s$. So from now on and as a general rule, all secondary representations are assumed to be minimal unless stated otherwise explicitly.

By convention, the zero R-module 0 is representable with $0 = 0$ being the unique secondary representation.

For concrete examples of secondary representation, see Examples 10.4 and 10.5.
Here is a theorem on the existence of secondary representation, cf. [6].

Theorem 7.2. *Every Artinian R-module is representable (over R).*

For any R-module M and any ideal $I \subseteq \mathrm{Ann}(M)$, the following is clear: M is representable over R if and only if M is representable over R/I.

Next, we state some useful results about secondary representations; compare with Theorem 2.1. We do not need to assume M is Artinian in Theorem 7.3, as long as M is representable. In case $U = R \setminus P$ with $P \in \mathrm{Spec}(R)$, we write $M^P := [U]M$.

Theorem 7.3 (Cf. [6]; compare with Theorem 2.1). *Let $M = Q_1 + \cdots + Q_s$ be a (minimal) secondary representation of an R-module M in which Q_i is P_i-secondary for each $i = 1, \ldots, s$. Then the following hold*

(i) $\{P_1, \ldots, P_s\} = \mathrm{Att}(M)$, *which is independent of the particular (minimal) secondary representation (cf.* [6, *Theorem* 2.2]*).*

i′ *We have* $\mathrm{Min}(M) \subseteq \{P_1, \ldots, P_s\}$*. In fact,* $\mathrm{Min}(M)$ *consists of the minimal members of* $\{P_1, \ldots, P_s\}$ *(under inclusion) precisely.*

(ii) *If P_i is minimal in* $\mathrm{Att}(M)$ *(i.e., $P_i \in \mathrm{Min}(M)$), then $Q_i = M^{P_i}$. See (iv).*

(iii) *Let $h\colon A \to R$ be a ring homomorphism, so that M is naturally an A-module. Let K be an A-submodule of M (e.g., $K = 0$). Then M/K is representable over A. In fact, if $M/K \neq 0$, then*

$$M/K = \sum_{Q_i \not\subseteq K} (Q_i + K)/K$$

is a (not necessarily minimal) secondary representation of M/K over A, in which $(Q_i + K)/K$ is $h^{-1}(P_i)$-secondary provided that $Q_i \not\subseteq K$.

iii′ *In particular,* $\mathrm{Att}_A(M) = h^*(\mathrm{Att}_R(M))$*, in which* $h^*\colon \mathrm{Spec}(R) \to \mathrm{Spec}(A)$ *is the continuous map naturally induced by h.*

(iv) *For any multiplicative subset U of R, $[U]M = \sum_{P_i \cap U = \varnothing} Q_i$ is a secondary representation over R (cf.* [6, *Theorem* 3.1]*).*

(v) *For any finitely generated ideal I of R, $\bigcap_{j \in \mathbb{N}}(I^j M) = \sum_{I \not\subseteq P_i} Q_i = I^n M$ for all $n \gg 0$ (cf.* [6, *Theorem* 3.3]*).*

v′ *For any non-empty subset I of R, $\bigcup_{r \in I}(\bigcap_{n \in \mathbb{N}}(r^n M)) = \sum_{I \not\subseteq P_i} Q_i$.*

Very much like $\mathrm{Ass}(-)$ (as well as Ass', Ass'' and $\mathrm{Ass}_{\mathrm{f}}$), the sets of attached primes are relatively well-behaved with exact sequences, as stated in the following well-known lemma. This will be referred to in the proof of Lemma 8.2.

Lemma 7.4 (Compare with Lemma 2.4). *Let $M_1 \to M_2 \to M_3 \to 0$ be an exact sequence of R-modules. Then* $\mathrm{Att}(M_3) \subseteq \mathrm{Att}(M_2) \subseteq \mathrm{Att}(M_1) \cup \mathrm{Att}(M_3)$.

Moreover, $\mathrm{Att}(\bigoplus_{i=1}^n K_i) = \bigcup_{i=1}^n \mathrm{Att}(K_i)$ *for R-modules $K_1, \ldots, K_n$ with $n \in \mathbb{N}$.*

8 Compatibility of Secondary Components

Throughout this section, we assume that R is a (not necessarily Noetherian) ring and M a (not necessarily Artinian) R-module. The reader should observe the similarity (or rather, "duality") between this section and Section 3.

The results in this section were obtained in [18].

Notation 8.1. Let M be a representable R-module and $X \subseteq \operatorname{Att}(M)$. Say $X = \{P_1, \ldots, P_r\} \subseteq \{P_1, \ldots, P_r, \ldots, P_s\} = \operatorname{Att}(M)$.

(i) If $M = Q_1 + \cdots + Q_r + \cdots + Q_s$ is a secondary representation of M with Q_i being P_i-secondary, then we say $Q = Q_1 + \cdots + Q_r$ is an X-secondary component (or a secondary component over X) of M. If $X = \varnothing$, then we agree that 0 is the only $\varnothing$-secondary component.

(ii) We call an X-secondary component of M minimal if it does not properly contain any X-secondary component of M.

(iii) Denote by $\Lambda_X^*(M)$ the set of all possible X-secondary components of M.

(iv) We use $\underset{\circ}{\Lambda}{}_X^*(M)$ to denote the set of all minimal X-secondary components of M. (Note that $\underset{\circ}{\Lambda}{}_X^*(M) \neq \varnothing$ if M is Artinian.)

(v) In case $X = \{P\} \subseteq \operatorname{Att}(M)$, we may simply write Λ_P^* and $\underset{\circ}{\Lambda}{}_P^*$ instead of $\Lambda_{\{P\}}^*$ and $\underset{\circ}{\Lambda}{}_{\{P\}}^*$ respectively.

Lemma 8.2 (Compare with Lemma 3.2). *Let M be a representable R-module and $X \subseteq \operatorname{Att}(M)$. For an R-module Q such that $Q \subseteq M$, the following are equivalent:*

(i) *Q is an X-secondary component of M, i.e., $Q \in \Lambda_X^*(M)$.*

(ii) *Q is representable, $\operatorname{Att}(Q) \subseteq X$ and $\operatorname{Att}(M/Q) \subseteq \operatorname{Att}(M) \setminus X$.*

(iii) *Q is representable, $\operatorname{Att}(Q) = X$ and $\operatorname{Att}(M/Q) = \operatorname{Att}(M) \setminus X$.*

Proof. Say $X = \{P_1, \ldots, P_r\} \subseteq \{P_1, \ldots, P_r, P_{r+1}, \ldots, P_s\} = \operatorname{Att}(M)$.

(i) $\Rightarrow$ (ii): Condition (i) means that there is a secondary representation $M = Q_1 + \cdots + Q_r + \cdots + Q_s$ with Q_i being P_i-secondary such that $Q = Q_1 + \cdots + Q_r$. Then evidently $\operatorname{Att}(Q) \subseteq X$ (since they are equal, see Theorem 7.3 (i)). Also, we have an R-linear isomorphism

$$\frac{M}{Q} = \frac{Q + \sum_{i=r+1}^{s} Q_i}{Q} \cong \frac{\sum_{i=r+1}^{s} Q_i}{Q \cap \sum_{i=r+1}^{s} Q_i},$$

which implies that $\operatorname{Att}(M/Q) \subseteq \operatorname{Att}(\sum_{i=r+1}^{s} Q_i) = \{P_{r+1}, \ldots, P_s\} = \operatorname{Att}(M) \setminus X$.

(ii) $\Rightarrow$ (iii): This is evident, since $\operatorname{Att}(M) \subseteq \operatorname{Att}(M/Q) \cup \operatorname{Att}(Q)$ by Lemma 7.4.

(iii) $\Rightarrow$ (i): As Q is representable and $\operatorname{Att}(Q) = \{P_1, \ldots, P_r\}$, we fix a secondary representation $Q = Q_1 + \cdots + Q_r$ in which Q_i is the P_i-secondary component for $i =$

$1, \ldots, r$. Next, we fix a secondary representation $M = Q'_1 + \cdots + Q'_r + \cdots + Q'_s$ of M with Q'_i being P_i-secondary and let $Q' = \bigcap_{i=r+1}^{s} Q'_i$, so that $Q' \in \Lambda^*_{\mathrm{Att}(M)\setminus X}(M)$. By the argument (i) $\Rightarrow$ (ii), $\mathrm{Att}(M/Q') \subseteq X$. Since $\frac{M}{Q+Q'}$ is a homomorphic image of both M/Q and M/Q', we know that $\mathrm{Att}(\frac{M}{Q+Q'}) \subseteq \mathrm{Att}(M/Q) \cap \mathrm{Att}(M/Q') = \varnothing$. Note that $\frac{M}{Q+Q'}$ is representable since M is so (cf. Theorem 7.3 (iii)). In light of this, the fact that $\mathrm{Att}(\frac{M}{Q+Q'}) = \varnothing$ necessarily implies $\frac{M}{Q+Q'} = 0$ (cf. Theorem 7.3 (i)), and hence $M = Q + Q'$. That is,

$$M = Q + Q' = Q_1 + \cdots + Q_r + Q'_{r+1} + \cdots Q'_s,$$

which is necessarily a (minimal) secondary representation of M. This implies that $Q = Q_1 + \cdots + Q_r$ is an X-secondary component of M, i.e., $Q \in \Lambda^*_X(M)$. □

As a consequence, we establish the following 'compatibility' property of secondary representation, as follows.

Theorem 8.3 (Compatibility). *Let M be a representable R-module. Then*

(i) *If $X_i \subseteq \mathrm{Att}(M)$ and $Q_{X_i} \in \Lambda^*_{X_i}(M)$ for $1 \le i \le n$. Then $\sum_{i=1}^{n} Q_{X_i} \in \Lambda^*_X(M)$, where $X = \bigcup_{i=1}^{n} X_i$.*

(ii) *In particular, suppose $\mathrm{Att}(M) = \{P_1, \ldots, P_s\}$ and $Q_i \in \Lambda^*_{P_i}(M)$ for each $i = 1, \ldots, s$. Then $M = Q_1 + \cdots + Q_s$, which is necessarily a minimal secondary representation of M.*

Proof. (i) By Lemma 8.2, we see $\mathrm{Att}(Q_{X_i}) = X_i$ and $\mathrm{Att}(M/Q_{X_i}) = \mathrm{Att}(M) \setminus X_i$ for $1 \le i \le n$. Therefore

$$\mathrm{Att}\Big(M/\Big(\sum_{i=1}^{n} Q_{X_i}\Big) \subseteq \bigcap_{i=1}^{n} \mathrm{Att}(M/Q_{X_i}\Big) = \mathrm{Att}(M) \setminus X$$

because of the natural surjections from M/Q_{X_i} onto $M/(\sum_{i=1}^{n} Q_{X_i})$. Also observe that $\mathrm{Att}(\sum_{i=1}^{n} Q_{X_i}) \subseteq \cup_{i=1}^{n} \mathrm{Att}(Q_{X_i}) = X$ (since there is an obvious surjection from $\oplus_{i=1}^{n} Q_{X_i}$ to $\sum_{i=1}^{n} Q_{X_i}$). Now Lemma 8.2 gives the desired result.

(ii) This is a special case of (i). By definition, M is the only $\mathrm{Att}(M)$-secondary component of M. (This can also be proved by "dualizing" the proof in [16, Theorem 1.1]; see [18, Theorem 4.1.2] for details.) □

9 Applying a Result of Sharp on Artinian Modules

Throughout this section, R is a ring and M is an Artinian R-module. Although R is not necessarily Noetherian, we are going to see that M can be naturally realized as an Artinian module over a Noetherian complete semi-local ring, thanks to a theorem of R. Y. Sharp in [9] (cf. Theorem 9.3). This would make the classic Matlis duality applicable, which then allows us to transform secondary representations to primary decompositions, as we are going to see in Observation 9.4. (Also see [10] for another result on Artinian modules.)

Notation 9.1. We will use the following notation in the sequel.

(i) Let $\operatorname{MSpec}(R) := \{\mathfrak{m} \in \operatorname{Spec}(R) \mid \mathfrak{m} \text{ is maximal in } R\}$.

(ii) For every $\mathfrak{m} \in \operatorname{MSpec}(R)$, denote $\Gamma_{\mathfrak{m}}(M) := \bigcup_{n \geq 0}(0 :_M \mathfrak{m}^n)$, which is isomorphic to $M_{\mathfrak{m}}$ since M is Artinian.

(iii) For every $\mathfrak{m} \in \operatorname{MSpec}(R)$, let $\widehat{R_{\mathfrak{m}}}$ be the $\mathfrak{m}$-adic completion of $R_{\mathfrak{m}}$ (or R), which is a quasi-local ring (i.e., a ring, not necessarily Noetherian, with a unique maximal ideal).

(iv) Let $\widehat{R} := \prod_{\mathfrak{m} \in \operatorname{MSpec}(R)} \widehat{R_{\mathfrak{m}}}$, which is a ring (not necessarily Noetherian).

(v) Let $\phi\colon R \to \widehat{R}$ be the natural ring homomorphism.

(vi) Let $\phi^*\colon \operatorname{Spec}(\widehat{R}) \to \operatorname{Spec}(R)$ denote the induced continuous map, that is, $\phi^*(P) = \phi^{-1}(P)$ for all $P \in \operatorname{Spec}(\widehat{R})$.

(vii) Let ϕ^*_M denote the resulting map if we restrict ϕ^* to $\operatorname{Att}_{\widehat{R}}(M) \to \operatorname{Att}_R(M)$. Thus, for $X \subseteq \operatorname{Att}_R(M)$, ${\phi^*_M}^{-1}(X) = \{P \in \operatorname{Att}_{\widehat{R}}(M) \mid \phi^{-1}(P) \in X\}$.

(viii) For each $\mathfrak{m} \in \operatorname{MSpec}(R)$, let $\operatorname{E}_{\widehat{R}}(R/\mathfrak{m})$ denote the injective hull of $R/\mathfrak{m}$ over $\widehat{R}$ (which is canonically isomorphic to its injective hull over $\widehat{R_{\mathfrak{m}}}$).

(ix) Let $E := \prod_{\mathfrak{m}} \operatorname{E}_{\widehat{R}}(R/\mathfrak{m})$, which is injective over $\widehat{R}$.

(x) Define the Matlis dualizing functor, denoted $\operatorname{D}(-)$, as follows: for every $\widehat{R}$-module N (e.g., N is an Artinian R-module), let $\operatorname{D}(N) := \operatorname{Hom}_{\widehat{R}}(N, E)$.

Let us recall the classic Matlis duality (over a Noetherian complete semi-local ring) and some consequences.

Theorem 9.2 (Matlis duality). *Let R be a Noetherian semi-local ring that is complete (with respect to its Jacobson radical) and M be an R-module that is Artinian or Noetherian. Say* $\operatorname{MSpec}(R) = \{\mathfrak{m}_1, \dots, \mathfrak{m}_n\}$, *so that* $R = \widehat{R} = \prod_{i=1}^n \widehat{R_{\mathfrak{m}_i}}$ *and* $E = \bigoplus_{i=1}^n \operatorname{E}_{R_{\mathfrak{m}_i}}(R/\mathfrak{m}_i)$. *Then*

(i) *If M is Artinian (resp. Noetherian), then* $\operatorname{D}(M)$ *is Noetherian (resp. Artinian).*

(ii) $\operatorname{D}(\operatorname{D}(M)) = M$ *and (hence)* $\operatorname{D}(\operatorname{D}(\operatorname{D}(M))) = \operatorname{D}(M)$.

(iii) *If* $\{N_i\}_{i\in\Delta}$ *is a family of (possibly infinitely many) R-submodules of M, then*

$$\mathrm{D}\Big(M/\sum_{i\in\Delta} N_i\Big) = \bigcap_{i\in\Delta} \mathrm{D}(M/N_i),$$

$$\sum_{i\in\Delta} N_i = \mathrm{D}\Big(\mathrm{D}(M)/\bigcap_{i\in\Delta} \mathrm{D}(M/N_i)\Big),$$

$$\mathrm{D}\Big(M/\bigcap_{i\in\Delta} N_i\Big) = \sum_{i\in\Delta} \mathrm{D}(M/N_i),$$

$$\bigcap_{i\in\Delta} N_i = \mathrm{D}\Big(\mathrm{D}(M)/\sum_{i\in\Delta} \mathrm{D}(M/N_i)\Big),$$

$$N_i \subseteq N_j \iff \mathrm{D}(M/N_i) \supseteq \mathrm{D}(M/N_j).$$

(iv) *For any R-submodule Q of M and* $P \in \mathrm{Spec}(R)$, *Q is P-secondary if and only if* $\mathrm{D}(Q)$ *is P-coprimary if and only if* $\mathrm{D}(M/Q)$ *is P-primary in* $\mathrm{D}(M)$.

(iv′) *For any submodule* Q' *of M and* $P \in \mathrm{Spec}(R)$, Q' *is P-primary in M if and only if* $\mathrm{D}(M/Q')$ *is P-secondary.*

(v) $M = \sum_{i=1}^{s} Q_i$ *is a (minimal) secondary representation of M if and only if* $0 = \bigcap_{i=1}^{s} \mathrm{D}(M/Q_i)$ *is a (minimal) primary decomposition of* 0 *in* $\mathrm{D}(M)$.

(v′) $0 = \bigcap_{i=1}^{s} Q'_i$ *is a (minimal) primary decomposition of* 0 *in M if and only if* $\mathrm{D}(M) = \sum_{i=1}^{s} \mathrm{D}(M/Q'_i)$ *is a (minimal) secondary representation of* $\mathrm{D}(M)$.

(vi) $\mathrm{Att}_R(M) = \mathrm{Ass}_R(\mathrm{D}(M))$ *and* $\mathrm{Ass}_R(M) = \mathrm{Att}_R(\mathrm{D}(M))$.

(By abuse of notation, we use "=" to denote natural isomorphisms, and also regard $\mathrm{D}(M/N_i)$, $\mathrm{D}(M/Q)$ *and* $\mathrm{D}(M/Q_i)$ *as R-submodules of* $\mathrm{D}(M)$ *via the natural injections.)*

Proof. Statements (i), (ii) and (iii) are standard results of the classic Matlis duality.

(iv) It is clear that $Q \neq 0 \iff \mathrm{D}(Q) \neq 0 \iff \mathrm{D}(M/Q) \subsetneq \mathrm{D}(M)$. So we assume $Q \neq 0$. Then we have that

$$Q \text{ is } P\text{-secondary} \iff f\colon Q \xrightarrow{r\cdot} Q \text{ is } \begin{cases} \text{surjective} & \text{if } r \in R \setminus P \\ \text{nilpotent} & \text{if } r \in P \end{cases}$$

$$\iff g\colon \mathrm{D}(Q) \xrightarrow{r\cdot} \mathrm{D}(Q) \text{ is } \begin{cases} \text{injective} & \text{if } r \in R \setminus P \\ \text{nilpotent} & \text{if } r \in P \end{cases}$$

$$\iff \mathrm{D}(Q) \text{ is } P\text{-coprimary}$$

$$\iff \mathrm{D}(M)/\mathrm{D}(M/Q) \text{ is } P\text{-coprimary}$$

$$\iff \mathrm{D}(M/Q) \text{ is } P\text{-primary in } \mathrm{D}(M).$$

(iv′) This can be proved in a similar way. (This also follows from (iv) in light of the duality results (i) and (ii).)

Finally, (v), (v′) and (vi) all follow from (i), (ii), (iii), (iv) and (iv′) directly. □

Let R be a general commutative ring (not necessarily Noetherian). Since M is Artinian, we see that $M = \bigoplus_{\mathfrak{m} \in \operatorname{MSpec}(R)} \Gamma_{\mathfrak{m}}(M)$ and $\Gamma_{\mathfrak{m}}(M) = 0$ for all but finitely many $\mathfrak{m}$. For each $\mathfrak{m} \in \operatorname{MSpec}(R)$, $\Gamma_{\mathfrak{m}}(M) = M_{\mathfrak{m}}$ is naturally a module over $\widehat{R_{\mathfrak{m}}}$. Thus M can be naturally viewed as a module over $\widehat{R}$ (via component-wise scalar multiplications). If we compose this derived $\widehat{R}$-module structure of M with ϕ, we recover the original R-module structure of M. Moreover, for a subset N of M, it is straightforward to see that

$$N \text{ is an } R\text{-submodule of } M \iff N \text{ is an } \widehat{R}\text{-submodule of } M.$$

So M must be Artinian over $\widehat{R}$, since M is Artinian over R. To study the R-module structure of M, one approach would be to study its $\widehat{R}$-module structure.

Let us study $\operatorname{Ann}_{\widehat{R}}(M)$, the annihilator of M over $\widehat{R}$. By the above, we see that

$$\operatorname{Ann}_{\widehat{R}}(M) = \prod_{\mathfrak{m} \in \operatorname{MSpec}(R)} \operatorname{Ann}_{\widehat{R_{\mathfrak{m}}}}(\Gamma_{\mathfrak{m}}(M)).$$

Thus M is naturally an Artinian module over the following quotient ring

$$\frac{\widehat{R}}{\operatorname{Ann}_{\widehat{R}}(M)} = \frac{\prod_{\mathfrak{m} \in \operatorname{MSpec}(R)} \widehat{R_{\mathfrak{m}}}}{\prod_{\mathfrak{m} \in \operatorname{MSpec}(R)} \operatorname{Ann}_{\widehat{R_{\mathfrak{m}}}}(\Gamma_{\mathfrak{m}}(M))} \cong \prod_{\mathfrak{m} \in \operatorname{MSpec}(R)} \frac{\widehat{R_{\mathfrak{m}}}}{\operatorname{Ann}_{\widehat{R_{\mathfrak{m}}}}(\Gamma_{\mathfrak{m}}(M))}.$$

As $\Gamma_{\mathfrak{m}}(M) = 0$ for all but finitely many $\mathfrak{m}$, say $\{\mathfrak{m} \in \operatorname{MSpec}(R) \mid \Gamma_{\mathfrak{m}}(M) \neq 0\} = \{\mathfrak{m}_1, \ldots, \mathfrak{m}_n\}$. Then $M = \bigoplus_{i=1}^{n} \Gamma_{\mathfrak{m}_i}(M)$ is naturally an Artinian module over

$$\prod_{\mathfrak{m} \in \operatorname{MSpec}(R)} \frac{\widehat{R_{\mathfrak{m}}}}{\operatorname{Ann}_{\widehat{R_{\mathfrak{m}}}}(\Gamma_{\mathfrak{m}}(M))} = \prod_{i=1}^{n} \frac{\widehat{R_{\mathfrak{m}_i}}}{\operatorname{Ann}_{\widehat{R_{\mathfrak{m}_i}}}(\Gamma_{\mathfrak{m}_i}(M))}.$$

So we study $\Gamma_{\mathfrak{m}_i}(M)$ over

$$\frac{\widehat{R_{\mathfrak{m}_i}}}{\operatorname{Ann}_{\widehat{R_{\mathfrak{m}_i}}}(\Gamma_{\mathfrak{m}_i}(M))}$$

for $i = 1, \ldots, n$. By construction, $\Gamma_{\mathfrak{m}_i}(M)$ is Artinian over the quasi-local ring

$$\frac{\widehat{R_{\mathfrak{m}_i}}}{\operatorname{Ann}_{\widehat{R_{\mathfrak{m}_i}}}(\Gamma_{\mathfrak{m}_i}(M))};$$

moreover, if we compose this module structure with the natural map $R \to \widehat{R_{\mathfrak{m}_i}}$, we recover the original R-module structure of $\Gamma_{\mathfrak{m}_i}(M)$.

Therefore, to study the secondary representations of an Artinian R-module M, it (usually) suffices to study them over $\widehat{R}$ (as secondary representations behave well under scalar restriction, see Theorem 7.3 (iii)). Then it suffices to regard M as an (Artinian) module over

$$\frac{\widehat{R}}{\operatorname{Ann}_{\widehat{R}}(M)} = \prod_{i=1}^{n} \frac{\widehat{R_{\mathfrak{m}_i}}}{\operatorname{Ann}_{\widehat{R_{\mathfrak{m}_i}}}(\Gamma_{\mathfrak{m}_i}(M))}.$$

The following theorem of R. Y. Sharp verifies that each of the rings $\frac{\widehat{R_{\mathfrak{m}_i}}}{\operatorname{Ann}_{\widehat{R_{\mathfrak{m}_i}}}(\Gamma_{\mathfrak{m}_i}(M))}$ is actually Noetherian. In the sequel, we say a ring is local if it is Noetherian with a unique maximal ideal. We say a ring is semi-local if it is Noetherian with finitely many maximal ideals.

Theorem 9.3 ([9]). *Let M be an Artinian R-module as above. Then for each $\mathfrak{m} \in \operatorname{MSpec}(R)$, $\frac{\widehat{R_{\mathfrak{m}}}}{(0:_{\widehat{R_{\mathfrak{m}}}}\Gamma_{\mathfrak{m}}(M))}$ is (either the zero ring or) a local (Noetherian) ring that is complete with respect to its maximal ideal. Therefore $\frac{\widehat{R}}{\operatorname{Ann}_{\widehat{R}}(M)}$ is a complete semi-local (Noetherian) ring (i.e., a direct product of finitely many complete local rings).*

Since each $\frac{\widehat{R_{\mathfrak{m}_i}}}{\operatorname{Ann}_{\widehat{R_{\mathfrak{m}_i}}}(\Gamma_{\mathfrak{m}_i}(M))}$ is complete local (Noetherian), the classic Matlis duality (Theorem 9.2) applies. It then follows that the functor $\mathrm{D}(-)$, which is defined over $\widehat{R}$, enjoys many of the properties of the classic Matlis duality, even though $\widehat{R}$ may not be Noetherian. Consequently, secondary representations of Artinian R-modules are in one-to-one correspondence with primary decompositions of Noetherian $\widehat{R}$-modules. (This is demonstrated in Observation 9.4 next.)

The following observations would show how the classic Matlis duality is applied, thanks to Theorem 9.3. This duality allows us to make a connection between the theory of secondary representation and the theory of primary decomposition.

Observation 9.4. Let R be a ring and M be an Artinian R-module. Keep all the above notation in this section. By abuse of notation, we may use "=" to denote natural isomorphisms. To further simplify the notation, let

$$T_{\mathfrak{m}} := \frac{\widehat{R_{\mathfrak{m}}}}{\operatorname{Ann}_{\widehat{R_{\mathfrak{m}}}}(\Gamma_{\mathfrak{m}}(M))} = \frac{\widehat{R_{\mathfrak{m}}}}{I_{\mathfrak{m}}},\ \forall\, \mathfrak{m} \text{ and } T := \prod_{\mathfrak{m}\in\operatorname{MSpec}(R)} T_{\mathfrak{m}} = \prod_{i=1}^{n} T_{\mathfrak{m}_i} = \frac{\widehat{R}}{I}$$

with $I := \operatorname{Ann}_{\widehat{R}}(M)$ and $I_{\mathfrak{m}} := \operatorname{Ann}_{\widehat{R_{\mathfrak{m}}}}(\Gamma_{\mathfrak{m}}(M))$. Then M is an Artinian T-module; and Theorem 9.3 says that T is a complete semi-local (Noetherian) ring. We make the following observations (many of them obvious):

(i) Although $\mathrm{D}(M)$ is defined as $\operatorname{Hom}_{\widehat{R}}(M, E)$ over $\widehat{R}$, $\mathrm{D}(M)$ is the same as taking the Matlis dual over the complete semi-local ring T, and it is also the same as taking the Matlis dual of each $\Gamma_{\mathfrak{m}_i}(M)$ individually over the complete local ring $T_{\mathfrak{m}_i}$ and then taking their direct sum. This is because, by Hom-⊗ adjointness,

$$\begin{aligned}
\operatorname{Hom}_{\widehat{R}}(M, E) &= \bigoplus_{i=1}^{n} \operatorname{Hom}_{\widehat{R_{\mathfrak{m}_i}}}\Big(\Gamma_{\mathfrak{m}_i}(M), \mathrm{E}_{\widehat{R_{\mathfrak{m}_i}}}(R/\mathfrak{m}_i)\Big) \\
&= \bigoplus_{i=1}^{n} \operatorname{Hom}_{T_{\mathfrak{m}_i}}\Big(\Gamma_{\mathfrak{m}_i}(M), \operatorname{Hom}_{\widehat{R_{\mathfrak{m}_i}}}(T_{\mathfrak{m}_i}, \mathrm{E}_{\widehat{R_{\mathfrak{m}_i}}}(R/\mathfrak{m}_i))\Big) \\
&= \bigoplus_{i=1}^{n} \operatorname{Hom}_{T_{\mathfrak{m}_i}}\Big(\Gamma_{\mathfrak{m}_i}(M), \mathrm{E}_{T_{\mathfrak{m}_i}}(R/\mathfrak{m}_i)\Big) \\
&= \operatorname{Hom}_T\Big(M, \bigoplus_{i=1}^{n} \mathrm{E}_{T_{\mathfrak{m}_i}}(R/\mathfrak{m}_i)\Big),
\end{aligned}$$

in which $\mathrm{E}_{T_{\mathfrak{m}_i}}(R/\mathfrak{m}_i)$ denotes the injective hull of $R/\mathfrak{m}_i$ over the ring $T_{\mathfrak{m}_i}$.

(ii) Thus $\mathrm{D}(M)$ is a Noetherian T-module, and hence a Noetherian $\widehat{R}$-module.

(iii) Therefore, $\mathrm{D}(\mathrm{D}(M)) = M$ and $\mathrm{D}(\mathrm{D}(\mathrm{D}(M))) = \mathrm{D}(M)$ (up to the canonical isomorphisms) as T-modules and hence as $\widehat{R}$-modules. This follows from the classic Matlis duality (cf. Theorem 9.2) over T.

(iv) If $\{N_i\}_{i\in\Delta}$ is a family of (possibly infinitely many) $\widehat{R}$-submodules of M and $\{K_i\}_{i\in\Delta}$ is a family of $\widehat{R}$-submodules of $\mathrm{D}(M)$, then

$$\begin{aligned}
&\mathrm{D}(M/\sum_{i\in\Delta} N_i) = \bigcap_{i\in\Delta} \mathrm{D}(M/N_i), \\
&\sum_{i\in\Delta} N_i = \mathrm{D}\Big(\mathrm{D}(M)/\bigcap_{i\in\Delta} \mathrm{D}(M/N_i)\Big), \\
&N_i \subseteq N_j \iff \mathrm{D}(M/N_i) \supseteq \mathrm{D}(M/N_j), \\
&N_i = N_j \iff \mathrm{D}(M/N_i) = \mathrm{D}(M/N_j), \\
&\mathrm{D}\Big(\mathrm{D}(M)/\bigcap_{i\in\Delta} K_i\Big) = \sum_{i\in\Delta} \mathrm{D}(\mathrm{D}(M)/K_i), \\
&\bigcap_{i\in\Delta} K_i = \mathrm{D}(M/\sum_{i\in\Delta} \mathrm{D}(\mathrm{D}(M)/K_i)), \\
&K_i \subseteq K_j \iff \mathrm{D}(\mathrm{D}(M)/K_i) \supseteq \mathrm{D}(\mathrm{D}(M)/K_j), \\
&K_i = K_j \iff \mathrm{D}(\mathrm{D}(M)/K_i) = \mathrm{D}(\mathrm{D}(M)/K_j).
\end{aligned}$$

(Indeed, the above equations and equivalences hold over T (cf. Theorem 9.2); hence they also hold over $\widehat{R}$.)

(v) For $Q \subseteq M$, Q is P-secondary if and only if $\mathrm{D}(Q)$ is P-coprimary if and only if $\mathrm{D}(M/Q)$ is P-primary in $\mathrm{D}(M)$ (over T or over $\widehat{R}$, no difference). This follows immediately from Theorem 9.2 over T (and hence over $\widehat{R}$).

(vi) $M = \sum_{i=1}^{s} Q_i$ is a (minimal) secondary representation of M over T (hence over $\widehat{R}$) if and only if $0 = \bigcap_{i=1}^{s} \mathrm{D}(M/Q_i)$ is a (minimal) primary decomposition of 0 in $\mathrm{D}(M)$ over T (hence over $\widehat{R}$). Thus $\mathrm{Att}_{\widehat{R}}(M) = \mathrm{Ass}_{\widehat{R}}(\mathrm{D}(M))$. Since every $\widehat{R}$-submodule of $\mathrm{D}(M)$ is of the form $\mathrm{D}(M/Q)$, the above is also a criterion of primary decompositions of 0 in $\mathrm{D}(M)$. Put directly, over T (and $\widehat{R}$), $0 = \bigcap_{i=1}^{s} Q'_i$ is a primary decomposition of 0 in $\mathrm{D}(M)$ if and only if $M = \sum_{i=1}^{s} \mathrm{D}(\mathrm{D}(M)/Q'_i)$ is a secondary representation of M. All these follow from Theorem 9.2 over T.

(vii) Thus, for any $Y \subseteq \mathrm{Att}_{\widehat{R}}(M)$ and $\widehat{R}$-submodules $Q \subseteq M$ and $Q' \subseteq \mathrm{D}(M)$,

$$Q \in \Lambda_Y^*(M) \iff \mathrm{D}(M/Q) \in \Lambda_Y(0 \subseteq \mathrm{D}(M)),$$

$$Q \in \underset{\circ}{\Lambda}{}_Y^*(M) \iff \mathrm{D}(M/Q) \in \mathring{\Lambda}_Y(0 \subseteq \mathrm{D}(M)),$$

$$\mathrm{D}(\mathrm{D}(M)/Q') \in \Lambda_Y^*(M) \iff Q' \in \Lambda_Y(0 \subseteq \mathrm{D}(M)),$$

$$\mathrm{D}(\mathrm{D}(M)/Q') \in \underset{\circ}{\Lambda}{}_Y^*(M) \iff Q' \in \mathring{\Lambda}_Y(0 \subseteq \mathrm{D}(M)).$$

Note that an R-submodule of M is the same as an $\widehat{R}$-submodule of M.

(viii) $\mathrm{D}(0 :_M J) = \mathrm{D}(M)/J\,\mathrm{D}(M)$ for every ideal J of T. This remains true if J is an ideal of R or $\widehat{R}$ (because of the natural maps $R \to \widehat{R} \to T$).

(ix) For convenience, we usually state the above results over $\widehat{R}$ rather than T, even though T (being Noetherian complete semi-local) is the reason why the results hold. This is because $\widehat{R}$ does not depend on the Artinian module M while T does, and sometimes we study several Artinian R-modules.

(x) Lastly, we make a summary as follows: For any Artinian R-module M, applying $\mathrm{D}(-)$ to M (over $\widehat{R}$ by construction) is the same as taking the Matlis dual of M over the complete semi-local (Noetherian) ring $T = \frac{\widehat{R}}{\mathrm{Ann}_{\widehat{R}}(M)}$. As a result, $\mathrm{D}(M)$ is a Noetherian module, $\mathrm{D}(\mathrm{D}(M)) = M$ and $\mathrm{D}(\mathrm{D}(\mathrm{D}(M))) = \mathrm{D}(M)$ over $\widehat{R}$; and studying the secondary representations of M over $\widehat{R}$ is equivalent to studying the primary decompositions of 0 in $\mathrm{D}(M)$ over $\widehat{R}$. All the above hold for every Artinian R-module M over $\widehat{R}$. In this sense, we (essentially) have the classic Matlis duality over $\widehat{R}$ for Artinian R-modules even though $\widehat{R}$ may not be Noetherian. For this reason, we also refer to $\mathrm{D}(-)$ (over $\widehat{R}$) as the Matlis functor.

(Again, the reader please be reminded that, by abuse of notation, we used "=" to denote natural isomorphisms in the above statements.)

In light of the above, we will frequently employ the following strategy in the remaining sections: To study the secondary representations of a given Artinian R-module M, we instead study the secondary representations of M over $\widehat{R}$ or, equivalently, over the complete semi-local ring $T = \frac{\widehat{R}}{\operatorname{Ann}_{\widehat{R}}(M)}$. Applying Matlis duality D(−), we obtain a Noetherian module $\mathrm{D}(M)$ (over the complete semi-local ring T). If we can show (or if we already know) certain properties of the primary decompositions of $\mathrm{D}(M)$, then, after applying Matlis duality D(−) again, we get corresponding properties of the secondary representations for $\mathrm{D}(\mathrm{D}(M)) = M$ (over the complete semi-local ring T). This in turn should reveal properties of secondary representation of the *original* Artinian R-module M that we intend to study (via the map $R \to \widehat{R} \to T$).

Next, we state a lemma concerning relations between the secondary representations of M as an R-module and the secondary representations of M as an $\widehat{R}$-module. To avoid confusion, we may use ${}_RM$ to indicate that the R-module structure of M is being considered; similarly, ${}_{\widehat{R}}M$ indicates the $\widehat{R}$-module structure.

Lemma 9.5. *Let R be a ring and M an Artinian R-module. Then the following hold:*

(i) $\Lambda^*_{\phi_M^{*\,-1}(X)}({}_{\widehat{R}}M) \subseteq \Lambda^*_X({}_RM)$ *for all* $X \subseteq \operatorname{Att}_R(M)$.

(ii) *For every* $X \subseteq \operatorname{Att}_R(M)$ *and every* $Q_X \in \Lambda^*_X({}_RM)$, *there exists* $Q_{\phi_M^{*\,-1}(X)} \in \Lambda^*_{\phi_M^{*\,-1}(X)}({}_{\widehat{R}}M)$ *such that* $Q_{\phi_M^{*\,-1}(X)} \subseteq Q_X$.

(iii) $\underset{\circ}{\Lambda}{}^*_X({}_RM) = \underset{\circ}{\Lambda}{}^*_{\phi_M^{*\,-1}(X)}({}_{\widehat{R}}M)$ *for all* $X \subseteq \operatorname{Att}_R(M)$.

Proof. Say $\operatorname{Att}_R(M) = \{\mathfrak{p}_1, \dots, \mathfrak{p}_s\}$. By Theorem 7.3 (iii′), we may write $\operatorname{Att}_{\widehat{R}}(M) = \{P_{i,j} \mid 1 \le i \le s;\ 1 \le j \le n(i)\}$ such that $\phi_M^{*\,-1}(\mathfrak{p}_i) = \{P_{i,j} \mid 1 \le j \le n(i)\}$.

(i) Let $Q_i \in \Lambda^*_{\phi_M^{*\,-1}(\mathfrak{p}_i)}({}_{\widehat{R}}M)$. Then there is a secondary representation $M = \sum_{i,j} Q_{i,j}$ of M with $Q_{i,j}$ being $P_{i,j}$-secondary such that $Q_i = \sum_{j=1}^{n(i)} Q_{i,j}$. Note that $M = \sum_i (\sum_{j=1}^{n(i)} Q_{i,j})$ is a secondary representation of M over R with $\sum_{j=1}^{n(i)} Q_{i,j}$ being the $\mathfrak{p}_i$-secondary component. Hence $Q_i = \sum_{j=1}^{n(i)} Q_{i,j} \in \Lambda^*_{\mathfrak{p}_i}({}_RM)$. This verifies the claim for $X = \{\mathfrak{p}_i\}$. The general claim follows, cf. Theorem 8.3 (1).

(ii) Let $M = \sum_i Q_i$ be any secondary representation of M over R with Q_i being the $\mathfrak{p}_i$-secondary component, so that $Q_i \in \Lambda^*_{\mathfrak{p}_i}({}_RM)$. Since each Q_i is Artinian, it is representable over $\widehat{R}$. Say $Q_i = \sum_{j=1}^{m(i)} Q_{i,j}$ is a secondary representation of Q_i over $\widehat{R}$ with $Q_{i,j}$ being $P'_{i,j}$-secondary (over $\widehat{R}$). After rearrangement, there is $k(i)$, $0 \le k(i) \le \min\{m(i), n(i)\}$, such that $P'_{i,j} = P_{i,j}$ for $1 \le j \le k(i)$ but $P'_{i,j} \notin \{P_{i,1}, \dots, P_{i,n(i)}\}$ for all $j > k(i)$. Then we have

$$M = \sum_{i=1}^{s}\sum_{j=1}^{m(i)} Q_{i,j} = Q_{1,1} + \dots + Q_{1,m(1)} + \dots + Q_{s,1} + \dots + Q_{s,m(s)}, \qquad (\dagger)$$

which is a not necessarily minimal secondary representation of M over $\widehat{R}$. We claim that, if we make (†) minimal, then $Q_{i,j}$ must be redundant for all $j > k(i)$. (Here is why: Suppose, for some $j > k(i)$, $Q_{i,j}$ remains in the minimized form of the above summation (†). Then we must have $P'_{i,j} \in \mathrm{Att}_{\widehat{R}}(M)$, cf. Theorem 7.3 (i). Because $j > k(i)$, we must have $P'_{i,j} = P_{a,b} \in \mathrm{Att}_{\widehat{R}}(M)$ for some $a \neq i$. But $\phi^*(P'_{i,j}) = \mathfrak{p}_i \neq \mathfrak{p}_a = \phi^*(P_{a,b})$, which is a contradiction.) Thus, we can throw out all the components $Q_{i,j}$ with $j > k(i)$, so that we get

$$M = \sum_{i=1}^{s}\sum_{j=1}^{k(i)} Q_{i,j} = Q_{1,1} + \cdots + Q_{1,k(1)} + \cdots + Q_{s,1} + \cdots + Q_{s,k(s)}. \quad (\ddagger)$$

But this implies $\mathrm{Att}_{\widehat{R}}(M) \subseteq \{P_{i,j} \mid 1 \le i \le s;\ 1 \le j \le k(i)\}$, which forces $k(i) = n(i)$ for all i in light of Theorem 7.3 (i). Consequently, (‡) must be a minimal secondary representation of M over $\widehat{R}$. Therefore, for each $i = 1, \dots, s$, we see

$$\Lambda^*_{\mathfrak{p}_i}({}_R M) \ni Q_i \supseteq \sum_{j=1}^{k(i)} Q_{i,j} \in \Lambda^*_{\phi_M^{*\,-1}(\mathfrak{p}_i)}({}_{\widehat{R}}M).$$

This verifies the claim for $X = \{\mathfrak{p}_i\}$. The general claim follows, cf. Theorem 8.3 (i).

(iii) This follows from (i) and (ii). □

Thus, when we study the minimal secondary components of an Artinian R-module, it suffices to do so over $\widehat{R}$, where Matlis duality applies.

We will frequently use Matlis duality to go between secondary representations of Artinian R-modules and primary decompositions of Noetherian $\widehat{R}$-modules. Most of the results in the following sections were obtained in [18].

10 Independence

Let R be a ring and M be a representable R-module. Note that $\mathrm{Att}(M)$ is finite, and $\mathrm{Att}(M)$ is a topological space because of the Zariski topology on $\mathrm{Spec}(R)$. As in Notation 4.1, for every $Y \subseteq \mathrm{Att}(M)$, we use $\mathrm{o}(Y)$ to denote the smallest superset of Y that is open in $\mathrm{Att}(M)$. The notation $\mathrm{o}(Y)$ depends on the ambient space, which should be made clear in the context.

If Y is an open subset of $\mathrm{Att}(M)$, then there is only one Y-secondary component in $\Lambda^*_Y(M)$, and it is $[U]M$ where $U = R \setminus \bigcup_{P \in Y} P$; see Theorem 7.3 (iv).

Definition 10.1. Let M be an R-module and $X \subseteq \mathrm{Att}_R(M)$. We say that the secondary representations of M are *independent* over X, or X-independent, if $\Lambda^*_X(M)$ consists of exactly one component, i.e., $|\Lambda^*_X(M)| = 1$.

Obviously, this definition is parallel to the definition of independence of primary decompositions (cf. Definition 4.2). In Theorem 4.4, it was shown that if K is a Noetherian R-module and $X \subseteq \operatorname{Ass}(K)$, then the primary decompositions of 0 in K are independent over X if and only if X is open in $\operatorname{Ass}(K)$.

Naturally, we ask the following question.

Question 10.2. Let R be a ring, M an Artinian R-module, and $X \subseteq \operatorname{Att}_R(M)$ such that the secondary representations of M are independent over X. Then is X an open subset of $\operatorname{Att}_R(M)$?

The next theorem indicates an answer of 'almost yes'. (The answer to the question is actually 'no', as explained in Example 10.4.)

Theorem 10.3. *Let M be an Artinian R-module and $X \subseteq \operatorname{Att}_R(M)$. Denote by ϕ^*_M the natural map from $\operatorname{Att}_{\widehat{R}}(M)$ to $\operatorname{Att}_R(M)$. If the secondary representations of M are independent over X, then ${\phi^*_M}^{-1}(X)$ is open in $\operatorname{Att}_{\widehat{R}}(M)$.*

Proof. As $\Lambda^*_{{\phi^*_M}^{-1}(X)}({}_{\widehat{R}}M) \subseteq \Lambda^*_X({}_RM)$ (by Lemma 9.5 (i)) and $|\Lambda^*_X({}_RM)| = 1$, we see $|\Lambda^*_{{\phi^*_M}^{-1}(X)}({}_{\widehat{R}}M)| = 1$. Now let us apply Matlis duality to M (over $\widehat{R}$). We see $|\Lambda_{{\phi^*_M}^{-1}(X)}(0 \subseteq \mathrm{D}(M))| = 1$ in light of the one-to-one correspondence in Observation 9.4 (vii). That is, the primary decompositions of 0 in $\mathrm{D}(M)$ over $\widehat{R}$ are independent over $\Lambda_{{\phi^*_M}^{-1}}(X)$. Since $\mathrm{D}(M)$ is Noetherian over $\widehat{R}$, we conclude that $\Lambda_{{\phi^*_M}^{-1}}(X)$ is open in $\operatorname{Ass}_{\widehat{R}}(\mathrm{D}(R))$ by Theorem 4.4. Now the proof is complete since $\operatorname{Ass}_{\widehat{R}}(\mathrm{D}(R)) = \operatorname{Att}_{\widehat{R}}(M)$, by Observation 9.4 (vi). □

The following example provides a negative answer to Question 10.2. (The ring in the example, i.e., $\mathbb{Z}$, is actually Noetherian.)

Example 10.4. Let $R = \mathbb{Z}$ and let $p \neq q$ be primes. Let $M := \Gamma_p(\frac{\mathbb{Q}}{\mathbb{Z}}) \oplus \frac{\mathbb{Z}}{(q)}$, which is Artinian over $\mathbb{Z}$. (Note that $\Gamma_p(\frac{\mathbb{Q}}{\mathbb{Z}})$ is the injective hull of $\mathbb{Z}/(p)$.) It is not hard to verify that the above direct sum is actually the unique secondary representation of M and $\operatorname{Att}_R(M) = \{(0), (q)\}$. In particular, the secondary representation of M is independent over $\{(q)\}$, but $\{(q)\}$ is not open in $\operatorname{Att}_R(M)$.

One might wonder whether the converse of Theorem 10.3 holds. It turns out that it fails, as shown in the following example.

Example 10.5. Let $(R, \mathfrak{m})$ be a Noetherian local ring that satisfies all the following conditions in relation with its completion $\widehat{R}$:

- There exist incomparable prime ideals $P_1,\ P_2 \in \operatorname{Spec}(\widehat{R})$ such that

$$P_1 \cap R =: \mathfrak{p}_1 \subsetneq \mathfrak{p}_2 := P_2 \cap R.$$

It follows that $\{P_2\}$ is an open subset of $\{P_1,\ P_2\}$.

- There are infinitely many (distinct) $\widehat{R}$-submodules $\frac{K_n}{P_1}$ of $\frac{\widehat{R}}{P_1}$, $n \geq 1$, such that $\frac{\widehat{R}}{K_n}$ are $\mathfrak{p}_2$-coprimary. (Thus, K_n are ideals of $\widehat{R}$, containing P_1.)

(Such a ring exists. For example, let $R := \mathbb{Q}[X, Y, Z]_{(X,Y,Z)}$, so that $\widehat{R} = \mathbb{Q}[\![X, Y, Z]\!]$. Let $e^Y := \sum_{k=0}^{\infty} \frac{Y^k}{k!}$, $P_1 := (X - e^Y + 1)\widehat{R}$ and $P_2 := Z\widehat{R}$. Then $\mathfrak{p}_1 = 0 \subsetneq ZR = \mathfrak{p}_2$; and $(\frac{\widehat{R}}{P_1+Z\widehat{R}})_{\mathfrak{p}_2} \neq 0$. Let $K_n := \operatorname{Ker}(\widehat{R} \to (\frac{\widehat{R}}{P_1+Z^n\widehat{R}})_{\mathfrak{p}_2})$, so that $\frac{\widehat{R}}{K_n}$ are $\mathfrak{p}_2$-coprimary as R-modules for all $n \geq 1$. Note that $K_n \supsetneq K_{n+1}$, since $(\frac{K_n}{K_{n+1}})_{\mathfrak{p}_2} \cong (\frac{P_1+Z^n\widehat{R}}{P_1+Z^{n+1}\widehat{R}})_{\mathfrak{p}_2} \cong (\frac{\widehat{R}}{P_1+Z\widehat{R}})_{\mathfrak{p}_2} \neq 0$, for all $n \geq 0$.)

It is straightforward to see that both

$$0 = \left(0 \oplus \frac{\widehat{R}}{P_2}\right) \cap \left(\frac{\widehat{R}}{P_1} \oplus 0\right) \quad \text{and} \quad 0 = \left(0 \oplus \frac{\widehat{R}}{P_2}\right) \cap \left(\frac{K_n}{P_1} \oplus 0\right), \quad n \geq 1,$$

are (minimal) primary decompositions of 0 in $\frac{\widehat{R}}{P_1} \oplus \frac{\widehat{R}}{P_2}$ over R. Let E be the injective hull of residue field $R/\mathfrak{m}$, and let $M := (0 :_E P_1) \oplus (0 :_E P_2)$.

Applying Matlis duality $\operatorname{Hom}_{\widehat{R}}(-, E)$ to the above primary decompositions, we see that both

$$\begin{aligned} M &= \big((0 :_E P_1) \oplus 0\big) + \big(0 \oplus (0 :_E P_2)\big) \qquad \text{and} \\ M &= \big((0 :_E P_1) \oplus 0\big) + \big((0 :_E K_n) \oplus (0 :_E P_2)\big), \quad n \geq 1, \end{aligned}$$

are (minimal) secondary representations of M over R. In the above, $0 \oplus (0 :_E P_2)$ and $(0 :_E K_n) \oplus (0 :_E P_2)$, $n \geq 1$, give rise to infinitely many (distinct) $\mathfrak{p}_2$-secondary components of ${}_RM$. Note that M is Artinian over R and over $\widehat{R}$, and the above secondary representations (over R) show that $\operatorname{Att}_R(M) = \{\mathfrak{p}_1, \mathfrak{p}_2\}$. It is also easy to see that

$$M = \big((0 :_E P_1) \oplus 0\big) + \big(0 \oplus (0 :_E P_2)\big) \qquad (*)$$

is a (minimal) secondary representation of M over $\widehat{R}$ and $\operatorname{Att}_{\widehat{R}}(M) = \{P_1,\ P_2\}$. (Thus $(*)$ is the unique secondary representation of M over $\widehat{R}$ by the choice of P_i.)

In summary, $(\phi_M^*)^{-1}(\{\mathfrak{p}_2\}) = \{P_2\}$ is open in $\operatorname{Att}_{\widehat{R}}(M)$, but the secondary representations of ${}_RM$ are not independent over $\{\mathfrak{p}_2\}$. In fact, $|\Lambda^*_{\mathfrak{p}_2}({}_RM)| = \infty$.

11 Minimal Secondary Components

Let M be an Artinian R-module. Using the notation introduced in Section 9, we present the following theorem about minimal secondary components. (The result was first obtained in [18].)

Theorem 11.1. *Let M be an Artinian R-module and $X \subseteq \mathrm{Att}_R(M)$. Say $X = \{P_1, \dots, P_r\}$. Then the following hold*

(i) $\mathring{\Lambda}^*_X(M) = \{\sum_{i=1}^r Q_i \mid Q_i \in \mathring{\Lambda}^*_{P_i}(M),\ 1 \le i \le r\}$.

(i′) $\mathring{\Lambda}^*_X({}_R M) = \mathring{\Lambda}^*_{\phi_M^{*\,-1}(X)}({}_{\widehat{R}}M) = \{\sum_{P \in \phi_M^{*\,-1}(X)} Q_P \mid Q_P \in \mathring{\Lambda}^*_P({}_{\widehat{R}}M)\}$.

(ii) *For all $Q \in \Lambda^*_{\phi_M^{*\,-1}(X)}({}_{\widehat{R}}M)$, $Q = \sum\{Q' \mid Q' \in \mathring{\Lambda}^*_{\phi_M^{*\,-1}(X)}({}_{\widehat{R}}M),\ Q' \subseteq Q\}$. In fact, every such Q is a sum of finitely many $Q' \in \mathring{\Lambda}^*_{\phi_M^{*\,-1}(X)}({}_{\widehat{R}}M)$.*

(iii) $\sum\{Q \mid Q \in \mathring{\Lambda}^*_X({}_R M)\} = \sum\{Q \mid Q \in \mathring{\Lambda}^*_{\phi_M^{*\,-1}(X)}({}_{\widehat{R}}M)\}$ *equals the unique $\mathrm{o}(\phi_M^{*\,-1}(X))$-secondary component of M over $\widehat{R}$, in which $\mathrm{o}(\phi_M^{*\,-1}(X))$ is the smallest open superset of $\phi_M^{*\,-1}(X)$ in $\mathrm{Att}_{\widehat{R}}(M)$.*

Proof. (i) and (i′): A direct proof will be given in Remark 11.2. But here we present a proof by the duality method described in Section 9. For (i′), the first equality follows from Lemma 9.5 (iii). To show the second equality, we regard M as an $\widehat{R}$-module. Then by Observation 9.4, it suffice to show

$$\mathring{\Lambda}_{\phi_M^{*\,-1}(X)}(0 \subseteq \mathrm{D}(M)) = \Big\{ \bigcap_{P \in \phi_M^{*\,-1}(X)} \mathrm{D}(\tfrac{M}{Q_P}) \mid \mathrm{D}(\tfrac{M}{Q_P}) \in \mathring{\Lambda}_P(0 \subseteq \mathrm{D}(M)) \Big\}.$$

But this holds by the virtue of Theorem 4.3 (i). Then (i) follows from (i′).

(ii) By Observation 9.4, it suffices to show that $\mathrm{D}(M/Q)$ equals the following

$$\bigcap\{\mathrm{D}(M/Q') \mid \mathrm{D}(M/Q') \in \mathring{\Lambda}_{\phi_M^{*\,-1}(X)}(0 \subseteq \mathrm{D}(M)),\ \mathrm{D}(M/Q') \supseteq \mathrm{D}(M/Q)\}.$$

But this follows from Theorem 4.3 (ii). The finiteness claim follows similarly.

(iii) The first equality follows from Lemma 9.5 (iii). To show the remaining claim, we use Matlis duality $\mathrm{D}(-)$. By Observation 9.4, it suffices to show that

$$\bigcap\{\mathrm{D}(M/Q) \mid \mathrm{D}(M/Q) \in \mathring{\Lambda}_{\phi_M^{*\,-1}(X)}(0 \subseteq \mathrm{D}(M))\}$$

is the unique $\mathrm{o}(\phi_M^{*\,-1}(X))$-primary component of 0 in $\mathrm{D}(M)$ as an $\widehat{R}$-module. But this follows from Theorem 4.3 (iii) as $\mathrm{D}(M)$ is a Noetherian $\widehat{R}$-module. □

Remark 11.2. We would like to present the following direct proofs of (i) and (i′) of Theorem 11.1 without using Matlis duality:

For (i), it is easy to show $\underset{\circ}{\Lambda}{}^*_X(M) \subseteq \{\sum_{i=1}^r Q_i \mid Q_i \in \underset{\circ}{\Lambda}{}^*_{P_i}(M),\ 1 \le i \le r\}$: For any $Q \in \underset{\circ}{\Lambda}{}^*_X(M)$, write $Q = Q_1' + \cdots + Q_r'$, where $Q_i' \in \Lambda^*_{P_i}$ for each $1 \le i \le r$. There exists $Q_i \in \underset{\circ}{\Lambda}{}^*_{P_i}$ such that $Q_i' \supseteq Q_i$ for each $i = 1, \ldots, r$, so that $Q = Q_1' + \cdots + Q_r' \supseteq Q_1 + \cdots + Q_r$. But $Q_1 + \cdots + Q_r \in \Lambda^*_X$ by the compatibility property (Theorem 8.3), which shows $Q = Q_1 + \cdots + Q_r$.

We show $\underset{\circ}{\Lambda}{}^*_X(M) \supseteq \{\sum_{i=1}^r Q_i \mid Q_i \in \underset{\circ}{\Lambda}{}^*_{P_i}(M),\ 1 \le i \le r\}$ by induction on $|X|$, the cardinality of X. If $|X| = 1$, there is nothing to prove. Assuming the containment holds for $|X| = r - 1$, we show the containment for $X = \{P_1, P_2, \ldots, P_r\}$. After rearrangement if necessary, we may assume that $P_r \not\subseteq P_i$ for $1 \le i \le r-1$. Set $U = R \setminus \bigcup_{i=1}^{r-1} P_i$. Let $Q = \sum_{i=1}^r Q_i$ such that $Q_i \in \underset{\circ}{\Lambda}{}^*_{P_i}(M)$ for $1 \le i \le r$. For any $Q' \in \Lambda^*_X$ such that $Q \supseteq Q'$, we need to show $Q = Q'$. Write $Q' = \sum_{i=1}^r Q_i'$ such that $Q_i' \in \underset{\circ}{\Lambda}{}^*_{P_i}$ for $1 \le i \le r$. Then we have

$$\sum_{i=1}^{r-1} Q_i = [U]Q \supseteq [U]Q' = \sum_{i=1}^{r-1} Q_i',$$

which forces $\sum_{i=1}^{r-1} Q_i = \sum_{i=1}^{r-1} Q_i'$ by the induction hypothesis. Therefore

$$\sum_{i=1}^{r-1} Q_i + (Q' \cap Q_r) = Q' \cap \sum_{i=1}^{r} Q_i = Q' \quad \Big(\text{since } \sum_{i=1}^{r-1} Q_i = \sum_{i=1}^{r-1} Q_i' \subset Q'\Big).$$

Hence we can derive a secondary representation $Q' = \sum_{i=1}^r Q_i''$ by putting together $\sum_{i=1}^{r-1} Q_i$ and any secondary representation of $(Q' \cap Q_r)$ (and then make it minimal). In this derived secondary representation $Q' = \sum_{i=1}^r Q_i''$, the P_r-secondary component, Q_r'', must come from the P_r-secondary component of $(Q' \cap Q_r)$, hence is contained in $Q' \cap Q_r$. Since $Q_r'' \in \Lambda^*_{P_r}(Q')$ and $Q' \in \Lambda^*_X(M)$, we see $Q_r'' \in \Lambda^*_{P_r}(M)$ (by compatibility, for example). This forces $Q_r'' = Q_r$ since Q_r is already a minimal P_r-secondary component of M. Hence $Q' \supseteq Q_r'' = Q_r$, which gives

$$Q = \sum_{i=1}^{r} Q_i = \sum_{i=1}^{r-1} Q_i + Q_r = \sum_{i=1}^{r-1} Q_i' + Q_r \subseteq \sum_{i=1}^{r-1} Q_i' + Q_r' = Q'.$$

Therefore $Q = Q'$, and the proof is complete.

Finally, the first equality of (i′) was done in Lemma 9.5 (iii); and the last equality follows from (i) applied to M as an Artinian module over $\widehat{R}$.

Because of Theorem 11.1, we can fine-tune Theorem 10.3 as follows.

Theorem 11.3. *Let M be an Artinian R-module and $X \subseteq \operatorname{Att}_R(M)$. Consider the following statements:*

(i) X is open in $\operatorname{Att}_R(M)$.	(i′) ${\phi_M^*}^{-1}(X)$ is open in $\operatorname{Att}_{\widehat{R}}(M)$.
(ii) $\lvert\Lambda^*_X(M)\rvert = 1$.	(ii′) $\lvert\Lambda^*_{{\phi_M^*}^{-1}(X)}({}_{\widehat{R}}M)\rvert = 1$.
(iii) $\Lambda^*_X(M)$ is finite.	(iii′) $\Lambda^*_{{\phi_M^*}^{-1}(X)}({}_{\widehat{R}}M)$ is finite.
(iv) $\underset{\circ}{\Lambda}{}^*_X(M)$ is finite.	(iv′) $\underset{\circ}{\Lambda}{}^*_{{\phi_M^*}^{-1}(X)}({}_{\widehat{R}}M)$ is finite.
(v) $\lvert\underset{\circ}{\Lambda}{}^*_X(M)\rvert = 1$.	(v′) $\lvert\underset{\circ}{\Lambda}{}^*_{{\phi_M^*}^{-1}(X)}({}_{\widehat{R}}M)\rvert = 1$.

Then (i) $\Rightarrow$ (ii) $\Rightarrow$ (iii) $\Rightarrow$ (iv) $\Leftrightarrow$ (v) $\Leftrightarrow$ (i′) $\Leftrightarrow$ (ii′) $\Leftrightarrow$ (iii′) $\Leftrightarrow$ (iv′) $\Leftrightarrow$ (v′).

Proof. The implications (i) $\Rightarrow$ (ii) $\Rightarrow$ (iii) $\Rightarrow$ (iv) are clear.

The implications (i′) $\Rightarrow$ (ii′) $\Rightarrow$ (iii′) $\Rightarrow$ (iv′) are clear.

The implications (ii′) $\Rightarrow$ (v′) $\Rightarrow$ (iv′) are clear.

(iv) $\Leftrightarrow$ (iv′) follows from Lemma 9.5 (iii), so does (v) $\Leftrightarrow$ (v′).

(iv) $\Rightarrow$ (i′): Say $\underset{\circ}{\Lambda}{}^*_X(M) = \{Q_1, \dots, Q_t\}$, so $\underset{\circ}{\Lambda}{}^*_{{\phi_M^*}^{-1}(X)}({}_{\widehat{R}}M) = \{Q_1, \dots, Q_t\}$. Let $Q = \sum_{i=1}^t Q_i$. Then $Q \in \Lambda^*_{{\phi_M^*}^{-1}(X)}({}_{\widehat{R}}M)$ by Theorem 8.3. On the other hand, Theorem 11.1 (iii) implies $Q \in \Lambda^*_{\mathrm{o}({\phi_M^*}^{-1}(X))}({}_{\widehat{R}}M)$, in which $\mathrm{o}({\phi_M^*}^{-1}(X))$ denotes the smallest open superset of ${\phi_M^*}^{-1}(X)$ in $\operatorname{Att}_{\widehat{R}}(M)$. By Lemma 8.2, we must have ${\phi_M^*}^{-1}(X) = \mathrm{o}({\phi_M^*}^{-1}(X))$, which is open in $\operatorname{Att}_{\widehat{R}}(M)$. □

By Examples 10.4, 10.5, implications (i) $\Leftarrow$ (ii) and (iii) $\Leftarrow$ (iv) are false in general.

12 Linear Growth of Secondary Components

Inspired by the linear growth of primary decomposition, and taking in account the duality between primary decomposition and secondary representation, we formulate a definition of the linear growth property of secondary representation as follows. We use the notation introduced in Notation 5.3. Let R be a ring.

Definition 12.1. Given a family $\mathcal{F} = \{M_a \mid a = (a_1, \dots, a_r) \in \mathbb{N}^r\}$ of R-modules, we say that $\mathcal{F}$ satisfies the linear growth property of secondary representation over R if there exists $k \in \mathbb{N}$ such that, for every $a = (a_1, \dots, a_r) \in \mathbb{N}^r$ such that $M_a \neq 0$, there exists a secondary representation of M_a

$$M_a = Q_{a,1} + \cdots + Q_{a,s(a)} \quad \text{(with } Q_{a,i} \text{ being } P_{a,i}\text{-secondary)}$$

such that $Q_{a,i} \subseteq (0 :_{M_a} (P_{a,i})^{k|a|})$ for all $i = 1, \dots, s(a)$, where $|a| = a_1 + \cdots + a_r$.

When the above occurs, we refer to k as a *slope* of $\mathcal{F}$.

Lemma 12.2. *Let $h\colon A \to R$ be a ring homomorphism, $\{M_n \mid n \in \mathbb{N}^t\}$ a family of R-modules, $\{K_n \mid n \in \mathbb{N}^t\}$ a family of A-modules such that $K_n \subseteq M_n$ as A-modules for all $n \in \mathbb{N}^t$.*

If $\{M_n \mid n \in \mathbb{N}^t\}$ satisfies the linear growth property of secondary representation over R with a slope k, then $\{M_n/K_n \mid n \in \mathbb{N}^t\}$ satisfies the linear growth property of secondary representation over A with the same slope k.

Proof. This follows (almost immediately) from Theorem 7.3 (iii). □

The next result is dual to Theorem 5.4.

Theorem 12.3. *Let A be a Noetherian ring, R an A-algebra, M a finitely generated A-module, N an Artinian R-module, and $J_1, \ldots, J_t$ ideals of R. Then each of the following families of R-modules has the linear growth property of secondary representation (over R):*

(i) *The family* $\{(0 :_N J^n) \mid n \in \mathbb{N}^t\}$.

(ii) *The family* $\{\mathrm{Ext}_A^c(M, (0 :_N J^n)) \mid n \in \mathbb{N}^t\}$.

(iii) *The family* $\{\mathrm{Tor}_c^A(M, (0 :_N J^n)) \mid n \in \mathbb{N}^t\}$.

Proof. Note that all the modules in all the families are Artinian R-modules. Apply the Matlis duality functor $\mathrm{D}(-)$ to the modules. By Observation 9.4 and Lemma 12.2, we only need to prove the linear growth property of primary decomposition for each of the following families over $\widehat{R}$:

(i*) The family $\{\mathrm{D}(N)/J^n\,\mathrm{D}(N) \mid n \in \mathbb{N}^t\}$.

(ii*) The family $\{\mathrm{Tor}_c^A(M, \mathrm{D}(N)/J^n\,\mathrm{D}(N)) \mid n \in \mathbb{N}^t\}$.

(iii*) The family $\{\mathrm{Ext}_A^c(M, \mathrm{D}(N)/J^n\,\mathrm{D}(N)) \mid n \in \mathbb{N}^t\}$.

Since $\mathrm{D}(N)$ is a Noetherian $\widehat{R}$-module and $\widehat{R}$ is clearly an A-algebra, the desired linear growth property of primary decomposition of the three families follows immediately from Theorem 5.4. □

Theorem 12.3 is a special case of the following dual statement of Theorem 5.5.

Theorem 12.4. *Let A be a ring and R an A-algebra. Let $J_1, J_2, \ldots, J_t$ be fixed ideals of R, N an Artinian R-module and $c \in \mathbb{Z}$. Fix a complex*

$$F_\bullet : \quad \cdots \to F_{c+1} \to F_c \to F_{c-1} \to \cdots \to F_i \to F_{i-1} \to \cdots$$

of finitely generated flat A-modules. For any $n \in \mathbb{N}^t$, let

$$T_n = \mathrm{H}_c(F_\bullet \otimes_A (0 :_N J^n)) \quad \text{and} \quad E_n = \mathrm{H}^c(\mathrm{Hom}_A(F_\bullet, (0 :_N J^n))).$$

Then the families $\{T_n \mid n \in \mathbb{N}^t\}$ and $\{E_n \mid n \in \mathbb{N}^t\}$, both consisting of Artinian R-modules, satisfy the linear growth property of secondary representation (over R).

Proof. By Observation 9.4 and Lemma 12.2, it suffices to show the linear growth property of primary decomposition for $\{\mathrm{D}(T_n) \mid n \in \mathbb{N}^t\}$ and $\{\mathrm{D}(E_n) \mid n \in \mathbb{N}^t\}$ over $\widehat{R}$. By Matlis duality, we have

$$\mathrm{D}(T_n) \cong \mathrm{H}^c\left(\mathrm{Hom}_A\left(F_\bullet, \frac{\mathrm{D}(N)}{J^n\,\mathrm{D}(N)}\right)\right) \quad \text{and} \quad \mathrm{D}(E_n) \cong \mathrm{H}_c\left(F_\bullet \otimes_A \frac{\mathrm{D}(N)}{J^n\,\mathrm{D}(N)}\right).$$

As $\mathrm{D}(N)$ is Noetherian over $\widehat{R}$, both $\{\mathrm{D}(T_n) \mid n \in \mathbb{N}^t\}$ and $\{\mathrm{D}(E_n) \mid n \in \mathbb{N}^t\}$ satisfy the linear growth property of primary decomposition by Theorem 5.5. □

The following may also be viewed as a dual of Theorem 5.5.

Theorem 12.5. *Let A be a ring, $J_1, J_2, \dots, J_t$ fixed ideals of A, and M a finitely generated A-module. Let R be an A-algebra and $c \in \mathbb{Z}$. Fix a complex*

$$F^\bullet: \quad \cdots \to F^i \to F^{i+1} \to \cdots \to F^{c-1} \to F^c \to F^{c+1} \to \cdots$$

of injective Artinian R-modules. Denote $E_n = \mathrm{H}^c(\mathrm{Hom}_A(\frac{M}{J^n M}, F^\bullet))$, the c-th cohomology, for all $n \in \mathbb{N}^t$. Then the family $\{E_n \mid n \in \mathbb{N}^t\}$, consisting of Artinian R-modules, satisfies the linear growth property of secondary representation over R.

Proof. Without affecting the claim, we assume $F^i = 0$ if $i \notin \{c-1, c, c+1\}$. Denote $I = \mathrm{Ann}_{\widehat{R}}(F^{c-1} \oplus F^c \oplus F^{c+1})$. Then $F^\bullet$ is naturally a complex over the complete semi-local Noetherian ring $\widehat{R}/I$ (cf. Observation 9.4). Clearly, each F_j remains injective and Artinian over $\widehat{R}/I$. Thus, replacing R with $\widehat{R}/I$, we may simply assume that R is Noetherian semi-local with $R = \widehat{R}$ (cf. Lemma 12.2).

Now the classic Matlis duality applies, which is still denoted $\mathrm{D}(-)$. What we observed in Observation 9.4 still holds (of course). For each $n \in \mathbb{N}^t$,

$$\begin{aligned} E_n &\cong \mathrm{H}^c\left(\mathrm{Hom}_A\left(\frac{M}{J^n M}, \mathrm{D}(\mathrm{D}(F^\bullet))\right)\right) \cong \mathrm{H}^c\left(\mathrm{D}\left(\frac{M}{J^n M} \otimes_A \mathrm{D}(F^\bullet)\right)\right) \\ &\cong \mathrm{D}\left(\mathrm{H}_c\left(\frac{M}{J^n M} \otimes_A \mathrm{D}(F^\bullet)\right)\right) \cong \mathrm{D}\left(\mathrm{H}_c\left(\frac{M \otimes_A R}{J^n(M \otimes_A R)} \otimes_R \mathrm{D}(F^\bullet)\right)\right). \end{aligned}$$

By Observation 9.4, it suffices to show that the family

$$\left\{\mathrm{D}(E_n) \cong \mathrm{H}_c\left(\frac{M \otimes_A R}{J^n(M \otimes_A R)} \otimes_R \mathrm{D}(F^\bullet)\right) \;\middle|\; n \in \mathbb{N}^t\right\}$$

has linear growth property of primary decomposition over $R = \widehat{R}$. Note that $\mathrm{D}(F^\bullet)$ is a complex of finitely generated projective R-modules; and $M \otimes_A R$ is Noetherian over R. By Theorem 5.5, $\{\mathrm{D}(E_n) \mid n \in \mathbb{N}^t\}$ satisfies the linear growth property of primary decomposition over $R = \widehat{R}$, which completes the proof. □

Now we prove the linear growth property of secondary representation for another family of Ext modules; compare with Theorem 12.3 (ii).

Theorem 12.6. *Let R be a Noetherian ring, $I_1, I_2, \ldots, I_s$ ideals of R, M a finitely generated R-module, N an Artinian R-module, and $c \in \mathbb{Z}$.*

Then the family $\{\mathrm{Ext}_R^c(\frac{M}{I^m M}, N) \mid m \in \mathbb{N}^s\}$, which consists of Artinian R-modules, satisfies the linear growth property of secondary representation over R.

Proof. Since R is Noetherian and N is Artinian, the minimal injective resolution of N consists of Artinian R-modules. Then the result follows from Theorem 12.5. □

Finally, we state a result (partially) dual to Theorem 6.2 and Corollary 6.5. It also contains Theorem 12.6 as a particular case.

Theorem 12.7. *Let R be a Noetherian ring, $I_1, \ldots, I_s$, $J_1, \ldots, J_t$ ideals of R, M a finitely generated R-module, N an Artinian R-module, and $c \in \mathbb{Z}$. For every $(\alpha, \beta) \in \mathbb{Z}^s \times \mathbb{Z}^t$, consider the following family (of Artinian R-modules)*

$$\mathcal{E}^{(\alpha,\beta)} := \left\{ \mathrm{Ext}_R^c\left(\frac{I^\alpha M}{I^{\alpha+m} M}, \frac{(0 :_N J^{\beta+n})}{(0 :_N J^\beta)}\right) \,\middle|\, (m,n) \in \mathbb{N}^s \times \mathbb{N}^t = \mathbb{N}^{s+t} \right\}.$$

Then there exists k such that for all $(\alpha, \beta) \in \mathbb{Z}^s \times \mathbb{Z}^t$, the family $\mathcal{E}^{(\alpha,\beta)}$ satisfies the linear growth property for secondary representation over R with the uniform slope k. That is, for every $(\alpha, \beta) \in \mathbb{Z}^{s+t}$ and for every $(m,n) \in \mathbb{N}^{s+t}$ such that $\mathrm{Ext}_R^c(\frac{I^\alpha M}{I^{\alpha+m} M}, \frac{(0:_N J^{\beta+n})}{(0:_N J^\beta)}) \neq 0$, there exists a secondary representation

$$\mathrm{Ext}_R^c\left(\frac{I^\alpha M}{I^{\alpha+m} M}, \frac{(0 :_N J^{\beta+n})}{(0 :_N J^\beta)}\right) = Q_{\alpha,\beta,m,n,1} + Q_{\alpha,\beta,m,n,2} + \cdots + Q_{\alpha,\beta,m,n,s(\alpha,\beta,m,n)},$$

with $Q_{\alpha,\beta,m,n,i}$ being $P_{\alpha,\beta,m,n,i}$-secondary, such that

$$Q_{\alpha,\beta,m,n,i} \subseteq \left(0 :_{\mathrm{Ext}_R^c\left(\frac{I^\alpha M}{I^{\alpha+m} M}, \frac{(0:_N J^{\beta+n})}{(0:_N J^\beta)}\right)} (P_{\alpha,\beta,m,n,i})^{k|(m,n)|}\right)$$

for all $i = 1, 2, \ldots, s(\alpha, \beta, m, n)$.

In particular, $\{\mathrm{Ext}_R^c(\frac{M}{I^m M}, (0 :_N J^n)) \mid (m,n) \in \mathbb{N}^s \times \mathbb{N}^t = \mathbb{N}^{s+t}\}$ satisfies the linear growth property of secondary representation over R.

Proof. As R is Noetherian, the minimal injective resolution of an Artinian R-module consists of Artinian R-modules. For the Artinian R-module N, there are only finitely many maximal ideals $\mathfrak{m}$ such that $\Gamma_{\mathfrak{m}}(N) \neq 0$; say $\mathfrak{m}_1, \ldots, \mathfrak{m}_r \in \mathrm{MSpec}(R)$ are all such maximal ideals. Let $B = \prod_{i=1}^r \widehat{R_{\mathfrak{m}_i}}$, which is a Noetherian flat R-algebra. Note that N, naturally a B-module, is Artinian over B.

Moreover, $\frac{(0:_N J^{\beta+n})}{(0:_N J^\beta)}$ are all naturally Artinian B-modules for all $\beta \in \mathbb{Z}^t$ and all $n \in \mathbb{N}^t$. Therefore, $\mathrm{Ext}_R^c(\frac{I^\alpha M}{I^{\alpha+m} M}, (0 :_N J^{\beta+n})(0 :_N J^\beta))$ are (naturally) Artinian

B-module for all $(\alpha, \beta) \in \mathbb{Z}^s \times \mathbb{Z}^t$ and $(m, n) \in \mathbb{N}^s \times \mathbb{N}^t$. By Lemma 12.2, it suffices to prove that the families $\mathcal{E}^{(\alpha,\beta)}$ satisfy the linear growth property for secondary representation over B with a uniform slope. Note that B is a complete semi-local ring.

Next, we apply the Matlis duality functor $\mathrm{D}(-)$ (over B) to the modules in the families $\mathcal{E}^{(\alpha,\beta)}$. By Observation 9.4, we only need to prove the linear growth property of primary decomposition, with a uniform slope, for the following families over B:

$$\mathrm{D}\,\mathcal{E}^{(\alpha,\beta)} = \left\{\mathrm{Tor}_c^R\left(\frac{I^\alpha M}{I^{\alpha+m}M}, \frac{J^\beta\,\mathrm{D}(N)}{J^{\beta+n}\,\mathrm{D}(N)}\right) \;\middle|\; (m,n) \in \mathbb{N}^s \times \mathbb{N}^t = \mathbb{N}^{s+t}\right\}.$$

Note that $\mathrm{D}(N)$ is Noetherian over B, while $B = \prod_{i-1}^r \widehat{R_{\mathfrak{m}_i}}$ is a Noetherian ring that is flat over R.

Now, by Theorem 6.2, all the families $\mathrm{D}\,\mathcal{E}^{(\alpha,\beta)}$ satisfy the linear growth property of primary decomposition over B with a uniform slope. The proof is complete. □

Acknowledgments. Some of the materials are from the author's thesis [18]. The author thanks his advisor, Craig Huneke, for help and support.

The author also thanks the anonymous referee for the suggestions, which improved the presentation of the paper.

Bibliography

[1] Atiyah, M. F., Macdonald, I. G., Introduction to commutative algebra. Cambridge, MA: Perseus Books Publishing, L.L.C.; 1969.

[2] Bourbaki, N., Elements of mathematics. Commutative algebra [translated from French]. Paris: Addison-Wesley Publishing Co. Hermann; 1972.

[3] Eisenbud, D., Commutative algebra with a view towards algebraic geometry. vol. 150 of Graduate texts in mathematics. New York: Springer-Verlag; 1995.

[4] Heinzer, W., Ratliff Jr, L. J., Shah, K., On the embedded primary components of ideals (I). J Algebra. 1994;167(3):724–744.

[5] Kirby, D., Coprimary decomposition of Artinian modules. J Lond Math Soc (2). 1973;6: 571–576.

[6] Macdonald, I. G., Secondary representation of modules over a commutative ring. Proceedings, Convegno di Algebra Commutativa, INDAM; 1971; Rome. vol. XI of Symposia mathematica. p. 23–43.

[7] Matsumura, H., Commutative ring theory. Cambridge University Press.

[8] Sharp, R. Y., Secondary representations for injective modules over commutative Noetherian rings. Proc Edinb Math Soc (2). 1976;20(2):143–151.

[9] Sharp, R. Y., A method for the study of Artinian modules, with an application to asymptotic behavior. Proceedings, Commutative Algebra; 1987; Berkeley, CA. vol. 15 of Mathematical Sciences Research Institute publications. New York: Springer; 1989. p. 443–465,

[10] Sharp, R. Y., Artinian modules over commutative rings. Math Proc Cambridge Philos Soc. 1992;111(1):25–33.

[11] Sharp, R. Y., Linear growth of primary decompositions of integral closures. J Algebra 1998;207(1):276–284.

[12] Sharp, R. Y., Injective modules and linear growth of primary decompositions. Proc Amer Math Soc. 2000;128(3):717–722.

[13] Stalvey, H., Weak primary decomposition of modules over a commutative ring [MSc thesis]. Georgia State University; 2010.

[14] Swanson, I., Powers of ideals: primary decompositions, Artin–Rees lemma and regularity. Math Ann. 1997:307(2):299–313.

[15] Swanson, I., Primary decompositions. Commutative algebra and combinatorics. vol 4 of Ramanujan Mathematical Society lecture notes series. Mysore: Ramanujan Mathematical Society; 2007. p. 117–155.

[16] Yao, Y., Primary decomposition: compatibility, independence and linear growth. Proc Amer Math Soc. 2002;130(6):1629–1637.

[17] Yao, Y., Primary decomposition II: primary components and linear growth. J Pure Appl Algebra 2006:205(1):226–242.

[18] Yao, Y., Finite F-representation type and primary decomposition [PhD thesis]. University of Kansas; 2002.

Author Information

Yongwei Yao, Department of Mathematics and Statistics, Georgia State University, Atlanta, GA, USA.
E-mail: yyao@gsu.edu

Progress in Commutative Algebra 1, 271–292

Recent Progress in Coherent Rings: a Homological Perspective

Livia Hummel

Abstract. Theories of coherent Cohen–Macaulay and Gorenstein rings have recently been developed by Hamilton and Marley, and Hummel and Marley, respectively. This work summarizes these theories after introducing the homological framework upon which they are built. We also explore recent developments in the theory of homological dimensions. These developments may provide further insight into the properties of coherent Cohen–Macaulay and Gorenstein rings, in addition to insight into the development of a characterization for coherent complete intersection rings.

Keywords. Gorenstein Dimension, FP-Injective Dimension, Coherent Ring, Gorenstein, Cohen–Macaulay, Complete Intersection.

2010 Mathematics Subject Classification. 13D05, 13D07, 13-02.

1 Introduction

Homological dimensions have been studied by Auslander and Bridger [4], Gerko [23], Avramov, Gasharov and Peeva [6], as well as many others, to create dimensions characterizing local *Noetherian* regular, complete intersection, Gorenstein, and Cohen–Macaulay rings. Others, including Bennis and Mahdou [9, 10], and Mao and Ding [37, 39], explored (global) Gorenstein dimensions in the coherent and Noetherian contexts. Through these explorations, a homological dimension introduced by Stenström [46] has been connected to flat and Gorenstein dimensions.

Concurrent with these activities is the exploration of the meaning of regular, complete intersection, Gorenstein and Cohen–Macaulay in the coherent context. A coherent ring is *regular* if every finitely generated ideal of the ring has finite projective dimension [11]. Glaz posed the question of whether there existed a theory of coherent Cohen Macaulay rings such that coherent regular rings are Cohen–Macaulay (see [26] and [27]). Hamilton and Marley [32] provided a positive answer to this question through homological methods. Hummel and Marley [34] extended the notion of Gorenstein dimension to lay the foundation for a theory of non-Noetherian, and even non-coherent, Gorenstein rings. This foundation has played a part in creating a rich theory of coherent rings where coherent regular, Gorenstein, and Cohen–Macaulay rings behave (mostly) like their Noetherian counterparts. Complete intersections are

This work was completed during a Triennial release granted by the University of Indianapolis.

thus far the missing character in this theory; their foundations likely still lie within the realm of homological dimensions.

The groundwork for the coherent theory comes from both homological dimensions and local cohomology. This work describes how the ever growing homological theory impacts the theory of coherent rings, and explores the homological methods that may lead to the expansion of this theory.

2 Coherent Rings and Grade

We say $(R, \mathfrak{m})$ is a local ring if it has a unique maximal ideal $\mathfrak{m}$. In this work all rings will be commutative, unless otherwise noted.

2.1 Coherent Rings and $(\mathrm{FP})_\infty^R$ Modules

A finitely generated module M of a ring R is *coherent* if every finitely generated submodule of M is finitely presented. A ring R is coherent if it is coherent as an R-module. Additional characterizations of coherent rings can be found in [25].

One important characterization of coherent rings is that any finitely presented module M over a coherent ring has an infinite resolution by finite free modules [25]. Moreover, $M^* = \mathrm{Hom}_R(M, R)$ has the same property. This property of modules in a coherent ring is denoted $(\mathrm{FP})_\infty^R$ by Bieri.

Definition 2.1 ([12]). Let R be a ring and let M be an R-module. M is $(\mathrm{FP})_\infty^R$ if M admits an infinite resolution of finitely generated free modules. If, instead, M admits a finite free resolution of length n, we say M is $(\mathrm{FP})_n^R$.

It follows that for any finitely presented R-module M over a coherent ring R, M and M^* are $(\mathrm{FP})_\infty^R$. Many of the properties of $(\mathrm{FP})_\infty^R$ modules have been explored in [12], [42], and [34]. Most of the interesting properties of $(\mathrm{FP})_\infty^R$ modules usually occur in the case where both the module and its dual are $(\mathrm{FP})_\infty^R$, that is, in the coherent-like case [34]. The assumption of M and M^* being in $(\mathrm{FP})_\infty^R$ carries the full force of coherence, without additional restrictions associated with the coherence assumption (see [34] and Section 4.2).

2.2 Non-Noetherian Grade

In the Noetherian case, the classical Noetherian notion of the *Depth*, or grade, of a module over an ideal I is defined as $\mathrm{Depth}_I\, M = \sup\{n | x_1, \dots, x_n \in I$ is an M-regular sequence$\}$. In the Noetherian case, Depth exhibits the following property.

Proposition 2.2 ([43]). *Let R be a local Noetherian ring, and let M be a finitely generated R-module. Then* $\mathrm{Depth}_I\, M > 0$ *if and only if* $(0 :_M I) = 0$.

However, there are examples of non-Noetherian rings where Proposition 2.2 does not hold (see, for instance [32]). To rectify this incongruity, Hochster extended Depth to non-Noetherian rings.

Definition 2.3 ([33]). Let R be a ring, let M be an R-module, and let I be an ideal of R. The depth of M with respect to I is defined as

$$\operatorname{depth}_I M = \sup\{\operatorname{Depth}_{IS}(M \otimes_R S) | S \text{ faithfully flat extension of R}\}.$$

If $(R, \mathfrak{m})$ is a local ring and $I = \mathfrak{m}$, then denote $\operatorname{depth}_{\mathfrak{m}} M = \operatorname{depth}_R M$, or depth M when the ring is unambiguous.

In the literature, depth has also been called polynomial, or p-depth (see [32, 42]). This definition of depth has most of the expected properties, which are summarized below.

Proposition 2.4 ([8; 33; 25, Chapter 7; 43, Chapter 5; 13, Section 9]). *Let M be an R-module and let I be an ideal of R such that $IM \neq M$.*

(i) $\operatorname{depth}_I M = \sup\{\operatorname{depth}_J M | J \subset I, J$ *finitely generated ideal*$\}$.

(ii) *Let $I = (x_1, \dots, x_n)$ and $H_j(\mathbf{x}, M)$ denote the jth Koszul homology of $\mathbf{x} = x_1, \dots, x_n$ on M, then* $\operatorname{depth}_I M = \inf\{i \geq 0 | H_{n-i}(\mathbf{x}, M) \neq 0\}$.

(iii) $\operatorname{depth}_I M = \operatorname{depth}_{IS}(M \otimes_R S)$ *for any faithfully flat R-algebra S.*

(iv) *If* $\operatorname{depth}_I M > 0$, *then* $\operatorname{Depth}_{IS}(M \otimes_R S) > 0$ *where $S = R[X]$ is a polynomial ring in one variable over R.*

(v) *If I is generated by n elements, then* $\operatorname{depth}_I M = \operatorname{Depth}_{IS}(M \otimes_R S)$ *where $S = R[X_1, \dots X_n]$.*

(vi) $\operatorname{depth}_I M = \operatorname{depth}_{\sqrt{I}} M$.

(vii) *If $x \in I$ is M-regular, then* $\operatorname{depth}_I M = \operatorname{depth}_I M/xM + 1$.

(viii) *Let $0 \longrightarrow L \longrightarrow M \longrightarrow N \longrightarrow 0$ be a short exact sequence of R-modules such that $IL \neq L$ and $IN \neq N$. If* $\operatorname{depth}_I M > \operatorname{depth}_I N$, *then* $\operatorname{depth}_I L = \operatorname{depth}_I N + 1$.

If in addition one assumes that R/I is $(\mathrm{FP})_{\infty}^{R}$, Hummel and Marley obtain the following homological characterization of depth for coherent rings.

Proposition 2.5 ([34]). *Let R be a ring, let M be an R-module, and let I be an ideal such that $IM \neq M$. If R/I is $(\mathrm{FP})_n^R$, then the following conditions are equivalent.*

(i) $\operatorname{depth}_I M \geq n$.

(ii) $\operatorname{Ext}_R^i(R/I, M) = 0$, *for* $0 \leq i < n$.

In particular, if $(R, \mathfrak{m})$ is a local coherent ring and $M \neq 0$ such that $\mathfrak{m}M \neq M$, then

$$\operatorname{depth} M = \sup\{n \geq 0 | \operatorname{Ext}_R^i(R/I, M) = 0 \textit{ for all } i < n \textit{ for some f.g. ideal } I \subset \mathfrak{m}\}.$$

This result connects the polynomial depth to the r-depth (as denoted by Barger [8]) based upon the vanishing of $\mathrm{Ext}(R/I, -)$-modules.

There are other notions of grade in the non-Noetherian case, based on the vanishing of the homology of the Hom of a Koszul a complex, of the Čech cohomology of a module, of the local cohomology of a module, of the $\mathrm{Ext}(R/I^n, -)$-modules, or of chain maps between complexes. Asgharzadeh and Tousi [3] explore the connections between these different grades to create additional characterizations of coherent Cohen–Macaulay modules (see Section 3).

3 Cohen–Macaulay Rings

Non-Noetherian Cohen–Macaulay rings were defined by Hamilton and Marley in [32]. Their homological approach to this question uses Schenzel's [45] notion of parameter sequences. Over non-Noetherian rings, parameter sequences play the role of systems of parameters over Noetherian rings. However parameter sequences are defined homologically rather than by height conditions.

Let $\mathbf{x} = x_1, \ldots, x_n$ be a finite sequence of elements in R. Given an R-module M, let $\check{H}^i_{\mathbf{x}}(M)$ be the ith Čech cohomology of M with respect to $\mathbf{x}$ and $H^i_{\mathbf{x}}(M)$ be the ith local cohomology of M. Schenzel gives the following definitions.

Definition 3.1 ([45]). Let R be a ring.

(i) The sequence $\mathbf{x}$ is weakly proregular if for all $i \geq 0$ and all R-modules M the natural map $H^i_{\mathbf{x}}(M) \longrightarrow \check{H}^i_{\mathbf{x}}(M)$ is an isomorphism.

(ii) The sequence $\mathbf{x} = x_1, \ldots, x_n$ is a parameter sequence if $\mathbf{x}$ is weakly proregular, $(\mathbf{x})R \neq R$, and $H^n_{\mathbf{x}}(R)_p \neq 0$ for all prime ideals p containing $\mathbf{x}$.

(iii) The sequence $\mathbf{x}$ is a strong parameter sequence if $x_1, \ldots, x_i$ is a parameter sequence for all $1 \leq i \leq n$.

Using these definitions, Hamilton and Marley define Cohen–Macaulay.

Definition 3.2 ([32]). A ring R is Cohen–Macaulay if every strong parameter sequence of R is a regular sequence.

Hamilton and Marley show that Cohen–Macaulay rings have the following properties.

Proposition 3.3 ([32]). *The following conditions are equivalent for a ring R.*

(i) *R is Cohen–Macaulay.*

(ii) $\mathrm{Depth}(\mathbf{x})R = \ell(x)$ *for every strong parameter sequence* $\mathbf{x}$ *of* R.

(iii) $\mathrm{depth}(\mathbf{x})R = \ell(x)$ *for every strong parameter sequence* $\mathbf{x}$ *of* R.

(iv) $H_i(\mathbf{x}; R) = 0$ *for all* $i \geq 1$ *and every strong parameter sequence* $\mathbf{x}$ *of* R.

(v) $H^i_{\mathbf{x}}(R) = 0$ *for all* $i < \ell(x)$ *and every strong parameter sequence* $\mathbf{x}$ *of* R.

The following results of Hamilton and Marley show the extent to which the Cohen–Macaulay property coincides with the Noetherian case.

Proposition 3.4 ([32]). *Let* R *be a ring.*

(i) *Let* $f : R \longrightarrow S$ *be a faithfully flat ring homomorphism, if* S *is Cohen–Macaulay then so is* R.

(ii) *If* $R[x]$ *is Cohen–Macaulay, then so is* R.

(iii) *If* $R_{\mathfrak{m}}$ *is Cohen–Macaulay for all maximal ideals* $\mathfrak{m}$ *of* R*, then so is* R.

It is unknown whether the converse of the statements above are true. In particular, since it is not known that the Cohen–Macaulay property localizes, Hamilton and Marley introduce the following modified definition of Cohen–Macaulay.

Definition 3.5 ([32]). A ring R is locally Cohen–Macaulay if R_p is Cohen–Macaulay for all $p \in \operatorname{Spec} R$.

Thus coherent regular and locally Cohen–Macaulay rings are related in the same way as their Noetherian counterparts.

Theorem 3.6 ([32]). *Coherent regular rings are locally Cohen–Macaulay.*

On the other hand, with the removal of the Noetherian assumption, the Cohen–Macaulay property is not retained after reduction by a non-zerodivisor, as seen in the following example.

Example 3.7 ([32, Example 4.9]). Let $S = \mathbb{C}[[x, y]]$ be the ring of formal power series in x and y over the field of complex numbers. Let $R = \mathbb{C} + x\mathbb{C}[[x, y]] \subseteq S$. R is a local Cohen–Macaulay domain, but R/xyR is not Cohen–Macaulay.

Towards the conclusion of their work, Hamilton and Marley [32] consider additional characterizations of Cohen–Macaulay rings. We begin with a few additional definitions. A prime ideal P is *weakly associated to* M if P is minimal over $(0 :_R x)$ for some $x \in R$. The set of weakly associated primes of M is denoted $\operatorname{wAss}_R M$. An ideal I of a ring is said to be *unmixed* if $\operatorname{wAss}_R R/I = \operatorname{Min}_R R/I$, the minimal primes of R/I over R. Using these definitions, additional properties of Cohen–Macaulay rings include the following.

Theorem 3.8 ([32]). *Let* R *be a ring.*

(i) *If every ideal of* R *generated by a strong parameter sequence is unmixed, then* R *is Cohen–Macaulay.*

(ii) *If* R *is an excellent Noetherian domain of characteristic* $p > 0$*, then* R^+ *is Cohen–Macaulay.*

(iii) *Let R be a coherent ring with* dim $R \leq 2$, *and let G be a finite group of automorphisms of R with* $|G|$ *a unit in R. Let* R^G *be the subring of invariants of R under the action of G and assume R is a finite* R^G*-module. Then* R^G *is a coherent locally Cohen–Macaulay module.*

Asgharzadeh and Tousi [3] also look at other characterization of Cohen–Macaulay rings in this context. However they use an approach used by Hamilton [31] to explore the question of coherent Cohen–Macaulay rings. Their work [3] compares characterizations of Cohen–Macaulay based upon height conditions of prime ideals to the definition of Hamilton and Marley. These other characterizations of Cohen–Macaulay use the notion of Koszul grade introduced by Alfonsi.

Definition 3.9 ([1]). Let R be a ring, let $I = (\underline{x}) = (x_1, \ldots, x_r)$ be an ideal of R, and let M be an R-module. If $\mathbb{K}_\bullet(\underline{x})$ is the Koszul complex of $(\underline{x})$, the Koszul grade is given by $\text{K.grade}_R(I, M) = \inf\{i \in \mathbb{N} \cup \{0\} | H^i(\text{Hom}_R(\mathbb{K}_\bullet(\underline{x}), M)) \neq 0\}$.

In the following definition, let $\mu(I)$ denote the minimal number of elements of a ring R needed to generate an ideal I of R. Denote the support of an R-module M by $\text{Supp}_R(M)$ and let $\text{Max}(R)$ denote the set of maximal ideals of the ring R.

Definition 3.10 ([3]). Let R be a ring and let M be an R-module.

(i) Hamilton–Marley Cohen–Macaulay [32]: R is Hamilton–Marley Cohen–Macaulay if R satisfies Definition 3.2.

(ii) Glaz Cohen–Macaulay [28]: M is Glaz Cohen–Macaulay if

$$\text{height}_R(p) = \text{K.grade}_{R_p}(pR_p, M_p).$$

(iii) WB Cohen–Macaulay [31]: R is WB Cohen–Macaulay if for each ideal I with height $I \geq \mu(I)$, then I is unmixed (also known as weak Bourbaki unmixed).

(iv) Spec Cohen–Macaulay [3]: M is Spec Cohen–Macaulay if

$$\text{height}_M(I) = \text{K.grade}_R(I, M) \text{ for all ideals } I \in \text{Supp}_R(M).$$

(v) Max ideals Cohen–Macaulay [3]: M is Max ideals Cohen–Macaulay if

$$\text{height}_M(I) = \text{K.grade}_R(I, M) \text{ for all ideals } I \in \text{Supp}_R(M) \cap \text{Max}(R).$$

(vi) f.g ideals Cohen–Macaulay [3]: M is f.g. ideals Cohen–Macaulay if

$$\text{height}_M(I) = \text{K.grade}_R(I, M) \text{ for all finitely generated ideals } I \text{ of } R.$$

(vii) ideals Cohen–Macaulay [3]: M is ideals Cohen–Macaulay if

$$\text{height}_M(I) = \text{K.grade}_R(I, M) \text{ for all ideals } I \text{ of } R.$$

Asgharzadeh and Tousi [3] show the following relations between these definitions of Cohen–Macaulay

$$\text{Max ideals} \Leftarrow \text{Spec} \Leftrightarrow \text{ideals} \Rightarrow \text{Glaz} \Rightarrow \text{f.g ideals} \Rightarrow \text{Hamilton-Marley} \Leftarrow \text{WB},$$

and provide examples of the non-existence of some of the missing implications above. See [3] for additional details.

4 Gorenstein Dimensions and the Auslander–Bridger Property

4.1 Gorenstein Dimensions

In [23], Gerko lists several properties that any generalized homological dimension should naturally fulfill. These are listed below for later reference.

Remark 4.1 ([23]). Let R be a ring, let $\mathcal{H}_R$ be a class of modules, and let $\mathcal{H}\text{-dim}_R$ be a homological dimension such that $\mathcal{H}\text{-dim}_R$ maps $\mathcal{H}_R$ into $\mathbb{Z}$. The following properties should hold for $\mathcal{H}$-dim.

(i) If $M \in \mathcal{H}_R$ then $\mathcal{H}\text{-dim}_R\, M + \operatorname{depth} M = \operatorname{depth} R$.

(ii) Let x be an R- and M-regular element. If $M \in \mathcal{H}_R$, then $M/xM \in \mathcal{H}_{R/xR}$ and $\mathcal{H}\dim_R M = \mathcal{H}\text{-dim}_{R/xR} M/xM$.

(iii) If $M \in \mathcal{H}_R$, then $M_\mathfrak{p} \in \mathcal{H}_{R_\mathfrak{p}}$ and $\mathcal{H}\text{-dim}_R\, M \geq \mathcal{H} - \dim_{R_\mathfrak{p}} M_\mathfrak{p}$.

(iv) Given an exact sequence $0 \longrightarrow L \longrightarrow M \longrightarrow N \longrightarrow 0$ of R-modules, if any two of the modules belongs to $\mathcal{H}_R$ then the third does as well.

If R is Noetherian, the following condition also holds.

(v) $k = R/\mathfrak{m} \in \mathcal{H}_R$ if and only if $M \in \mathcal{H}_R$ for all R-modules.

In this section, as well as in Sections 5 and 6, we discuss homological dimensions that have been explored in the coherent context. We begin the discussion with Gorenstein dimensions.

Definition 4.2 ([4]). Let R be a ring and let M be a finitely generated R-module.

(i) M is in the class $G(R)$ if

(a) $\operatorname{Ext}^i_R(M, R) = \operatorname{Ext}^i_R(M^*, R) = 0$ for all $i \geq 0$.

(b) $M \cong M^{**}$.

(ii) M has Gorenstein dimension n, denoted $\operatorname{Gdim} M = n$, if there exists a minimal length exact resolution $0 \longrightarrow G_n \longrightarrow \cdots \longrightarrow G_0 \longrightarrow M \longrightarrow 0$ such that $G_i \in G(R)$ for each i. If no finite resolution exists, then $\operatorname{Gdim} M = \infty$.

Projective modules are in $G(R)$, and have Gorenstein dimension zero; modules with finite projective dimension thus have finite Gorenstein dimension. Gorenstein dimension satisfies the properties of Remark 4.1 as shown in [4] and is thoroughly summarized by Christensen in [14] and extended to the finitely presented modules over a coherent ring by [41], [19], and [34].

McDowell [41] extends most of the results of Gorenstein dimensions to finitely generated modules over local coherent rings. Taking a different approach, Hummel and Marley [34] simply modify the assumption on the module, providing the following definition of Gorenstein dimension for coherent-like modules.

Definition 4.3 ([34]). Let R be a ring and let M be an R-module.

(i) M is in the class $\tilde{\mathrm{G}}(R)$ if

(a) M and M^* are $(\mathrm{FP})_\infty^R$.

(b) $\mathrm{Ext}_R^i(M, R) = \mathrm{Ext}_R^i(M^*, R) = 0$ for all $i \geq 0$.

(c) $M \cong M^{**}$.

(ii) M has $\tilde{\mathrm{G}}$-dimension n, denoted $\tilde{\mathrm{G}}\dim M = n$, if there exists an exact resolution $0 \longrightarrow G_n \longrightarrow \cdots \longrightarrow G_0 \longrightarrow M \longrightarrow 0$ of minimal length such that $G_i \in \tilde{\mathrm{G}}(R)$ for each i. If no finite resolution exists, $\tilde{\mathrm{G}}\dim M = \infty$.

The characteristics of modules in coherent rings leads to the following result.

Proposition 4.4 ([34]). *If* $(R, \mathfrak{m})$ *is a local coherent ring, then* $\mathrm{Gdim}\, M = \tilde{\mathrm{G}}\dim M$ *for every finitely presented R-module M.*

The properties of $\tilde{\mathrm{G}}\dim$, explored in [34] and [42], satisfy the first four properties of Remark 4.1 and are analogous to Gorenstein dimension.

Taking a different approach to generalizing Gorenstein dimension over finitely presented modules, Enochs and Jenda developed Gorenstein projective dimension.

Definition 4.5 ([19]). A complex $\mathbf{E} : \cdots \longrightarrow P_1 \longrightarrow P_0 \longrightarrow P^0 \longrightarrow P^1 \longrightarrow \cdots$ of modules is called acyclic if $H_i(\mathbf{E}) = 0$, where $H_i(\mathbf{E})$ is the ith homology module of $\mathbf{E}$.

Definition 4.6 ([19]). Let R be a ring.

(i) An R-module M is Gorenstein projective if there is an acyclic complex $\mathbf{P}$ of projective R-modules with $\mathrm{Coker}\, P^0 \longrightarrow P^1 \cong M$ and $\mathrm{Hom}(\mathbf{P}, Q) = 0$ for every projective R-module Q.

(ii) The Gorenstein projective dimension of a module M, denoted $\mathrm{Gpd}\, M$, is n if

$$0 \longrightarrow G_n \longrightarrow G_{n-1} \longrightarrow \cdots \longrightarrow G_1 \longrightarrow G_0 \longrightarrow M \longrightarrow 0$$

is an exact resolution of minimal length such that G_i is Gorenstein projective.

Again, projective modules are Gorenstein projective, and modules with finite projective dimension have finite Gorenstein projective dimension. Gorenstein projective dimensions also satisfy the properties of Remark 4.1. Over Noetherian rings $\operatorname{Gdim} M = \operatorname{Gpd} M$ [7]; Hummel and Marley [34] show the equality holds in the coherent case for finitely presented modules. Analogous definitions can be given for Gorenstein injective and Gorenstein flat modules.

Definition 4.7 ([19]). Let R be a ring.

(i) An R-module M is Gorenstein injective if there is an acyclic complex

$$\mathbf{E} : \cdots \longrightarrow E^1 \longrightarrow E^0 \longrightarrow E_0 \longrightarrow E_1 \longrightarrow \cdots$$

of injective R-modules with

$$\operatorname{Coker}(E^0 \longrightarrow E^1) \cong M \quad \text{and} \quad \operatorname{Hom}(N, \mathbf{E}) = 0$$

for every injective R-module N.

(ii) An R-module M is Gorenstein flat if there is an acyclic complex

$$\mathbf{F} : \cdots \longrightarrow F^1 \longrightarrow F^0 \longrightarrow F_0 \longrightarrow F_1 \longrightarrow \cdots$$

of flat R-modules with

$$\operatorname{Coker}(F^0 \longrightarrow F^1) \cong M \quad \text{and} \quad N \otimes \mathbf{F} = 0$$

for every injective R-module N.

Gorenstein flat and injective dimensions are defined analogously to Gorenstein projective dimension. All three Gorenstein dimensions are being actively studied by many authors including Bennis and Mahdou [9, 10], and Mao and Ding [37, 39], among others.

More recently Iyengar and Krause [35], Christensen and Veliche [16], Sather-–Wagstaff, Sharif and White [44], and others have studied Gorenstein projective, injective, and flat modules in the context of totally acyclic complexes. A *totally acyclic* complex $\mathbf{M}$ is an acyclic complex that satisfies the following equivalent conditions.

Proposition 4.8 ([14, 15]). *Let R be a ring, and let $\mathbf{M}$ be an acyclic complex of finitely generated projective R-modules. Then the following conditions are equivalent.*

(i) $\operatorname{Hom}_R(\mathbf{M}, R)$ *is acyclic.*

(ii) $\operatorname{Hom}_R(\mathbf{M}, F)$ *is acyclic for every flat R-module F.*

(iii) $E \otimes_R \mathbf{M}$ *is acyclic for every injective R-module E.*

Using the definition above, we see that Gorenstein projective, injective and flat modules are kernels of totally acyclic complexes.

In any ring R, these three homological dimensions are related in the following way for any R-module M

$$\operatorname{Gdim}_R M \leq \operatorname{Gpd} M \leq \operatorname{pd}_R M,$$

where equality holds in the first spot if R is coherent and M finitely presented; equality holds in the second spot if M has finite projective dimension.

4.2 The Auslander–Bridger Formula

This section explores the several iterations of Remark 4.1(i) for Gorenstein dimension. We start with the Auslander–Buchsbaum formula, which relates projective dimension to Depth.

Theorem 4.9 ([5]). *Let R be a local Noetherian ring and let M be an R-module with* $\operatorname{pd} M < \infty$. *Then*

$$\operatorname{pd} M + \operatorname{Depth} M = \operatorname{Depth} R.$$

The Auslander–Bridger formula provides a link between Depth and Gorenstein dimension.

Theorem 4.10 ([4]). *Let R be a local Noetherian ring and let M be an R-module with* $\operatorname{Gdim} M < \infty$. *Then*

$$\operatorname{Gdim} M + \operatorname{Depth} M = \operatorname{Depth} R.$$

The Auslander–Bridger formula was first extended to coherent rings by McDowell [41], who considered a subclass of coherent rings called pseudo-Noetherian rings.

Definition 4.11 ([41]). Let R be a ring and let M be a nonzero R-module. R is pseudo-Noetherian if

(i) R is coherent, and

(ii) if for any finitely generated ideal I contained in set of zero-divisors of M, there exists a nonzero m in M with $Im = 0$.

Any R-module M satisfying the second condition is called pseudo-Noetherian.

Note that the property held by pseudo-Noetherian modules is a characteristic of modules over Noetherian rings (see [36]). However, not all coherent rings are pseudo-Noetherian, as seen in the following example.

Example 4.12 ([41]). Let K be the quotient field of $\mathbb{Z}[x]_{(2,z)}$, and let R be the power series ring $K[[t]]$. R is a coherent domain, but R/tR is not a pseudo-Noetherian R-module.

On the other hand, a ring whose modules are all pseudo-Noetherian may not be coherent; this is the case for any non-coherent generalized valuation ring [41].

McDowell [41] used the characterization of depth over coherent local rings from Proposition 2.5 as the definition of depth over pseudo-Noetherian rings. Over local pseudo-Noetherian rings, if M is a finitely presented module, $\operatorname{depth} M$ is the length of a maximal M-regular sequence [41]. In conjunction with McDowell's extension of Gorenstein dimension to coherent rings, the Auslander–Bridger formula can be generalized to pseudo-Noetherian rings.

Theorem 4.13 ([41]). *Let R be a local pseudo-Noetherian ring, and let M be a nonzero finitely presented R-module with* $\operatorname{Gdim} M < \infty$. *Then,*

$$\operatorname{Gdim} M + \operatorname{depth} M = \operatorname{depth} R.$$

Generalizing the Auslander–Bridger formula to any coherent rings is problematic. A key step in the proof of the coherent result requires both Gorenstein dimension and coherence to pass through faithfully flat extensions. However it is well known that coherence is not maintained under faithfully flat extensions; for instance $R[x]$ is not necessarily coherent even if R is coherent. In fact, much attention has been concentrated on the question of what conditions guarantee that coherence is maintained under faithfully flat extensions, and requires the assumption that the ring has finite weak dimension. The *weak dimension* of a ring R is defined as $\sup\{\operatorname{fd} M \mid M \text{ an } R\text{-module}\}$ (see [25] and [1]). As shown in [34], $(\mathrm{FP})_\infty^R$ is preserved under faithfully flat extensions. Thus the $(\mathrm{FP})_\infty^R$ assumption in the next result serves as a stepping stone to the coherent result desired.

Theorem 4.14 (Generalized Auslander–Bridger Formula [34]). *Let R be a local ring and let M be an R-module with* $\tilde{\operatorname{G}}\operatorname{dim} M < \infty$. *Then*

$$\tilde{\operatorname{G}}\operatorname{dim} M + \operatorname{depth} M = \operatorname{depth} R.$$

As finitely presented modules over a coherent ring are $(\mathrm{FP})_\infty^R$, the coherent case follows easily.

Corollary 4.15 ([34]). *Let R be a local coherent ring and let M be an R-module with* $\operatorname{Gdim} M < \infty$. *Then*

$$\operatorname{Gdim} M + \operatorname{depth} M = \operatorname{depth} R.$$

5 Gorenstein Rings and Injective Dimensions

Recall that over Noetherian and coherent rings, Gorenstein dimension and Gorenstein projective dimension coincide. Over Noetherian rings, Gorenstein rings have been characterized via Gorenstein dimensions.

Proposition 5.1 ([4, 19]). *Let $(R, \mathfrak{m}, k)$ be a Noetherian ring. The following conditions are equivalent.*

(i) *R is Gorenstein.*

(ii) $\operatorname{Gdim} M < \infty$ *for all R-modules M.*

(iii) $\operatorname{Gdim} k < \infty$.

(iv) $\operatorname{Gpd} M < \infty$ *for all R-modules M.*

(v) $\operatorname{Gpd} k < \infty$.

In light of the above characterization, Hummel and Marley [34] define Gorenstein as follows.

Definition 5.2 ([34]). A local ring R is Gorenstein if $\operatorname{Gdim} R/I < \infty$ for every finitely generated ideal I. An arbitrary ring R is Gorenstein if $R_{\mathfrak{m}}$ is Gorenstein for every maximal ideal $\mathfrak{m}$.

While Gorenstein projective dimension and G-dimension are equivalent in coherent rings, note that all the (coherent) Gorenstein results below were first proved in the context of $(\mathrm{FP})_\infty$-modules and $\tilde{\mathrm{G}}$-dimensions. Since these results make use of the defining characteristics of G-dimension, they will be stated in those terms instead of Gorenstein projective dimension.

By the inequality $\operatorname{Gdim} M \leq \operatorname{pd} M$, it follows immediately that

Theorem 5.3 ([34]). *Coherent regular rings are Gorenstein.*

Using Corollary 4.15, it follows that

Theorem 5.4 ([34]). *A coherent Gorenstein ring is locally Cohen–Macaulay.*

While most Gorenstein results focus on coherent Gorenstein rings, the following example from [34] constructs a non-coherent Gorenstein ring. For the following example we define an R-module M to be linearly compact if every collection $\{N_i \mid i \in \mathcal{I}\}$ of cosets of M having the finite intersection property satisfies $\bigcap_{i \in \mathcal{I}} N_i \neq \emptyset$ [22]. A valuation ring R is *almost maximal* if for every ideal $I \neq 0$, R/I is linearly compact in the discrete topology.

Example 5.5 ([34, Example 5.2]). Let V be an almost maximal valuation domain with value group $\mathbb{R}$ (see Section II.6 of [22] for details). Let $\mathfrak{m}$ be the maximal ideal of V and let $a \in \mathfrak{m}$ be a nonzero element. Then $R = V/a\mathfrak{m}$ is a non-coherent Gorenstein ring.

Additional characterizations of Gorenstein rings include the following.

Proposition 5.6 ([34]). *Let R be a ring.*

(i) *If R is Gorenstein, then R_p is Gorenstein for any prime ideal p.*

(ii) *R is a local coherent Gorenstein ring if and only if $R/(x)$ is Gorenstein for any R-regular element x.*

(iii) *R is Gorenstein if and only if $R[x]$ is coherent and Gorenstein.*

(iv) *If $\{R_i\}$ is a family of coherent Gorenstein rings, then $R = \varinjlim R_i$ is coherent Gorenstein.*

While the definition of Gorenstein rings rests upon a characterization using G-dimension, one may ask how well this definition behaves in light the following characterization of Gorenstein rings:

Proposition 5.7 ([40]). *Let $(R, \mathfrak{m}, k)$ be an n-dimensional Noetherian local ring. The following conditions are equivalent.*

(i) *R is Gorenstein.*

(ii) $\operatorname{id} R \leq n$.

(iii) $\operatorname{Ext}_R^i(k, R) = 0$ *for* $i < n$ *and* $\operatorname{Ext}_R^n(k, R) \cong k$.

(iv) *R is a Cohen–Macaulay ring and* $\operatorname{Ext}_R^n(k, R) \cong k$.

(v) *R is a Cohen–Macaulay ring and every parameter ideal is irreducible.*

(vi) *R is a Cohen–Macaulay ring and there exists an irreducible parameter ideal.*

Hummel and Marley [34] provide a one-directional analogy of Proposition 5.7(v) and (vi) over coherent rings.

Proposition 5.8 ([34]). *If R is a local coherent Gorenstein ring with* $\operatorname{depth} R = n < \infty$, *then every n-generated ideal generated by a regular sequence is irreducible.*

Notice that while regular sequences of length depth R may not exist, if coherence can be preserved one may pass to a faithfully flat ring to obtain the necessary regular sequence [34]. It is currently unknown whether the reverse of Proposition 5.8 is true.

Hummel and Marley [34] have also made connections with characterization (ii), replacing injective dimension with FP-injective dimension. FP-injective modules, introduced by Stenström arises from a modification of the definition of injective modules.

Definition 5.9 ([46]). Let R be a ring and let M be an R-module.

(i) M is called FP-injective if $\operatorname{Ext}_R^1(F, M) = 0$ for all finitely presented modules F.

(ii) The FP-injective dimension of M is defined as

$$\text{FP-id}_R\, M = \inf\{n \geq 0 \mid \operatorname{Ext}_R^{n+1}(F, M) = 0 \forall \text{ finitely presented } R\text{-module } F\}.$$

FP-injective modules have also appeared in the literature as absolutely pure modules (for instance see [25] and [22]). FP-injective modules were connected to other homological dimensions by Ding and Chen [18] who explored FP-injectivity in conjunction with coherent rings. Below, some of the more salient properties of FP-injective dimension from Lemma 3.1 of [46] are summarized in the context of coherent rings.

Proposition 5.10 ([46]). *Let R be a coherent ring, let M be an R-module, and let n be a non-negative integer. The following conditions are equivalent.*

(i) $\text{FP-id}_R\, M \leq n$.

(ii) $\text{Ext}^i_R(F, M) = 0$ *for all $i > n$ and all finitely presented R-modules F.*

(iii) $\text{Ext}^{n+1}_R(R/I, M) = 0$ *for all finitely generated ideals I of R.*

(iv) *Given an exact sequence* $0 \longrightarrow M \longrightarrow E^0 \longrightarrow E^1 \longrightarrow \cdots \longrightarrow E^{n-1} \longrightarrow E^n \longrightarrow 0$ *with E^i an FP-injective module for $0 \leq i \leq n-1$, then E^n is FP-injective.*

Additional details on FP-injective modules and coherent rings can be found in [34], [46], and [18].

The first characterization of Gorenstein rings via FP-injective dimension was by Ding and Chen in [18], who characterized local coherent rings of finite self-FP-injective dimension. They called local coherent rings with FP-id $R \leq n$ n-FC rings, later denoted Ding–Chen rings by Gillespie in [24]. The following theorem of Ding and Chen makes the initial connection between Gorenstein and n-FC rings.

Theorem 5.11 ([18]). *Let R be a local coherent ring. The following conditions are equivalent for $n \geq 1$.*

(i) *R is n-FC.*

(ii) Gpd $M \leq n$ *for all finitely presented R-modules M, that is, R is Gorenstein.*

However, there is an example of a local coherent Gorenstein ring with infinite FP-injective dimension.

Example 5.12 ([41]). Let k be a field, and let $R = k[[(x_n)_{n \in \mathbb{N}}]]$ be the power series ring in a countable infinite number of indeterminates over k such that only a finite number of indeterminates occur in the expansion of any element of R. R is a local coherent regular ring (and hence Gorenstein), but FP-id $R = \infty$.

Thus additional assumptions are needed for a Gorenstein ring to have finite FP-injective dimension. In light of the bound

$$\begin{aligned} \text{depth}\, R &\leq \sup\{\text{Gdim}(R/I) \mid I \text{ finitely generated ideal}\} \\ &= \sup\{n \mid \text{Ext}^i_R(R/I, R) = 0, i \geq n, I \text{ finitely generated ideal}\}, \end{aligned}$$

for any ring R, the work of Hummel and Marley and Theorem 5.11 leads to the following result.

Theorem 5.13 ([34]). *Let R be a local coherent ring. The following conditions are equivalent for $n \geq 0$.*

(i) *R is n-FC.*

(ii) *R is Gorenstein with* depth $R = n$.

Rings satisfying this theorem will be denoted n-FC Gorenstein rings. In the case of n-FC Gorenstein rings, a result of Ding and Chen also provides a coherent equivalent to the following Noetherian result.

Proposition 5.14 ([13, Exercise 3.1.25]). *A Noetherian ring R is Gorenstein if and only if the set of modules with finite projective dimension is equal to the set of modules with finite injective dimension.*

In the coherent case, flat modules play the role of projective modules.

Proposition 5.15 ([17]). *Let R be a coherent ring with* $\text{FP-id}\, R \leq n$. *The following conditions are equivalent.*

(i) $\text{fd}\, M < \infty$.

(ii) $\text{fd}\, M \leq n$.

(iii) $\text{FP-id}\, M < \infty$.

(iv) $\text{FP-id}\, M \leq n$.

Thus, the following unpublished generalization of Proposition 5.14 follows easily.

Proposition 5.16. *Let R be a local coherent ring; the following conditions are equivalent.*

(i) *R is Gorenstein with* $\text{depth}\, R < \infty$.

(ii) *For any module M,* $\text{fd}\, M < \infty$ *if and only if* $\text{FP-id}\, M < \infty$.

Proof. The forward direction holds by Proposition 5.15. The reverse direction holds trivially, as R is flat, and hence $\text{FP-id}\, R < \infty$. □

Note that Foxby [21] extended Proposition 5.14 by showing that a Noetherian ring with a single module of both finite projective and injective dimension is Gorenstein. It is unknown whether this result carries over to coherent rings.

The connection between FP-injective and flat modules is natural in light of the following duality between FP-injective dimension and flat dimension. In the following, the character module of M is denoted $M^+ = \text{Hom}_R(M, \mathbb{Q}/\mathbb{Z})$.

Lemma 5.17 ([20]). *Let R be a ring and let M be an R-module.*

(i) $\text{fd}M = \text{id}M^+ = \text{FP-id}\, M^+$.

(ii) *If R is right coherent and M is a right R-module then* $\text{fd}M^+ = \text{FP-id}\, M$.

This relation is analogous to the relation between injective and flat modules over Noetherian rings.

Recall that an R-module M has *weak dimension n*, denoted $\text{w.dim}_R\, M = n$, if there is a minimal length exact resolution of M, $0 \longrightarrow F_n \longrightarrow \cdots \longrightarrow F_0 \longrightarrow M \longrightarrow 0$, consisting of flat modules. The *weak dimension* of a ring R is defined $\text{w.dim}\, R = \sup\{\text{w.dim}\, M \mid M \text{ an } R\text{-module}\}$. By [29], coherent rings of finite weak dimension are regular coherent rings. One may ask whether there are Gorenstein rings with infinite weak dimension. The following result of Gillespie [24] yields a positive answer by providing examples of n-FC Gorenstein rings with infinite weak dimension.

Proposition 5.18 ([24]). *If R is an n-FC Gorenstein ring, then the group ring $R[G]$ is an n-FC Gorenstein ring for any finite group G.*

More work has been done recently with FP-injective dimension that has lead to further characterizations of n-FC, and hence coherent Gorenstein, rings. Mao and Ding [37–39] explore the existence of FP-injective (pre-) covers and flat (pre-)envelopes. Yang and Liu [48] extend the notion of FP-injectivity to complexes.

Given the connection between Gorenstein dimension and FP-injective dimension, and the fact that Gorenstein projective, injective and flat modules are kernels of totally acyclic complexes, a natural question arises of whether FP-injective dimension can also be viewed in terms of totally acyclic complexes. Mao and Ding [39] do this through the following definition of Gorenstein FP-injective modules, which are FP-injective modules that approximate the properties of Gorenstein injective modules.

Definition 5.19 ([39]). Let R be a ring, and let M be a left R-module. M is Gorenstein FP-injective if there is an exact sequence

$$\mathbf{E}: \quad \cdots \longrightarrow E_1 \longrightarrow E_0 \longrightarrow E^0 \longrightarrow E^1 \longrightarrow \cdots$$

of injective left R-modules with $M = \ker(E^0 \longrightarrow E^1)$ and $\operatorname{Hom}(F, \mathbf{E})$ exact for every FP-injective module F.

Clearly any kernel or cokernel of the sequence $\mathbf{E}$ above is Gorenstein FP-injective. Gorenstein FP-injectives are also closed under direct products [39]. In particular, Gorenstein FP-injective modules satisfy the properties of homological dimensions in Remark 4.1; see [39] for details. In addition, Gorenstein FP-injective and Gorenstein flat modules are related in the same way as FP-injective and flat modules are in Lemma 5.17.

Proposition 5.20 ([39]). *Let R be a coherent ring and let M be a right R-module. Then M is Gorenstein flat if and only if M^+ is Gorenstein FP-injective.*

Note that [39] shows the forward direction of this result holds for any ring.

Another variation on FP-injective and flat modules are FI-injective and FI-flat modules introduced by Mao and Ding.

Definition 5.21 ([38]). Let R be a ring.

(i) A left R-module is FI-injective if $\operatorname{Ext}^1_R(F, M) = 0$ for any FP-injective left R-module F.

(ii) A right R-module is FI-flat if $\operatorname{Tor}^R_1(N, F) = 0$ for any FP-injective left R-module F.

Note that Gorenstein FP-injective modules are also FI-injective.

Gorenstein FP-injective modules fit between injective and Gorenstein injective modules. Over a Noetherian ring, the classes of Gorenstein FP-injective and Gorenstein injective modules are identical. The following result illuminates the link between Gorenstein FP-injective and FP-injective modules.

Proposition 5.22 ([38]). *Let R be a coherent ring.*

(i) *R is left Noetherian if and only if every FP-injective left R-module is Gorenstein FP-injective.*

(ii) *If the class of Gorenstein FP-injective left R-modules is closed under direct sums, then R is left Noetherian.*

(iii) *If* FP-id $R \leq n < \infty$ *the following conditions are equivalent.*

(a) w. dim $R \leq n$.

(b) *Every Gorenstein flat right R-module is flat.*

(c) *Every Gorenstein FP-injective left R-module is FP-injective.*

(d) *Every Gorenstein FP-injective left F-module is injective.*

The relation between these modules can be summarized as follows with the arrows indicating containment under the indicated conditions:

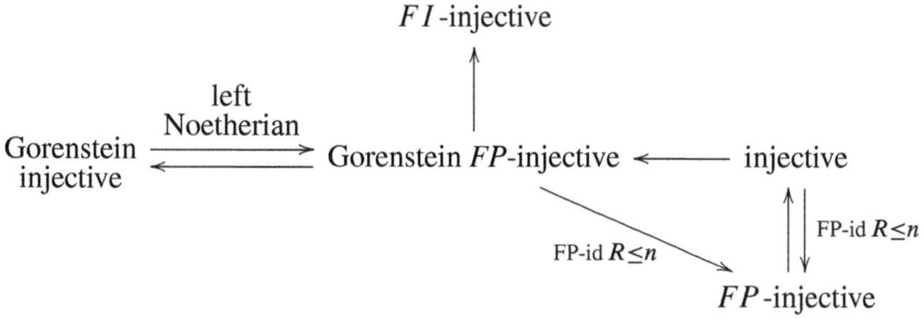

Using Gorenstein FP-injective dimension, Theorem 3.4 in [39] provides an additional characterization of Gorenstein n-FC rings over perfect rings.

Theorem 5.23 ([39]). *Let R be a coherent perfect ring. The following conditions are equivalent.*

(i) *R is an n-FC ring.*

(ii) *For every exact sequence $0 \longrightarrow M \longrightarrow F^0 \longrightarrow \cdots \longrightarrow F^{n-1} \longrightarrow F^n \longrightarrow 0$ with F^i Gorenstein FP-injective for $0 \leq i \leq n-1$, then F^n is Gorenstein FP-injective.*

(iii) *For every exact sequence $0 \longrightarrow F_n \longrightarrow F_{n-1} \longrightarrow \cdots \longrightarrow F_0 \longrightarrow M \longrightarrow 0$ with each F_i Gorenstein flat for $0 \leq i \leq n-1$, then F_n is Gorenstein flat.*

With multiple characterizations of Gorenstein rings that are compatible with the Noetherian case, we move on to a discussion of potential candidates for a theory of coherent complete intersection rings.

6 Foundations for Coherent Complete Intersections

Let $(R, \mathfrak{m}, k)$ be a local ring, and let M be any R-module. Define the ith Betti number of R to be $\beta_i(R) = \dim_R \operatorname{Tor}_i^R(k, k)$, and the ith Betti number of M to be $\beta_i(M) = \dim_k \operatorname{Tor}_i^R(M, k)$. The first formal definition of non-Noetherian complete intersections found by this author is by André [2]. This definition uses André–Quillen homology theory to extend the following "well-known" characterization of Noetherian complete intersection rings given by André.

Proposition 6.1 ([2]). *A local Noetherian ring is a complete intersection if its Poincaré series has the following form*

$$\sum \beta_i x^i = \frac{(1+x)^r}{(1-x^2)^s}$$

with the integer $r - s = \dim R > 0$.

In [2], André characterizes the rings satisfying the Poincaré equality given above without the restrictions on r and s, and defines rings satisfying this equality to be complete intersections. Since then, several authors have worked to characterize complete intersections via homological dimensions. The work of Avramov, Vesselin, Gasharov, and Peeva introduced complete intersection dimension.

Definition 6.2 ([6]). Let R and R' be local rings, and let M be an R-module.

(i) The map $R \longrightarrow R'$ is a (codimension c) deformation if it is a surjective local homomorphism with kernel generated by a (length c) regular sequence.

(ii) A quasi-deformation of R is a diagram of local homomorphisms $R \longrightarrow R' \leftarrow Q$, with $R \longrightarrow R'$ a flat extension and $R' \leftarrow Q$ a (codimension c) deformation. Given a quasi-deformation $R \longrightarrow R' \leftarrow Q$ and an R-module M, set $M' = M \otimes_R R'$.

(iii) For a nonzero R-module M, denote the complete intersection dimension of M to be $\text{CI-dim}_R M = \inf\{\operatorname{pd}_Q M' - \operatorname{pd}_Q R' \mid R \longrightarrow R' \leftarrow Q \text{ is a quasi-deformation}\}$. For a module M over a Noetherian ring R,

$$\text{CI-dim}_R M = \sup\{\text{CI-dim}_{R_\mathfrak{m}} M_\mathfrak{m} \mid \mathfrak{m} \text{ a maximal ideal of } R\}.$$

Complete intersection dimension characterizes Noetherian complete intersection rings.

Theorem 6.3 ([6]). *Let* $(R, \mathfrak{m}, k)$ *be a local Noetherian ring. If* R *is a complete intersection, then every* R*-module has finite CI-dimension. If* $\text{CI-dim}_R k < \infty$, *then* R *is a complete intersection.*

Gerko [23] uses Gorenstein dimension and complexity to characterize Noetherian complete intersection rings. The *complexity* of an R-module is defined as $\mathrm{cx}_R M = \sup\{n | \beta_i^R(M) \leq \alpha x^{n-1}\}$.

In the Noetherian case, Gulliksen provides the following connection between complexity and complete intersection rings.

Proposition 6.4 ([30]). *If R is a local Noetherian complete intersection ring, then* $\mathrm{cx}\ M < \infty$ *for every R-module M.*

Gerko's investigation in [23] with PCI-dimension, also denoted lower CI dimension (*CI_*-dimension*), yields an easy definition of coherent complete intersection rings.

Definition 6.5 ([23]). Let R be a ring and let M be an R-module.

(i) Define PCI-dim $M = 0$ if Gdim $M = 0$ and cx $M < \infty$.

(ii) Define

$$\text{PCI-}\dim M = \inf\{n | 0 \longrightarrow P_n \longrightarrow \cdots \longrightarrow P_0 \longrightarrow M \longrightarrow 0, \text{PCI-}\dim P_i = 0\ \forall i\}.$$

With this definition Gerko makes the following connection, which mirrors Theorem 6.3.

Proposition 6.6 ([23]). *If R is a Noetherian complete intersection,* PCI-dim $M < \infty$ *for every R-module M. Conversely, if* PCI-dim $k < \infty$ *then R is a complete intersection.*

Gerko also shows that PCI-dimension is related to CI-dimension.

Proposition 6.7 ([23]). *Let R be a ring, and let M be an R-module. Then* PCI-dim $M \leq$ CI-dim M*, with equality if and only if* CI-dim $M < \infty$.

However, Veliche showed the classes of modules with finite PCI-dimension and finite CI-dimension are not the same.

Proposition 6.8 ([47]). *Let R be a local Noetherian ring containing a field, with* depth $R \geq 4$. *Then there exists a perfect ideal I in R with* grade $I = 4$*, and a module M over R/I such that* PCI-dim$_{R/I}$ $M = 0$*, but* CI-dim$_{R/I}$ $M = \infty$.

Hence, using the following definition of complete intersection would allow complete intersections to be immediately Gorenstein.

Definition 6.9. Let $(R, \mathfrak{m})$ be a local ring. Define R to be a complete intersection if PCI-dim $R/I < \infty$ for all finitely generated ideals I. If R is a local coherent ring, R is a complete intersection if PCI-dim $M < \infty$ for all finitely presented modules M.

However this definition is unsatisfying in that it a priori assumes finite Gorenstein dimension, in particular that the ring is Gorenstein. Instead, via a suggestion to the author by Avramov, a preferable definition for coherent complete intersections may be the following.

Definition 6.10. A local coherent ring R is a complete intersection if cx $M < \infty$ for every finitely presented R-module.

While $(\mathrm{FP})_\infty$-modules certainly have finite complexity, it is unclear whether finite complexity is sufficient to imply finite Gorenstein dimension. More work needs to be done to discover if complexity is a sufficient condition for finite Gorenstein dimension.

Acknowledgments. The author would like to thank Lucho Avramov and Sean Sather-Wagstaff for discussions about complete intersection rings, and Diana White for support and discussions related to Gorenstein rings and FP-injective dimension. The author also thanks the referee for providing thoughtful comments and suggestions, as well as an additional reference, that helped to improve this work.

Bibliography

[1] Alfonsi, B., Grade non-noethérien. Comm Algebra. 1981;9:811–840.

[2] André, M., Nonnoetherian complete intersections. Bull Amer Math Soc. 1972;78:724–729.

[3] Asgharzadeh, M., Tousi, M., On the notion of Cohen–Macaulayness for non-Noetherian rings. J Algebra. 2009;322:2297–2320.

[4] Auslander, M., Bridger, M., Stable module theory. vol. 94 of Memoirs of the American Mathematical Society. Providence, RI: American Mathematical Society; 1969.

[5] Auslander, M., Buchsbaum, D. A., Homological dimension in local rings. Trans Amer Math Soc. 1957;85:390–405.

[6] Avramov, L. L., Gasharov, V. N., Peeva, I. V., Complete intersection dimension. Publ Math Inst Hautes Études Sci 1997:67–114.

[7] Avramov, L. L., Martsinkovsky, A., Absolute, relative, and Tate cohomology of modules of finite Gorenstein dimension. Proc Lond Math Soc 3. 2002;85:393–440.

[8] Barger, S. F., A theory of grade for commutative rings. Proc Amer Math Soc. 1972;36: 365–368.

[9] Bennis, D., Mahdou, N., Global Gorenstein dimensions of polynomial rings and of direct products of rings. Houston J Math. 2009;35:1019–1028.

[10] Bennis, D., Mahdou, N., Global Gorenstein dimensions. Proc Amer Math Soc. 2010;138: 461–465.

[11] Bertin, J., Anneaux cohérents réguliers. C R Acad Sci Paris Sér A–B. 1971;273:A590–A591.

[12] Bieri, R., (Mathematics Department, Queen Mary College). Homological dimension of discrete groups. Queen Mary College Mathematics Notes. London; 1976.

[13] Bruns, W., Herzog, J., Cohen–Macaulay rings. vol. 39 of Cambridge studies in advanced mathematics. Cambridge: Cambridge University Press; 1993.

[14] Christensen, L. W., Gorenstein dimensions. vol. 1747 of Lecture notes in mathematics. Berlin: Springer-Verlag; 2000.

[15] Christensen, L. W., Foxby, H.-B., Holm, H., Beyond totally reflexive modules and back. In: Fontana, M., Kabbaj, S.-E., Olberding, B., Swanson, I., editors, Commutative Algebra. New York: Springer; 2011. p. 101–143.

[16] Christensen, L. W., Veliche, O., A test complex for Gorensteinness. Proc Amer Math Soc. 2008;136:479–487 [electronic].

[17] Ding, N., Chen, J., The flat dimensions of injective modules. Manuscripta Math. 1993; 78:165–177.

[18] Ding, N., Chen, J., Coherent rings with finite self-FP-injective dimension. Comm Algebra. 1996;24:2963–2980.

[19] Enochs, E. E., Jenda, O. M. G., Relative homological algebra. vol. 90 of De Gruyter expositions in mathematics, Berlin: Walter de Gruyter & Co.; 2000.

[20] Fieldhouse, D. J., Character modules, dimension and purity. Glasg Math J 1972;13:144–146.

[21] Foxby, H.-B., Isomorphisms between complexes with applications to the homological theory of modules. Math Scand. 1977;40:5–19.

[22] Fuchs, L., Salce, L., Modules over non-Noetherian domains. vol. 84 of Mathematical surveys and monographs. Providence, RI: American Mathematical Society; 2001.

[23] Gerko, A. A., On homological dimensions. Mat Sb. 2001;192:79–94.

[24] Gillespie, J., Model structures on modules over Ding–Chen rings. Homology, Homotopy Appl. 2010;12:61–73.

[25] Glaz, S., Commutative coherent rings. vol. 1371 of Lecture notes in mathematics, Berlin: Springer-Verlag; 1989.

[26] Glaz, S., Commutative coherent rings: historical perspective and current developments. Nieuw Arch Wiskd (4). 1992;10:37–56.

[27] Glaz, S., Coherence, regularity and homological dimensions of commutative fixed rings. Proceedings, Commutative Algebra; 1992; Trieste. River Edge, NJ: World Sci. Publ.; 1994. p. 89–106.

[28] Glaz, S., Homological dimensions of localizations of polynomial rings. Proceedings, Zero-dimensional Commutative Rings; 1994; Knoxville, TN. vol. 171 of Lecture notes in pure and applied mathemaics, New York: Dekker; 1995. p. 209–222.

[29] Glaz, S., Homological characterizations of rings: the commutative case. In: The concise handbook of algebra. Kluwer Publ.; 2002.

[30] Gulliksen, T. H., A change of ring theorem with applications to Poincaré series and intersection multiplicity. Math Scand. 1974;34:167–183.

[31] Hamilton, T. D., Weak Bourbaki unmixed rings: a step towards non-Noetherian Cohen–Macaulayness. Rocky Mountain J Math. 2004;34:963–977.

[32] Hamilton, T. D., Marley, T., Non-Noetherian Cohen–Macaulay rings. J Algebra. 2007; 307:343–360.

[33] Hochster, M., Grade-sensitive modules and perfect modules. Proc Lond Math Soc (3). 1974;29:55–76.

[34] Hummel, L., Marley, T., The Auslander–Bridger formula and the Gorenstein property for coherent rings. J Commut Algebra. 2009;1:283–314.

[35] Iyengar, S., Krause, H., Acyclicity versus total acyclicity for complexes over Noetherian rings. Doc Math. 2006;11:207–240 [electronic].

[36] Kaplansky, I., Commutative rings. Revised ed. Chicago, IL, London: The University of Chicago Press; 1974.

[37] Mao, L., Ding, N., Envelopes and covers by modules of finite FP-injective and flat dimensions. Comm Algebra. 2007;35:833–849.

[38] Mao, L., Ding, N., FI-injective and FI-flat modules. J Algebra. 2007;309:367–385.

[39] Mao, L., Ding, N., Gorenstein FP-injective and Gorenstein flat modules. J Algebra Appl. 2008;7:491–506.

[40] Matsumura, H., Commutative ring theory [translated from Japanese by Reid M]. vol. 8 of Cambridge studies in advanced mathematics. Cambridge: Cambridge University Press; 1986.

[41] McDowell, K. P., Commutative coherent rings [PhD thesis]. McMaster University (Canada). Ann Arbor, MI: ProQuest LLC; 1974.

[42] Miller, L. M., A theory of non-Noetherian Gorenstein rings [PhD thesis]. The University of Nebraska, Lincoln. Ann Arbor, MI: ProQuest LLC; 2008.

[43] Northcott, D. G., Finite free resolutions. vol. 71 of Cambridge tracts in mathematics. Cambridge: Cambridge University Press; 1976.

[44] Sather-Wagstaff, S., Sharif, T., White, D., Stability of Gorenstein categories. J Lond Math Soc (2). 2008;77:481–502.

[45] Schenzel, P., Proregular sequences, local cohomology, and completion. Math Scand. 2003;92:161–180.

[46] Stenström, B., Coherent rings and FP-injective modules. J Lond Math Soc (2). 1970;2: 323–329.

[47] Veliche, O., Construction of modules with finite homological dimensions. J Algebra. 2002;250:427–449.

[48] Yang, X., Liu, Z., FP-injective complexes. Comm Algebra. 2010;38:131–142.

Author Information

Livia Hummel, Department of Mathematics and Computer Science, University of Indianapolis, Indianapolis, IN, United States.
E-mail: hummell@uindy.edu

Progress in Commutative Algebra 1, 293–361

Non-commutative Crepant Resolutions: Scenes From Categorical Geometry

Graham J. Leuschke

Abstract. Non-commutative crepant resolutions are algebraic objects defined by Van den Bergh to realize an equivalence of derived categories in birational geometry. They are motivated by tilting theory, the McKay correspondence, and the minimal model program, and have applications to string theory and representation theory. In this expository article I situate Van den Bergh's definition within these contexts and describe some of the current research in the area.

Keywords. Non-commutative Algebraic Geometry, Categorical Geometry, Derived Categories, Non-commutative Crepant Resolution, NCCR, McKay Correspondence, Minimal Model Program, Tilting.

2010 Mathematics Subject Classification. Primary: 13-02, 13C14, 14A22, 14E15, 16S38; Secondary: 13C14, 13D09.

1 Introduction

A resolution of singularities replaces a singular algebraic variety by a non-singular one that is isomorphic on a dense open set. As such, it is a great boon to the algebraic geometer, allowing the reduction of many calculations and constructions to the case of a smooth variety. To the pure commutative algebraist, however, this process can seem like the end of a story rather than the beginning: it replaces a well-understood thing, the spectrum of a ring, with a much more mysterious thing glued together out of other spaces. Put simply, a resolution of singularities of an affine scheme Spec R is almost never another affine scheme (but see Section 13). One cannot in general resolve singularities and stay within the categories familiar to commutative algebraists.

The usual solution, of course, is to expand one's landscape on the geometric side to include more complicated schemes. There are plenty of good reasons to do this other than resolving singularities, and it has worked well for a century. Locally, the more convoluted objects are built out of affine schemes/commutative algebra, so one has not strayed too far.

Here is another alternative: expand the landscape on the algebraic side instead, to include non-commutative rings as well as commutative ones. This suggestion goes by the name "non-commutative algebraic geometry" or, my preference, "categorical geometry". For some thoughts on the terminology, see Section 7. Whatever the name,

The author was partly supported by NSF grant DMS-0902119.

the idea is to treat algebraic objects, usually derived categories, as coming from geometric objects even when no such geometric things exist. Since one trend in algebraic geometry in the last forty years has been to study algebraic varieties indirectly, by studying their (derived) categories of quasicoherent sheaves, one can try to get along without the variety at all. Given a category of interest C, one can postulate a "non-commutative" space X such that the (derived) category of quasicoherent sheaves on X is C, and write $\mathsf{C} = \mathsf{D}^b(\mathsf{Qch}\, X)$. In this game, the derived category *is* the geometry and the symbol X simply stands in as a grammatical placeholder; the mathematical object in play is C.

Of course, such linguistic acrobatics can only take you so far. The bonds between algebra and geometry cannot be completely severed: the "non-commutative spaces" must be close enough to the familiar commutative ones to allow information to pass back and forth. This article is about a particular attempt to make this program work.

The idea of non-commutative resolutions of singularities appeared around the same time in physics [42, 27, 30, 69] and in pure mathematics, notably in [40, 50]. In 2002, inspired by Bridgeland's proof [56] of a conjecture due to Bondal and Orlov, Van den Bergh [205] proposed a definition for a *non-commutative crepant resolution* of a ring R. This is an R-algebra Λ which is (i) finitely generated as an R-module, (ii) generically Morita equivalent to R, and (iii) has finite global dimension. These three attributes are supposed to stand in for the components of the definition of a resolution of singularities: it is (i) proper, (ii) birational, and (iii) non-singular. The additional "crepancy" condition is a certain symmetry hypothesis on Λ which is intended to stand in for the condition that the resolution of singularities not affect the canonical sheaf. See Section 12 for details.

My main goal for this article is to motivate the definition of a non-commutative crepant resolution (Definition 12.4). In order to do that effectively, I will attempt to describe the contexts out of which it arose. These are several, including Morita theory and tilting, the McKay correspondence, the minimal model program of Mori and Reid, and especially work of Bondal and Orlov on derived categories of coherent sheaves. Of course, the best motivation for a new definition is the proof of a new result, and I will indicate where the new concepts have been applied to problems in "commutative" geometry. Finally, the article contains a healthy number of examples, both of existence and of non-existence of non-commutative crepant resolutions. Since it is not at all clear yet that the definitions given below are the last word, we can hope that reasoning by example will point the way forward.

As Miles Reid writes in [174],

> It is widely appreciated that mathematicians usually treat history in a curiously dishonest way, rewriting the history of the subject as it should have been discovered [...] The essential difficulty seems to be that the story in strictly chronological order will not make sense to anyone; the writer wants to give an explanation based on the logical layout of the subject, whatever violence it does to historical truth.

This article will be guilty of the dishonesty Reid suggests, intentionally in some places and, I fear, unintentionally in others. I do intend to build a certain logical layout around the ideas below, and I sincerely apologize to any who feel that violence has been done to their ideas.

Here is a thumbnail sketch of the contents. The first few sections consider, on both the algebraic side and the geometric, the reconstruction of the underlying ring or space from certain associated categories. The obstructions to this reconstruction – and even to reconstruction of the commutative property – are explained by Morita equivalence (Section 2) and tilting theory (Sections 4–5). Section 6 contains a central example: Beilinson's "tilting description" of the derived category of coherent sheaves on projective space.

This is not intended to be a comprehensive introduction to non-commutative algebraic geometry; for one thing, I am nowise competent to write such a thing. What I cannot avoid saying is in Section 7.

The next two sections give synopses of what I need from the geometric theory of resolutions of singularities and the minimal model program, followed in Section 10 by some remarks on purely category-theoretic replacements for resolutions of singularities. Another key example, the McKay correspondence, appears in Section 11.

At last in Section 12 I define non-commutative crepant resolutions. The definition I give is slightly different from Van den Bergh's original, but agrees with his in the main case of interest. The next few sections, 13–16, focus on particular aspects of the definition, recapping some related research and focusing on obstructions to existence. In particular I give several more examples of existence and non-existence of non-commutative crepant resolutions. Two more families of examples take up Section 17 and Section 18: rings of finite representation type and the generic determinantal hypersurface. Here tilting returns, now as a source of non-commutative crepant resolutions. I investigate a potential theory of "non-commutative blowups" in Section 19, and give very quick indications of some other examples in Section 20. That section also lists a few open questions and gestures at some topics that were omitted for lack of space, energy, or expertise.

Some results are simplified from their published versions for expository reasons. In particular I focus mostly on local rings, allowing some cleaner statements at the cost of generality, even though such generality is in some cases necessary for the proofs. In any case I give very few proofs, and sketchy ones at that. The only novel contribution is a relatively simple proof, in Section 18, of the $m = n$ case of the main theorem of [44].

The reader I have in mind has a good background in commutative algebra, but perhaps less in non-commutative algebra, algebraic geometry, and category theory. Thus I spend more time on trivialities in these latter areas than in the first. I have tried to make the references section comprehensive, though it surely is out of date already.

Conventions. All modules will be left modules, so for a ring Λ I will denote by $\Lambda\text{-mod}$ the category of finitely generated left Λ-modules, and by $\text{mod-}\Lambda = \Lambda^{\text{op}}\text{-mod}$ the category of finitely generated left Λ^{op}-modules. Other categories of modules will be defined on the fly. Capitalized versions of names, namely Λ-Mod, etc., will denote the same categories without any hypothesis of finite generation.

Throughout R and S will be commutative Noetherian rings, usually local, while Λ and its Greek-alphabet kin will not necessarily be commutative.

2 Morita Equivalence

To take a representation-theoretic view is to replace the study of a ring by the study of its (Abelian category of) modules. Among other advantages, this allows to exploit the tools of homological algebra. A basic question is: How much information do we lose by becoming representation theorists? In other words, when are two rings indistinguishable by their module categories, so that Λ-mod and Γ-mod are the same Abelian category for different rings Λ and Γ?

To fix terminology, recall that a functor $\mathsf{F}\colon \mathsf{A} \longrightarrow \mathsf{B}$ between Abelian categories is *fully faithful* if it induces an isomorphism on Hom-sets, and *dense* if it is surjective on objects up to isomorphism. If F is both fully faithful and dense, then it is an *equivalence* [153, Section IV.4], that is, there is a functor $\mathsf{G}\colon \mathsf{B} \longrightarrow \mathsf{A}$ such that both compositions are isomorphic to the respective identities. In this case write $\mathsf{A} \simeq \mathsf{B}$. Equivalences preserve and reflect essentially all "categorical" properties and attributes: mono- and epimorphisms, projectives, injectives, etc. Thus the question above asks when two module categories are equivalent.

Morita's theorem on equivalences of module categories [157, Section 3] completely characterizes the contexts in which Λ-mod $\simeq$ Γ-mod for rings Λ and Γ. First I define some of the necessary terms.

Definition 2.1. Let Λ be a ring and $M \in \Lambda$-Mod.

(i) Denote by add M the full subcategory of Λ-Mod containing all direct summands of finite direct sums of copies of M.

(ii) Say M is a *generator* (for Λ-mod) if every finitely generated left Λ-module is a homomorphic image of a finite direct sum of copies of M. Equivalently, $\Lambda \in$ add M.

(iii) Say M is a *progenerator* if M is a finitely generated projective module and a generator. Equivalently, add Λ = add M.

Theorem 2.2 (Morita equivalence, see e.g. [78, Chapter V]). *The following are equivalent for rings Λ and Γ.*

(i) *There is an equivalence of Abelian categories* Λ-mod $\simeq$ Γ-mod.

(ii) *There exists a progenerator* $P \in \Lambda$-mod *such that* $\Gamma \cong \operatorname{End}_\Lambda(P)^{op}$.

(iii) *There exists a* $(\Lambda\text{-}\Gamma)$*-bimodule* ${}_\Lambda P_\Gamma$ *such that* $\operatorname{Hom}_\Lambda(P,-)\colon \Lambda$-mod $\longrightarrow$ Γ-mod *is an equivalence.*

In this case, say that Λ and Γ are Morita equivalent.

Interesting bits of the history of Morita's theorem, as well as his other work, can be found in [3].

An immediate corollary of Morita's theorem will be useful later.

Corollary 2.3. *Let* Λ *be a ring and* M, N *two* Λ*-modules such that* $\mathsf{add}\, M = \mathsf{add}\, N$, *equivalently* M *is a direct summand of* N^s *for some* s *and* N *is a direct summand of* M^t *for some* t. *Then* $\mathrm{End}_\Lambda(M)$ *and* $\mathrm{End}_\Lambda(N)$ *are Morita equivalent via the functors* $\mathrm{Hom}_\Lambda(M,N) \otimes_{\mathrm{End}_\Lambda(M)} -$ *and* $\mathrm{Hom}_\Lambda(N,M) \otimes_{\mathrm{End}_\Lambda(N)} -$.

Among the consequences of Theorem 2.2, the most immediately relevant to our purposes are those related to commutativity. One can show [143, Corollary 18.42] that if Λ and Γ are Morita equivalent, then the centers $Z(\Lambda)$ and $Z(\Gamma)$ are isomorphic. It follows that two *commutative* rings R and S are Morita equivalent if and only if they are isomorphic. On the other hand, in general Morita equivalence is blind to the commutative property. Indeed, the free module Λ^n is a progenerator for any $n \geq 1$, so that Λ and the matrix ring $\mathrm{End}_\Lambda(\Lambda^n) \cong \mathrm{Mat}_n(\Lambda)$ are Morita equivalent. Even if Λ is commutative, $\mathrm{Mat}_n(\Lambda)$ will not be for $n \geq 2$.

The fact that commutativity is invisible to the module category is a key motivation for categorical geometry. It is interesting to observe that this idea, and even the connection with endomorphism rings, is already present in the Freyd–Mitchell Theorem [77, Theorem 7.34] classifying Abelian categories as categories of modules. In detail, the Freyd–Mitchell Theorem says that if C is a category whose objects form a set (as opposed to a proper class) which is closed under all set-indexed direct sums, and C has a progenerator P such that $\mathrm{Hom}_{\mathsf{C}}(P,-)$ commutes with all set-indexed direct sums, then $\mathsf{C} \simeq \Lambda\text{-}\mathsf{Mod}$ for $\Lambda = \mathrm{End}_{\mathsf{C}}(P)$. Different choices of P obviously give potentially non-commutative rings Λ, even if $\mathsf{C} = R\text{-}\mathsf{mod}$ for some commutative ring R.

The property of P referred to above will recur later: say that P is *compact* if $\mathrm{Hom}_{\mathsf{C}}(P,-)$ commutes with all (set-indexed) direct sums.

The following cousin of Morita equivalence will be essential later on.

Proposition 2.4 (Projectivization [7, Proposition II.2.1]). *Let* Λ *be a ring and* M *a finitely generated* Λ*-module which is a generator. Set* $\Gamma = \mathrm{End}_\Lambda(M)^{\mathrm{op}}$. *Then the functor*

$$\mathrm{Hom}_\Lambda(M,-)\colon \Lambda\text{-}\mathsf{mod} \longrightarrow \Gamma\text{-}\mathsf{mod}$$

is fully faithful, and restricts to an equivalence

$$\mathrm{Hom}_\Lambda(M,-)\colon \mathsf{add}\, M \longrightarrow \mathsf{add}\, \Gamma\,.$$

In particular, the indecomposable projective Γ*-modules are precisely the modules of the form* $\mathrm{Hom}_\Lambda(M,N)$ *for* N *an indecomposable module in* $\mathsf{add}\, M$.

3 (Quasi)coherent Sheaves

On the geometric side, it has also long been standard operating procedure to study a variety or scheme X in a representation-theoretic mode by investigating the sheaves on X, particularly those of algebraic origin, the quasicoherent sheaves.

Let X be a Noetherian scheme. Recall that an $\mathcal{O}_X$-module is *quasicoherent* if it locally can be represented as the cokernel of a homomorphism between direct sums of copies of $\mathcal{O}_X$. A quasicoherent sheaf is *coherent* if those direct sums can be chosen to be finite. Write $\mathsf{Qch}\, X$ for the category of quasicoherent sheaves and $\mathsf{coh}\, X$ for that of coherent sheaves. Since X is assumed to be Noetherian, these are both Abelian categories (for the quasicoherent sheaves, quasi-compact and quasi-separated is enough [59]).

The category $\mathsf{Qch}\, X$ is a natural environment for homological algebra over schemes; for example, computations of cohomology naturally take place in $\mathsf{Qch}\, X$. Now one may ask the same question as in the previous section: What information, if any, is lost in passage from the geometric object X to the category $\mathsf{coh}\, X$ or $\mathsf{Qch}\, X$? In this case, the kernel is even smaller: we lose essentially nothing. Indeed, it is not hard to show that for arbitrary complex varieties X and Y, the categories $\mathsf{coh}\, X$ and $\mathsf{coh}\, Y$ are equivalent if and only if X and Y are isomorphic. The key idea is to associate to a coherent sheaf the closed subset of X on which it is supported; for example, the points of X correspond to the simple objects in $\mathsf{coh}\, X$. See for example [41, Section 8]. More generally, Gabriel [78] taught us how to associate to any Abelian category A a *geometric realization*: a topological space Spec A, together with a sheaf of rings $\mathcal{O}_{\mathsf{A}}$. (In fact, the sheaf $\mathcal{O}_{\mathsf{A}}$ is the endomorphism sheaf of the identity functor on A, reminiscent of the Freyd–Mitchell theorem mentioned in the previous section. The space Spec A is nothing but the set of isomorphism classes of indecomposable injective objects of A, with a base for the topology given by $\operatorname{Supp} M = \{[I] \mid \text{there is a nonzero arrow } M \longrightarrow I\}$ for Noetherian objects M.) In the case $\mathsf{A} = \mathsf{Qch}\, X$ for a Noetherian scheme X, the pair $(\operatorname{Spec} \mathsf{A},\ \mathcal{O}_{\mathsf{A}})$ is naturally isomorphic to $(X,\ \mathcal{O}_X)$. This construction has been generalized to arbitrary schemes by Rosenberg [181, 182], giving the following theorem.

Theorem 3.1 (Gabriel–Rosenberg Reconstruction). *A scheme X can be reconstructed up to isomorphism from the Abelian category* $\mathsf{Qch}\, X$.

Theorem 3.1 implies that there is no interesting Morita theory for (quasi)coherent sheaves. This is not all that surprising, given that Morita-equivalent commutative rings are necessarily isomorphic. The well-known equivalence between modules over a ring R and quasicoherent sheaves over the affine scheme Spec R strongly suggests the same sort of uniqueness on the geometric side as on the algebraic.

For *projective* schemes, Serre's fundamental construction [188] describes the quasicoherent sheaves on X in terms of the graded modules over the homogeneous coordinate ring. Explicitly, let A be a finitely generated graded algebra over a field, and set $X = \operatorname{Proj} A$, the associated projective scheme. Let $\mathsf{GrMod}\,A$, resp. $\mathsf{grmod}\,A$, denote the category of graded, resp. finitely generated graded, A-modules. The graded modules annihilated by $A_{\geq n}$ for $n \gg 0$ form a subcategory $\mathsf{Tors}\,A$, resp. $\mathsf{tors}\,A$, and

$$\mathsf{Tails}\,A = \mathsf{GrMod}\,A/\,\mathsf{Tors}\,A \quad \text{and} \quad \mathsf{tails}\,A = \mathsf{grmod}\,A/\,\mathsf{tors}\,A$$

are defined to be the quotient categories. This means that two graded modules M and N are isomorphic in $\mathsf{Tails}\,A$ if and only if $M_{\geq n} \cong N_{\geq n}$ as graded modules for large enough n.

Theorem 3.2 (Serre). *Let A be a commutative graded algebra generated in degree one over $A_0 = k$, a field, and set $X = \operatorname{Proj} A$. Then the functor $\Gamma_*\colon \mathsf{coh}\,X \longrightarrow \mathsf{tails}\,A$, which sends a coherent sheaf $\mathcal{F}$ to the image in $\mathsf{tails}\,A$ of $\bigoplus_{n=-\infty}^{\infty} H^0(X, \mathcal{F}(n))$, defines an equivalence of categories $\mathsf{coh}\,X \simeq \mathsf{tails}\,A$.*

Serre's theorem is the starting point for "non-commutative projective geometry," as we shall see in Section 7 below. From the point of view of categorical geometry, it is the first instance of a purely algebraic description of the (quasi)coherent sheaves on a space, and thus opens the possibility of "doing geometry" with only a category in hand.

4 Derived Categories of Modules

Originally introduced as technical tools for organizing homological (or "hyperhomological" [208]) information, derived categories have in the last 30 years been increasingly viewed as a basic invariant of a ring or variety. Passing from an Abelian category to an associated derived category not only tidies the workspace by incorporating the non-exactness of various natural functors directly into the notation, but in some cases it allows a "truer description" [57] of the underlying algebra or geometry than the Abelian category does. For example, there are varieties with non-trivial derived auto-equivalences $\mathsf{D}^b(X) \simeq \mathsf{D}^b(X)$ that do not arise from automorphisms; one might think of these as additional symmetries that were invisible from the geometric point of view. Another example is Kontsevich's Homological Mirror Symmetry conjecture [139], which proposes an equivalence of certain derived categories related to "mirror pairs" of Calabi–Yau manifolds.

Let us fix some notation. Let A be an Abelian category. The *homotopy category* $\mathsf{K}(\mathsf{A})$ has for objects the complexes over A, and for morphisms homotopy-equivalence classes of chain maps. The *derived category* $\mathsf{D}(\mathsf{A})$ is obtained by formally inverting those morphisms in $\mathsf{K}(\mathsf{A})$ which induce isomorphisms on cohomology, i.e. the quasi-isomorphisms.

We decorate $\mathsf{K}(\mathsf{A})$ and $\mathsf{D}(\mathsf{A})$ in various ways to denote full subcategories. For the moment I need only $\mathsf{K}^b(\mathsf{A})$, the full subcategory composed of complexes C having only finitely many non-zero components, and $\mathsf{D}^b(\mathsf{A})$, the corresponding *bounded derived category* of A.

The homotopy category $\mathsf{K}(\mathsf{A})$, the derived category $\mathsf{D}(\mathsf{A})$ and their kin are no longer Abelian categories, but they have a *triangulated structure*, consisting of a *shift functor* $(-)[1]$ shifting a complex one step against its differential and changing the sign of that differential, and a collection of *distinguished triangles* taking the place occupied by the short exact sequences in Abelian categories. A functor between triangulated categories is said to be a *triangulated functor* if it preserves distinguished triangles and intertwines the shift operators.

Homomorphisms $\varphi\colon M \longrightarrow N$ in $\mathsf{D}(A)$ are diagrams $M \xleftarrow{f} P \xrightarrow{g} N$ of homotopy classes of chain maps, where $f\colon P \longrightarrow M$ is a quasi-isomorphism and we think of φ as $f^{-1}g$. Much more usefully,

$$\operatorname{Hom}_{\mathsf{D}(A)}(M, N[i]) = \operatorname{Ext}^i_A(M, N)$$

for all $i \in \mathbb{Z}$ and all M, N in $\mathsf{D}(A)$.

Let us say that two rings Λ and Γ are *derived equivalent* if there is an equivalence of triangulated categories $\mathsf{D}^b(\Lambda\text{-}\mathsf{Mod}) \longrightarrow \mathsf{D}^b(\Gamma\text{-}\mathsf{Mod})$.

It is nearly obvious that a Morita equivalence between rings Λ and Γ gives rise to a derived equivalence $\mathsf{D}^b(\Lambda\text{-}\mathsf{Mod}) \simeq \mathsf{D}^b(\Gamma\text{-}\mathsf{Mod})$. (Any equivalence between Abelian categories preserves short exact sequences.) In general, derived equivalence is a much weaker notion. It does, however, preserve some essential structural information. For example, if Λ and Γ are derived equivalent, then their Grothendieck groups $K_0(\Lambda)$ and $K_0(\Gamma)$ are isomorphic [177, Proposition 9.3], as are the Hochschild homology and cohomology groups [178] and the cyclic cohomologies [93]. If Λ and Γ are derived-equivalent finite-dimensional algebras over a field k, then they have the same number of simple modules, and simultaneously have finite global dimension [91, 92].

Most importantly for this article, derived-equivalent rings have isomorphic centers [177, Proposition 9.2]. In particular, if R and S are commutative rings, then $\mathsf{D}^b(R\text{-}\mathsf{Mod}) \simeq \mathsf{D}^b(S\text{-}\mathsf{Mod})$ if and only if $R \cong S$. Thus there is at most one commutative ring in any derived-equivalence class, another sign that one should look at non-commutative rings for non-trivial derived equivalences.

All these facts follow from Rickard's Morita theory of derived equivalences, in which the progenerator of Theorem 2.2 is replaced by a *tilting object.* Here is the main result of Rickard's theory.

Theorem 4.1 (Rickard [177]). *Let Λ and Γ be rings. The following conditions are equivalent.*

(i) $\mathsf{D}^b(\Lambda\text{-}\mathsf{Mod})$ *and* $\mathsf{D}^b(\Gamma\text{-}\mathsf{Mod})$ *are equivalent as triangulated categories.*

(ii) $\mathsf{K}^b(\operatorname{add}\Lambda)$ *and* $\mathsf{K}^b(\operatorname{add}\Gamma)$ *are equivalent as triangulated categories.*

(iii) *There is an object $T \in \mathsf{K}^b(\operatorname{add}\Lambda)$ satisfying*

  (a) $\operatorname{Ext}^i_\Lambda(T, T) = 0$ *for all $i > 0$, and*

  (b) $\operatorname{add} T$ *generates* $\mathsf{K}^b(\operatorname{add}\Lambda)$ *as a triangulated category,*

such that $\Gamma \cong \operatorname{End}_\Lambda(T)$.

If Λ and Γ are finite dimensional algebras over a field, then these are all equivalent to $\mathsf{D}^b(\Lambda\text{-mod}) \simeq \mathsf{D}^b(\Gamma\text{-mod})$.

A complex T as in condition (iii) is called a *tilting complex* for Λ, and Γ is *tilted from* Λ. Tilting complexes appeared first in the form of *tilting modules*, as part of Brenner and Butler's [26] study of the reflection functors of Bernšteĭn, Gelfand, and Ponomarev [33]. (The word was chosen to illustrate their effect on the vectors in a root system, namely a change of basis that tilts the axes relative to the positive roots.) Their properties were generalized, formalized, and investigated subsequently by Happel and Ringel [103], Bongartz [52], Cline–Parshall–Scott [64], Miyashita [152], and others. Happel seems to have been the first to realize [92] that if T is a Λ-module of finite projective dimension, having no higher self-extensions $\operatorname{Ext}^{>0}_{\Lambda}(T, T) = 0$, and Λ has a finite co-resolution $0 \longrightarrow \Lambda \longrightarrow T_1 \longrightarrow \cdots \longrightarrow T_r \longrightarrow 0$ with each $T_i \in \operatorname{add} T$, then the functor $\operatorname{Hom}_{\Lambda}(T, -)\colon \mathsf{D}^b(\Lambda\text{-Mod}) \longrightarrow \mathsf{D}^b(\operatorname{End}_{\Lambda}(T)\text{-Mod})$ is an equivalence. In their earliest incarnation, tilting modules were defined to have projective dimension one, but this more general version has become standard.

Morita equivalence is a special case of tilting. Indeed, any progenerator is a tilting module. However, see Section 6 below for a pair of derived-equivalent algebras which are not Morita equivalent.

5 Derived Categories of Sheaves

The first real triumphs of the derived category came in the geometric arena: the construction by Grothendieck and coauthors of a global intersection theory and the theorem of Riemann–Roch [32] are the standard examples [60]. The idea of the (bounded) derived category of a scheme as a geometric invariant first emerged around 1980 in the work of Beilinson, Mukai, and others. I will describe some of Beilinson's observations in the next section. Mukai found the first example of non-isomorphic varieties which are derived equivalent [160]; he showed that an Abelian variety X and its dual $X^{\vee}$ always have equivalent derived categories of quasicoherent sheaves. His construction is modeled on a Fourier transform and is now called a Fourier–Mukai transform [105, 106]. It would draw us too far afield from our subject to discuss Fourier–Mukai transforms in any depth here. Several examples will appear later in the text: see the end of Section 6 and Theorems 9.4, 11.7, and 11.8. It is an important result of Orlov [162] that *any* equivalence $\mathsf{D}^b(\operatorname{coh} X) \longrightarrow \mathsf{D}^b(\operatorname{coh} Y)$, for X and Y connected smooth projective varieties, is given by a Fourier–Mukai transform.

The existence of non-trivial derived equivalences for categories of sheaves means that one cannot hope for a general reconstruction theorem, even for smooth varieties. However, under an assumption on the *canonical sheaf* ω_X, the variety X can be reconstructed from its derived category. Recall that for X a smooth complex variety over $\mathbb{C}$, $\omega_X = \bigwedge^{\dim X} \Omega_{X/\mathbb{C}}$ is the sheaf of top differential forms on X where $\Omega_{X/\mathbb{C}}$ is the cotangent bundle, a.k.a. the sheaf of 1-forms on X. It is an invertible sheaf. Recall further [94, Section II.7] that an invertible sheaf $\mathcal{L}$ is *ample* if for every coherent sheaf $\mathcal{F}$, $\mathcal{F} \otimes \mathcal{L}^n$ is generated by global sections for $n \gg 0$.

Theorem 5.1 (Bondal–Orlov [49]). *Let X and Y be smooth connected projective varieties over $\mathbb{C}$. Assume that either the canonical sheaf ω_X or the anticanonical sheaf ω_X^{-1} is ample. If $\mathsf{D}^b(\mathsf{coh}\, X) \simeq \mathsf{D}^b(\mathsf{coh}\, Y)$, then X is isomorphic to Y.*

Note that the result is definitely false for Abelian varieties by the result of Mukai mentioned above; in this case $\omega_X \cong \mathcal{O}_X$ is trivial, so not ample. Calabi–Yau varieties are another example where $\omega_X \cong \mathcal{O}_X$ is not ample, and the conclusion does not hold.

One consequence of this theorem is that, under the same hypotheses, the group of *auto-equivalences* $\mathsf{D}^b(\mathsf{coh}\, X) \xrightarrow{\simeq} \mathsf{D}^b(\mathsf{coh}\, X)$ of X is generated by the obvious suspects: $\operatorname{Aut}(X)$, the shift $(-)[1]$, and the tensor products $- \otimes_{\mathcal{O}_X} \mathcal{L}$ with fixed line bundles $\mathcal{L}$.

Triangulated categories arising in nature like $\mathsf{D}^b(\mathsf{coh}\, X)$ generally have a lot of additional structure: there is a tensor (symmetric monoidal) structure induced from the derived tensor product, among other things. Taking this into account gives stronger results. To give an example, recall that a *perfect complex* on a scheme X is one which locally is isomorphic in the derived category to a bounded complex of locally free sheaves of finite rank. Perfect complexes form a subcategory $\mathsf{D}^{\mathsf{perf}}(\mathsf{Qch}\, X)$. As long as X is quasi-compact and separated (Noetherian is enough), $\mathsf{D}^{\mathsf{perf}}(\mathsf{Qch}\, X)$ contains precisely the *compact* objects of $\mathsf{D}(\mathsf{Qch}\, X)$, that is, the complexes C such that $\operatorname{Hom}_{\mathcal{O}_X}(C, -)$ commutes with set-indexed direct sums. See [59, Theorem 3.1.1]. Balmer [23, 24] shows that a Noetherian scheme X can be reconstructed up to isomorphism from $\mathsf{D}^{\mathsf{perf}}(\mathsf{Qch}\, X)$, as long as the natural tensor structure is taken into account, and that two reduced Noetherian schemes X and X' are isomorphic if and only if $\mathsf{D}^{\mathsf{perf}}(\mathsf{Qch}\, X)$ and $\mathsf{D}^{\mathsf{perf}}(\mathsf{Qch}\, X')$ are equivalent as tensor triangulated categories.

The theory of tilting sketched in the previous section has a geometric incarnation as well, which signals the first appearance of non-commutative rings on the geometric side of our story.

Definition 5.2. Let X be a Noetherian scheme and T an object of $\mathsf{D}(\mathsf{Qch}\, X)$. Say that T is a *tilting object* if it is compact, is a classical generator for $\mathsf{D}^{\mathsf{perf}}(\mathsf{Qch}\, X)$, and has no non-trivial self-extensions. Explicitly, this is to say:

(i) T is a perfect complex;

(ii) The smallest triangulated subcategory of $\mathsf{D}(\mathsf{Qch}\, X)$ containing T and closed under direct summands is $\mathsf{D}^{\mathsf{perf}}(\mathsf{Qch}\, X)$; and

(iii) $\operatorname{Ext}^i_{\mathcal{O}_X}(T, T) = 0$ for $i > 0$.

If T is quasi-isomorphic to a complex consisting of a locally free sheaf in a single degree, it is sometimes called a *tilting bundle*.

The generating condition (ii) is sometimes replaced by the requirement that T *generates* $\mathsf{D}(\mathsf{Qch}\,X)$, i.e. that if an object N in $\mathsf{D}(\mathsf{Qch}\,X)$ satisfies $\mathrm{Ext}^i_{\mathcal{O}_X}(T, N) = 0$ for all $i \in \mathbb{Z}$, then $N = 0$. If an object T classically generates $\mathsf{D}^{\mathsf{perf}}(\mathsf{Qch}\,X)$ as in the definition, then it generates $\mathsf{D}(\mathsf{Qch}\,X)$; the converse holds in the presence of the assumption (i) that T is compact. This is a theorem due to Ravenel and Neeman [59, Theorem 2.1.2].

A class of schemes particularly well-suited for geometric tilting theory consists of those which are projective over a scheme Z, which in turn is affine of finite type over an algebraically closed field k. This generality allows a wide range of interesting examples, but also ensures, by [87, Théorème 2.4.1(i)], that if T is a tilting object on X then the endomorphism ring $\Lambda = \mathrm{End}_{\mathcal{O}_X}(T)$ is a *finitely generated* algebra over the field k. In particular, Λ is finitely generated as a module over its center.

The next result is fundamental for everything that follows. It has origins in the work of Beilinson presented in the next section, with further refinements in [22, 53, 59].

Theorem 5.3 (Geometric Tilting Theory [106, Theorem 7.6]). *Let X be a scheme, projective over a finite-type affine scheme over an algebraically closed field k. Let T be a tilting object in $\mathsf{D}(\mathsf{Qch}\,X)$, and set $\Lambda = \mathrm{End}_{\mathcal{O}_X}(T)$. Then*

(i) *The functor $\mathbf{R}\mathrm{Hom}_{\mathcal{O}_X}(T, -)$ induces an equivalence of triangulated categories between $\mathsf{D}(\mathsf{Qch}\,X)$ and $\mathsf{D}(\Lambda\text{-}\mathsf{Mod})$, with inverse $- \overset{\mathbf{L}}{\otimes}_\Lambda T$.*

(ii) *If T is in $\mathsf{D}^b(\mathsf{coh}\,X)$, then this equivalence restricts to give an equivalence between $\mathsf{D}^b(\mathsf{coh}\,X)$ and $\mathsf{D}^b(\Lambda\text{-}\mathsf{mod})$.*

(iii) *If X is smooth, then Λ has finite global dimension.*

It is not at all clear from this result when tilting objects exist, though it does impose some necessary conditions on X. For example, assume that in addition X is projective over k and T is a tilting object in $\mathsf{D}(\mathsf{Qch}\,X)$. Then $\Lambda = \mathrm{End}_{\mathcal{O}_X}(T)$ is a finite-dimensional algebra over k. The Grothendieck group $K_0(\Lambda)$ is thus a free Abelian group of finite rank, equal to the number of simple Λ-modules. This implies that $K_0(X)$ is free Abelian as well. Thus any torsion in K_0 rules out the existence of a tilting object.

6 Example: Tilting on Projective Space

In this section I illustrate Theorem 5.3 via Beilinson's tilting description of the derived category of projective space. The techniques have been refined and are now standard; they have been used, most notably by Kapranov, to construct explicit descriptions of the derived category of coherent sheaves on several classes of varieties. For example, there are tilting bundles on smooth projective quadrics [127], on Grassmannians [43, 126], on flag manifolds [128], on various toric varieties [101, 102, 131], and on weighted projective spaces [22, 81]. Here we stick to projective space.

Let k be a field, V a k-vector space of dimension $n \geq 2$, and $\mathbb{P} = \mathbb{P}^{n-1} = \mathbb{P}(V)$ the projective space on V. We consider two families of n locally free sheaves on $\mathbb{P}$. First let

$$\mathcal{E}_1 = \{\mathcal{O}, \mathcal{O}(-1), \dots, \mathcal{O}(-n+1)\}$$

where $\mathcal{O} = \mathcal{O}_{\mathbb{P}}$ is the structure sheaf. Also let $\Omega = \Omega_{\mathbb{P}}$ be the cotangent sheaf, so that $\Omega^i = \bigwedge^i \Omega$ is the $\mathcal{O}$-module of differential i-forms on $\mathbb{P}$, and set

$$\mathcal{E}_2 = \left\{\Omega^0(1) = \mathcal{O}(1), \Omega^1(2), \dots, \Omega^{n-1}(n)\right\}.$$

Let T_1 and T_2 be the corresponding direct sums,

$$T_1 = \bigoplus_{a=0}^{n-1} \mathcal{O}(-a) \quad \text{and} \quad T_2 = \bigoplus_{a=1}^{n} \Omega^{a-1}(a).$$

The constituent sheaves of $\mathcal{E}_1$ and $\mathcal{E}_2$ are related by the tautological Koszul complex on $\mathbb{P}$. Indeed, the Euler derivation $e\colon V \otimes_k \mathcal{O}(-1) \longrightarrow \mathcal{O}$, which corresponds to the identity on V under $\operatorname{Hom}_{\mathcal{O}}(V \otimes_k \mathcal{O}(-1), \mathcal{O}) \cong \operatorname{Hom}_{\mathcal{O}}(V \otimes_k \mathcal{O}, \mathcal{O}(1)) \cong \operatorname{Hom}_k(V, V)$, gives rise to a complex

$$0 \longrightarrow \textstyle\bigwedge^n V \otimes_k \mathcal{O}(-n) \longrightarrow \cdots \longrightarrow \bigwedge^1 V \otimes_k \mathcal{O}(-1) \longrightarrow \mathcal{O} \longrightarrow 0 \tag{6.1}$$

on $\mathbb{P}$. In fact it is acyclic [74, Example 17.20], and the cokernels are exactly the sheaves Ω^i, which decompose the Koszul complex into short exact sequences

$$0 \longrightarrow \Omega^a \longrightarrow \textstyle\bigwedge^a V \otimes_k \mathcal{O}(-a) \longrightarrow \Omega^{a-1} \longrightarrow 0.$$

Together with the well-known calculation of the cohomologies of the $\mathcal{O}(-a)$ [94, Theorem III.5.1], the identification $\mathcal{H}om_{\mathcal{O}}(\mathcal{O}(-a), \mathcal{O}(-b)) = \mathcal{O}(a-b)$, and the fact that $\operatorname{Ext}^i_{\mathcal{O}}(-,-) = H^i(\mathcal{H}om_{\mathcal{O}}(-,-))$ on vector bundles, this produces the following data. (See [44] for a jazzed-up version which holds over any base ring k.)

Lemma 6.1. *Keep the notation established so far in this section.*

(i) *We have* $\operatorname{Ext}^i_{\mathcal{O}}(\mathcal{O}(-a), \mathcal{O}(-b)) = 0$ *for all* $i > 0$, *and*

$$\operatorname{Hom}_{\mathcal{O}}(\mathcal{O}(-a), \mathcal{O}(-b)) \cong \operatorname{Sym}_{a-b}(V)$$

for $0 \leq a, b \leq n-1$.

(ii) *We have* $\operatorname{Ext}^i_{\mathcal{O}}(\Omega^{a-1}(a), \Omega^{b-1}(b)) = 0$ *for all* $i > 0$, *and*

$$\operatorname{Hom}_{\mathcal{O}}(\Omega^{a-1}(a), \Omega^{b-1}(b)) \cong \textstyle\bigwedge^{a-b}(V^*)$$

for $1 \leq a, b \leq n$, *where* V^* *is the dual of* V.

The lemma in particular implies that the endomorphism rings of T_1 and T_2,

$$\Lambda_1 = \mathrm{End}_{\mathcal{O}}\Big(\bigoplus_{a=0}^{n-1} \mathcal{O}(-a)\Big) \cong \bigoplus_{a,b=0}^{n-1} \mathrm{Sym}_{a-b}(V)$$

and

$$\Lambda_2 = \mathrm{End}_{\mathcal{O}}\Big(\bigoplus_{a=1}^{n} \Omega^{a-1}(a)\Big) \cong \bigoplus_{a,b=0}^{n-1} \bigwedge^{a-b}(V^*)$$

are "spread out" versions of the truncated symmetric and exterior algebras, respectively. This can be made more precise by viewing Λ_1 and Λ_2 as *quiver algebras.* Consider a quiver on n vertices labeled, say, $0, 1, \dots, n-1$, and having n arrows from each vertex to its successor, corresponding to a basis of V, resp. of V^*. Introduce quadratic relations $v_i v_j = v_j v_i$ corresponding to the kernel of the natural map $V \otimes_k V \longrightarrow \mathrm{Sym}_2(V)$, respectively $v_i v_j = -v_j v_i$ corresponding to the kernel of $V^* \otimes_k V^* \longrightarrow \bigwedge^2(V^*)$. The resulting path algebras with relations are isomorphic to Λ_1 and Λ_2, respectively. In [44] we call these "quiverized" symmetric and exterior algebras.

I have not yet proven that Λ_1 and Λ_2 are derived equivalent to $\mathbb{P}$. For this, it remains to show that the collections $\mathcal{E}_1$ and $\mathcal{E}_2$ generate the derived category $\mathsf{D}^b(\mathsf{coh}\,\mathbb{P})$. This is accomplished via Beilinson's "resolution of the diagonal" argument. Let $\Delta \subset \mathbb{P}\times\mathbb{P}$ denote the diagonal, and $p_1, p_2 \colon \mathbb{P}\times\mathbb{P} \longrightarrow \mathbb{P}$ the projections onto the factors. For sheaves $\mathcal{F}$ and $\mathcal{G}$ on $\mathbb{P}$, set

$$\mathcal{F} \boxtimes \mathcal{G} = p_1^*\mathcal{F} \otimes_{\mathbb{P}\times\mathbb{P}} p_2^*\mathcal{G}\,,$$

a sheaf on $\mathbb{P}\times\mathbb{P}$. One can show that the structure sheaf of the diagonal $\mathcal{O}_\Delta$ is resolved over $\mathcal{O}_{\mathbb{P}\times\mathbb{P}}$ by a Koszul-type resolution

$$0 \longrightarrow \mathcal{O}(-n) \boxtimes \Omega^n(n) \longrightarrow \cdots \longrightarrow \mathcal{O}(-1) \boxtimes \Omega^1(1) \longrightarrow \mathcal{O}_{\mathbb{P}\times\mathbb{P}} \longrightarrow \mathcal{O}_\Delta \longrightarrow 0\,.$$

In particular, $\mathcal{O}_\Delta$ is in the triangulated subcategory of $\mathsf{D}^b(\mathsf{coh}(\mathbb{P}\times\mathbb{P}))$ generated by sheaves of the form $\mathcal{O}(-i) \boxtimes Y$ for Y in $\mathsf{D}^b(\mathsf{coh}\,\mathbb{P})$. The same goes for any object of the form $\mathcal{O}_\Delta \boxtimes \mathbf{L}p_1^* X$ with X in $\mathsf{D}^b(\mathsf{coh}\,\mathbb{P})$ as well. Push down now by p_2 and use the projection formula to see that $X = \mathbf{R}p_{2*}(\mathcal{O}_\Delta \overset{\mathbf{L}}{\otimes} \mathbf{L}p_1^* X)$ belongs to the triangulated subcategory of $\mathsf{D}^b(\mathsf{coh}\,\mathbb{P})$ generated by $\mathcal{O}(-i) \otimes \mathbf{R}p_{2*}p_1^* Y$. The factor $\mathbf{R}p_{2*}p_1^* Y$ is represented by the complex of k-vector spaces with zero differential $\mathbf{R}\Gamma(Y)$, and hence $\mathcal{E}_1 = \{\mathcal{O}, \mathcal{O}(-1), \dots, \mathcal{O}(-n+1)\}$ generates $\mathsf{D}^b(\mathsf{coh}\,\mathbb{P})$. On the other hand, reversing the roles of p_1 and p_2 gives the result for $\mathcal{E}_2 = \big\{\mathcal{O}(1), \Omega^1(2), \dots, \Omega^{n-1}(n)\big\}$ as well.

This discussion proves the following theorem.

Theorem 6.2 (Beilinson). *Let k be a field, V a vector space of dimension $n \geq 2$ over k, and $\mathbb{P} = \mathbb{P}(V)$. The vector bundles*

$$T_1 = \bigoplus_{a=0}^{n-1} \mathcal{O}_{\mathbb{P}}(-a) \quad \text{and} \quad T_2 = \bigoplus_{a=1}^{n} \Omega_{\mathbb{P}}^{a-1}(a)$$

are tilting bundles on $\mathbb{P}$. In particular, there are equivalences of triangulated categories

$$\mathsf{D}^b(\Lambda_1\text{-mod}) \simeq \mathsf{D}^b(\mathsf{coh}\,\mathbb{P}) \simeq \mathsf{D}^b(\Lambda_2\text{-mod})$$

defined by $\mathbf{R}\mathrm{Hom}_{\mathcal{O}_{\mathbb{P}}}(T_i, -)$ for $i = 1, 2$, where $\Lambda_i = \mathrm{End}_{\mathcal{O}_{\mathbb{P}}}(T_i)$.

By the way, the construction $\mathbf{R}p_{2*}(\mathcal{O}_\Delta \overset{\mathbf{L}}{\otimes}_{\mathcal{O}_{\mathbb{P}}} \mathbf{L}p_1^*(-))$, which accepts sheaves on $\mathbb{P}$ and returns sheaves on $\mathbb{P}$, is an example of a Fourier–Mukai transform, the definition of which was gracefully avoided in Section 5. Replacing $\mathcal{O}_\Delta$ by any other fixed complex in $\mathsf{D}^b(\mathsf{coh}(\mathbb{P} \times \mathbb{P}))$ would give another.

7 The Non-existence of Non-commutative Spaces

As mentioned in the Introduction, I personally am reluctant to use the phrase "non-commutative algebraic geometry" to describe results like Beilinson's in Section 6. While the phrase is certainly apposite on a word-by-word basis, given that the ideas are a natural blend of algebraic geometry and non-commutative algebra, I find that using it in public leads immediately to being asked awkward questions like, "What on earth is non-commutative geometry?" While many people have offered thoughtful, informed answers to this question – [79, 121, 124, 125, 130, 144, 158, 189, 192, 210] are some of my personal favorites – I find the whole conversation distracting from the more concrete problems at hand. I propose instead that results like Beilinson's and those to follow in later sections should be considered as part of "categorical geometry". The name seems unclaimed, apart from an online book from 1998.

In this section I say a few words about a couple of approaches to building a field called non-commutative algebraic geometry. I have chosen a deliberately provocative title for the section, so that there can be no question that these are opinionated comments. The reader who is intrigued by the ideas mentioned here would do well to seek out a less idiosyncratic, more comprehensive introduction such as those cited in the previous paragraph.

One potential pitfall for the prospective student of non-commutative geometry is that there are several disparate approaches. For one thing, the approach of Connes and his collaborators [120], which some hope will explain aspects of the Standard Model of particle physics or even prove the Riemann Hypothesis, is based on differential geometry and C^*-algebras, and is, as far as I can tell, completely separate from most of the considerations in this article. More subtly, even within non-commutative *algebraic* geometry, there are a few different points of view. I do not consider myself competent even to give references, for fear of giving offense by omission.

So what is the problem here? Why cannot one simply do algebraic geometry, say at the level of [94], over non-commutative rings [156]? There have been several sustained attempts to do exactly this, starting in the 1970s. There are a couple of immediate obstacles to a naïve approach.

The first problem is to mimic the fact that a ring R can be recovered from the Zariski topology on the prime spectrum Spec R and the structure sheaf $\mathcal{O}_{\operatorname{Spec} R}$. (One finds, of course, $H^0(\operatorname{Spec} R, \mathcal{O}_{\operatorname{Spec} R}) = R$.) Both of these sets of information depend essentially on *localization.* For non-commutative rings, the prime spectrum is rather impoverished; for example, the Weyl algebra $\mathbb{C}\langle x, y\rangle/(yx - xy - 1)$ has trivial two-sided prime spectrum. Even ignoring this difficulty, localization for non-commutative rings [118, 172] only functions well for Øre sets, and the complement of a prime ideal need not be an Øre set.

One possible resolution of the problem would be to focus on the quotient modules $\Lambda/\mathfrak{p}$ instead of the prime ideals $\mathfrak{p}$. The points of a commutative affine variety X (over $\mathbb{C}$, say) are in one-one correspondence with the simple modules over the coordinate ring $\mathbb{C}[X]$. Furthermore, a point $x \in X$ is a non-singular point if and only if the corresponding simple module $\mathbb{C}[X]/\mathfrak{m}_x$ has finite projective dimension.

Unfortunately, here there is a second problem: finite projective dimension, even finite *global* dimension, is a very weak property for non-commutative rings. For example, there is no Auslander–Buchsbaum Theorem giving a uniform upper bound on finite projective dimensions over a given ring; the existence of such a bound over an Artin algebra is called the *finitistic dimension conjecture,* and has been open since at least 1960 [25, 107]. There are a host of additional technical problems to be overcome. It's unknown, for instance, whether finite global dimension implies primeness (as regularity implies domain for a commutative local ring); the Jacobson radical might fail to satisfy the Artin–Rees property [36], derailing the standard proof. Pathologies abound: for example, there is a local Noetherian domain Λ of global dimension 3 such that every quotient ring other than Λ itself, 0, and $\Lambda/\operatorname{rad}(\Lambda)$ has infinite global dimension [36, Example 7.3].

Restricting to a smaller class of rings solves some of these problems. For example, the class of rings Λ which are finitely generated modules over their center $Z(\Lambda)$ are much better-behaved than the norm. For example, Λ is left and right Noetherian if $Z(\Lambda)$ is, so that $\operatorname{gldim} \Lambda = \operatorname{gldim} \Lambda^{\mathrm{op}}$. The "lying over", "incomparability", and "going up" properties hold for prime ideals along the extension $Z(\Lambda) \hookrightarrow \Lambda$ [168, Theorem 16.9]. Furthermore, the following reassuring results hold [110, Section 2].

Proposition 7.1. *Let $(R, \mathfrak{m})$ be a local ring and Λ a module-finite R-algebra. Let M be a finitely generated Λ-module.*

(i) *The* dimension *of M, defined by* $\dim M = \dim(R/\operatorname{Ann}_R(M))$, *is independent of the choice of central subring R over which Λ is a finitely generated module.*

(ii) *The* depth *of* M, *defined by* $\operatorname{depth} M = \inf\{i \mid \operatorname{Ext}^i_R(R/\mathfrak{m}, M) \neq 0\}$, *is also independent of the choice of* R.

(iii) *(Ramras* [171]*) We have*

$$\operatorname{depth} M \leq \dim M \leq \operatorname{injdim}_\Lambda M .$$

If in particular Λ *is a torsion-free* R*-module and* $\operatorname{gldim} \Lambda < \infty$, *then* $\operatorname{injdim}_\Lambda \Lambda = \operatorname{gldim} \Lambda$ [17, *Lemma* 1.3], *so that*

$$\operatorname{depth}_R \Lambda \leq \dim R \leq \operatorname{gldim} \Lambda .$$

(iv) *(*[170] *or* [83]*) The global dimension of* Λ *is the supremum of* $\operatorname{pd}_\Lambda L$ *over all* Λ*-modules* L *of finite length.*

Restricting still further, one arrives at a very satisfactory class of rings. Recall that for $(R, \mathfrak{m})$ a local ring, a finitely generated R-module M is *maximal Cohen–Macaulay (MCM)* if $\operatorname{depth} M = \dim R$. Equivalently, there is a system of parameters $x_1, \dots, x_d$, with $d = \dim R$, which is an M-regular sequence. In the special case where R is Gorenstein, this condition is equivalent to $\operatorname{Ext}^i_R(M, R) = 0$ for all $i > 0$.

Definition 7.2. Let $(R, \mathfrak{m})$ be a local ring and Λ a module-finite R-algebra. Say that Λ is an R*-order* if Λ is maximal Cohen–Macaulay as an R-module.

The terminology is imperfect: there are several other definitions of the word "order" in the literature, going back decades. Here we follow [16]. See Section 12 for a connection to the classical theory of hereditary and classical orders over Dedekind domains.

Localization is still problematic, even for orders. In order to get a workable theory, a condition stronger than finite global dimension is sometimes needed.

Definition 7.3. Let R be a commutative ring and let Λ be a module-finite R-algebra. Say that Λ is *non-singular* if $\operatorname{gldim} \Lambda_\mathfrak{p} = \dim R_\mathfrak{p}$ for every prime ideal $\mathfrak{p}$ of R.

Non-singular orders have a very satisfactory homological theory, especially over Gorenstein local rings. A non-singular order over a local ring satisfies a version of the Auslander–Buchsbaum Theorem [110, Proposition 2.3]: If Λ is an R-order with $\operatorname{gldim} \Lambda = d < \infty$, then for any Λ-module M the equality $\operatorname{pd}_\Lambda M + \operatorname{depth} M = d$ holds. Furthermore, the following characterization of non-singularity holds for orders [113, Proposition 2.13].

Proposition 7.4. *Let* R *be a CM ring with a canonical module* ω, *and let* Λ *be an* R*-order. Then the following are equivalent.*

(i) Λ *is non-singular.*

(ii) $\operatorname{gldim} \Lambda_\mathfrak{m} = \dim R_\mathfrak{m}$ *for all maximal ideals* $\mathfrak{m}$ *of* R.

(iii) *The finitely generated* Λ*-modules which are MCM as* R*-modules are precisely the finitely generated projective* Λ*-modules.*

(iv) $\mathrm{Hom}_R(\Lambda, \omega)$ *is a projective* Λ*-module and* $\mathrm{gldim}\,\Lambda < \infty$.

The definitions above represent an attempt to force classical algebraic geometry, or equivalently commutative algebra, to work over a class of non-commutative rings. Here is a different approach, more consonant with the idea of "categorical geometry." Rather than focusing attention on the rings, concentrate on an Abelian or triangulated category C, which we choose to think of as $\mathsf{Qch}\,X$ or $\mathsf{D}^b(\mathsf{Qch}\,X)$ for some space X *about which we say nothing further.* In this approach, the space X is nothing but a notational placeholder, and the geometric object is the category C.

This idea has had particular success in taking Serre's Theorem 3.2 as a template and writing $\mathsf{D}^b(\mathsf{Qch}\,X)$ for a quotient category of the form $\mathsf{tails}\,\Lambda = \mathsf{grmod}\,\Lambda/\,\mathsf{tors}\,\Lambda$. One thus obtains what is called *non-commutative projective geometry.* To describe these successes, let us make the following definition, which is based on the work of Geigle–Lenzing [81], Verëvkin [207, 206], Artin–Zhang [21] and Van den Bergh [203].

Definition 7.5. A *quasi-scheme* (over a field k) is a pair $X = (X\text{-}\mathsf{mod}, \mathcal{O}_X)$ where X-mod is a (k-linear) Abelian category and $\mathcal{O}_X \in X\text{-}\mathsf{mod}$ is an object. Two quasi-schemes X and Y are isomorphic (over k) if there exists a (k-linear) equivalence $\mathsf{F}\colon X\text{-}\mathsf{mod} \longrightarrow Y\text{-}\mathsf{mod}$ such that $\mathsf{F}(\mathcal{O}_X) \cong \mathcal{O}_Y$.

The obvious first example is that a (usual, commutative) scheme X is a quasi-scheme $(\mathsf{coh}\,X, \mathcal{O}_X)$. For any ring Λ, commutative or not, one can define the affine quasi-scheme associated to Λ to be $\mathrm{Spec}\,\Lambda := (\Lambda\text{-}\mathsf{mod}, \Lambda)$. One checks that if R is commutative and $X = \mathrm{Spec}\,R$ is the usual prime spectrum, then the global section functor $\Gamma(X, -)\colon \mathsf{coh}\,X \longrightarrow R\text{-}\mathsf{mod}$ induces an isomorphism of quasi-schemes $(\mathsf{coh}\,X, \mathcal{O}_X) \longrightarrow (R\text{-}\mathsf{mod}, R)$.

The basic example of a quasi-scheme in non-commutative projective geometry is the following, which mimics the definition of tails from Section 3 precisely. Let Λ be a Noetherian graded algebra over a field k. For simplicity, assume that $A_0 = k$. Let $\mathsf{GrMod}\,\Lambda$ and $\mathsf{grmod}\,\Lambda$ be the categories of graded Λ-modules, resp. finitely generated graded Λ-modules. Let $\mathsf{Tors}\,\Lambda$, resp. $\mathsf{tors}\,\Lambda$, be the subcategory of graded modules annihilated by $\Lambda_{\geq n}$ for $n \gg 0$. Then define the quotient categories

$$\mathsf{Tails}\,\Lambda = \mathsf{GrMod}\,\Lambda/\,\mathsf{Tors}\,\Lambda \quad \text{and} \quad \mathsf{tails}\,\Lambda = \mathsf{grmod}\,\Lambda/\,\mathsf{tors}\,\Lambda\,,$$

and set

$$\mathsf{Proj}\,\Lambda = (\mathsf{Tails}\,\Lambda, \mathcal{O}) \quad \text{and} \quad \mathsf{proj}\,\Lambda = (\mathsf{tails}\,\Lambda, \mathcal{O})$$

where $\mathcal{O}$ is the image of Λ in $\mathsf{tails}\,\Lambda$. Call $\mathsf{Proj}\,\Lambda$ and $\mathsf{proj}\,\Lambda$ the *(Noetherian) projective quasi-scheme* determined by Λ. The *dimension* of the projective quasi-scheme is $\mathrm{GKdim}\,\Lambda - 1$, where $\mathrm{GKdim}\,\Lambda$ is the Gelfand–Kirillov dimension; this means that $\dim \mathsf{proj}\,\Lambda$ is the polynomial rate of growth of $\{\dim_k \Lambda_n\}_{n\geq 0}$.

One can define sheaf cohomology $H^j(\mathsf{Tails}\,\Lambda, -)$ in $\mathsf{Tails}\,\Lambda$ to directly generalize the commutative definition. In particular the global sections functor is $\Gamma(-) = \mathrm{Hom}_{\mathsf{Tails}\,\Lambda}(\mathcal{O}, -)$. For M in $\mathsf{Tails}\,\Lambda$, then, one would like versions of two basic results in algebraic geometry: Serre-finiteness ($H^j(\mathsf{Tails}\,\Lambda, M) = 0$ for $j \gg 0$) and Serre-vanishing ($H^j(\mathsf{Tails}\,\Lambda, M(i)) = 0$ for all $j \geq 1$ and $i \gg 0$). These results turn out only to be true under a technical condition called χ (see [21]), which is automatic in the commutative case. There is also an analogue of Serre's Theorem 3.2 due to Artin and Van den Bergh [20], which gives the same sort of purely algebraic description of $\mathsf{Qch}\,X$ as $\mathsf{Tails}\,\Lambda$, where Λ is defined to be a *twisted homogeneous coordinate ring.* For details, see [192].

The classification of projective quasi-schemes of small dimension, i.e. categories of the form $\mathsf{tails}\,\Lambda$ where Λ is a graded algebra with small rate of growth, is an ongoing program. The case of non-commutative curves (where $\dim_k A_n$ grows linearly) was completed by Artin and Stafford [10]. There is a conjectural classification of non-commutative surfaces due to Artin, but it is still open. The important special case of *non-commutative projective planes,* that is, $\mathsf{tails}\,\Lambda$ where Λ is a so-called Artin–Schelter regular algebra of Gelfand–Kirillov dimension 3 with Hilbert series $(1-t)^{-3}$, has been completely understood [9, 12, 13, 54, 202].

8 Resolutions of Singularities

So far I have considered only "absolute" situations, that is, constructions applied to individual rings or categories in isolation. In the sections to come, I will want to understand certain relative situations, particularly analogues of resolutions of singularities. In this section I collect a few definitions and facts about resolutions of singularities, for easy reference later. Begin with the definition.

Definition 8.1. Let X be an algebraic variety over a field k. A *resolution of singularities* of X is a proper, birational morphism $\pi: \widetilde{X} \longrightarrow X$ with $\widetilde{X}$ a non-singular algebraic variety.

Resolutions of singularities are also sometimes called "smooth models," indicating that the non-singular variety $\widetilde{X}$ is not too different from X: the map is an isomorphism on a dense open set and is proper, hence surjective. For curves, construction of resolution of singularities is easy, as every irreducible curve is birational to a unique smooth projective curve, namely the normalization (see Section 13). For surfaces, resolutions of singularities still exist in any characteristic, but now an irreducible surface is birational to infinitely many smooth surfaces. This observation is the beginning of the minimal model program, cf. Section 9.

Of course existence of resolutions of singularities in any dimension is a theorem due to Hironaka for k an algebraically closed field of characteristic zero; in this case the morphism $\pi: \widetilde{X} \longrightarrow X$ can be taken to be an isomorphism over the smooth locus of X, and even to be obtained as a sequence of blowups of non-singular subvarieties of the singular locus followed by normalizations. We will not need this.

As an aside, I mention here that a proper map between affine schemes is necessarily finite [94, Example II.4.6]. It follows immediately that a resolution of singularities of a singular normal affine scheme is *never* an affine scheme.

Our other definitions require the canonical sheaf of a singular variety. The canonical sheaf ω_Y of a smooth variety Y has already appeared, as the sheaf of top differential forms $\bigwedge^{\dim Y} \Omega_Y$ (see the discussion before Theorem 5.1). If Y is merely normal, then define ω_Y to be $j_* \omega_{Y_{\text{reg}}}$, where j is the open immersion $Y_{\text{reg}} \hookrightarrow Y$ of the smooth locus. When Y is Cohen–Macaulay, ω_Y is also a dualizing sheaf [94, Section III.7]; in other words, if the local rings of Y are CM, then the stalks of ω_Y are canonical modules in the sense of [98]. Similarly, ω_Y is an invertible sheaf (line bundle) if and only if Y is Gorenstein. The Weil divisor K_Y such that $\omega_Y = \mathcal{O}_Y(K_Y)$ is called the *canonical divisor*.

The behavior of the canonical sheaf/divisor under certain morphisms is of central interest. For example, the Grauert–Riemenschneider Vanishing theorem describes the higher direct images of ω.

Theorem 8.2 (GR Vanishing [84]). *Let $\pi\colon \widetilde{X} \longrightarrow X$ be a resolution of singularities of a variety X over $\mathbb{C}$. Then $\mathbf{R}^i \pi_* \omega_{\widetilde{X}} = 0$ for all $i > 0$.*

Now I come to a pair of words which will be central for the rest of the article.

Definition 8.3. Let $\pi\colon \widetilde{X} \longrightarrow X$ be a resolution of singularities of a normal variety X.

(i) Say that π is a *rational* resolution if $\mathbf{R}^i \pi_* \mathcal{O}_{\widetilde{X}} = 0$ for $i > 0$. Equivalently, since X is normal, $\mathbf{R}\pi_* \mathcal{O}_{\widetilde{X}} = \mathcal{O}_X$. In this case X is said to have *rational singularities.*

(ii) Say that π is a *crepant*[1] resolution if $\pi^* \omega_X = \omega_{\widetilde{X}}$.

Crepancy is a condition relating the two ways of getting a sheaf on $\widetilde{X}$ from one on X, namely via Hom and via $\otimes$. To get an idea what this condition is, consider a homomorphism of CM local rings $R \longrightarrow S$ such that S is a finitely generated R-module. Let ω_R be a canonical module for R. Then one knows that the "co-induced" module $\operatorname{Ext}_R^t(S, \omega_R)$, where $t = \dim R - \dim S$, is a canonical module for S [35, Theorem 3.3.7]. The "induced" module $S \otimes_R \omega_R$ is not necessarily a canonical module. Back in the geometric world, $\pi^* \omega_{\operatorname{Spec} R}$ corresponds to $S \otimes_R \omega_R$, so the assumption that this is equal to ω_S is locally a condition of the form $\operatorname{Ext}_R^t(S, \omega_R) \cong S \otimes_R \omega_R$. When X is Gorenstein, i.e. $\omega_X \cong \mathcal{O}_X$, a crepant resolution $\widetilde{X}$ is also Gorenstein.

One of the main motivations for considering crepant resolutions of singularities comes from the study of Calabi–Yau varieties, which in particular have trivial canon-

[1] Obligatory comment on the terminology: the word "crepant" is due to Miles Reid. He describes it [174, page 330] as a pun meaning "non-discrepant", in that the *discrepancy divisor* $K_{\widetilde{X}} - \pi^* K_X$ vanishes.

ical sheaves. In this case, if one wants a resolution $\pi\colon \widetilde{X} \longrightarrow X$ in which $\widetilde{X}$ is also Calabi–Yau, then π needs to be crepant.

A *small* resolution, that is, one for which the exceptional locus has codimension at least two, is automatically crepant. This is a very useful sufficient condition.

The next proposition follows from GR vanishing [132, page 50].

Proposition 8.4. *Let X be a complex algebraic variety and let $\pi\colon \widetilde{X} \longrightarrow X$ be a resolution of singularities.*

(i) *X has rational singularities if and only if X is CM and $\pi_*\omega_{\widetilde{X}} = \omega_X$.*

(ii) *If X is Gorenstein and has a crepant resolution of singularities, then X has rational singularities.*

Not every rational singularity has a crepant resolution. Here are two examples.

Example 8.5. Let R be the diagonal hypersurface ring $\mathbb{C}[x,y,z,t]/(x^3+y^3+z^3+t^2)$. Then R is quasi-homogeneous with the variables given weights 2, 2, 2, and 3. The a-invariant of R is thus $6-(2+2+2+3) = -3 < 0$, and R has rational singularities by Fedder's criterion [104, Example 3.9]. However, Lin [146] shows that a diagonal hypersurface defined by $x_0^r + x_1^d + \cdots + x_d^d$ has a crepant resolution of singularities if and only if r is congruent to 0 or 1 mod d.

Example 8.6. Quotient singularities $X = Y/G$, where Y is smooth and G is a finite group of automorphisms, have rational singularities [209]. Consider quotient singularities $\mathbb{C}^n/G$, where $G \subset \mathrm{SL}(n,\mathbb{C})$ is finite. These are by [211] the Gorenstein quotient singularities.

If $n = 2$, the results are the *rational double points*, also known as Kleinian singularities or Du Val singularities, which are the quotient singularities $X = \mathbb{C}^2/G = \mathrm{Spec}(\mathbb{C}[u,v]^G)$, where $G \subset \mathrm{SL}(2,\mathbb{C})$ is a finite subgroup. These are also described as ADE hypersurface rings $\mathbb{C}[x,y,z]/(f(x,y,z))$ with explicit equations as follows.

$$\begin{aligned}
&(A_n): && x^2 + y^{n+1} + z^2\,, && n \geq 1\\
&(D_n): && x^2y + y^{n-1} + z^2\,, && n \geq 4\\
&(E_6): && x^3 + y^4 + z^2 &&\\
&(E_7): && x^3 + xy^3 + z^2 &&\\
&(E_8): && x^3 + y^5 + z^2 &&
\end{aligned} \tag{8.1}$$

For these singularities, a crepant resolution always exists and is unique. In fact, a normal affine surface singularity R over $\mathbb{C}$ admits a crepant resolution if and only if every local ring of R is (at worst) a rational double point. I will return to the rational double points in Section 11 below.

If $n = 3$, $\mathbb{C}^3/G$ always has a crepant resolution as well, though they are no longer unique, thanks to the existence of *flops* (see the next section). There is a classification of the finite subgroups of SL(3, $\mathbb{C}$) up to conjugacy, and existence of crepant resolutions was verified on a case-by-case basis by Markushevich [150], Roan [179, 180], Ito [111, 112], and Ito–Reid [109]. See Theorem 11.8 below for a unified statement.

For $n \geq 4$, quotient singularities need not have crepant resolutions of singularities. For example, the quotient of $\mathbb{C}^4$ by the involution $(x, y, z, w) \mapsto (-x, -y, -z, -w)$ admits no crepant resolution [175, Example 5.4].

9 The Minimal Model Program

A key motivation for categorical desingularizations in general, and non-commutative crepant resolutions in particular, is the *minimal model program* of Mori and Reid. This is an attempt to find a unique "best" representative for the birational equivalence class of any algebraic variety. For curves, this is obvious, since there is in each equivalence class a unique smooth projective representative.

It is also the case that every surface is birationally equivalent to a smooth projective surface, but now matters are complicated by the fact that the blowup of a smooth surface at a point is again smooth. However, every birational morphism of surfaces factors as a sequence of blowups, so must have a (-1)-curve, that is, a rational curve $C \cong \mathbb{P}^1$ with self-intersection -1, lying over a smooth point. One can compute that if C is a (-1)-curve on a surface X, then $K_X \cdot C = -1$, where K_X is the canonical divisor.

By Castelnuovo's criterion, a (-1)-curve can always be blown down, essentially undoing the blowup. The algorithm for obtaining a minimal model is thus to contract all the (-1)-curves, and one obtains the classification of minimal models for surfaces [94, Section V.5]: the result of the algorithm is a smooth projective surface S which is either $\mathbb{P}^2$, a ruled surface over a curve (the "Fano" case), or such that $K_S \cdot C \geq 0$ for every curve C in S. In this last case say that K_S is *nef.*

The minimal model program is a framework for extending this simple-minded algorithm to one that will work for threefolds and higher-dimensional varieties. The theory turns out to be much richer, in part because it turns out that one must allow minimal models to be a little bit singular. Here "a little bit" means in codimension ≥ 2. Precisely, a projective variety X is a *minimal model* if every birational map $Y \dashrightarrow X$ is either a contraction of a divisor to a set of codimension at least two, or is an isomorphism outside sets of codimension at least two [138]. There are compelling reasons to allow singular minimal models; for example, there exists a three-dimensional smooth variety which is not birational to any smooth variety with nef canonical divisor [155]. Mori and Reid realized that this meant minimal models need not be smooth; they can be taken to be *terminal* instead.

I will not worry about the technical definitions of terminal and canonical singularities here, but only illustrate with a class of examples. A diagonal hypersurface singularity defined by $x_0^{a_0} + x_1^{a_1} + \cdots + x_d^{a_d}$ is

(i) *canonical* if and only if $a_1 + \cdots + a_d > 1$, and

(ii) *terminal* if and only if $a_1 + \cdots + a_d > 1 + \frac{1}{\operatorname{lcm}(a_i)}$.

For Gorenstein singularities, canonical singularities are the same as rational singularities, so Proposition 8.4 (ii) says that the existence of a crepant resolution implies canonical singularities.

In this language, a projective variety X is a minimal model if and only if it is $\mathbb{Q}$-factorial (i.e. the divisor class group of every local ring is torsion), has nef canonical divisor, and has terminal singularities.

In dimension two, minimal models are unique up to isomorphism by definition. Terminal surface singularities are smooth, and the canonical surface singularities are the rational double points of Example 8.6 [137, (2.6.2)].

In dimension three, terminal singularities are well-understood, cf. [137, 2.7] or [173]. The Gorenstein ones are precisely the isolated compound Du Val (cDV) singularities. (Recall that a cDV singularity is a hypersurface defined by $f(x, y, z) + tg(x, y, z, t)$, where f is a simple singularity as in (8.1) and g is arbitrary.) However, minimal models of threefolds are no longer unique [63]. Here is the simplest example.

Example 9.1 (The "classic flop"). Let X be the three-dimensional (A_1) singularity over $\mathbb{C}$, so $X = \operatorname{Spec} \mathbb{C}[u, v, x, y]/(uv - xy)$. Consider the blowup $f\colon Y \longrightarrow X$ of the plane $u = x = 0$. It's easy to check that Y is smooth, and that $f\colon Y \longrightarrow X$ is a birational map which contracts a line $L \cong \mathbb{P}^1$ to the origin. Thus f is a small resolution, whence crepant. Furthermore Y is a minimal model.

One could also have considered the plane $u = y = 0$ and its blowup $f'\colon Y' \longrightarrow X$. Symmetrically, Y' is smooth, f' contracts a line $L' \cong \mathbb{P}^1$ and is crepant, and Y' is a minimal model.

The resolutions Y and Y' are almost indistinguishable, but they are not isomorphic over X. One can check that the birational transforms of the plane $u = x = 0$ to Y and Y' have intersection number $+1$ with L and -1 with L'.

On the other hand, the induced birational map $\varphi\colon Y \dashrightarrow Y'$ is an isomorphism once one removes L from Y and L' from Y'. This φ is called a (or "the classic") *flop*. It is also sometimes called the "Atiyah flop" after [11], though Reid traces it back through work of Zariski in the 1930s, and assigns it a birthdate of around 1870.

Let Z be the blowup of the origin of X. Then Z is in fact the closed graph of φ and there is a diagram

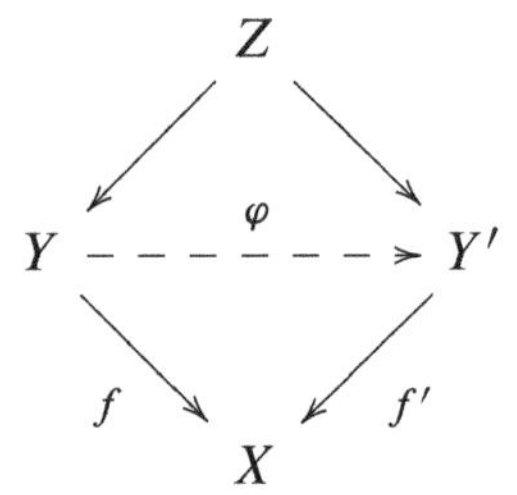

The exceptional surface of $Z \longrightarrow X$ is the quadric $Q = \mathbb{P}^1 \times \mathbb{P}^1$, which is cut out by two families of lines. The lines L and L' are the contractions of Q along these two rulings, and conversely Q is the blowup of $L \subset Y$, resp. $L' \subset Y'$.

The next definition is a special case of the usual definition of a flop [134, Definition 6.10] (in general, one need not assume Y and Y' are smooth, nor that X is Gorenstein).

Definition 9.2. Let Y and Y' be smooth projective varieties. Say that a birational map $\varphi\colon Y \dashrightarrow Y'$ is a *flop* if there is a diagram

$$\begin{array}{ccc} Y & \overset{\varphi}{\dashrightarrow} & Y' \\ & {\scriptstyle f}\searrow \quad \swarrow{\scriptstyle f'} & \\ & X & \end{array}$$

where X is a normal projective Gorenstein variety, f and f' are small resolutions of singularities, and there is a divisor D on Y such that, if D' is the strict transform of D on Y', then $-D'$ is ample.

Say $\varphi\colon Y \dashrightarrow Y'$ is a *generalized flop* if for some (equivalently, for every) diagram

$$\begin{array}{ccc} & Z & \\ & {\scriptstyle \pi}\swarrow \quad \searrow{\scriptstyle \pi'} & \\ Y & \overset{\varphi}{\dashrightarrow} & Y' \end{array}$$

with Z smooth, there is an equality $\pi^* K_Y = \pi'^* K_{Y'}$.

It is known that the existence of a crepant resolution forces canonical singularities, so that in particular if X participates in a flop as above, it has canonical singularities. On the other hand, if X is $\mathbb{Q}$-factorial and has terminal singularities, then it can have no crepant resolution of singularities [136, Corollary 4.11] (this is one explanation of the name "terminal").

Bondal and Orlov [50] observed that one ingredient of the minimal model program, namely the blowup $\widetilde{X}$ of a smooth variety X at a smooth center, induces a fully faithful functor on derived categories $\mathsf{D}^b(\mathsf{coh}\, X) \longrightarrow \mathsf{D}^b(\mathsf{coh}\, \widetilde{X})$. They propose that each of the operations of the program should induce such fully faithful embeddings. In particular, they make the following conjecture.

Conjecture 9.3 (Bondal–Orlov). For a generalized flop $\varphi\colon Y \dashrightarrow Y'$ between smooth varieties, there is an induced exact equivalence $\mathsf{F}\colon \mathsf{D}^b(\mathsf{coh}\, Y') \longrightarrow \mathsf{D}^b(\mathsf{coh}\, Y)$.

Notice that even though there always exists a natural Fourier–Mukai type functor $\mathbf{R}\pi_*\mathbf{L}\pi'^*(-)\colon \mathsf{D}^b(\operatorname{coh} Y') \longrightarrow \mathsf{D}^b(\operatorname{coh} Y)$, this is known not to be fully faithful in general, so some new idea is needed.

Bondal and Orlov proved Conjecture 9.3 in some special cases in dimension three, and Bridgeland [56] gave a complete proof for threefolds. Here is Bridgeland's result.

Theorem 9.4 (Bridgeland). *Let X be a projective complex threefold with terminal singularities. Let $f\colon Y \longrightarrow X$ and $f'\colon Y' \longrightarrow X$ be crepant resolutions of X. Then $\mathsf{D}^b(\operatorname{coh} Y) \simeq \mathsf{D}^b(\operatorname{coh} Y')$.*

The equivalence in this theorem is a Fourier–Mukai type functor $\mathbf{R} f_*(\mathcal{P} \overset{\mathbf{L}}{\otimes} f'^*(-))$, where $\mathcal{P}$ is a well-chosen object of $\mathsf{D}^b(\operatorname{coh}(Y \times_X Y'))$. In fact the construction of $\mathcal{P}$ is very difficult and is the heart of the proof.

10 Categorical Desingularizations

Now let us combine the philosophical ramblings of Section 7 with the concrete problems of Sections 8 and 9. Treating commutative and non-commutative varieties – in the form of their derived categories – on equal footing, one can entertain the notion of a *resolution of a commutative algebraic variety by a non-commutative one.* Bondal and Orlov [50] seem to have been the first to articulate such a possibility in pure mathematics. Other authors have considered modified or specialized versions, e.g. [31, 141, 147].

To begin, let us consider resolutions of singularities from a categorical point of view. Let X be a normal algebraic variety, and let $\pi\colon \widetilde{X} \longrightarrow X$ be a resolution of singularities. There are two natural functors between derived categories, namely the derived pushforward $\mathbf{R}\pi_*\colon \mathsf{D}^b(\operatorname{coh} \widetilde{X}) \longrightarrow \mathsf{D}^b(\operatorname{coh} X)$ and the derived pullback $\mathbf{L}\pi^*\colon \mathsf{D}(\operatorname{coh} X) \longrightarrow \mathsf{D}(\operatorname{coh} \widetilde{X})$. The derived pullback may not take bounded complexes to right-bounded ones, so does not generally give a functor on D^b. One could restrict $\mathbf{L}\pi^*$ to the perfect complexes over X and write instead $\mathbf{L}\pi^*\colon \mathsf{D}^{\mathrm{perf}}(\operatorname{coh} X) \longrightarrow \mathsf{D}^{\mathrm{perf}}(\operatorname{coh} \widetilde{X}) = \mathsf{D}^b(\operatorname{coh} X)$.

The pullback and pushforward form an adjoint pair. If X is assumed to have rational singularities, much more can be said. For an object $\mathcal{E}$ in $\mathsf{D}^b(\operatorname{coh} \widetilde{X})$ and a perfect complex $\mathcal{P}$ over X, the derived projection formula gives

$$\mathbf{R}\pi_*(\mathcal{E} \overset{\mathbf{L}}{\otimes}_{\mathcal{O}_{\widetilde{X}}} \mathbf{L}\pi^*\mathcal{P}) = \mathbf{R}\pi_*\mathcal{E} \overset{\mathbf{L}}{\otimes}_{\mathcal{O}_X} \mathcal{P}\,.$$

In particular, setting $\mathcal{E} = \mathcal{O}_{\widetilde{X}}$ and taking into account $\mathbf{R}\pi_*\mathcal{O}_{\widetilde{X}} = \mathcal{O}_X$, this yields

$$\mathbf{R}\pi_*\mathbf{L}\pi^*\mathcal{P} = \mathcal{P}$$

for every perfect complex $\mathcal{P}$ in $\mathsf{D}^b(\operatorname{coh} X)$. Otherwise said, $\mathbf{R}\pi_*\colon \mathsf{D}^b(\operatorname{coh} \widetilde{X}) \longrightarrow \mathsf{D}^b(\operatorname{coh} X)$ identifies the target with the quotient of the source by the kernel of $\mathbf{R}\pi_*$. Bondal and Orlov propose to take this as a template:

Definition 10.1 (Bondal–Orlov). A *categorical desingularization* of a triangulated category D is an Abelian category C of finite homological dimension and a triangulated subcategory K of $\mathsf{D}^b(\mathsf{C})$, closed under direct summands, such that $\mathsf{D}^b(\mathsf{C})/\mathsf{K} \simeq \mathsf{D}$.

One problem with this definition is the assumption that C have finite homological dimension. As observed in Section 7, this is a very weak condition when C is the category of modules over a non-commutative ring. There are a number of proposals for a better – that is, more restrictive – notion of *smoothness* for a (triangulated) category, but as far as I can tell, no consensus on a best candidate [140, 141, 147, 198].

As an aside, I note here that the condition for $\pi: \widetilde{X} \longrightarrow X$ to be crepant can be translated into categorical language as "the right adjoint functor $\pi^!$, which is locally represented by $\mathrm{Hom}_{\mathcal{O}_X}(\mathcal{O}_{\widetilde{X}}, -)$, is isomorphic to π^*." We will not need this.

Let us reconsider Example 9.1 from the point of view of categorical geometry. This can be thought of as a warmup for Section 18.

Example 10.2. Set $R = \mathbb{C}[u, v, x, y]/(uv - xy)$, so that $X = \operatorname{Spec} R$ is the three-dimensional ordinary double point as in Example 9.1. Let $I = (u, x)$ and $I' = (u, y)$. Then in fact $I' = I^{-1} = I^* = \mathrm{Hom}_R(I, R)$ is the dual of I. Notice too that $\mathrm{End}_R(I) = R$, either by direct computation or by Theorem 13.1 below.

Let $f: Y \longrightarrow X$ and $f': Y' \longrightarrow X$ be the blowups of I and I' as before. On Y, consider the locally free sheaf $\mathcal{E} = \mathcal{O}_Y \oplus \mathcal{O}_Y(1)$, which is the pullback of $\mathcal{O}_X \oplus \mathcal{I}$, where $\mathcal{I}$ is the ideal sheaf of I. Straightforward calculations (or see Section 18) show that $\mathcal{E}$ is a *tilting bundle* on Y (Definition 5.2), and hence $\mathbf{R}\mathrm{Hom}_{\mathcal{O}_Y}(\mathcal{E}, -): \mathsf{D}^b(\mathsf{coh}\, Y) \longrightarrow \mathsf{D}^b(\Lambda\text{-}\mathsf{mod})$ is an equivalence, where $\Lambda = \mathrm{End}_{\mathcal{O}_Y}(\mathcal{E})$. Furthermore, we have

$$\Lambda \cong f_* \mathcal{E}nd_{\mathcal{O}_Y}(\mathcal{E}) = \mathrm{End}_R(R \oplus I),$$

which can also be written as a block-matrix ring

$$\Lambda = \begin{pmatrix} R & I \\ I^{-1} & \mathrm{End}_R(I) = R \end{pmatrix}.$$

The induced functor $\mathsf{D}^b(\Lambda\text{-}\mathsf{mod}) \longrightarrow \mathsf{D}^b(\mathsf{coh}\, X)$ is then obviously a categorical desingularization.

Repeating the construction above with $\mathcal{E}' = \mathcal{O}_{Y'} \oplus \mathcal{O}_{Y'}(1)$ on Y', one obtains $\Lambda' = \mathrm{End}_R(R \oplus I')$. But since $I' = I^{-1}$, Λ' is isomorphic to Λ. This implies equivalences

$$\mathsf{D}^b(\mathsf{coh}\, Y) \simeq \mathsf{D}^b(\Lambda\text{-}\mathsf{mod}) \simeq \mathsf{D}^b(\mathsf{coh}\, Y').$$

Inspired by the example above and others from the minimal model program, Bondal and Orlov expect that for a singular variety X, the category $\mathsf{D}^b(\mathsf{coh}\, X)$ should have a *minimal* categorical desingularization, i.e. one embedding in any other. Such a category would be unique up to derived equivalence. They propose in particular the following conjecture.

Conjecture 10.3 (Bondal–Orlov [50]). Let X be a complex algebraic variety with canonical singularities and let $f\colon Y \longrightarrow X$ be a finite morphism with Y smooth. Then $\mathcal{A} = \mathcal{E}nd_{\mathcal{O}_X}(f_*\mathcal{O}_Y)$ gives a minimal categorical desingularization, in the sense that $\mathcal{A}$-mod has finite global dimension and if $\widetilde{X} \longrightarrow X$ is any other resolution of singularities of X, then there exists a fully faithful embedding $\mathsf{D}^b(\mathcal{A}\text{-mod}) \longrightarrow \mathsf{D}^b(\operatorname{coh}\widetilde{X})$. Moreover, if $\widetilde{X} \longrightarrow X$ is crepant, then the embedding is an equivalence.

In the next section I will consider another family of examples providing strong evidence for this conjecture.

11 Example: the McKay Correspondence

In this section I sketch a main motivating example, already foreshadowed in Example 8.6. The finite subgroups of SL(2, $\mathbb{C}$) were carefully studied by Klein in the 1880s, and the resolutions of the corresponding singularities $\mathbb{C}^2/G = \operatorname{Spec}\mathbb{C}[u,v]^G$ were understood by Du Val in the 1930s. The structure of the resolution faithfully reflects the representation theory of the group G, as observed by McKay [151], and the correspondence naturally extends to the reflexive modules over the (completed) coordinate ring $\mathbb{C}[\![u,v]\!]^G$. Even more, there is a natural resolution of singularities of the quotient singularity, built from the group G, which is derived equivalent to a certain non-commutative ring built from these reflexive modules. Thus the group G already knows the geometry of $\mathbb{C}^2/G$ and its resolution of singularities.

This section is about this circle of ideas, which together go by the name "McKay correspondence." I consider first, more generally, finite subgroups $G \subset \mathrm{GL}(n,k)$ with $n \geq 2$ and k a field of characteristic relatively prime to $|G|$. Then I specialize to $n = 2$ and subgroups of SL, where the strongest results hold. See [148] or [218] for proofs.

Let $S = k[\![x_1,\dots,x_n]\!]$ be a power series ring over an algebraically closed field k with $n \geq 2$. Let $G \subset \mathrm{GL}(n,k)$ be a finite subgroup with order invertible in k. Make G act on S by linear changes of variables, and set $R = S^G$, the ring of invariants. The ring R is Noetherian, local, and complete, of dimension n. It is even CM by the Hochster–Eagon theorem [95]. Furthermore, S is a module-finite R-algebra, and is a maximal Cohen–Macaulay R-module.

The central character in the story is the skew, or twisted, group algebra $S\#G$. As an S-module, $S\#G$ is free on the elements of G, and the product of two elements $s \cdot \sigma$ and $t \cdot \tau$, with $s,t \in S$ and $\sigma,\tau \in G$, is defined by $(s\cdot\sigma)(t\cdot\tau) = s\sigma(t)\cdot\sigma\tau$. Thus moving σ past t "twists" the ring element.

Left modules over $S\#G$ are precisely S-modules with a compatible action of G, and one computes that $\operatorname{Hom}_{S\#G}(M,N) = \operatorname{Hom}_S(M,N)^G$ for $S\#G$-modules M and N. Since the order of G is invertible, taking invariants is an exact functor, whence $\operatorname{Ext}^i_{S\#G}(M,N) = \operatorname{Ext}^i_S(M,N)^G$ for all $i > 0$ as well. It follows that an $S\#G$-module P is projective if and only if it is free over S. This, together with a moment's contemplation of the (G-equivariant) Koszul complex over S on $x_1, \dots, x_n$, gives the following observation.

Proposition 11.1. *The twisted group ring $S\#G$, where $S = k[\![x_1, \dots, x_n]\!]$ and G is a finite group of linear automorphisms of S with order invertible in k, has finite global dimension equal to n.*

The "skew" multiplication rule in $S\#G$ is cooked up precisely so that the homomorphism $\gamma\colon S\#G \longrightarrow \operatorname{End}_R(S)$, defined by $\gamma(s\cdot\sigma)(t) = s\sigma(t)$, is a ring homomorphism extending the group homomorphism $G \longrightarrow \operatorname{End}_R(S)$ defining the action of G on S. In general, γ is neither injective nor surjective, but under an additional assumption on G, it is both. Recall that a *pseudo-reflection* is an element $\sigma \in \mathrm{GL}(n,k)$ of finite order which fixes a hyperplane.

Theorem 11.2 (Auslander [14, 18]). *Let $S = k[\![x_1, \dots, x_n]\!]$, $n \geq 2$, let $G \subset \mathrm{GL}(n,k)$ be a finite group acting on S, and assume $|G|$ is invertible in S. Set $R = S^G$. If G contains no non-trivial pseudo-reflections then the homomorphism $\gamma\colon S\#G \longrightarrow \operatorname{End}_R(S)$ is an isomorphism.*

Consequently, in this case $\operatorname{End}_R(S)$ has finite global dimension and as an R-module is isomorphic to a direct sum of copies of S, so in particular is a MCM R-module.

The condition that G contain no non-trivial pseudo-reflections is equivalent to the extension $R \hookrightarrow S$ being unramified in codimension one [218, Lemma 10.7].

Let $\varrho\colon G \longrightarrow \mathrm{GL}(W)$ be a representation of G on the finite-dimensional k-vector space W. Then $S \otimes_k W$, with the diagonal action of G, is a finitely generated $S\#G$-module. It is free over S, whence projective over $S\#G$. The submodule of fixed points, $M_\varrho = (S \otimes_k W)^G$, is naturally an R-module. If ϱ is irreducible, then one can show that M_ϱ is a direct summand of S as an R-module. Conversely, given any R-direct summand of S, the corresponding idempotent in $\operatorname{End}_R(S)$ defines an $S\#G$-direct summand P of $S\#G$, whence a representation $P/(x_1, \dots, x_n)P$ of G.

Corollary 11.3. *These operations induce equivalences between the categories $\mathsf{add}_R(S)$ of R-direct summands of S, $\mathsf{add}\operatorname{End}_R(S)$ of finitely generated projective $\operatorname{End}_R(S)$-modules, and $\mathsf{add}\, S\#G$ of finitely generated projective $S\#G$-modules, and a $1-1$ correspondence with the objects of the category $\mathsf{rep}_k\, G$ of finite-dimensional representations of G.*

As a final ingredient, define a quiver from the data of the representation theory of G, or equivalently – given the correspondences above – of the R-module structure of S.

Definition 11.4 (McKay [151]). The *McKay quiver* of $G \subset \mathrm{GL}(n,k)$ has vertices $\varrho_0, \dots, \varrho_d$, a complete set of the non-isomorphic irreducible k-representations of G, with ϱ_0 the trivial irrep. Denote by ϖ the given n-dimensional representation of G as a subgroup of $\mathrm{GL}(n,k)$. Then draw m_{ij} arrows $\varrho_i \longrightarrow \varrho_j$ if the multiplicity of ϱ_i in $\varpi \otimes_k \varrho_j$ is equal to m_{ij}.

Now let us specialize to the case $n = 2$. Here the MCM R-modules are precisely the reflexive ones. This case is unique thanks to the following result, which fails badly for $n \geq 3$.

Lemma 11.5 (Herzog [96]). *Let $S = k[\![u, v]\!]$, let $G \subset \mathrm{GL}(2, k)$ be a finite group of order invertible in k, and let $R = S^G$. Then every finitely generated reflexive R-module is a direct summand of a direct sum of copies of S as an R-module. In particular, the MCM R-modules coincide with $\mathsf{add}_R(S)$, and there are only finitely many indecomposable ones.*

The one-one correspondences that hold for arbitrary n can thus be augmented in dimension two, giving a correspondence between the irreducible representations of G and the indecomposable MCM R-modules.

Specialize one last time, to assume now that $G \subset \mathrm{SL}(2, k)$. Note that then G automatically contains no non-trivial pseudo-reflections. Furthermore, $R = S^G$ is Gorenstein by a result of Watanabe [211]; in fact, it is classical [133] that $\operatorname{Spec} R$ embeds as a hypersurface in k^3, so $R \cong k[\![x, y, z]\!]/f(x, y, z)$ for some polynomial f. As long as k has characteristic not equal to 2, 3, or 5, the polynomials arising are precisely the ADE polynomials of (8.1) defining the rational double points.

The rational double points are distinguished among normal surface singularities by the fact that their local rings have unique crepant resolutions of singularities, which are the minimal resolutions of singularities. They are particularly easy to compute, being achieved by a sequence of blowups of points (no normalization required). The preimage of the singular point is a bunch of rational curves $E_1, \dots, E_n$ on the resolution. These curves define the *dual graph of the desingularization*: it has for vertices the irreducible components $E_1, \dots, E_n$, with an edge joining E_i to E_j if $E_i \cap E_j \neq 0$. This graph is related to the other data as follows.

Theorem 11.6 (Classical McKay Correspondence). *Let k be an algebraically closed field of characteristic not 2, 3, or 5, and let $G \subset \mathrm{SL}(2, k)$ be a finite subgroup of order invertible in k. Set $S = k[\![u, v]\!]$, with a natural linear action of G, set $R = S^G$, and let $\pi\colon \widetilde{X} \longrightarrow \operatorname{Spec} R$ be the minimal resolution of singularities with exceptional curves $E_1, \dots, E_n$. Then*

(i) *There is a one-one correspondence between*

  (a) *the exceptional curves E_i;*

  (b) *the irreducible representations of G; and*

  (c) *the indecomposable MCM R-modules.*

(ii) *(McKay) The dual graph of the desingularization is isomorphic to the McKay quiver after deleting the trivial vertex and replacing pairs of opposed arrows by edges. It is an ADE Coxeter–Dynkin diagram.*

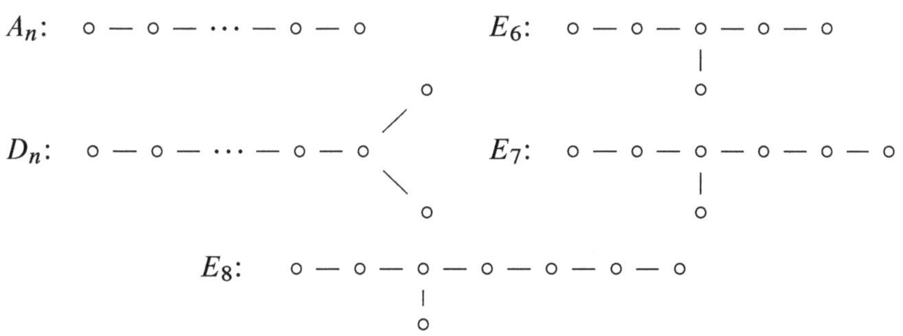

Shortly after McKay's original observation [151] of the isomorphism of graphs above, Gonzalez-Sprinberg and Verdier [90] gave, in characteristic zero, a geometric construction linking the representation theory of G and the resolution of singularities $\widetilde{X}$. Later constructions by Artin–Verdier [19], Esnault [75], and Knörrer [135] made explicit the correspondences between the exceptional curves E_i, the indecomposable reflexive R-modules, and the irreducible representations of G.

The first intimation of a "higher geometric McKay correspondence" appeared in string theory in the mid-1980s. Dixon, Harvey, Vafa, and Witten [71] observed that for certain $G \subseteq \mathrm{SL}(3,\mathbb{C})$, and a certain crepant resolution $\widetilde{X} \longrightarrow \mathbb{C}^3/G$, there is an equality between the Euler characteristic $\chi(\widetilde{X})$ and the number of conjugacy classes (= number of irreducible representations) of G. There followed a great deal of work on the existence of crepant resolutions of singularities for quotient singularities of the form Y/G, where Y is an arbitrary smooth variety of dimension two or more. Specifically, one can ask for the existence of a crepant resolution $\widetilde{X} \longrightarrow Y/G$ and a derived equivalence between $\widetilde{X}$ and the G-equivariant coherent sheaves on Y. Let $\mathsf{D}^b_G(Y)$ denote the bounded derived category of the latter.

Such an equivalence was first constructed by Kapranov and Vasserot in the setting of Theorem 11.6. In this case, the minimal resolution of singularities $\widetilde{X}$ has an alternative construction, as *Nakamura's G-Hilbert scheme* $\mathrm{Hilb}^G(\mathbb{C}^2)$ [108, 161]. This is an irreducible component of the subspace of the Hilbert scheme of points in $\mathbb{C}^2$ given by the ideal sheaves $\mathcal{J} \subseteq \mathcal{O}_{\mathbb{C}^2}$ such that the $\mathcal{O}_{\mathbb{C}^2}/\mathcal{J} \cong \mathbb{C}[G]$ as G-modules.

Theorem 11.7 (Kapranov–Vasserot [142]). *Let $G \subset \mathrm{SL}(2,\mathbb{C})$ be a finite group, $S = \mathbb{C}[u,v]$, $R = S^G$, and $X = \operatorname{Spec} R$. Set $H = \mathrm{Hilb}^G(\mathbb{C}^2)$. Then there is a commutative triangle*

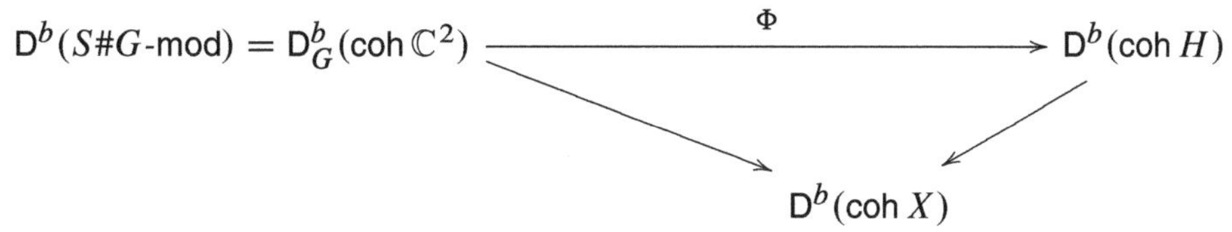

in which Φ is an equivalence of triangulated categories.

The equivalence Φ is given by an explicit "equivariant" Fourier–Mukai type functor, $\Phi(-) = (\mathbf{R}p_*\mathbf{L}q^*(-))^G$, where $Z \subseteq X \times \mathbb{C}^2$ is the incidence variety and p, q are the projections onto the factors.

In dimension greater than two, there is no minimal resolution of singularities. However, Nakamura's G-Hilbert scheme is still a candidate for a crepant resolution of singularities in dimension three. Bridgeland, King, and Reid proved the following general result about the G-Hilbert scheme.

Theorem 11.8 (Bridgeland–King–Reid [40]). *Suppose that Y is a smooth and quasi-projective complex variety, and that $G \subseteq \operatorname{Aut} Y$ is a finite group of automorphisms such that the quotient Y/G has Gorenstein singularities. Let $H = \operatorname{Hilb}^G(Y)$. If*

$$\dim(H \times_{Y/G} H) \le \dim H + 1\,,$$

then H is a crepant resolution of singularities of Y/G and there is an equivalence (explicitly given by a Fourier–Mukai functor) of derived categories $\mathsf{D}^b(\operatorname{Hilb}^G(Y)) \longrightarrow \mathsf{D}^b_G(Y)$, where $\mathsf{D}^b_G(Y)$ is the bounded derived category of G-equivariant coherent sheaves on Y.

The assumption on the fiber product $H \times_{Y/G} H$ is automatic if $\dim Y \le 3$, so this result implies a derived McKay correspondence for three-dimensional quotient singularities $\mathbb{C}^3/G$ with $G \subset \mathrm{SL}(3, \mathbb{C})$. In particular, such singularities have a crepant resolution, which had been verified on a case-by-case basis using the classification of finite subgroups of $\mathrm{SL}(3, \mathbb{C})$. The full details of the correspondences in dimension three are still being worked out [62].

In dimension four, the hypothesis on $H \times_{Y/G} H$ need not hold if $H \longrightarrow Y/G$ contracts a divisor to a point. Indeed, we have seen in Example 8.6 that some quotients $\mathbb{C}^4/G$ have *no* crepant resolutions of singularities. Furthermore, even when a crepant resolution exists, the G-Hilbert scheme may be singular, or non-crepant, or both [175, Example 5.4]. In general, the following conjecture is due to Reid.

Conjecture 11.9 (Derived McKay Correspondence Conjecture). For a crepant resolution of singularities $\widetilde{X} \longrightarrow \mathbb{C}^n/G$, should one exist, there is an equivalence between $\mathsf{D}^b(\operatorname{coh} \widetilde{X})$ and $\mathsf{D}^b_G(\mathbb{C}^n)$.

Compare with Conjecture 9.3 above. The derived McKay correspondence conjecture is known when G preserves a complex symplectic form on $\mathbb{C}^n$ [39], and when G is Abelian [129].

Notice, for a last comment, that the "resolution" $S\#G \cong \operatorname{End}_R(S)$ of Theorem 11.2 exists in any dimension for $G \subset \mathrm{GL}(n, k)$ having no non-trivial pseudo-reflections, and delivers a derived equivalence $\mathsf{D}^b(S\#G) \simeq \mathsf{D}^b_G(\mathbb{C}^n)$ by definition. In dimension two, it is even derived equivalent to the "preferred" desingularization $\operatorname{Hilb}^G(\mathbb{C}^2)$. As we shall see in the next section, it is even in a certain sense "crepant," so represents a potential improvement on the geometric situation.

12 Non-commutative Crepant Resolutions

Now I come to the title character of this article. It is an attempt, due to Van den Bergh, to define a concrete algebraic object whose derived category will realize a categorical desingularization in the sense of Definition 10.1, and which will also verify Conjecture 10.3. The main motivations are Examples 9.1 and 10.2, and Section 11.

Let R be a commutative ring. Recall from Definitions 7.2 and 7.3 that a ring Λ is an R-order if it is finitely generated and MCM as an R-module, and is non-singular if $\operatorname{gldim} \Lambda_{\mathfrak{p}} = \dim R_{\mathfrak{p}}$ for all $\mathfrak{p} \in \operatorname{Spec} R$. Let us also agree that a module-finite algebra Λ over a domain R is *birational* to R if $\Lambda \otimes_R K \cong M_n(K)$ for some n, where K is the quotient field. If Λ is torsion-free as an R-module, this is equivalent to asking that $\Lambda \subseteq M_n(K)$ and that Λ spans $M_n(K)$ when scalars are extended to K. The terminology is consistent with our determination to identify objects that are Morita equivalent; the birationality of Λ should mean that $\Lambda \otimes_R K$ is Morita equivalent to K, and the only candidates are the matrix rings $M_n(K)$.

Here is a provisional definition, to be improved shortly.

Provisional Definition 12.1. Let R be a CM normal domain with quotient field K. A *non-commutative desingularization* of R is a non-singular birational R-order Λ.

There is also a natural candidate for a "crepancy" condition.

Definition 12.2. Let R be a local ring, and let Λ be a module-finite R-algebra. Let us say that Λ is a *symmetric* R-algebra if $\operatorname{Hom}_R(\Lambda, R) \cong \Lambda$ as a $(\Lambda$-$\Lambda)$-bimodule.

Notice immediately that if Λ is a symmetric R-algebra, then for any left Λ-module M, there are natural isomorphisms

$$\operatorname{Hom}_\Lambda(M, \Lambda) \cong \operatorname{Hom}_\Lambda(M, \operatorname{Hom}_R(\Lambda, R)) \cong \operatorname{Hom}_R(\Lambda \otimes_\Lambda M, R) \cong \operatorname{Hom}_R(M, R).$$

We also have the following direct consequence of Proposition 7.4:

Corollary 12.3. *Let R be a Gorenstein local ring. If Λ is a symmetric R-order of finite global dimension, then* $\operatorname{gldim} \Lambda = \dim R$. *In particular, Λ is non-singular.*

Notice that this corollary fails badly for non-Gorenstein R; a counterexample is Example 17.3 below.

Here finally is the definition [205].

Definition 12.4. Let $(R, \mathfrak{m})$ be a CM local normal domain with quotient field K. A *non-commutative crepant resolution* of R (or of $\operatorname{Spec} R$) is a symmetric, birational, R-order Λ having finite global dimension.

I first observe that the definition is Morita-invariant, i.e. if Λ and Γ are Morita-equivalent R-algebras and Λ is a symmetric birational R-order of finite global dimension, then so is Γ. Indeed, global dimension is known to pass across Morita equivalence. Suppose Λ is MCM over R and $\Gamma = \operatorname{End}_\Lambda(P)$ for some Λ-progenerator P. Since P is a progenerator, P is a direct summand of Λ^n for some n, and it follows that Γ is a direct summand of $\operatorname{End}_\Lambda(\Lambda^n) \cong M_n(\Lambda)$ as an R-module. Thus Γ is a MCM R-module as well. Symmetry is similarly easy to verify.

Before considering other possible definitions and addressing the examples from previous sections, I point out a connection with the classical theory of orders [2, 176], following [110]. Let R be a domain with quotient field K. Recall that a module-finite R-algebra Λ, contained in a finite-dimensional division algebra D over K, is called a *classical order* in D if Λ spans D over K, and is called *maximal* in D if it is maximal among classical orders in D with respect to containment. Maximal orders over Dedekind domains have been completely understood for many years; the following facts are well-known.

- Every finite-dimensional division K-algebra D contains a unique maximal order Δ_D.
- A classical order is maximal if and only if it is Morita equivalent to $\Delta_{D_1} \times \cdots \times \Delta_{D_k}$ for finite-dimensional division algebras $D_1, \dots, D_k$ over K.
- A classical order is *hereditary*, that is, has global dimension at most one, if and only if it is Morita equivalent to a ring of the form $T_{n_1}(\Delta_{D_1}) \times \cdots \times T_{n_k}(\Delta_{D_k})$, where $T_n(\Delta)$ denotes the subring of $M_n(\Delta)$ containing matrices (a_{ij}) with $a_{ij} \in \operatorname{rad}(\Delta)$ for $i > j$.

With these facts in mind, let R be a complete discrete valuation ring and Λ a module-finite R-algebra. If Λ is a symmetric R-algebra of global dimension 1, then it follows that Λ is a maximal order. Indeed, Λ is hereditary, so Morita equivalent to $T_{n_1}(\Delta_{D_1}) \times \cdots \times T_{n_k}(\Delta_{D_k})$ as above. One can check, however, that $T_n(\Delta)$ is symmetric only for $n = 1$. Thus Λ is maximal.

Now, a classical order Λ over a normal domain R is maximal if and only if Λ is reflexive as an R-module and $\Lambda_\mathfrak{p}$ is a maximal order for all primes $\mathfrak{p}$ of height one in R [2, Theorem 1.5] and [176, 11.5]. Combining this with the discussion above gives the following result.

Proposition 12.5. *Let R be a normal domain with quotient field K, and Λ a symmetric birational R-order of finite global dimension. Then Λ is a maximal order.*

The connection with the classical theory of orders gives a structure theorem for symmetric non-singular orders, via the following results of Auslander–Goldman [2, Lemma 4.2] and Auslander [18, Lemma 5.4].

Theorem 12.6. *Let R be a normal domain with quotient field K.*

(i) *Let Λ be a classical order over R in $M_n(K)$. Then Λ is a maximal order if and only if there exists a finitely generated reflexive R-module M such that $\Lambda \cong \operatorname{End}_R(M)$.*

(ii) *Let M be a reflexive R-module, and set $\Lambda = \operatorname{End}_R(M)$. Then Λ is reflexive as an R-module and the map $\alpha\colon \Lambda \longrightarrow \operatorname{Hom}_R(\Lambda, R)$ defined by $\alpha(f)(g) =$* $\operatorname{tr}(fg)$, *where* $\operatorname{tr}\colon \operatorname{End}_K(K \otimes_R M) \longrightarrow \operatorname{End}_R(M)$ *is the usual trace map, is an isomorphism of $(\Lambda\text{-}\Lambda)$-bimodules. Hence Λ is a symmetric R-algebra.*

Here are a few definitions which, at least under certain hypotheses, are equivalent to Definition 12.4. Part (ii) of the next proposition is the original definition of a non-commutative crepant resolution [204, 205]. For that definition, say that Λ is *homologically homogeneous* over the central subring R if it is finitely generated as an R-module and every simple Λ-module has the same projective dimension, equal to $\dim R$ [34, 36]. This condition seems first to have been introduced by Vasconcelos [201] under the name "moderated algebras." If R is equidimensional, it is equivalent to asking that for every $\mathfrak{p} \in \operatorname{Spec} R$ the localization $\Lambda_{\mathfrak{p}}$ is MCM as an $R_{\mathfrak{p}}$-module and $\operatorname{gldim} \Lambda_{\mathfrak{p}} = \dim R_{\mathfrak{p}}$ [37].

Proposition 12.7. *Let R be a Gorenstein local normal domain and let Λ be a module-finite R-algebra. Then the following sets of conditions on Λ are equivalent.*

(i) *Λ is a symmetric birational R-order and has finite global dimension.*

(ii) *$\Lambda \cong \operatorname{End}_R(M)$ for some reflexive R-module M, and Λ is homologically homogeneous.*

(iii) *$\Lambda \cong \operatorname{End}_R(M)$ for some reflexive R-module M, Λ is MCM as an R-module, and* $\operatorname{gldim} \Lambda < \infty$.

Proof. Assume first that Λ satisfies (i). Then Λ is MCM over R by definition, and this localizes well. By Theorem 12.6 (i), Λ is an endomorphism ring of a reflexive module M, and by Corollary 12.3 Λ is non-singular, giving (ii). Clearly (ii) implies (iii). Finally, if $\Lambda \cong \operatorname{End}_R(M)$ for a reflexive R-module M, then Λ is birational to R, and is symmetric by Theorem 12.6 (ii). □

The implication (iii) $\Longrightarrow$ (ii) fails if R is not Gorenstein. Again see Example 17.3 below.

Now it is clear that Auslander's Theorem 11.2 proves that, for any $n \geq 2$ and any finite group $G \subset \mathrm{SL}(n, k)$ with order invertible in k, the ring of invariants $R = k[\![x_1, \dots, x_n]\!]^G$ has a non-commutative crepant resolution. Namely, with S denoting the power series ring, the endomorphism ring $\operatorname{End}_R(S)$ has finite global dimension and, since S is a MCM R-module, is an R-order. Thus $\operatorname{End}_R(S) \cong S\#G$ is a non-commutative crepant resolution. One can also prove directly that the twisted group ring $S\#G$ is symmetric over R.

Similarly, the three-dimensional ordinary double point in Example 10.2 admits the non-commutative crepant resolution $\Lambda = \operatorname{End}_R(R \oplus I)$, which is derived equivalent to the resolutions of singularities Y and Y'.

For the next equivalent definition we need the notion of a *d-Calabi–Yau* algebra. There are a few approaches to topics with this name. I follow [55, 110]; see also [51, 80]. I will always assume that the base ring is local, which eases the exposition considerably.

Definition 12.8. Let R be a local ring and let Λ be a module-finite R-algebra. Write $D(-) = \operatorname{Hom}_R(-, E)$ for Matlis duality over R, where E is the injective hull of the residue field. Say that Λ is *d-Calabi–Yau (d-CY)* if there is a functorial isomorphism

$$\operatorname{Hom}_{\mathsf{D}(\Lambda\text{-}\mathsf{Mod})}(X, Y[d]) \cong D\operatorname{Hom}_{\mathsf{D}(\Lambda\text{-}\mathsf{Mod})}(Y, X)$$

for all X and Y in $\mathsf{D}^b(\Lambda\text{-}\mathsf{fl})$, the bounded derived category of finite-length Λ-modules. Similarly, Λ is *d-CY⁻* if an isomorphism as above holds for all X in $\mathsf{D}^b(\Lambda\text{-}\mathsf{fl})$ and all Y in $\mathsf{K}^b(\operatorname{\mathsf{add}}\Lambda)$.

These definitions are perhaps a bit much to swallow all at once. Here are some basic facts about the Calabi–Yau conditions. Let R be a local ring and Λ a module-finite R-algebra. Then Λ is n-CY for some integer n if and only if Λ is n-CY$^-$ and has finite global dimension. Indeed, if $\operatorname{gldim}\Lambda < \infty$ then $\mathsf{D}^b(\Lambda\text{-}\mathsf{fl}) \subset \mathsf{D}^b(\Lambda\text{-}\mathsf{mod}) = \mathsf{K}^b(\operatorname{\mathsf{add}}\Lambda)$. The "only if" part is proved by completing and considering the finite-length Λ-module $Y/\operatorname{rad}(\Lambda)^n Y$ for Y in $\mathsf{K}^b(\operatorname{\mathsf{add}}\Lambda)$.

Calabi–Yau algebras are best-behaved when R is Gorenstein. In that case [110, Theorem 3.2], if Λ is n-CY or n-CY$^-$ for some n, then $n = \dim R$. Furthermore, Λ is d-CY$^-$ if and only if Λ is a symmetric R-order. (This is one point where life is easier because R is local. Iyama and Reiten give an example, which they credit to J. Miyachi, of a d-CY$^-$ algebra over a non-local Gorenstein ring which is not symmetric, even though $R \longrightarrow \Lambda$ is injective. It is *locally* symmetric.) More precisely, the following equivalent conditions hold.

Proposition 12.9 ([110]). *Let $(R, \mathfrak{m}, k)$ be a Gorenstein local ring with* $\dim R = d$, *and Λ a module-finite R-algebra. The following are equivalent for any integer n.*

(i) *Λ is n-CY⁻.*

(ii) *As functors on Λ-fl, $\operatorname{Ext}^n_\Lambda(-, \Lambda)$ is isomorphic to the Matlis duality $D(-) = \operatorname{Hom}_R(-, E)$, and $\operatorname{Ext}^i_\Lambda(-, \Lambda) = 0$ for $i \neq n$.*

(iii) *$\mathbf{R}\operatorname{Hom}_R(\Lambda, R) \cong \Lambda[n-d]$ in the bounded derived category of $(\Lambda$-$\Lambda)$-bimodules.*

(iv) *Λ is a CM R-module of dimension n and $\operatorname{Ext}^{d-n}_R(\Lambda, R) \cong \Lambda$ as $(\Lambda$-$\Lambda)$-bimodules.*

In particular, a birational module-finite algebra Λ is d-CY if and only if it is symmetric and has finite global dimension.

Of course the value of a definition, even one as motivated as this one has been, is in the theorems. Here is the main result of [205].

Theorem 12.10. *Let R be a Gorenstein normal $\mathbb{C}$-algebra, set $X = \operatorname{Spec} R$, and let $\pi\colon \widetilde{X} \longrightarrow X$ be a crepant resolution of singularities. Assume that the fibers of π have dimension at most one. Then there exists a MCM R-module M such that the endomorphism ring $\Lambda = \operatorname{End}_R(M)$ is homologically homogeneous. In particular, Λ is a non-commutative crepant resolution of R. Furthermore, $\widetilde{X}$ and Λ are derived equivalent: $\mathsf{D}^b(\operatorname{coh} \widetilde{X}) \simeq \mathsf{D}^b(\Lambda\text{-mod})$.*

Here is a sketch of the proof of Theorem 12.10. We know that existence of a crepant resolution implies that X has rational singularities. Let $\mathcal{L}$ be an ample line bundle on the smooth variety $\widetilde{X}$ generated by global sections. Then by the hypothesis on the fibers of π ([59, Lemma 4.2.4]), $\mathcal{O}_{\widetilde{X}} \oplus \mathcal{L}$ generates $\mathsf{D}(\operatorname{Qch} \widetilde{X})$, that is, if $\mathcal{N}$ in $\mathsf{D}(\operatorname{Qch} \widetilde{X})$ satisfies $\operatorname{Hom}_{\mathsf{D}(\operatorname{Qch} \widetilde{X})}(\mathcal{O}_{\widetilde{X}} \oplus \mathcal{L}, \mathcal{N}[i]) = 0$ for $i \neq 0$, then $\mathcal{N} = 0$ (see the discussion after Definition 5.2). Take an extension $0 \longrightarrow \mathcal{O}_{\widetilde{X}}^r \longrightarrow \mathcal{M}' \longrightarrow \mathcal{L} \longrightarrow 0$ corresponding to a set of r generators for $\operatorname{Ext}^1_{\mathcal{O}_{\widetilde{X}}}(\mathcal{L}, \mathcal{O}_{\widetilde{X}})$ as an R-module. Set $\mathcal{M} = \mathcal{M}' \oplus \mathcal{O}_{\widetilde{X}}$. Then $\mathcal{M}$ also generates $\mathsf{D}(\operatorname{Qch} \widetilde{X})$. One can show that $\operatorname{Ext}^i_{\mathcal{O}_{\widetilde{X}}}(\mathcal{M}, \mathcal{M}) = 0$ for $i > 0$ (this takes a good bit of work). Thus $\mathcal{M}$ is a tilting bundle on $\widetilde{X}$. Set $\Lambda = \operatorname{End}_{\mathcal{O}_{\widetilde{X}}}(\mathcal{M})$; then the vanishing of the derived pushforwards $\mathbf{R}^i\pi_*\mathcal{E}nd_{\mathcal{O}_{\widetilde{X}}}(\mathcal{M}) = \operatorname{Ext}^i_{\mathcal{O}_{\widetilde{X}}}(\mathcal{M}, \mathcal{M})$ implies that $\Lambda \cong \operatorname{End}_R(M)$, where $M = \Gamma(\widetilde{X}, \mathcal{M})$. The proofs that Λ and M are both MCM are more involved.

Van den Bergh also proves a converse to Theorem 12.10, constructing a geometric crepant resolution $\pi\colon \widetilde{X} \longrightarrow \operatorname{Spec} R$ from a non-commutative one under certain assumptions [204, §6]. The method is roughly as follows: let Λ be a non-commutative crepant resolution of R, and take for $\widetilde{X}$ a moduli space of certain stable representations of Λ. Then he proves that if $\dim(\widetilde{X} \times_{\operatorname{Spec} R} \widetilde{X}) \leq \dim R + 1$, then $\widetilde{X} \longrightarrow \operatorname{Spec} R$ is a crepant resolution and there is an equivalence of derived categories $\mathsf{D}^b(\operatorname{coh} \widetilde{X}) \simeq \mathsf{D}^b(\Lambda\text{-mod})$. Observe that the hypothesis is exactly similar to that of Theorem 11.8. In particular, the hypothesis holds if $\dim R \leq 3$, giving the following theorem.

Theorem 12.11. *Let R be a three-dimensional Gorenstein normal $\mathbb{C}$-algebra with terminal singularities.*

(i) *There is a non-commutative crepant resolution of R if and only if $X = \operatorname{Spec} R$ has a crepant resolution of singularities.*

(ii) *All crepant resolutions of R – geometric as well as non-commutative – are derived equivalent.*

The second statement verifies Conjecture 9.3 of Bondal and Orlov in this case. Iyama and Reiten [110] have recently shown that, even without the assumption on the singularities of R being terminal, all non-commutative crepant resolutions of R are derived equivalent. Even more recently, Iyama and Wemyss [114] have announced a sufficient criterion for the existence of a derived equivalence between the non-commutative crepant resolutions of R. When $d \leq 3$ this criterion is always satisfied, recovering the Iyama–Reiten result.

Van den Bergh suggests the following extension of Theorem 12.11(ii).

Conjecture 12.12 (Van den Bergh). Let R be a Gorenstein normal $\mathbb{C}$-algebra and $X = \operatorname{Spec} R$. Then all crepant resolutions of R – geometric as well as non-commutative – are derived equivalent.

13 Example: Normalization

The appearance of endomorphism rings as ersatz resolutions of singularities may initially be unsettling. It does, however, have a precedent. One can think of the *normalization* $\overline{R}$ of an integral domain R, i.e. the integral closure in its quotient field, as a partial resolution of singularities, one that is especially tractable since it does not leave the category of Noetherian rings. This result of Grauert and Remmert [85, 86] interprets the normalization as an endomorphism ring.

Theorem 13.1 (Grauert–Remmert). *Let R be an integral domain and I a non-zero integrally closed ideal of R such that $R_{\mathfrak{p}}$ is normal for every $\mathfrak{p} \not\supseteq I$. Then the following are equivalent.*

(i) *R is normal;*

(ii) *For all non-zero fractional ideals J of R,* $\operatorname{Hom}_R(J, J) = R$*;*

(iii) *For all non-zero ideals J of R,* $\operatorname{Hom}_R(J, J) = R$*;*

(iv) $\operatorname{Hom}_R(I, I) = R$.

For any fractional ideal J, the containments $R \subseteq \operatorname{Hom}_R(J, J) \subseteq \overline{R}$ always hold. The latter inclusion sends $\varphi\colon J \longrightarrow J$ to the fraction $\varphi(r)/r$ for any fixed non-zerodivisor $r \in J$; this is well-defined. In particular, $\operatorname{Hom}_R(J, J)$ is a commutative(!) ring.

Theorem 13.1 was used by de Jong [72] to give an algorithm for computing the normalization $\overline{R}$ of an affine domain over a perfect field, or slightly more generally. Let R be a local domain such that its normalization $\overline{R}$ is a finitely generated R-module; equivalently, the completion $\widehat{R}$ is reduced. One needs to determine a non-zero integrally closed ideal I such that $V(I)$ contains the non-normal locus of R. If R is affine over a perfect field, then the Jacobian criterion implies that the radical of the Jacobian ideal will work; there are other choices in other cases. Set $R' = \operatorname{Hom}_R(I, I)$. If $R' = R$, then R is normal, so stop. Otherwise, replace R by R' and repeat. The algorithm has been refined and extended since [68, 82].

It follows from Serre's criterion for normality that if R is the coordinate ring of an irreducible curve singularity, then the normalization $\overline{R}$ is regular, whence is the coordinate ring of a resolution of singularities of $\operatorname{Spec} R$. Thus in this situation, desingularization can be achieved as an iterated endomorphism ring. In fact, as long as R is affine over a perfect field, one actually has $\overline{R} = \mathrm{Hom}_R(\overline{R}, \overline{R})$, a single endomorphism ring of a finitely generated module giving resolution of singularities.

14 MCM Endomorphism Rings

The requirement that a non-commutative crepant resolution of singularities should be an order, i.e. a MCM module, raises a basic question: Does the depth of $\mathrm{Hom}_R(M, M)$ depend in any predictable way on the depth of M? The short answer is No. In this section we look at some examples.

First, observe that there is at least a lower bound on the depth of a Hom module: If R is any local ring and M, N are finitely generated modules with depth $N \geq 2$, then $\mathrm{Hom}_R(M, N)$ has depth ≥ 2 as well. Indeed, applying $\mathrm{Hom}_R(-, N)$ to a free presentation of M displays $\mathrm{Hom}_R(M, N)$ as the kernel of a map between direct sums of copies of N, so the Depth Lemma gives the conclusion. That's about the end of the good news.

Next notice that the depth of $\mathrm{Hom}_R(M, M)$ can be strictly greater than that of M. Indeed, let R be a CM normal domain and let J be any non-zero ideal of R. Then $\mathrm{Hom}_R(J, J) = R$ by Theorem 13.1, even though $\mathrm{depth}_R\, J$ can take any value between 1 (if, say, J is a maximal ideal) and d (if for example J is principal). Furthermore, R can be taken to be Gorenstein, or even a hypersurface ring, so finding a class of rings that avoids this problem seems hopeless.

One might hope at least that if M is MCM then $\mathrm{Hom}_R(M, M)$ is MCM as well. This question was raised by Vasconcelos [200] for R a Gorenstein local ring. It also has a negative answer, though it is at least harder. A counterexample is given by Jinnah [119], based on [100, Example 5.9].

Example 14.1. Let k be a field and set $A = k[x, y, z]/(x^3 + y^3 + z^3)$, $B = k[u, v]$. Let R be the Segre product of A and B, the graded ring defined by $R_n = A_n \otimes_k B_n$. Then R is the subring of $A[u, v]$ generated by xu, xv, yu, yv, zu, zv, a three-dimensional normal domain of depth 2. The ideal $I = vA[u, v] \cap R$ has depth 3 over R.

Write R as a quotient of a graded complete intersection ring S of dimension 3. Then I has depth 3 over S as well, but $\mathrm{Hom}_S(I, I) = \mathrm{Hom}_R(I, I) = R$ has depth two as an S-module. Localizing S at its irrelevant ideal gives a local example.

Here is a useful characterization of the depth of $\mathrm{Hom}_R(M, N)$.

Lemma 14.2 ([117, 66]). *Let R be a CM local ring and let M and N be finitely generated R-modules. Fix $n \geq 2$ and consider the following properties.*

(i) $\mathrm{Hom}_R(M, N)$ *satisfies* (S_{n+1})*; and*

(ii) $\mathrm{Ext}^i_R(M, N) = 0$ *for* $i = 1, \dots, n-1$.

If M is locally free in codimension n and N satisfies (S_n), then (i) $\implies$ (ii). *If N satisfies (S_{n+1}) then* (ii) $\implies$ (i).

Since this result is used in some later sections, I'll sketch the proof. First assume that M is locally free in codimension n and that N satisfies (S_n). If $n \geq \dim R$ then M is free and there is nothing to prove, so one may localize at a prime ideal of height $n + 1$ to assume by induction that M is locally free on the punctured spectrum, and so $\operatorname{Ext}^i_R(M, N)$ has finite length for $i \geq 1$. Take a free resolution $P_\bullet$ of M and consider the first n terms of the complex $\operatorname{Hom}_R(P_\bullet, N)$. Since the cohomologies of this complex, namely $\operatorname{Ext}^i_R(M, N)$ for $i = 1, \ldots, n - 1$, all have finite length, $\operatorname{Hom}_R(P_\bullet, N)$ is exact by the Acyclicity Lemma [35, Example 1.4.23]. For the second statement, take once again the free resolution $P_\bullet$ of M and consider the first n terms of $\operatorname{Hom}_R(P_\bullet, N)$, which form an exact sequence by the assumption. The Depth Lemma then implies $\operatorname{depth} \operatorname{Hom}_R(M, N)_\mathfrak{p} \geq \min\{n + 1, \operatorname{depth} N_\mathfrak{p}\}$ for every $\mathfrak{p} \in \operatorname{Spec} R$, which gives the conclusion.

The homological consequences of Lemma 14.2 are even stronger than is immediately apparent. To describe these, recall that module N over a commutative ring R is called Tor-*rigid* if whenever $\operatorname{Tor}^R_i(M, N) = 0$ for some $i \geq 0$ and some finitely generated R-module M, necessarily $\operatorname{Tor}^R_j(M, N) = 0$ for all $j \geq i$. Deciding whether a given module is Tor-rigid is generally a delicate problem, as Dao observes [66]. The following result of Jothilingam [123] (see also [122]) gives a very useful necessary condition.

Proposition 14.3 (Jothilingam). *Let R be a local ring and let M, N be finitely generated R-modules. Assume that N is* Tor-*rigid. If* $\operatorname{Ext}^1_R(M, N) = 0$, *then the natural map $\Phi_{M,N} \colon M^* \otimes_R N \longrightarrow \operatorname{Hom}_R(M, N)$ is an isomorphism. In particular, if* $\operatorname{Ext}^1_R(N, N) = 0$ *then N is free.*

It follows immediately that if R is a local ring satisfying (R_2) and (S_3), and M is a reflexive R-module with a non-free direct summand which is Tor-rigid, then $\Lambda = \operatorname{End}_R(M)$ is *not* MCM, whence is not a non-commutative crepant resolution. Indeed, let N be a Tor-rigid summand of M which is not free. Then N is reflexive, so satisfies (S_2), and is free in codimension two as R is regular on that locus. Moreover, $\operatorname{Hom}_R(N, N)$ is a direct summand of $\operatorname{Hom}_R(M, M)$. If $\operatorname{Hom}_R(M, M)$ were MCM, then $\operatorname{Hom}_R(N, N)$ would also be, so would satisfy (S_3). But then $\operatorname{Ext}^1_R(N, N) = 0$ by Lemma 14.2, contradicting Proposition 14.3.

It is now easy to bolster Example 14.1 by constructing, over any CM local ring $(R, \mathfrak{m}, k)$ of dimension 3 or more, a MCM module M such that $\operatorname{Hom}_R(M, M)$ is not MCM[2]. Indeed, take M to be a high enough syzygy of the residue field k; since k is Tor-rigid, the same is true of M, and it is locally free on the punctured spectrum. By Lemma 14.2 and Proposition 14.3, then, $\operatorname{Hom}_R(M, M)$ has depth at most 2.

[2] I'm grateful to Hailong Dao for pointing this out to me.

From Proposition 14.3 and progress on understanding Tor-rigid modules over hypersurface rings, Dao derives the next theorem, which identifies obstructions to the existence of non-commutative crepant resolutions.

Theorem 14.4 ([65, 66, 67]). *Let $R = S/(f)$ be a local hypersurface ring with S an equicharacteristic or unramified regular local ring and $f \in S$ a non-zero non-unit. Assume that R is regular in codimension two.*

(i) *If* $\dim R = 3$ *and R is $\mathbb{Q}$-factorial, then every finitely generated R-module is* Tor-*rigid, so R admits no non-commutative crepant resolution.*

(ii) *If R has an isolated singularity and* $\dim R$ *is an even number greater than 3, then* $\operatorname{Hom}_R(M, M)$ *satisfies (S_3) only if M is free, so R admits no non-commutative crepant resolution.*

Recall from Example 8.5 that the isolated hypersurface singularity defined by $x_0^r + x_1^d + \cdots + x_d^d = 0$ has a crepant resolution of singularities only if $r \equiv 0$ or 1 modulo d. Part (ii) of Dao's theorem thus implies that the extension of Van den Bergh's Theorem 12.11(i) to higher dimensions has a negative answer, at least without some further assumptions.

Example 14.5 ([66, Example 3.6; 38, §2]). Theorem 14.4 allows some progress toward deciding which of the three-dimensional ADE singularities (see (8.1)) have non-commutative crepant resolutions. Let k be an algebraically closed field of characteristic zero. The 3-dimensional versions of $(A_{2\ell})$, (E_6), and (E_8) are factorial, so do not admit a non-commutative crepant resolution at all.

Let $R = k[\![x, y, z, w]\!]/(xy + z^2 - w^{2\ell+2})$, an $(A_{2\ell+1})$ singularity, with $\ell \geq 1$. (Observe that the case $\ell = 0$ is the ordinary double point of Example 9.1.) Then I claim that R has a non-commutative crepant resolution $\Lambda = \operatorname{End}_R(M)$ in which M is MCM. Indeed, the indecomposable MCM R-modules are completely known [218, Example 5.12]; they are the free module R, the ideal $I = (x, z + w^{\ell+1})$, the dual ideal $I^* = (y, z - w^{\ell+1})$, and ℓ indecomposables $M_1, \dots, M_\ell$ of rank two.

Each M_i is its own Auslander–Reiten translate, so in particular $\operatorname{Ext}^1_R(M_i, M_i) \neq 0$ for each $i = 1, \dots, \ell$. By Lemma 14.2, no M_i can be a constituent in a non-commutative crepant resolution. On the other hand, I and I^* satisfy $\operatorname{Hom}_R(I, I) \cong \operatorname{Hom}_R(I^*, I^*) \cong R$ by Theorem 13.1. Thus at least $\operatorname{End}_R(R \oplus I)$ and $\operatorname{End}_R(R \oplus I^*)$ are symmetric R-orders; it will follow from the results in the next section that since $R \oplus I$ and $R \oplus I^*$ are cluster tilting modules (Theorem 15.5), the endomorphism rings have global dimension equal to 3, so are non-commutative crepant resolutions.

15 Global Dimension of Endomorphism Rings

The tendency for endomorphism rings to have finite global dimension was first observed by Auslander [15, §III.3]. Recall that Λ is an *Artin algebra* if the center of Λ is a commutative Artin ring and Λ is a finitely generated module over its center.

Theorem 15.1 (Auslander). *Let Λ be an Artin algebra with radical $\mathfrak{r}$ and assume that $\mathfrak{r}^n = 0$, $\mathfrak{r}^{n-1} \neq 0$. Set $M = \bigoplus_{i=0}^n \Lambda/\mathfrak{r}^i$. Then $\Gamma = \operatorname{End}_\Lambda(M)$ is a coherent Artin algebra of global dimension at most $n+1$.*

Based on this result Auslander was led to define the *representation dimension* of an Artin algebra Λ as the least value of gldim $\operatorname{End}_\Lambda(M)$ as M runs through all finitely generated Λ-modules which are generators-cogenerators for Λ, that is, M contains as direct summands all indecomposable projective and injective Λ-modules. Observe that Theorem 15.1 does not prove finiteness of the representation dimension; while M has a non-zero free direct summand, it need not be a cogenerator unless Λ is self-injective.

Auslander proved in [15] that repdim $\Lambda \leq 2$ if and only if Λ has finite representation type (see Section 17), but it was not until 2003 that Rouquier constructed the first examples with representation dimension greater than 3 [184]. Incidentally, Rouquier's proof uses the notion of the *dimension* [59, 185] of the derived category $\mathsf{D}^b(\Lambda\text{-}\mathsf{mod})$. The dimension of a triangulated category is a measure of how many steps are required to obtain it starting from a single object and inductively taking the closure under shifts, direct sums and summands, and distinguished triangles. Rouquier proved that if Λ is a finite-dimensional algebra over a perfect field k, then $\dim \mathsf{D}^b(\Lambda\text{-}\mathsf{mod}) \leq \operatorname{repdim} \Lambda$.

Iyama showed in 2003 that the representation dimension of a finite-dimensional algebra is always finite [115]. He also extended the definition of representation dimension to CM local rings of positive Krull dimension.

Definition 15.2. Let R be a complete CM local ring with canonical module ω. Set

$$\operatorname{repdim} R = \inf_M \{\operatorname{gldim} \operatorname{End}_R(R \oplus \omega \oplus M)\} ,$$

where the infimum is taken over all MCM R-modules M.

Iyama's techniques involved *maximal n-orthogonal modules,* now called *cluster tilting modules* [117]. Here I will not say anything about cluster algebras or cluster categories; see [46] for an exposition. Here is a direct definition of cluster tilting modules [38].

Definition 15.3. Let R be a CM local ring and M a MCM R-module. Fix $n \geq 1$.

(i) Set

$$M^{\perp_n} = \{X \mid X \text{ is MCM and } \operatorname{Ext}^i_R(M, X) = 0 \text{ for } 1 \leq i \leq n\}$$

and symmetrically

$${}^{\perp_n}M = \{X \mid X \text{ is MCM and } \operatorname{Ext}^i_R(X, M) = 0 \text{ for } 1 \leq i \leq n\}.$$

(ii) Say that M is *cluster tilting* if

$$M^{\perp_1} = \operatorname{\mathsf{add}} M = {}^{\perp_1}M .$$

There are some isolated results about cluster tilting in small dimension. For example, [38] constructs and classifies cluster tilting modules for the one-dimensional ADE hypersurface singularities. When R is two-dimensional and Gorenstein the Auslander–Reiten translate τ is the identity, so $\operatorname{Ext}^1_R(M, M)$ is never zero for MCM M; this rules out cluster tilting in this case.

To describe the connection between cluster tilting modules and non-commutative crepant resolutions, let's consider the following theorem of Dao–Huneke [70, Theorem 3.2].

Theorem 15.4. *Let R be a CM local ring of dimension $d \geq 3$. Let M be a MCM R-module with a non-zero free direct summand, and set $\Lambda = \operatorname{End}_R(M)$. Assume that Λ is MCM as an R-module. Consider the following conditions.*

(i) $M^{\perp_{d-2}} = \mathsf{add}\, M$.

(ii) *There exists an integer n with $1 \leq n \leq d-2$ such that* $M^{\perp_n} = \mathsf{add}\, M$.

(iii) $\operatorname{gldim} \Lambda \leq d$.

(iv) $\operatorname{gldim} \Lambda = d$.

Then (i) $\implies$ (ii) $\implies$ (iii) $\iff$ (iv). *If R has an isolated singularity, then all four are equivalent.*

The main assertion here is (ii) $\implies$ (iii). Everything else is relatively straightforward or follows from Lemma 14.2. To prove (ii) $\implies$ (iii), Dao and Huneke use Proposition 2.4 to get, for any R-module N satisfying (S_2), a long exact sequence

$$\sigma: \qquad \cdots \longrightarrow M^{n_{j+1}} \longrightarrow M^{n_j} \longrightarrow \cdots \longrightarrow M^{n_0} \longrightarrow N \longrightarrow 0$$

such that $\operatorname{Hom}_R(M, \sigma)$ is exact. Let N_j be the kernel at the jth spot; then one shows by induction on j that $\operatorname{Ext}^1_R(M, N_j) \subseteq \operatorname{Ext}^1_R(M, M)^{n_{j+1}}$, so that $N_{d-2} \in M^{\perp_{d-2}} = \mathsf{add}\, M$. It follows that $\operatorname{Hom}_R(M, N_{d-2})$ is Λ-projective. Thus every Λ-module of the form $\operatorname{Hom}_R(M, N)$ has projective dimension at most $d-2$, so that $\operatorname{gldim} \Lambda \leq d$.

As a corollary of Theorem 15.4, Dao and Huneke obtained another proof of the following result of Iyama, which nicely encapsulates the significance of cluster tilting modules to non-commutative crepant resolutions.

Theorem 15.5 (Iyama [116, Theorem 5.2.1]). *Let R be a CM local ring of dimension $d \geq 3$ and with canonical module ω. Assume that R has an isolated singularity. Let M be a MCM R-module and set $\Lambda = \operatorname{End}_R(M)$. The following conditions are equivalent.*

(i) *M contains R and ω as direct summands, Λ is MCM, and* $\operatorname{gldim} \Lambda = d$.

(ii) $M^{\perp_{d-2}} = \mathsf{add}\, M = {}^{\perp_{d-2}}M$.

In particular, if $d = 3$ and R is a Gorenstein isolated singularity, then a MCM R-module M gives a non-commutative crepant resolution if and only if it is a cluster tilting module.

For dimension 3, this result gives a very clear picture of the landscape of non-commutative crepant resolutions. In higher dimension, however, the assumption of isolated singularity becomes more restrictive. Moreover, as Dao and Huneke observe, for $d \geq 4$ the condition $\mathsf{add}\, M = M^{\perp_{d-2}}$ rules out a large class of examples. Specifically, if $\mathrm{Ext}^2_R(M, M) = 0$ for a MCM module M over a complete intersection ring R, then M is necessarily free, since one can complete and lift M to a regular local ring [1].

Back in dimension 3, one can obtain even stronger results, and address possible extensions of Theorem 12.11 (i), by imposing geometric hypotheses. Recall that a cDV singularity (see Section 9) is a three-dimensional hypersurface singularity defined by a polynomial $f(x, y, z) + tg(x, y, z, t)$, where f is ADE and g is arbitrary. A cDV singularity is called cA_n if the generic hyperplane section is a surface singularity of type (A_n).

Theorem 15.6 ([38, Theorem 5.5]). *Let $(R, \mathfrak{m})$ be a local isolated cDV singularity. Then* Spec R *has a crepant resolution of singularities if and only if R has a non-commutative crepant resolution, and these both occur if and only if there is a cluster tilting module in the stable category $\underline{\mathrm{CM}}(R)$. If R is a cA_n singularity defined by $g(x, y) + zt$, then these are equivalent to the number of irreducible power series in a prime decomposition of $g(x, y)$ being $n + 1$.*

16 Rational Singularities

As we saw in Proposition 8.4, GR Vanishing implies that any complex algebraic variety with a crepant resolution of singularities has rational singularities. Furthermore, the idea of a categorical, or non-commutative, desingularization is really only well-behaved for rational singularities. It would therefore be most satisfactory if existence of a non-commutative crepant resolution – a symmetric birational order of finite global dimension – implied rational singularities. This is true by work of Stafford and Van den Bergh [193]. Their result is somewhat more general. Recall from the discussion preceding Proposition 12.7 that Λ is homologically homogeneous if every simple Λ-module has the same projective dimension.

Theorem 16.1 (Stafford–Van den Bergh). *Let k be an algebraically closed field of characteristic zero, and let Λ be a prime affine k-algebra which is finitely generated as a module over its center R. If Λ is homologically homogeneous then the center R has rational singularities. In particular, if R is a Gorenstein normal affine domain* *and has a non-commutative crepant resolution of singularities, then it has rational singularities.*

Van den Bergh gave a proof of the final sentence in case R is graded in [204, Proposition 3.3]. (This argument in the published version of [204] is not quite correct; see the updated version online for a corrected proof.)

Here are a few brief comments on the proof, only in the case where Λ is a non-commutative crepant resolution of its Gorenstein center R, so is symmetric, birational and of finite global dimension. The first step is a criterion for rational singularities, which is an algebraicization of the criterion $\pi_*\omega_{\widetilde{X}} = \omega_X$ of Proposition 8.4.

Lemma 16.2. *Let R be a CM normal affine k-algebra, where k is an algebraically closed field of characteristic zero. Let K be the quotient field of R and let ω_R be the canonical module for R. Then R has rational singularities if and only if for every regular affine S with $R \subseteq S \subseteq K$, one has $\omega_R \subseteq \operatorname{Hom}_R(S, \omega_R)$ inside $\operatorname{Hom}_R(K, \omega_R)$.*

Given the lemma, the derivation of the theorem is somewhat technical. Here I simply note that one key idea is to show ([193, Proposition 2.6]) that if Λ is homologically homogeneous of dimension d then $\omega_\Lambda = \operatorname{Hom}_R(\Lambda, R)$ is an invertible Λ-module, and furthermore the shift $\omega_\Lambda[d]$ is a dualizing complex for Λ in the sense of Yekutieli [217]. This result has been extended [149, Theorem 5.1.12] to remove the hypothesis of finite global dimension (so Λ is assumed to be "injectively homogeneous") and the hypotheses on the field k.

The theorem of Stafford and Van den Bergh does require an assumption on the characteristic of k, as they observe [193, page 671]: there is a homologically homogeneous ring in characteristic 2 with CM center R for which R fails to have rational singularities (in any reasonable sense). The root cause of this bad behavior seems to be the failure of a fixed ring S^G to be a direct summand of S in bad characteristic. It is reasonable to ask, then, as Stafford and Van den Bergh do: Suppose Λ is a homologically homogeneous ring whose center R is an affine k-algebra for a field k of characteristic $p > 0$, and assume that R is an R-module direct summand of Λ. Must R have rational singularities?

One application of Theorem 16.1 is to rule out overly optimistic thoughts on the existence of "generalized" non-commutative desingularizations. For example, one might remove the assumption that Λ be an R-order and simply say that a *weak non-commutative desingularization* is an R-algebra $\Lambda = \operatorname{End}_R(M)$, where M is a reflexive R-module, such that $\operatorname{gldim} \Lambda < \infty$. One might then hope that such things exist quite generally, for, say, every Gorenstein normal domain [154]. However, in dimension two this definition would coincide with that of a non-commutative crepant resolution since endomorphism rings of reflexive R-modules have depth at least two, so would only exist for rational singularities by Theorem 16.1. Therefore a counterexample to the hope would be something like $\mathbb{C}[x, y, z]/(x^3 + y^3 + z^3)$, which is a Gorenstein normal domain but does not have rational singularities.

17 Examples: Finite Representation Type

Let Λ be an Artin algebra of *finite representation type*, i.e. there are only a finite number of non-isomorphic indecomposable finitely generated Λ-modules. Auslander defined what is now called the *Auslander algebra* of Λ to be $\Gamma = \operatorname{End}_\Lambda(M_1 \oplus \cdots \oplus M_t)$, where $M_1, \ldots, M_t$ is a complete set of non-isomorphic indecomposable finitely generated Λ-modules. By Corollary 2.3, Γ is Morita equivalent to any other algebra of the form $\operatorname{End}_\Lambda(N)$, where N is a *representation generator* for Λ, that is, contains every indecomposable finitely generated Λ-module as a direct summand. These algebras are distinguished by the following result.

Theorem 17.1 (Auslander [15]). *Let Λ be an Artin algebra of finite representation type with representation generator M. Assume that Λ is not semisimple. Set $\Gamma = \operatorname{End}_\Lambda(M)$. Then* $\operatorname{gldim} \Gamma = 2$.

The proof of this theorem is quite direct from proposition 2.4 and the left-exactness of $\operatorname{Hom}_\Lambda(M, -)$. Indeed, assume that Λ is not semisimple and let X be a finitely generated Γ-module, with projective presentation $P_1 \xrightarrow{\varphi} P_0 \longrightarrow X \longrightarrow 0$. The projective modules P_i are each of the form $\operatorname{Hom}_\Lambda(M, M_i)$ for Λ-modules M_1 and M_0, both in $\operatorname{add} M$. Similarly, $\varphi = \operatorname{Hom}_\Lambda(M, f)$ for some $f \colon M_1 \longrightarrow M_0$. Put $M_2 = \ker f$. Then

$$0 \longrightarrow \operatorname{Hom}_\Lambda(M, M_2) \longrightarrow \operatorname{Hom}_\Lambda(M, M_1) \xrightarrow{\operatorname{Hom}_\Lambda(M,f)} \operatorname{Hom}_\Lambda(M, M_0) \longrightarrow X \longrightarrow 0$$

is a projective resolution of X of length two.

Auslander and Roggenkamp [5] proved a version of this theorem in Krull dimension one, specifically for (classical) orders over complete discrete valuation rings. For their result, define an order Λ over a complete DVR T to have finite representation type if there are only a finite number of non-isomorphic indecomposable finitely generated Λ-modules which are free over T; these are called Λ-*lattices*. If M contains all indecomposable Λ-lattices as direct summands, then $\Gamma = \operatorname{End}_\Lambda(M)$ is proven to have global dimension at most two; the proof is nearly identical to the one sketched above. One need only observe that the kernel M_2 of a homomorphism between Λ-lattices $f \colon M_1 \longrightarrow M_0$ is again a Λ-lattice.

In general, say that a (commutative) local ring R has *finite representation type*, or finite Cohen–Macaulay type, if there are only a finite number of non-isomorphic indecomposable maximal Cohen–Macaulay (MCM) R-modules. Recall that when R is complete, a finitely generated R-module M is MCM if and only if it is free over a Noether normalization of R.

We have already met, in Section 11, the two-dimensional complete local rings of finite representation type, at least over $\mathbb{C}$. By results of Auslander and Esnault [18, 75], they are precisely the quotient singularities $R = \mathbb{C}[\![u, v]\!]^G$, where $G \subset \mathrm{GL}(2, \mathbb{C})$ is a finite group. Moreover, Herzog's Lemma 11.5 implies that in that case the power series ring $S = \mathbb{C}[\![u, v]\!]$ is a representation generator for (the MCM modules over) R. Once again the proof above applies nearly verbatim to show (redundantly, cf. Proposition 11.1) that $\mathrm{End}_R(S)$ has global dimension two.

In dimension three or greater, the kernel $M_2 = \ker(M_1 \longrightarrow M_0)$ is no longer a MCM module. When the ring R is CM, however, one can replace it by a high syzygy to obtain the following result.

Theorem 17.2 (Iyama [116], Leuschke [145], Quarles [169]). *Let R be a CM local ring of finite representation type and let M be a representation generator for R. Set $\Lambda = \mathrm{End}_R(M)$. Then Λ has global dimension at most* $\max\{2, \dim R\}$, *and equality holds if* $\dim R \geq 2$. *More precisely,* $\mathrm{pd}_\Lambda S = 2$ *for every simple Λ-module S except the one corresponding to R, which has projective dimension equal to* $\dim R$.

Recall that the projective module corresponding to an indecomposable direct summand N of M is $P_N = \mathrm{Hom}_R(M, N)$, and the corresponding simple module is $S_N = P_N / \operatorname{rad} P_N$.

The proof of the assertion $\operatorname{gldim} \Lambda \leq \max\{2, \dim R\}$ is exactly similar to the argument sketched above.[3] For the more precise statement about the projective dimensions of the simple modules, recall that over a CM local ring of finite representation type, every non-free indecomposable MCM module X has an *AR (or almost split) sequence*. This is a non-split short exact sequence of MCM modules, $0 \longrightarrow Y \longrightarrow E \longrightarrow X \longrightarrow 0$, such that every homomorphism $Z \longrightarrow X$ from a MCM module Z to X, which is not a split surjection, factors through E. In particular, one can show that if M is a representation generator, then applying $\mathrm{Hom}_R(M, -)$ to the AR sequence ending in X yields the exact sequence

$$0 \longrightarrow \mathrm{Hom}_R(M, Y) \longrightarrow \mathrm{Hom}_R(M, E) \longrightarrow \mathrm{Hom}_R(M, X) \longrightarrow S_X \longrightarrow 0,$$

where S_X is the simple $\mathrm{End}_R(M)$-module corresponding to X. In particular, this displays a projective resolution of S_X for every non-free indecomposable MCM module X. The simple S_R corresponding to R is thus very special, and has projective dimension equal to $\dim R$ by Proposition 7.1 (iii). Observe that this argument relies essentially on the fact that R has a representation generator; below is an example where $\mathrm{pd}_\Lambda S > \dim R$ for a simple S even though Λ has finite global dimension.

[3] In the published version of [145], I gave an incorrect argument for the equality $\operatorname{gldim} \Lambda = \dim R$ if $\dim R \geq 2$, pointed out to me by C. Quarles and I. Burban. I claimed that if S is a simple Λ-module, then a Λ-projective resolution of S consists of MCM R-modules, so has length at least $\dim R$ by the depth lemma. That's not true, since Λ is not MCM. The equality can be rescued by appealing to Proposition 7.1 (iii).

Among other things, the statement about simple modules implies that when dim R is at least 3, the endomorphism ring of a representation generator is never homologically homogeneous, so is never a non-commutative crepant resolution. A concrete example of this failure has already appeared in Example 14.5. Here is another example in the non-Gorenstein case.

Example 17.3 ([144, Example 12; 189]). Let k be an infinite field and let R be the complete scroll of type (2, 1), that is, $R = k[\![x,y,z,u,v]\!]/I$, where I is generated by the 2×2 minors of the matrix $\left(\begin{smallmatrix} x & y & u \\ y & z & v \end{smallmatrix}\right)$. Then R is a three-dimensional CM normal domain which is not Gorenstein, and has finite representation type [6]. The only non-free indecomposable MCM modules are, up to isomorphism,

- the canonical module $\omega \cong (u,v)R$;
- the first syzygy of ω, isomorphic to $\omega^* = \operatorname{Hom}_R(\omega, R)$ and to $(x,y,u)R$;
- the second syzygy N of ω, rank two and 6-generated; and
- the dual $L = \operatorname{Hom}_R(\omega^*, \omega)$ of ω^*, isomorphic to $(x,y,z)R$.

By Theorem 17.2, $\Lambda = \operatorname{End}_R(R \oplus \omega \oplus \omega^* \oplus N \oplus L)$ has global dimension 3. However, Λ is not MCM as an R-module, since none of L^*, N^*, and $\operatorname{Hom}_R(\omega, \omega^*)$ is MCM. One can check with, say, `Macaulay2` [88] that $\operatorname{End}_R(R \oplus \omega)$ and $\operatorname{End}_R(R \oplus \omega^*)$ are up to Morita equivalence the only endomorphism rings of the form $\operatorname{End}_R(D)$, with D non-free MCM, that are themselves MCM. In fact $\operatorname{End}_R(R \oplus \omega) \cong \operatorname{End}_R(R \oplus \omega^*)$ as rings.

Set $\Gamma = \operatorname{End}_R(R \oplus \omega)$. Then Γ has two simple modules S_ω and S_R. Using Lemma 14.2 and the known structure of the AR sequences over R, Smith and Quarles [190] show that $\operatorname{pd}_\Gamma S_\omega = 4$ and $\operatorname{pd}_\Gamma S_R = 3$. Thus Γ has global dimension equal to 4 by Proposition 7.1 (iii), but is not a non-commutative crepant resolution of R.

Example 17.4. There is only one other known example of a non-Gorenstein CM complete local ring of finite representation type in dimension three or more. It is the (completion of the) homogeneous coordinate ring of the cone over the Veronese embedding $\mathbb{P}^2 \hookrightarrow \mathbb{P}^5$. Explicitly, set $R = \mathbb{C}[\![x^2, xy, xz, y^2, yz, z^2]\!] \subset \mathbb{C}[\![x,y,z]\!] = S$. Then the indecomposable non-free MCM R-modules are the canonical module $\omega = (x^2, xy, xz)R$ and its first syzygy N. Observe that $S \cong R \oplus \omega$ as R-modules, so by Theorem 11.2, $\operatorname{End}_R(R \oplus \omega) \cong S\#(\mathbb{Z}_2)$ has finite global dimension. Since $\operatorname{End}_R(S) \cong S \oplus S$, $\Lambda = \operatorname{End}_R(S)$ is a non-commutative crepant resolution for R.

By Theorem 17.2, $\Gamma = \operatorname{End}_R(R \oplus \omega \oplus N)$ has global dimension 3. But $\operatorname{Hom}_R(N, R)$ and $\operatorname{Hom}_R(N, N)$ have depth 2, so Γ is not a non-commutative crepant resolution.

18 Example: the Generic Determinant

The most common technique thus far for constructing non-commutative crepant resolutions has been to exploit a known (generally crepant) resolution of singularities and a tilting object on it. In fact, the basic technique is already present in Van den Bergh's proof of Theorem 12.11. This has been used in several other families of examples. This section is devoted to describing a particular example of this technique in action, namely the generic determinantal hypersurface ring.

Let k be a field and $X = (x_{ij})$ the generic square matrix of size $n \geq 2$, whose entries x_{ij} are thus a family of n^2 indeterminates over k. Set $S = k[X] = k[\{x_{ij}\}]$ and let R be the hypersurface ring $S/(\det X)$ defined by the determinant of X. Then R is a normal Gorenstein domain of dimension $n^2 - 1$.

Fix a free S-module $\mathcal{F}$ of rank n. Left-multiplication with the matrix X naturally defines the generic S-linear map $\mathcal{F} \longrightarrow \mathcal{F}$. The exterior powers $\bigwedge^a X : \bigwedge^a \mathcal{F} \longrightarrow \bigwedge^a \mathcal{F}$ define natural S-modules

$$M_a = \operatorname{cok} \textstyle\bigwedge^a X$$

for $a = 1, \ldots, n$. In fact each M_a is annihilated by $\det X$, so is naturally an R-module. The pair $(\bigwedge^a X, \bigwedge^{n-a} X^T)$ forming a *matrix factorization* of $\det X$, the M_a are even MCM modules over R [73]. They are in particular reflexive, of rank $\binom{n-1}{a-1}$.

Set $M = \bigoplus_{a=1}^n M_a$ and $\Lambda = \operatorname{End}_R(M)$. The crucial result of [44], in this case, is then

Theorem 18.1. *The R-algebra Λ provides a non-commutative crepant resolution of R.*

The proof in [44] proceeds by identifying the M_a as geometric objects with tilting in their ancestries, as follows. Let F be a k-vector space of dimension n, and set $\mathbb{P} = \mathbb{P}(F^\vee) \cong \mathbb{P}_k^{n-1}$ be the projective space over R, viewed as equivalence classes $[\lambda]$ of linear forms $\lambda : F \longrightarrow k$. Put

$$Y = \mathbb{P} \times \operatorname{Spec} S\,,$$

with canonical projections $\widetilde{p} : Y \longrightarrow \mathbb{P}$ and $\widetilde{q} : Y \longrightarrow \operatorname{Spec} S$. Identify $\operatorname{Spec} S$ with the space of $(n \times n)$ matrices A over the field k, with coordinate functions given by the indeterminates x_{ij}. Then the *incidence variety*

$$Z = \{([\lambda], A) \mid \text{image}\, A \subseteq \ker \lambda\}$$

is a resolution of singularities of $\operatorname{Spec} R$. (Compare with Example 9.1, which is the case $n = 2$.) Indeed, the image of Z under $\widetilde{q} : Y \longrightarrow \operatorname{Spec} S$ is precisely the locus of matrices A with $\operatorname{rank} A < n$, that is, $\operatorname{Spec} R$. Furthermore, the singular locus of $\operatorname{Spec} R$ consists of the matrices of rank $< n - 1$, and $q := \widetilde{q}|_Z : Z \longrightarrow \operatorname{Spec} R$ is an isomorphism away from these points. One can explicitly write down the equations cutting Z out of Y, and verify that Z is smooth, and is a complete intersection in Y; if in particular $j : Z \longrightarrow Y$ is the inclusion, then this implies that $j_* \mathcal{O}_Z$ is resolved over $\mathcal{O}_Y$ by a Koszul complex on the Euler form $F \otimes_k \mathcal{O}_Y(-1) \longrightarrow \mathcal{O}_Y$.

Here is a pictorial description of the situation.

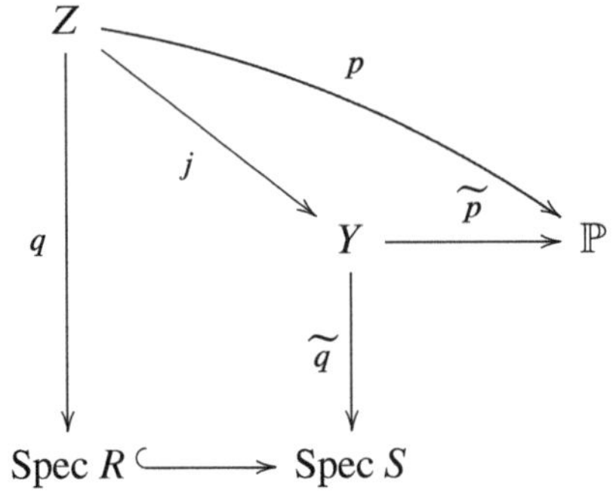

(18.1)

Recall from Section 6 that $T = \bigoplus_{a=1}^{n} \Omega^{a-1}(a)$, where $\Omega = \Omega_{\mathbb{P}/k}$ is the sheaf of differential forms on $\mathbb{P}$ and $\Omega^j = \bigwedge^j \Omega$, is a tilting bundle on $\mathbb{P}$. Set $\mathcal{M}_a = p^*\Omega^{a-1}(a)$ for $a = 1, \dots, n$, a locally free sheaf on the resolution Z. As the typography hints, $\mathcal{M}_a$ is a geometric version of M_a, in the following sense.

Proposition 18.2. *As R-modules, $\mathbf{R}q_*\mathcal{M}_a = M_a$. More precisely, $\mathbf{R}^j q_*\mathcal{M}_a = 0$ for $j > 0$ and $q_*\mathcal{M}_a = M_a$ for all a.*

The proof of the proposition involves juggling two Koszul complexes. Tensoring (6.1) with $\mathcal{O}_{\mathbb{P}}(a)$ and truncating gives an exact sequence

$$0 \longrightarrow \Omega^{a-1}(a) \longrightarrow \textstyle\bigwedge^{a-1} F \otimes_k \mathcal{O}_{\mathbb{P}}(1) \longrightarrow \cdots \longrightarrow F \otimes_k \mathcal{O}_{\mathbb{P}}(a-1) \longrightarrow \mathcal{O}_{\mathbb{P}}(a) \longrightarrow 0.$$

The projection p being flat, the pullback p^* is exact, yielding

$$0 \longrightarrow \mathcal{M}_a \longrightarrow \textstyle\bigwedge^{a-1} F \otimes_{\mathcal{O}_Z} (1) \longrightarrow \cdots \longrightarrow F \otimes_k \mathcal{O}_Z(a-1) \longrightarrow \mathcal{O}_Z(a) \longrightarrow 0\,.$$

Compute $\mathbf{R}q_*$ as $\mathbf{R}\widetilde{q}_* j_*$. As $j_*\mathcal{O}_Z$ is resolved over $\mathcal{O}_Y$ by a Koszul complex, we may replace the former with the latter and obtain a double complex in the fourth quadrant, with $\bigwedge^{a-1} F \otimes_k \mathcal{O}_Y(1)$ at the origin and $\bigwedge^{a-i+1} F \otimes_k \bigwedge^{-j} F \otimes_k \mathcal{O}_Y(i+j+1)$ in the (i, j) position. Now apply $\mathbf{R}q_*$. By [94, Example III.8.4], the higher direct images of the projective bundle $q \colon Y \longrightarrow \operatorname{Spec} S$ are completely known,

$$\mathbf{R}^j\widetilde{q}_*\mathcal{O}_Y(t) = \begin{cases} 0 & \text{if } t < 0 \text{ or } 1 < j < n-1; \\ \operatorname{Sym}_t(F) \otimes_k S = \operatorname{Sym}_t(\mathcal{F}) & \text{for } j = 0; \text{ and} \\ 0 & \text{for } j = n-1 \text{ if } t \geq -n. \end{cases}$$

This already proves $\mathbf{R}^j q_* \mathcal{M}_a = 0$ for $j > 0$, and allows one to represent $q_* \mathcal{M}_a$ by the homology of the total complex of the following double complex of free S-modules. (For notational simplicity write $\bigwedge^i$ and Sym_j instead of $\bigwedge^i \mathcal{F}$ and $\mathrm{Sym}_j \mathcal{F}$.)

$$
\begin{array}{ccccccccc}
 & & 0 & & 0 & & 0 & & \\
 & & \uparrow & & \uparrow & & \uparrow & & \\
0 & \to & \bigwedge^{a-1} \otimes \mathrm{Sym}_1 & \to \cdots \to & \bigwedge^1 \otimes \mathrm{Sym}_{a-1} & \longrightarrow & \mathrm{Sym}_a & \longrightarrow & 0 \\
 & & \uparrow & & \uparrow & & \uparrow & & \\
0 & \to & \bigwedge^{a-1} \otimes \bigwedge^1 & \to \cdots \to & \bigwedge^1 \otimes \bigwedge^1 \otimes \mathrm{Sym}_{a-2} & \to & \bigwedge^1 \otimes \mathrm{Sym}_{a-1} & \to & 0 \\
 & & \uparrow & & \uparrow & & \uparrow & & \\
 & & 0 & \longrightarrow \cdots \to & \bigwedge^1 \otimes \bigwedge^2 \otimes \mathrm{Sym}_{a-3} & \to & \bigwedge^2 \otimes \mathrm{Sym}_{a-2} & \to & 0 \\
 & & & & \uparrow & & \uparrow & & \\
 & & & & \vdots & & \vdots & & \\
 & & & & \uparrow & & \uparrow & & \\
 & & & 0 \longrightarrow & \bigwedge^1 \otimes \bigwedge^{a-1} & \longrightarrow & \bigwedge^{a-1} \otimes \mathrm{Sym}_1 & \to & 0 \\
 & & & & \uparrow & & \uparrow & & \\
 & & & & 0 & & \bigwedge^a & & \\
 & & & & & & \uparrow & & \\
 & & & & & & 0 & &
\end{array}
$$

Here the jth column is obtained by tensoring the strand of degree j in the Koszul complex with $\bigwedge^{a-j-1} \mathcal{F}$, so is acyclic [74, Theorem A2.10]. Similarly, the $(-i)^{\text{th}}$ row is the degree a strand in a Koszul complex tensored with $\bigwedge^i \mathcal{F}$, and so is exact with the exceptions of the top and bottom rows. The top row has homology equal to $\bigwedge^a \mathcal{F}$ at the leftmost end, while the bottom row has homology $\bigwedge^a \mathcal{F}$ on the right. One checks from the explicit nature of the maps that the total complex is thus reducible to $\bigwedge^a X \colon \bigwedge^a \mathcal{F} \longrightarrow \bigwedge^a \mathcal{F}$, whence $q_* \mathcal{M}_a = M_a$, as claimed.

Now it is relatively easy to prove that

$$
\mathbf{R}^j q_* \mathcal{H}om_{\mathcal{O}_Z}(\mathcal{M}_b, \mathcal{M}_a) = \begin{cases} \mathrm{Hom}_R(M_b, M_a) & \text{if } j = 0, \text{ and} \\ 0 & \text{otherwise,} \end{cases}
$$

and to establish the rest of the assertions in the next theorem.

Theorem 18.3. *The object* $\mathbf{R}q_*\mathcal{H}om_{\mathcal{O}_Z}(\mathcal{M}_b, \mathcal{M}_a)$ *is isomorphic in the bounded derived category* $\mathsf{D}^b(S\text{-mod})$ *to a single morphism between free* S*-modules situated in (cohomological) degrees* -1 *and* 0. *Therefore the* R*-module* $q_*\mathcal{H}om_{\mathcal{O}_Z}(\mathcal{M}_b, \mathcal{M}_a) = \operatorname{Hom}_R(M_b, M_a)$ *is a MCM* R*-module and the higher direct images vanish, so that in particular*

$$\mathbf{R}^1 q_*\mathcal{H}om_{\mathcal{O}_Z}(\mathcal{M}_b, \mathcal{M}_a) = \operatorname{Ext}^1_R(M_b, M_a) = 0\,.$$

It remains to see that $\bigoplus_{a=1}^n \mathcal{M}_a$ is a tilting object on the resolution Z, so that $\Lambda = \operatorname{End}_R(\bigoplus_a M_a) = q_*\mathcal{E}nd_{\,\mathcal{O}_Z}(\bigoplus_a \mathcal{M}_a)$ has finite global dimension, whence is a non-commutative crepant resolution of R. It suffices for this to compute the cohomology

$$H^i(\mathbb{P}, \mathcal{H}om_{\mathcal{O}_{\mathbb{P}}}(\Omega^{b-1}(b), \Omega^{a-1}(a))(c))\,;$$

since p is flat, this will compute $\operatorname{Ext}^i_{\mathcal{O}_Z}(\mathcal{M}_b, \mathcal{M}_a)$ as well. In [44] we gave a characteristic-free proof of this vanishing, and another appears in the appendix by Weyman to [76]. In characteristic 0, one can compute the cohomology with Bott vanishing [215, Chapter 4]. This allows the following proposition and theorem.

Proposition 18.4. *The* $\mathcal{O}_Z$*-module* $\mathcal{M} = \bigoplus_a \mathcal{M}_a = \bigoplus_{a=1}^n p^*\Omega^{a-1}(a)$ *is a tilting bundle in* $\mathsf{D}^b(\mathsf{coh}\,Z)$. *In detail, with* $\mathcal{A} = \operatorname{End}_{\mathsf{D}^b(\mathsf{coh}\,Z)}(\mathcal{M})$,

(i) $\operatorname{Ext}^i_{\mathcal{O}_Z}(\mathcal{M}, \mathcal{M}) := \operatorname{Hom}_{\mathsf{D}^b(\mathsf{coh}\,Z)}(\mathcal{M}, \mathcal{M}[i]) = 0$ *for* $i > 0$;

(ii) $\mathbf{R}\operatorname{Hom}_{\mathcal{O}_Z}(\mathcal{M}, -) \colon \mathsf{D}^b(\mathsf{coh}\,Z) \longrightarrow \mathsf{D}^b(\mathcal{A}\text{-mod})$ *is an equivalence of triangulated categories, with* $- \overset{\mathbf{L}}{\otimes}_{\mathcal{A}} \mathcal{M}$ *as inverse;*

(iii) $\mathcal{A}$ *has finite global dimension.*

(iv) $\mathcal{A} \cong \Lambda = \operatorname{End}_R(M)$.

Theorem 18.5. *Let* k *be a field,* X *an* $(n \times n)$ *matrix of indeterminates,* $n \geq 2$, *and* $R = k[X]/(\det X)$ *the generic determinantal hypersurface ring. Let* $M_a = \operatorname{cok} \bigwedge^a X$ *for* $a = 1, \ldots, n$, *and put* $M = \bigoplus_a M_a$. *Then the* R*-algebra* $\Lambda = \operatorname{End}_R(M)$ *has finite global dimension and is MCM as an* R*-module. It is in particular a non-commutative crepant resolution of* R.

In [44] we replace the square matrix X by an $(m \times n)$ matrix with $n \geq m$ and R with the quotient by the maximal minors $k[X]/I_m(X)$, which defines the locus in $\operatorname{Spec} k[X]$ of matrices with non-maximal rank. The same construction $M_a = \operatorname{cok} \bigwedge^a X$ yields an algebra $\Lambda = \operatorname{End}_R(\bigoplus_{a=1}^m M_a)$ which is still MCM as an R-module and still has finite global dimension. In this case, however, Λ is not a non-singular R-algebra, so not a non-commutative crepant resolution according to our definition. This is directly attributable to the fact that quotients by minors are Gorenstein if and only if $n = m$, so that Corollary 12.3 fails for non-square matrices. In a forthcoming paper [45], we establish the same result for the quotient by arbitrary minors

$k[X]/I_t(X)$, with $1 \le t \le m$, using a tilting bundle on the Grassmannian [43]. In particular we obtain non-commutative crepant resolutions when the matrix is square.

Similar techniques, i.e. constructions using tilting objects on known resolutions of singularities, are used by Kuznetsov [141] to give non-commutative desingularizations for several more classes of examples, including cones over Veronese/Segre embeddings and Grassmannians, as well as Pfaffian varieties.

19 Non-commutative Blowups

It was clear from early on in the development of non-commutative (projective) geometry that it would be most desirable to have a non-commutative analogue of the most basic birational transformation, the blowup. This section sketches a few approaches to the problem.

First recall that if R is a commutative ring and I is an ideal of R, the *blowup* $\mathcal{B}_I(X)$ of $X = \operatorname{Spec} R$ at (the closed subscheme defined by) I is $\operatorname{Proj} R[It]$, where $R[It]$ is the Rees algebra $R \oplus I \oplus I^2 \oplus \cdots$. The exceptional locus of the blowup is the *fiber cone* $\operatorname{Proj} R[It]/IR[It] = R/I \oplus I/I^2 \oplus \cdots$.

One might hope to mimic this definition for sufficiently nice non-commutative rings. This turns out to give unsatisfactory results. For example ([8; 203, page 2]) set $\Lambda = k\langle x, y\rangle/(yx - xy - y)$, and consider the ideal $\mathfrak{m} = (x, y)$ "corresponding" to the origin of this non-commutative surface. Then $\mathfrak{m}^n = (x^n, y)$ for all y, so the fiber cone $R[\mathfrak{m}t]/\mathfrak{m}R[\mathfrak{m}t]$ is one-dimensional in each degree, and is isomorphic to $k[z]$. This means that the exceptional locus is in some sense zero-dimensional, whereas one should expect the exceptional divisor of a blowup of a point in a surface to have dimension 1.

Van den Bergh [203] constructs an analogue of the Rees algebra directly over projective quasi-schemes $\mathsf{Proj}\,\Lambda$ (see Section 7). Specifically, if $X = \mathsf{Proj}\,\Lambda$ is a quasi-scheme, he gives a construction of the blowup of a smooth point p in a commutative curve Y contained in X. (This means that $\mathsf{Qch}\,Y \simeq \mathsf{Proj}(\Lambda/x\Lambda)$ for some $x \in \Lambda$.) Using this construction, Van den Bergh considers blowups of quantum projective planes at small numbers of points, in particular non-commutative deformations of the del Pezzo surfaces obtained by blowing up in ≤ 8 points. I will not go into the details of the construction or the applications here.

There is a more recent proposal for a definition of the phrase "non-commutative blowup," which is inspired by the classic flop of Example 10.2 and by Theorem 13.1. In general, the idea is that for an ideal I of a ring Λ, the *non-commutative blowup* of Λ in I is the ring

$$\mathcal{B}_I^{\mathrm{nc}}(\Lambda) = \operatorname{End}_\Lambda(\Lambda \oplus I)\,.$$

In the situation of Example 10.2, we saw that $\mathcal{B}_I^{\mathrm{nc}}(R)$ was derived equivalent to the usual blowup $\mathcal{B}_I(\operatorname{Spec} R)$. Thus suggests the following question, a version of which I first heard from R.-O. Buchweitz.

Question 19.1. Can one generalize or imitate the normalization algorithm of Section 13 to show that there is a sequence of non-commutative blowups starting with $\Lambda = \Lambda_0$ and continuing with $\Lambda_{i+1} = \mathcal{B}^{\mathrm{nc}}_{I_i}(\Lambda_i) = \mathrm{End}_{\Lambda_i}(\Lambda_i \oplus I_i)$ for some ideals $I_i \subset \Lambda_i$, such that Λ_i eventually has finite global dimension?

One might try to follow Hironaka and blow up only in "smooth centers," i.e. assume that Λ_i/I_i is non-singular, as is the case in Example 10.2.

Very recent work of Burban–Drozd [28] confirms that the non-commutative blowup as above is a sort of categorical desingularization whenever X is a reduced algebraic curve singularity having only nodes and cusps for singular points, and I is the conductor ideal. They observe that $\mathcal{A} = \mathcal{E}nd_{\mathcal{O}_X}(\mathcal{O}_X \oplus I)$ has global dimension equal to 2, that $(\mathcal{O}_X \oplus I) \otimes_{\mathcal{O}_X} -\colon \operatorname{coh} X \longrightarrow \mathcal{A}\text{-mod}$ is fully faithful, and that $\mathrm{Hom}_{\mathcal{A}}(\mathcal{O}_X \oplus I, -)\colon \mathcal{A}\text{-mod} \longrightarrow \operatorname{coh} X$ is exact.

In his Master's thesis, Quarles constructs a direct connection between blowups and non-commutative blowups, the only one I know of. Let $(R, \mathfrak{m}, k)$ be a Henselian local k-algebra, with k an algebraically closed field. Let I be an ideal of R which is MCM and reflexive as an R-module, and set $\Lambda = \mathcal{B}^{\mathrm{nc}}_{I^*}(R) = \mathrm{End}_R(R \oplus I^*)$. Then Quarles defines [169, Section 7] a bijection between the closed points of $\mathcal{B}_I(\operatorname{Spec} R) = \operatorname{Proj} R[It]$ and the set of indecomposable Λ-modules X arising as extensions $0 \longrightarrow S_R \longrightarrow X \longrightarrow S_{I^*} \longrightarrow 0$ of the two simple modules S_R and S_I. The bijection is just as sets, and carries no known algebraic information; in particular, it is not known to be a moduli space.

There are some immediate problems. For example, in Example 17.3 $\mathrm{End}_R(R \oplus \omega) \cong \mathrm{End}_R(R \oplus \omega^*)$, but $R[\omega t] \ncong R[\omega^* t]$, since one is regular and the other is not. The associated projective schemes are isomorphic, of course. It is not clear how to reconcile this.

A similar approach has been suggested in prime characteristic [195–197, 199, 216]. For the rest of this section let k be an algebraically closed field of characteristic $p > 0$. Let X and Y be normal algebraic schemes over k, and let $f\colon Y \longrightarrow X$ be a finite dominant morphism. Then Yasuda [216] proposes to call the endomorphism ring $\mathcal{E}nd_{\mathcal{O}_X}(f_*\mathcal{O}_Y)$ the non-commutative blowup attached to f.

In particular, consider the non-commutative blowup of the Frobenius. For every $e \geq 1$, set $X_e = X$ and let $F^e_X\colon X_e \longrightarrow X$ be the eth iterate of the Frobenius morphism. Assume that F_X is finite. Then the non-commutative blowup of the eth Frobenius, $\mathcal{E}nd_{\mathcal{O}_X}(F^e_{X*}\mathcal{O}_{X_e})$ is locally given by $\mathrm{End}_R(R^{1/p^e})$, where R^{1/p^e} is the ring of $(p^e)^{\mathrm{th}}$ roots of elements of R. It is isomorphic to $\mathrm{End}_{R^{p^e}}(R)$, where now R^{p^e} is the subring of $(p^e)^{\mathrm{th}}$ powers. The ring $\mathrm{End}_{R^{p^e}}(R)$ consists of differential operators on R [191] and is sometimes a non-commutative crepant resolution of R.

Theorem 19.2 (Toda–Yasuda [199]). *Let R be a complete local ring of characteristic p which is one of the following.*

(i) *a one-dimensional domain;*

(ii) *the ADE hypersurface singularity of type* (A_1) *(and* $p \neq 2$*); or*

(iii) *a ring of invariants* $k[\![x_1, \dots, x_n]\!]^G$, *where* $G \subset \mathrm{GL}(n, k)$ *is a finite subgroup with order invertible in* k.

Then for $e \gg 0$, $\mathrm{End}_R(R^{1/p^e})$ *has finite global dimension. However it is not generally MCM as an* R*-module, so is not a non-commutative crepant resolution.*

Let me make a few comments on the proofs. For (i), consider the integral closure $S \cong k[\![x]\!]$ of R. Then for any $e \geq 1$, one checks that $\mathrm{End}_{R^{p^e}}(S^{p^e}) = \mathrm{End}_{S^{p^e}}(S^{p^e}) = S^{p^e}$. Take e large enough that $S^{p^e} \subseteq R$. Then R is free over S^{p^e} of rank p^e, so $\mathrm{End}_{R^{p^e}}(R) \cong M_{p^e}(\mathrm{End}_{R^{p^e}}(S^{p^e})) = M_{p^e}(S^{p^e})$. This is Morita equivalent to $S^{p^e} \cong S$, so has global dimension equal to 1. It is also clearly MCM.

For (ii), assume $p \neq 2$ and set $R = k[\![x_1, \dots, x_d]\!]/(x_1^2 + \cdots + x_d^2)$. Then one can show that for all $e \geq 1$, R is a representation generator for R^{p^e}. (This requires separate arguments for d odd/even.) By Theorem 17.2, $\mathrm{End}_{R^{p^e}}(R)$ has finite global dimension. It is not a non-commutative crepant resolution by Theorem 14.4.

Finally, for (iii), Toda and Yasuda use results of Smith and Van den Bergh to show that if $S = k[\![x_1, \dots, x_d]\!]$ and $R = S^G$ as in the statement, then for $e \gg 0$ every module of covariants $(S \otimes_k W)^G$ appears as an R^{p^e}-direct summand in R, in S, and in S^{p^e}. Thus $\mathrm{End}_{R^{p^e}}(S^{p^e}) \cong S^{p^e} \# G$ (Theorem 11.2) is Morita equivalent to $\mathrm{End}_{R^{p^e}}(R)$ by Corollary 2.3, and they simultaneously have finite global dimension.

In general there are known non-trivial obstructions to $\mathrm{End}_R(R^{1/p^e})$ being a non-commutative crepant resolution. For example, Dao points out [66] that when R is a complete intersection ring, R^{1/p^e} is known to be Tor-rigid [4], so if R satisfies (R_2) then $\mathrm{End}_R(R^{1/p^e})$ is not MCM for any $e \geq 1$ by the discussion following Proposition 14.3.

20 Omissions and Open Questions

In addition to the examples already mentioned in previous section, there is a large and growing array of examples of non-commutative crepant resolutions and related constructions. Lack of space and expertise prevent me from describing them in full, but here are a few references and comments.

Deformations of the Kleinian singularities $\mathbb{C}^2/G$, with $G \subset \mathrm{SL}(2, \mathbb{C})$, have non-commutative crepant resolutions [89], which are identified as deformed preprojective algebras in the sense of [61].

In a different direction, Wemyss has considered the non-Gorenstein case of the classical McKay correspondence, where $G \not\subset \mathrm{SL}(2, \mathbb{C})$ [212–214]; much of Theorem 11.6 breaks down, but much can be recovered by restricting to the so-called "special" representations. This leads to the *reconstruction algebra,* which is the endomorphism ring of the special MCM modules.

Beil [29] shows that *square superpotential algebras,* which are certain quiver algebras with relations coming from cyclic derivatives of a superpotential, are non-commutative crepant resolutions of their centers (which are three-dimensional toric Gorenstein normal domains). In fact, Broomhead gives a construction of a non-commutative crepant resolution for every Gorenstein affine toric threefold, from superpotential algebras called dimer models [58]. Similar algebras associated to brane tilings have non-commutative crepant resolutions as well [47, 159].

Finally, in [31] Bezrukavnikov constructs a non-commutative version of the Springer resolution Z from (18.1), which is different from that in [44].

Many other topics have been omitted that could have played a role. For example, I have said nothing about (semi-)orthogonal decompositions of triangulated categories and exceptional sequences. These grew out of Beilinson's result in Section 6, via the Rudakov seminar [186]. See [105, Chapter 1] or [48].

Connections of this material with string theory appear at every turn [194]. For example, the derived category $\mathsf{D}(X)$ appears in string theory as the category of branes propagating on the space X. Non-commutativity arises naturally in this context from the fact that *open* strings can be glued together in two different ways, unlike *closed* strings [42]. Furthermore, the Calabi–Yau condition of Section 12 is essential to the string-theoretic description of spacetime [58, 167]. Most obviously, the field of high energy physics has been a driving force in non-commutative desingularizations and the higher geometric McKay correspondence. I am not competent to do more than gesture at these connections.

I end the article with a partial list of open problems. Some of these are mentioned in the text, while others are implicit.

(i) Conjecture 9.3 of Bondal and Orlov, that a generalized flop between smooth varieties induces a derived equivalence, is still largely open outside of dimension three. The related Conjecture 12.12 of Van den Bergh, which asks for derived equivalence of both geometric and non-commutative crepant resolutions, is similarly open. See [114] for some very recent progress on the non-commutative side.

(ii) Existence of a non-commutative crepant resolution is *not* equivalent to existence of a crepant resolution of singularities in dimension four or higher. See the end of Example 8.6 for examples with non-commutative resolutions but no geometric ones, and Example 14.5 for failure of the other direction. However, it still may hold in general in dimension three. One might also be optimistic and ask for additional hypotheses to rescue the case of dimension four.

(iii) Various results in the text fail for rings that are not Gorenstein, notably Corollary 12.3 and Proposition 12.7. Is there a better definition of non-commutative crepant resolutions which would satisfy these statements over non-Gorenstein Cohen–Macaulay rings? Perhaps we should not expect one, since crepant resolutions of singularities exist only in the Gorenstein case.

On a related note, is symmetry (Definition 12.2) too strong a condition? The relevant property in [193] is that $\mathrm{Hom}_R(\Lambda, R)$ is an *invertible* (Λ-Λ)-bimodule, rather than insisting that $\mathrm{Hom}_R(\Lambda, R) \cong \Lambda$. This would, unfortunately, rule out endomorphism rings $\mathrm{End}_R(M)$, since they are automatically symmetric by Theorem 12.6 (ii). Or perhaps the appropriate generalization to non-Gorenstein rings is that $\mathrm{Hom}_R(\Lambda, \omega_R) = \Lambda$.

(iv) Crepant resolutions of singularities are very special: they exist only for canonical singularities, not in general for terminal singularities. The non-commutative version is more general. One might therefore hope that Theorem 12.11 is true for canonical threefolds as well. Van den Bergh's proof of Theorem 12.11 applies verbatim for any canonical threefold admitting a crepant resolution of singularities with one-dimensional fibers.

(v) In nearly all of the examples of non-commutative crepant resolutions, the module M such that $\Lambda = \mathrm{End}_R(M)$ can be taken maximal Cohen–Macaulay. Lemma 14.2 indicates one obstruction to M having high depth. Are there general situations where a non-commutative crepant resolution exists, but no MCM module will suffice? Or situations (other than surfaces) where every non-commutative crepant resolution is given by a MCM module? See [113, Proposition 5.11] for one result in this direction.

(vi) Van den Bergh points out in [204] that one might try to build a theory of rational singularities for non-commutative rings, extending the results of Section 16. It would be essential to have a non-commutative analogue of the Grauert–Riemenschneider Vanishing theorem (8.2), but none seems to be known. There is an algebraic reformulation of GR Vanishing due to Sancho de Salas [187], cf. [104, Chapter 5]: Let R be a reduced CM local ring essentially of finite type over an algebraically closed field of characteristic zero, and let I be an ideal of R such that $\mathrm{Proj}\, R[It]$ is smooth; then the associated graded ring $\mathrm{gr}_{I^n}(R)$ is Cohen–Macaulay for $n \gg 0$. It would be very interesting to have a purely algebraic proof of this result, particularly if it encompassed some non-commutative rings. The proof of Sancho de Salas uses results from [84], so relies on complex analysis; see [99] for some progress toward an algebraic proof in dimension two.

(vii) In a similar direction, Question 19.1 asks for an algorithm to resolve singularities via a sequence of "non-commutative blowups." For a start, one needs any non-trivial connection between $\mathsf{D}^b(\mathsf{coh}\,\mathrm{Proj}\, R[It])$ and $\mathsf{D}^b(\mathrm{End}_R(R \oplus I)\text{-}\mathsf{mod})$; other than Quarles' bijection, none seems to be known.

(viii) Even given a very strong result along the lines of (vii), an enormous amount of work would still be needed to obtain applications of non-commutative desingularizations analogous to those of resolutions of singularities. For example, can one define an "arithmetic genus" in a non-commutative context, and show, as Hironaka does, that it is a "birational" invariant?

Acknowledgments. I am grateful to Jesse Burke, Hailong Dao, Kos Diveris, and Michael Wemyss for insightful comments on earlier drafts, and to Ragnar-Olaf Buchweitz and Michel Van den Bergh for encouragement in this project, as well as for years of enjoyable mathematical interaction.

Bibliography

[1] Auslander, M., Ding, S., Solberg, Ø., Liftings and weak liftings of modules. J Algebra. 1993;156(2):273–317. MR 1216471

[2] Auslander, M., Goldman, O., Maximal orders. Trans Amer Math Soc. 1960;97:1–24. MR 0117252

[3] Arhangel′skii, A. V., Goodearl, K. R., Huisgen-Zimmermann, B., Kiiti Morita (1915–1995). Notices Amer Math Soc. 1997;44(6):680–684. MR 1452070

[4] Avramov, L. L., Miller, C., Frobenius powers of complete intersections. Math Res Lett. 2001;8(1-2):225–232. MR 1825272

[5] Auslander, M., Roggenkamp, K. W., A characterization of orders of finite lattice type. Invent Math. 1972;17:79–84. MR 0320053

[6] Auslander, M., Reiten, I., The Cohen–Macaulay type of Cohen–Macaulay rings. Adv Math. 1989;73(1):1–23. MR 979585

[7] Auslander, M., Reiten, I., Smalø, S. O., Representation theory of Artin algebras. vol. 36 of Cambridge studies in advanced mathematics. Corrected reprint of the 1995 original. Cambridge: Cambridge University Press; 1997. MR 1476671

[8] Artin, M., Some problems on three-dimensional graded domains. Proceedings, Representation Theory and Algebraic Geometry; 1995; Waltham, M. A. vol. 238 of London Mathematical Society lecture note series. Cambridge: Cambridge University Press; 1997. p. 1–19. MR 1477464

[9] Artin, M., Schelter, W. F., Graded algebras of global dimension 3. Adv Math. 1987;66(2):171–216. MR 917738

[10] Artin, M., Stafford, J. T., Noncommutative graded domains with quadratic growth. Invent Math. 1995;122(2):231–276. MR 1358976

[11] Atiyah, M. F., On analytic surfaces with double points. Proc R Soc London Ser A Math Phys Eng Sci. 1958;247:237–244. MR 0095974

[12] Artin, M., Tate, J., Van den Bergh, M., Some algebras associated to automorphisms of elliptic curves. In: The Grothendieck Festschrift, Vol. I. vol. 86 of Progress in Mathematics. Boston, MA: Birkhäuser Boston; 1990. p. 33–85. MR 1086882

[13] Artin, M., Tate, J., Van den Bergh, M., Modules over regular algebras of dimension 3. Invent Math. 1991;106(2):335–388. MR 1128218

[14] Auslander, M., On the purity of the branch locus. Amer J Math. 1962;84:116–125. MR 0137733

[15] Auslander, M., Representation dimension of Artin algebras. In: Selected works of Maurice Auslander, vol. 1. American Mathematical Society; 1999 [originally Queen Mary College mathematics notes. University of London; 1971]. p. 505–574.

[16] Auslander, M., Functors and morphisms determined by objects. Proceedings, Representation Theory of Algebras; 1976; Temple University Philadelphia, P. A. vol. 37 of Lecture notes in pure and applied mathematics. New York: Dekker; 1978. p. 1–244. MR 0480688

[17] Auslander, M., Isolated singularities and existence of almost split sequences. Proceedings, Representation Theory, II; 1984; Ottawa, ON. vol. 1178 of Lecture notes in mathematics. Berlin: Springer; 1986. p. 194–242. MR 842486

[18] Auslander, M., Rational singularities and almost split sequences. Trans Amer Math Soc. 1986;293(2):511–531. MR 816307

[19] Artin, M., Verdier, J.-L., Reflexive modules over rational double points. Math Ann. 1985;270(1):79–82. MR 769609

[20] Artin, M., Van den Bergh, M., Twisted homogeneous coordinate rings. J Algebra. 1990;133(2):249–271. MR 1067406

[21] Artin, M., Zhang, J. J., Noncommutative projective schemes. Adv Math. 1994;109(2):228–287. MR 1304753

[22] Baer, D., Tilting sheaves in representation theory of algebras. Manuscripta Math. 1988;60(3):323–347. MR 928291

[23] Balmer, P., Presheaves of triangulated categories and reconstruction of schemes. Math Ann. 2002;324(3):557–580. MR 1938458

[24] Balmer, P., The spectrum of prime ideals in tensor triangulated categories. J Reine Angew Math. 2005;588:149–168. MR 2196732

[25] Bass, H., Finitistic dimension and a homological generalization of semi-primary rings. Trans Amer Math Soc. 1960;95:466–488. MR 0157984

[26] Brenner, S., Butler, M. C. R., Generalizations of the Bernstein–Gel′fand–Ponomarev reflection functors. Proceedings of the Second International Conference on Representation Theory; 1979; Carleton University, Ottawa, ON. vol. 832 of Lecture notes in mathematics. Berlin: Springer; 1980. p. 103–169. MR 607151

[27] Berenstein, D., Douglas, M. R., Seiberg duality for quiver gauge theories. Available from: http://arxiv.org/abs/hep-th/0207027. 2002.

[28] Burban, I., Drozd, Y., Tilting on non-commutative rational projective curves. Available from: http://arxiv.org/abs/1103.3936. April 2011.

[29] Beil, C., The noncommutative geometry of square superpotential algebras. Available from: http://arxiv.org/abs/0811.2439. November 2008.

[30] Berenstein, D., Reverse geometric engineering of singularities. J High Energy Phys. 2002;(4):Paper 18, 52. MR 1911393

[31] Bezrukavnikov, R. V., Noncommutative counterparts of the Springer resolution. Proceedings, International Congress of Mathematicians, Vol. II. Zürich: European Mathematical Society; 2006. p. 1119–1144. MR 2275638

[32] Berthelot, P., Grothendieck, A., Illusie, L., editors. Théorie des intersections et théorème de Riemann-Roch. vol. 225 of Lecture notes in mathematics. New York: Springer-Verlag; 1971. MR 0354655

[33] Bernšteĭn, J. N., Gel′fand, I. M., Ponomarev, V. A., Coxeter functors, and Gabriel's theorem. Uspekhi Mat Nauk. 1973;38(2):19–33. MR 0393065

[34] Brown, K. A., Hajarnavis, C. R., Homologically homogeneous rings. Trans Amer Math Soc. 1984;281(1):197–208. MR 719665

[35] Bruns, W., Herzog, J., Cohen–Macaulay rings. vol. 39 of Cambridge studies in advanced mathematics. Cambridge: Cambridge University Press; 1993. MR 1251956

[36] Brown, K. A., Hajarnavis, C. R., MacEacharn, A. B., Noetherian rings of finite global dimension. Proc Lond Math Soc (3). 1982;44(2):349–371. MR 647437

[37] Brown, K. A., Hajarnavis, C. R., MacEacharn, A. B., Rings of finite global dimension integral over their centres. Comm Algebra. 1983;11(1):67–93. MR 687406

[38] Burban, I., Iyama, O., Keller, B., Reiten, I., Cluster tilting for one-dimensional hypersurface singularities. Adv Math. 2008;217(6):2443–2484. MR 2397457

[39] Bezrukavnikov, R. V., Kaledin, D. B., McKay equivalence for symplectic resolutions of quotient singularities. Tr Mat Inst Steklova. 2004;246[Algebr. Geom. Metody: Svyazi i Prilozh.]:20–42. MR 2101282

[40] Bridgeland, T., King, A., Reid, M., The McKay correspondence as an equivalence of derived categories. J Amer Math Soc. 2001;14(3):535–554 [electronic]. MR 1824990

[41] Buan, A. B., Krause, H., Solberg, Ø., Support varieties: an ideal approach. Homology, Homotopy Appl. 2007;9(1):45–74. MR 2280286

[42] Berenstein, D., Leigh, R. G., Resolution of stringy singularities by non-commutative algebras. J High Energy Phys. 2001;(6):Paper 30, 37. MR 1849725

[43] Buchweitz, R.-O., Leuschke, G. J., Van den Bergh, M. A characteristic free tilting bundle for Grassmannians. Available from: http://arxiv.org/abs/1006.1633. June 2010.

[44] Buchweitz, R.-O., Leuschke, G. J., Van den Bergh, M., Non-commutative desingularization of determinantal varieties, I. Invent Math. 2010;182(1):47–115. MR 2672281

[45] Buchweitz, R.-O., Leuschke, G. J., Van den Bergh, M., Non-commutative desingularization of determinantal varieties, II: Arbitrary minors. Submitted 2011.

[46] Buan, A. B., Marsh, R., Cluster-tilting theory. Trends in representation theory of algebras and related topics, vol. 406 of Contemporary mathematics. Providence, RI: American Mathematical Society; 2006. p. 1–30. MR 2258039

[47] Bender, M., Mozgovoy, S., Crepant resolutions and brane tilings II: Tilting bundles. Available from: http://arxiv.org/abs/0909.2013. September 2009.

[48] Bondal, A. I., Orlov, D., Semiorthogonal decompositions for algebraic varieties. Available from: http://arxiv.org/abs/alg-geom/9506012. 1995.

[49] Bondal, A. I., Orlov, D., Reconstruction of a variety from the derived category and groups of autoequivalences. Compos Math. 2001;125(3):327–344. MR 1818984

[50] Bondal, A. I., Orlov, D., Derived categories of coherent sheaves. Proceedings of the International Congress of Mathematicians, Vol. II.; 2002, Beijing; Higher Ed. Press; 2002. p. 47–56. MR 1957019

[51] Bocklandt, R., Calabi–Yau algebras and weighted quiver polyhedra. Available from: http://arxiv.org/abs/0905.0232. 2009.

[52] Bongartz, K., Tilted algebras. Proceedings, Representations of Algebras; 1980; Puebla. vol. 903 of Lecture notes in mathematics. Berlin: Springer; 1981. p. 26–38. MR 654701

[53] Bondal, A. I., Representations of associative algebras and coherent sheaves. Izv Akad Nauk SSSR Ser Mat. 1989;53(1):25–44. MR 992977

[54] Bondal, A. I., Polishchuk, A. E., Homological properties of associative algebras: the method of helices. Izv Ross Akad Nauk Ser Mat. 1993;57(2):3–50. MR 1230966

[55] Braun, A., On symmetric, smooth and Calabi–Yau algebras. J Algebra. 2007;317(2):519–533. MR 2362929

[56] Bridgeland, T., Flops and derived categories. Invent Math. 2002;147(3):613–632. MR 1893007

[57] Bridgeland, T., Derived categories of coherent sheaves. Proceedings of the International Congress of Mathematicians, Vol. II. Zürich: European Mathematical Society; 2006. p. 563–582. MR 2275610

[58] Broomhead, N., Dimer models and Calabi–Yau algebras. Available from: http://arxiv.org/abs/0901.4662. January 2009.

[59] Bondal, A. I., Van den Bergh, M., Generators and representability of functors in commutative and noncommutative geometry. Mosc Math J. 2003;3(1):1–36, 258. MR 1996800

[60] Căldăraru, A., Derived categories of sheaves: a skimming. In: Snowbird lectures in algebraic geometry. vol. 388 of Contemporary mathematics. Providence, RI: American Mathematical Society; 2005. p. 43–75. MR 2182889

[61] Crawley-Boevey, W. W., Holland, M. P., Noncommutative deformations of Kleinian singularities. Duke Math J. 1998;92(3):605–635. MR 1620538

[62] Cautis, S., Logvinenko, T., A derived approach to geometric McKay correspondence in dimension three. J Reine Angew Math. 2009;636:193–236. MR 2572250

[63] Corti, A., What is ... a flip?. Notices Amer Math Soc. 2004;51(11):1350–1351. MR 2105240

[64] Cline, E. T., Parshall, B. J., Scott, L. L., Derived categories and Morita theory. J Algebra. 1986;104(2):397–409. MR 866784

[65] Dao, H., Some observations on local and projective hypersurfaces. Math Res Lett. 2008;15(2):207–219. MR 2385635

[66] Dao, H., Remarks on non-commutative crepant resolutions of complete intersections. Adv Math. 2010;224(3):1021–1030. MR 2628801

[67] Dao, H., Decent intersection and Tor-rigidity for modules over local hypersurfaces. To appear in Trans Amer Math Soc. Available from: http://arxiv.org/abs/math/0611568. February 2011.

[68] Decker, W., de Jong, T., Greuel, G.-M., Pfister, G., The normalization: a new algorithm, implementation and comparisons. Proceedings, Computational Methods for Representations of Groups and Algebras; 1997; Essen. vol. 173 of Progress in mathematics. Basel: Birkhäuser; 1999. p. 177–185. MR 1714609

[69] Douglas, M. R., Greene, B. R., Morrison, D. R., Orbifold resolution by D-branes. Nuclear Phys B. 1997;506:84–106.

[70] Dao, H., Huneke, C., Vanishing of Ext, cluster tilting modules and finite global dimension of endomorphism rings. Available from: http://arxiv.org/abs/1005.5359. May 2010.

[71] Dixon, L. J., Harvey, J. A., Vafa, C., Witten, E., Strings on orbifolds. Nuclear Phys B. 1985;261(4):678–686. MR 818423

[72] de Jong, T., An algorithm for computing the integral closure. J Symbolic Comput. 1998;26(3):273–277. MR 1633931

[73] Eisenbud, D., Homological algebra on a complete intersection, with an application to group representations. Trans Amer Math Soc. 1980;260(1):35–64. MR 570778

[74] Eisenbud, D., Commutative algebra with a view toward algebraic geometry. vol. 150 of Graduate texts in mathematics. New York: Springer-Verlag; 1995. MR 1322960

[75] Esnault, H., Reflexive modules on quotient surface singularities. J Reine Angew Math. 1985;362:63–71. MR 809966

[76] Eisenbud, D., Schreyer, F.-O., Weyman, J., Resultants and Chow forms via exterior syzygies. J Amer Math Soc. 2003;16(3):537–579 [electronic]. MR 1969204

[77] Freyd, P., Abelian categories. An introduction to the theory of functors. Harper's series in modern mathematics. New York: Harper & Row Publishers; 1964. MR 0166240

[78] Gabriel, P., Des catégories abéliennes. Bull Soc Math France. 1962;90:323–448. MR 0232821

[79] Ginzburg, V., Lectures on noncommutative geometry. Available from: http://arXiv.org/abs/0506603. June 2005.

[80] Ginzburg, V., Calabi–Yau algebras. Available from: http://arXiv.org/abs//0612139. December 2006.

[81] Geigle, W., Lenzing, H., A class of weighted projective curves arising in representation theory of finite-dimensional algebras. Proceedings, Singularities, Representation of Algebras, and Vector Bundles; 1985; Lambrecht. vol. 1273 of Lecture notes in mathematics. Berlin: Springer; 1987. p. 265–297. MR 915180

[82] Greuel, G.-M., Laplagne, S., Seelisch, F., Normalization of rings. J Symbolic Comput. 2010;45(9):887–901. MR 2661161

[83] Goodearl, K. R., Global dimension of differential operator rings. IV. Simple modules. Forum Math. 1989;1(2):185–200. MR 990143

[84] Grauert, H., Riemenschneider, O., Verschwindungssätze für analytische Kohomologiegruppen auf komplexen Räumen. Invent Math. 1970;11:263–292. MR 0302938

[85] Grauert, H., Remmert, R., Riemenschneider, O., Analytische Stellenalgebren. vol. 176 of Grundlehren der mathematischen Wissenschaften. Berlin: Springer-Verlag; 1971. MR 0316742

[86] Grauert, H., Remmert, R., Riemenschneider, O., Coherent analytic sheaves. vol. 265 of Grundlehren der Mathematischen Wissenschaften [Fundamental principles of mathematical sciences]. Berlin: Springer-Verlag; 1984. MR 755331

[87] Grothendieck, A., Éléments de géométrie algébrique. III. In: Étude cohomologique des faisceaux cohérents. I. vol. 11 of Publications Mathématiques Institute Hautes Études Scientifiques. 1961. p. 5–167. MR 0217085

[88] Grayson, D., Stillman, M., Macaulay 2, a software system for research in algebraic geometry. Available from: http://www.math.uiuc.edu/Macaulay2/.

[89] Gordon, I., Smith, S. P., Representations of symplectic reflection algebras and resolutions of deformations of symplectic quotient singularities. Math Ann. 2004;330(1):185–200. MR 2091684

[90] González-Sprinberg, G., Verdier J-L., Points doubles rationnels et représentations de groupes. C R Math Acad Sci Paris Sér I. 1981;293(2):111–113. MR 637103

[91] Happel, D., On the derived category of a finite-dimensional algebra. Comment Math Helv. 1987;62(3):339–389. MR 910167

[92] Happel, D., Triangulated categories in the representation theory of finite-dimensional algebras. vol. 119 of London Mathematical Society lecture note series. Cambridge: Cambridge University Press; 1988. MR 935124

[93] Happel, D., Hochschild cohomology of finite-dimensional algebras. Proceedings, Séminaire d'Algèbre Paul Dubreil et Marie-Paul Malliavin, 39ème année; 1987/1988; Paris. vol. 1404 of Lecture notes in mathematics. Berlin: Springer; 1989, p. 108–126. MR 1035222

[94] Hartshorne, R., Algebraic geometry. vol. 52 of Graduate texts in mathematics. New York: Springer-Verlag; 1977. MR 0463157

[95] Hochster, M., Eagon, J. A., Cohen–Macaulay rings, invariant theory, and the generic perfection of determinantal loci. Amer J Math. 1971;93:1020–1058. MR 0302643

[96] Herzog, J., Ringe mit nur endlich vielen Isomorphieklassen von maximalen, unzerlegbaren Cohen–Macaulay-Moduln. Math Ann. 1978;233(1):21–34. MR 0463155

[97] Holm, T., Jørgensen, P., Triangulated categories: definitions, properties, and examples. In: Triangulated categories. vol. 375 of London Mathematical Society lecture note series. Cambridge: Cambridge University Press; 2010. p. 1–51. MR 2681706

[98] Herzog, J., Kunz, E., editors. Der kanonische Modul eines Cohen–Macaulay-Rings. Seminar über die lokale Kohomologietheorie von Grothendieck; Wintersemester 1970/1971; Universität Regensburg. vol. 238 of Lecture notes in mathematics. Berlin: Springer-Verlag; 1971. MR 0412177

[99] Huckaba, S., Marley, T., On associated graded rings of normal ideals. J Algebra. 1999;222(1):146–163. MR 1728165

[100] Hochster, M., Cohen–Macaulay modules. Proceedings of the Conference on Commutative Algebra; 1972; University of Kansas, Lawrence, KS. vol. 311 of Lecture notes in mathematics. Berlin: Springer; 1973. p. 120–152. MR 0340251

[101] Hille, L., Perling, M., Exceptional sequences of invertible sheaves on rational surfaces. Available from: http://arXiv.org/abs/0810.1936.

[102] Hille, L., Perling, M., A counterexample to King's conjecture. Compos Math. 2006;142(6):1507–1521. MR 2278758

[103] Happel, D., Ringel, C. M., Tilted algebras. Trans Amer Math Soc. 1982;274(2):399 443. MR 675063

[104] Huneke, C., Tight closure and its applications [with an appendix by Hochster M]. vol. 88 of CBMS regional conference series in mathematics. Washington, DC: Conference Board of the Mathematical Sciences; 1996. MR 1377268

[105] Huybrechts, D., Fourier–Mukai transforms in algebraic geometry. Oxford Mathematical Monographs. Oxford: The Clarendon Press Oxford University Press; 2006. MR 2244106

[106] Hille, L., Van den Bergh, M., Fourier–Mukai transforms. In: Handbook of tilting theory. vol. 332 of London Mathematical Society lecture note series. Cambridge: Cambridge University Press; 2007. p. 147–177. MR 2384610

[107] Huisgen-Zimmermann, B., The finitistic dimension conjectures – a tale of 3.5 decades. Proceedings, Abelian Groups and Modules; 1994; Padova. vol. 343 of Mathematics and its applications. Dordrecht: Kluwer Acad. Publ.; 1995. p. 501–517. MR 1378224

[108] Ito, Y., Nakamura, I., Hilbert schemes and simple singularities. Proceedings, New Trends in Algebraic Geometry; 1996; Warwick. vol. 264 of London Mathematical Society lecture note series. Cambridge: Cambridge University Press; 1999. p. 151–233. MR 1714824

[109] Ito, Y., Reid, M., The McKay correspondence for finite subgroups of $SL(3, \mathbb{C})$. Proceedings, Higher-dimensional Complex Varieties; 1994; Trento. Berlin: De Gruyter; 1996. p. 221–240. MR 1463181

[110] Iyama, O., Reiten, I., Fomin–Zelevinsky mutation and tilting modules over Calabi–Yau algebras. Amer J Math. 2008;130(4):1087–1149. MR 2427009

[111] Ito, Y., Crepant resolution of trihedral singularities and the orbifold Euler characteristic. Internat J Math. 1995;6(1):33–43. MR 1307302

[112] Ito, Y., Gorenstein quotient singularities of monomial type in dimension three. J Math Sci Univ Tokyo. 1995;2(2):419–440. MR 1366564

[113] Iyama, O., Wemyss, M., Auslander–Reiten duality and maximal modifications for non-isolated singularities. Math Z. 2010;265(1):41–83. MR 2606949

[114] Iyama, O., Wemyss, M., On the noncommutative Bondal–Orlov conjecture. Available from: http://arXiv.org/abs/1101.3642. January 2011.

[115] Iyama, O., Finiteness of representation dimension. Proc Amer Math Soc. 2003;131(4):1011–1014 [electronic]. MR 1948089

[116] Iyama, O., Auslander correspondence. Adv Math. 2007;210(1):51–82. MR 2298820

[117] Iyama, O., Higher-dimensional Auslander–Reiten theory on maximal orthogonal subcategories. Adv Math. 2007;210(1):22–50. MR 2298819

[118] Vinayak Jategaonkar, A., Localization in Noetherian rings. vol. 98 of London Mathematical Society lecture note series. Cambridge: Cambridge University Press; 1986. MR 839644

[119] Jinnah, M. I., Reflexive modules over regular local rings. Arch Math (Basel). 1975;26(4):367–371. MR 0376662

[120] Jones, V., Moscovici, H., Review of "Noncommutative geometry" by Alain Connes [San Diego, CA: Academic Press; 1994; 1303779 (95j:46063)]. Notices Amer Math Soc. 1997;44(7):792–799. MR 1460207

[121] Jørgensen, P., Non-commutative projective geometry [lecture notes]. Available from: http://www.staff.ncl.ac.uk/peter.jorgensen/manus/opgteach.ps.

[122] Jorgensen, D. A., Finite projective dimension and the vanishing of $\mathrm{Ext}_R(M, M)$. Comm Algebra. 2008;36(12):4461–4471. MR 2473341

[123] Jothilingam, P., A note on grade. Nagoya Math J. 1975;59:149–152. MR 0392979

[124] Kaledin, D. B., Homological methods in non-commutative geometry [lecture notes]. University of Tokyo; Oct 2007–Mar 2008. Available from: http://imperium.lenin.ru/~kaledin/tokyo/. 2008.

[125] Kaledin, D. B., Non-commutative geometry from the homological point of view [lecture notes]. KIAS, Seoul; Oct 2009. Available from: http://imperium.lenin.ru/~kaledin/seoul/. 2009.

[126] Kapranov, M. M., Derived category of coherent sheaves on Grassmann manifolds. Izv Akad Nauk SSSR Ser Mat. 1984;48(1):192–202. MR 733363

[127] Kapranov, M. M., Derived category of coherent bundles on a quadric. Funktsional Anal i Prilozhen. 1986;20(2):67. MR 847146

[128] Kapranov, M. M., On the derived categories of coherent sheaves on some homogeneous spaces. Invent Math. 1988;92(3):479–508. MR 939472

[129] Kawamata, Y., Derived categories of toric varieties. Michigan Math J. 2006;54(3):517–535. MR 2280493

[130] Keeler, D. S., The rings of noncommutative projective geometry. Proceedings, Advances in Algebra and Geometry; 2001; Hyderabad. New Delhi: Hindustan Book Agency; 2003, p. 195–207. MR 1988956

[131] King, A., Tilting bundles on some rational surfaces [unpublished]. Available from: http://www.maths.bath.ac.uk/~masadk/papers/. 1997.

[132] Kempf, G., Knudsen, F. F., Mumford, D., Saint-Donat, B., Toroidal embeddings. vol. 339 of Lecture notes in mathematics. Berlin: Springer-Verlag; 1973. MR 0335518

[133] Klein, F., Vorlesungen über das Ikosaeder und die Auflösung der Gleichungen vom fünften Grade [Lectures on the icosahedron and the solution of the 5th degree equations]. Basel: Birkhäuser Verlag. Stuttgart: B. G. Teubner Verlagsgesellschaft; 1993 [German].

[134] Kollár, J., Mori, S., Birational geometry of algebraic varieties [with the collaboration of Clemens, C. H., Corti, A., translated from the 1998 Japanese original]. vol. 134 of Cambridge tracts in mathematics. Cambridge: Cambridge University Press; 1998. MR 1658959

[135] Knörrer, H., Group representations and the resolution of rational double points. Proceedings, Finite Groups – Coming of Age; 1982; Montreal, QC. vol. 45 of Contemporary mathematics. Providence, RI: American Mathematical Society; 1985. p. 175–222. MR 822239

[136] Kollár, J., Flops. Nagoya Math J. 1989;113:15–36. MR 986434

[137] Kollár, J., Minimal models of algebraic threefolds: Mori's program. Séminaire Bourbaki, Vol. 712. Astérisque 1989;177–178:303–326. MR 1040578

[138] Kollár, J., What is ... a minimal model?. Notices Amer Math Soc. 2007;54(3):370–371. MR 2292143

[139] Kontsevich, M., Homological algebra of mirror symmetry. Proceedings of the International Congress of Mathematicians, Vol. 1, 2; 1994; Zürich. Basel: Birkhäuser; 1995. p. 120–139. MR 1403918

[140] Kontsevich, M., Soibelman, Y., Notes on A_∞-algebras, A_∞-categories and noncommutative geometry. In: Homological mirror symmetry, vol. 757 of Lecture notes in physics. Berlin: Springer; 2009. p. 153–219. MR 2596638

[141] Kuznetsov, A., Lefschetz decompositions and categorical resolutions of singularities. Selecta Math (N S). 2008;13(4):661–696. MR 2403307

[142] Kapranov, M. M., Vasserot, E., Kleinian singularities, derived categories and Hall algebras. Math Ann. 2000;316(3):565–576. MR 1752785

[143] Lam, T. Y., Lectures on modules and rings. vol. 189 of Graduate texts in mathematics. New York: Springer-Verlag; 1999. MR 1653294

[144] Laudal, O. A., Noncommutative algebraic geometry. Proceedings of the International Conference on Algebraic Geometry and Singularities, vol. 19 [Spanish]; 2001; Sevilla. 2003. p. 509–580. MR 2023198

[145] Leuschke, G. J., Endomorphism rings of finite global dimension. Canad J Math. 2007;59(2):332–342. MR 2310620

[146] Lin, H.-W., On crepant resolution of some hypersurface singularities and a criterion for UFD., Trans Amer Math Soc. 2002;354(5):1861–1868 [electronic]. MR 1881020

[147] Lunts, V. A., Categorical resolution of singularities. J Algebra. 2010;323(10):2977–3003. MR 2609187

[148] Leuschke, G. J., Wiegand, R., Cohen–Macaulay representations [in progress]. Available from: http://www.leuschke.org/Research/MCMBook.

[149] Macleod, M. J., Generalizing the Cohen–Macaulay condition and other homological properties [PhD thesis]. University of Glasgow; 2010. Available from: http://theses.gla.ac.uk/2004/.

[150] Markushevich, D., Resolution of $\mathbb{C}^3/H_{168}$. Math Ann. 1997;308(2):279–289. MR 1464903

[151] McKay, J., Graphs, singularities, and finite groups. Proceedings of The Santa Cruz Conference on Finite Groups; 1979; University of California, Santa Cruz, CA. vol. 37 of Proceedings of symposia in pure mathematics. Providence, RI: American Mathematical Society; 1980. p. 183–186. MR 604577

[152] Miyashita, Y., Tilting modules of finite projective dimension. Math Z. 1986;193(1):113–146. MR 852914

[153] Mac Lane, S., Categories for the working mathematician. 2nd ed. vol. 5 of Graduate texts in mathematics. New York: Springer-Verlag; 1998. MR 1712872

[154] Existence of non-commutative desingularizations. MathOverflow, Available from: http://mathoverflow.net/questions/14350 [version: 2010-02-06].

[155] Flips in the minimal model program. MathOverflow, Available from: http://mathoverflow.net/questions/5321 [version: 2009-11-16].

[156] Non-commutative algebraic geometry. MathOverflow, Available from: http://mathoverflow.net/questions/7917 [version: 2009-12-14].

[157] Morita, K., Duality for modules and its applications to the theory of rings with minimum condition. Sci Rep Tokyo Kyoiku Daigaku Sect A. 1958;6:83–142. MR 0096700

[158] Mori, I., An introduction to noncommutative algebraic geometry. Proceedings of the 40th Symposium on Ring Theory and Representation Theory. Yamaguchi: Symp. Ring Theory Represent. Theory Organ. Comm.; 2008. p. 53–59. MR 2427915

[159] Mozgovoy, S., Crepant resolutions and brane tilings I: Toric realization. Available from: http://arxiv.org/abs/0908.3475. August 2009.

[160] Mukai, S., Duality between $D(X)$ and $D(\hat{X})$ with its application to Picard sheaves. Nagoya Math J. 1981;81:153–175. MR 607081

[161] Nakamura, I., Hilbert schemes of Abelian group orbits. J Algebraic Geom. 2001;10(4):757–779. MR 1838978

[162] Orlov, D. O., Equivalences of derived categories and $K3$ surfaces. J Math Sci (N Y). 1997;84(5):1361–1381 [Algebraic geometry, 7]. MR 1465519

[163] Orlov, D. O., Derived categories of coherent sheaves and equivalences between them. Uspekhi Mat Nauk. 2003;58(3(351)):89–172. MR 1998775

[164] Orlov, D. O., Quasicoherent sheaves in commutative and noncommutative geometry. Izv Ross Akad Nauk Ser Mat. 2003;67(3):119–138. MR 1992195

[165] Orlov, D. O., Derived categories of coherent sheaves and triangulated categories of singularities. In: Algebra, arithmetic, and geometry: in honor of Yu. I. Manin. Vol. II. vol. 270 of Progress in mathematics. Boston, MA: Birkhäuser Boston Inc.; 2009. p. 503–531. MR 2641200

[166] Orlov, D. O., Remarks on generators and dimensions of triangulated categories. Mosc Math, J. 2009;9(1):153–159. MR 2567400

[167] Ooguri, H., Yamazaki, M., Crystal melting and toric Calabi–Yau manifolds. Comm Math Phys. 2009;292(1):179–199. MR 2540074

[168] Passman, D. S., Infinite crossed products. vol. 135 of Pure and applied mathematics. Boston, MA: Academic Press Inc.; 1989. MR 979094

[169] Quarles, C. L., Krull–Schmidt Rings & noncommutative resolutions of singularities [MSc thesis]. University of Washingon; 2005.

[170] Rainwater, J., Global dimension of fully bounded Noetherian rings. Comm Algebra. 1987;15(10):2143–2156. MR 909958

[171] Ramras, M., Maximal orders over regular local rings of dimension two. Trans Amer Math Soc. 1969;142:457–479. MR 0245572

[172] Ranicki, A., editor. Non-commutative localization in algebra and topology. vol. 330 of London Mathematical Society lecture note series. Cambridge: Cambridge University Press; 2006. MR 2222649

[173] Reid, M., Minimal models of canonical 3-folds. Proceedings, Algebraic Varieties and Analytic Varieties; 1981; Tokyo. vol. 1 of Advanced studies in pure mathematics. Amsterdam: North-Holland; 1983, p. 131–180. MR 715649

[174] Reid, M., Twenty-five years of 3-folds – an old person's view. In: Explicit birational geometry of 3-folds, vol. 281 of London Mathematical Society lecture note series. Cambridge: Cambridge University Press; 2000, p. 313–343. MR 1798985

[175] Reid, M., La correspondance de McKay. Séminaire Bourbaki, Vol. 867; 1999/2000. Astérisque 2002;276:53–72. MR 1886756

[176] Reiner, I., Maximal orders [with a foreword by Taylor MJ]. vol. 28 of London Mathematical Society monographs. New series. Corrected reprint of the 1975 original. Oxford: The Clarendon Press Oxford University Press; 2003. MR 1972204

[177] Rickard, J., Morita theory for derived categories. J Lond Math Soc (2). 1989;39(3):436–456. MR 1002456

[178] Rickard, J., Derived equivalences as derived functors. J Lond Math Soc (2). 1991;43(1):37–48. MR 1099084

[179] Roan, S.-S., On $c_1 = 0$ resolution of quotient singularity. Internat J Math. 1994;5(4):523–536. MR 1284568

[180] Roan, S.-S., Minimal resolutions of Gorenstein orbifolds in dimension three. Topology. 1996;35(2):489–508. MR 1380512

[181] Rosenberg, A. L., Noncommutative schemes. Compos Math. 1998;112(1):93–125. MR 1622759

[182] Rosenberg, A. L., The spectrum of Abelian categories and reconstruction of schemes. Proceedings, Rings, Hopf Algebras, and Brauer Groups; 1996; Antwerp/Brussels. vol. 197 of Lecture notes in pure and applied mathematics. New York: Dekker; 1998, p. 257–274. MR 1615928

[183] Rouquier, R., Catégories dérivées et géométrie birationnelle [d'après Bondal, Orlov, Bridgeland, Kawamata et al.]. Séminaire Bourbaki. Vol. 946; 2004/2005. Astérisque 2006;307:283–307. MR 2296422

[184] Rouquier, R., Representation dimension of exterior algebras. Invent Math. 2006;165(2):357–367. MR 2231960

[185] Rouquier, R., Dimensions of triangulated categories. J K-Theory. 2008;1(2):193–256. MR 2434186

[186] Rudakov, A. N., Exceptional collections, mutations and helices. In: Helices and vector bundles. vol. 148 of London Mathmatical Society lecture note series. Cambridge: Cambridge University Press; 1990, p. 1–6. MR 1074777

[187] Sancho de Salas, J. B., Blowing-up morphisms with Cohen–Macaulay associated graded rings. Proceedings, Géométrie algébrique et applications, I; 1984; La Rábida. vol. 22 of Travaux en cours. Paris: Hermann; 1987. p. 201–209. MR 907914

[188] Serre, J.-P., Faisceaux algébriques cohérents. Ann of Math (2). 1955;61:197–278. MR 0068874

[189] Smith, S. P., Non-commutative algebraic geometry [draft]. 2000. Available from: http://www.math.washington.edu/~smith/Research/nag.pdf.

[190] Smith, S. P., Quarles, C., On an example of Leuschke [private communication]. 2005.

[191] Smith, K. E., Van den Bergh, M., Simplicity of rings of differential operators in prime characteristic. Proc Lond Math Soc (3). 1997;75(1):32–62. MR 1444312

[192] Stafford, J. T., Van den Bergh, M., Noncommutative curves and noncommutative surfaces. Bull Amer Math Soc (N S). 2001;38(2):171–216. MR 1816070

[193] Stafford, J. T., Noncommutative resolutions and rational singularities. Michigan Math J. 2008;57(special volume in honor of Melvin Hochster):659–674. MR 2492474

[194] Seiberg, N., Witten, E., String theory and noncommutative geometry. J High Energy Phys. 1999;(9):Paper 32. MR 1720697

[195] Toda, Y., Birational Calabi–Yau threefolds and BPS state counting. Commun Number Theory Phys. 2008;2(1):63–112. MR 2417847

[196] Toda, Y., Stability conditions and crepant small resolutions. Trans Amer Math Soc. 2008;360(11):6149–6178. MR 2425708

[197] Toda, Y., Stability conditions and Calabi–Yau fibrations. J Algebraic Geom. 2009;18(1):101–133. MR 2448280

[198] Toën, B., Vaquié M., Algébrisation des variétés analytiques complexes et catégories dérivées. Math Ann. 2008;342(4):789–831. MR 2443764

[199] Toda, Y., Yasuda, T., Noncommutative resolution, F-blowups and D-modules. Adv Math. 2009;222(1):318–330. MR 2531377

[200] Vasconcelos, W. V., Reflexive modules over Gorenstein rings. Proc Amer Math Soc. 1968;19:1349–1355. MR 0237480

[201] Vasconcelos, W. V., On quasi-local regular algebras. Proceedings, Convegno di Algebra Commutativa, INDAM; 1971, Rome. vol. XI of Symposia Mathematica. London: Academic Press; 1973, p. 11–22. MR 0330159

[202] Van den Bergh, M., Regular rings of dimension three. Proceedings, Séminaire d'algèbre Paul Dubreil et Marie-Paule Malliavin; 1986; Paris. vol. 1296 of Lecture notes in mathematics. Berlin: Springer; 1987. p. 228–234. MR 932058

[203] Van den Bergh, M., Blowing up of non-commutative smooth surfaces. Mem Amer Math Soc. 2001;154(734):x+140. MR 1846352

[204] Van den Bergh, M., Non-commutative crepant resolutions (with some corrections). In: The legacy of Niels Henrik Abel. Berlin: Springer; 2004. p. 749–770. [the updated (2009) version on arXiv has some minor corrections over the published version]. MR 2077594

[205] Van den Bergh, M., Three-dimensional flops and noncommutative rings. Duke Math J. 2004;122(3):423–455. MR 2057015

[206] Verëvkin, A. B., Injective Serre sheaves. Mat Zametki. 1992;52(4):35–41, 128. MR 1203950

[207] Verëvkin, A. B., On a noncommutative analogue of the category of coherent sheaves on a projective scheme. Proceedings, Algebra and Analysis; 1989; Tomsk. vol. 151 of American mathematical society translations, series 2. Providence, RI: American Mathematical Society; 1992. p. 41–53. MR 1191171

[208] Verdier, J.-L., Des catégories dérivées des catégories abéliennes [with a preface by Illusie, L., edited and with a note by Maltsiniotis G]. Astérisque 1996;239. MR 1453167

[209] Viehweg, E., Rational singularities of higher dimensional schemes. Proc Amer Math Soc. 1977;63(1):6–8. MR 0432637

[210] Verschoren, A., Willaert, L., Noncommutative algebraic geometry: from pi-algebras to quantum groups. Bull Belg Math Soc Simon Stevin. 1997;4(5):557–588. MR 1600288

[211] Watanabe, K., Certain invariant subrings are Gorenstein. I, II. Osaka J Math. 1974;11:1–8, 379–388. MR 0354646

[212] Wemyss, M., Reconstruction algebras of type A. Available from: http://arXiv.org/abs/0704.3693. April 2007.

[213] Wemyss, M., Reconstruction algebras of type D (I). Available from: http://arXiv.org/abs/0905.1154. May 2009.

[214] Wemyss, M., Reconstruction algebras of type D (II). Available from: http://arXiv.org/abs/0905.1155. May 2009.

[215] Weyman, J., Cohomology of vector bundles and syzygies. vol. 149 of Cambridge tracts in mathematics. Cambridge: Cambridge University Press; 2003. MR 1988690

[216] Yasuda, T., Frobenius morphisms of noncommutative blowups. Available from: http://arXiv.org/abs/0906.1662. June 2009.

[217] Yekutieli, A., Dualizing complexes over noncommutative graded algebras. J Algebra. 1992;153(1):41–84. MR 1195406

[218] Yoshino, Y., Cohen–Macaulay modules over Cohen–Macaulay rings. vol. 146 of London Mathematical Society lecture note series. Cambridge: Cambridge University Press; 1990. MR 1079937

Author Information

Graham J. Leuschke, Mathematics Department, Syracuse University, Syracuse NY 13244, USA.
E-mail: gjleusch@math.syr.edu

Bei Fragen zur Produktsicherheit wenden Sie sich bitte an:
If you have any questions regarding product safety,
please contact:

Walter de Gruyter GmbH
Genthiner Straße 13
10785 Berlin
productsafety@degruyterbrill.com